Advances in Wireless Communication: Applications and Developments

Advances in Wireless Communication: Applications and Developments

Guest Editor

Mario E. Rivero-Angeles

MDPI

Basel • Beijing • Wuhan • Barcelona • Belgrade • Novi Sad • Cluj • Manchester

Guest Editor
Mario E. Rivero-Angeles
Computer Research Center (CIC)
Instituto Politécnico Nacional (IPN)
Mexico City
Mexico

Editorial Office
MDPI AG
Grosspeteranlage 5
4052 Basel, Switzerland

This is a reprint of the Special Issue, published open access by the journal *Telecom* (ISSN 2673-4001), freely accessible at: https://www.mdpi.com/journal/telecom/special_issues/R2QDC44TRK.

For citation purposes, cite each article independently as indicated on the article page online and as indicated below:

Lastname, A.A.; Lastname, B.B. Article Title. *Journal Name* **Year**, *Volume Number*, Page Range.

ISBN 978-3-7258-6938-1 (Hbk)
ISBN 978-3-7258-6939-8 (PDF)
https://doi.org/10.3390/books978-3-7258-6939-8

Contents

About the Editor

Mario E. Rivero-Angeles

Mario E. Rivero-Angeles was born in Mexico City, Mexico, in 1976. He received his BSc degree from the Metropolitan Autonomous University (UAM), Mexico, in 1998, and M.Sc. and Ph.D. degrees in Electrical Engineering from CINVESTAV-IPN in 2000 and 2006, respectively. He has been a professor at the National Polytechnic Institute, currently working at the Center of Research in Computation (CIC-IPN), Mexico, since 2002. He was a Postdoctoral Fellow at the Dyonisos research project at INRIA (Institut National de Recherche en Informatique et en Automatique), Rennes, France, from 2007 to 2010. His research interests include random access protocols and data transmission in cellular networks, P2P networks, and wireless sensor networks.

Preface

Wireless communications have evolved from user-centered services, where the system design was primarily focused on providing an adequate quality of service (QoS) for voice services, to a world where things are connected, providing an entire ecosystem in support of a user's everyday life. Indeed, communication systems are no longer focused on connecting people, but rather on allowing everything that surrounds the user to perform context-based, automatic routines that improve quality of life. For instance, modern and future communication systems must allow massive packet transmissions between user wearables (ranging from smart watches and augmented reality glasses to medical devices like prosthetics and pacemakers) and the smartphone, but also to the user's vehicle and domestic devices (such as surveillance cameras, smart appliances, robot vacuums, among others). In this regard, a machine learning algorithm must provide nodes with sufficient intelligence to scan the environment and learn users' patterns in a way that distinguishes everyday life situations from emergency events and even unusual activities, such as vacations and social/sporting/cultural events. This capacity will be necessary to address data packets inside the user ecosystem. Additionally, it will aid other parties (mainly when the user is unable to communicate), like vehicles, to directly communicate with the police in case of a robbery or with a hospital in case of an accident. Indeed, to date, most devices communicate and transmit all the available data to the user's smartphone, leaving the user to discard, attend to, or ignore these notifications or address them by configuring the notification settings. However, this is no longer sustainable due to high energy consumption, high bandwidth usage, and the highly cumbersome procedure of configuring each node, especially for applications in the near future, where more and more things become smart and indiscriminately transmit packets. This will be the next step in the design and development of the Internet of Things (IoT). In the same context, due to the environmental impact of the batteries used in these devices, the search for clean and sustainable options to power the nodes will have a significant impact on wireless communication devices. This is because alternative means of providing electricity (such as solar, organic, and bio-based electricity, among others) generate significantly less energy than common lithium-based batteries. As such, communication systems must now consider the limited energy and charging times of energy-harvesting modules to determine when and what information to transmit, thereby minimizing the impact on the user's personal ecosystem.

In this Special Issue, we delve into the details of some key topics that will be crucial in the design of present and future communication systems.

Mario E. Rivero-Angeles
Guest Editor

telecom

Editorial

Recent Advances in Applications and Performance Improvement Schemes in Wireless Communication

Mario Eduardo Rivero-Ángeles * and Izlian Yolanda Orea-Flores

Computer Research Center (CIC), Instituto Politécnico Nacional (IPN), Mexico City 07738, Mexico; iorea@ipn.mx
* Correspondence: mriveroa@ipn.mx

1. Introduction

To this date, the Fifth Generation (5G) of mobile communications has been deployed and has opened a great number of opportunities by increasing transmission rates (partially through the use of MIMO systems), decreasing latency, providing the amount of bandwidth required for video services, Virual and Augmented Reality applications, and social media and providing a solid ground for the massive implementation of the Internet of Things (IoT), which we believe is still in its initial phases of development. Indeed, the IoT will operate autonomously and with a high degree of intelligence, recognizing different users, routines, and special events in the lives of device owners, and communicating among different devices to provide seamless operation in people's everyday lives. Furthermore, devices used by users will also communicate among themselves and with devices from cities, highways, rivers, and forests, enabling the monitoring of ecosystems and the prevention of flooding and fires, among many other applications. To this day, the IoT operates mainly from a user's point of view, requiring manual configuration of many devices across various applications. Communication among devices and between different companies is extremely limited, not to mention communication with public devices.

From the works presented in this issue, we can confidently see that the design of wireless communications will require the use of one or more of these aspects in the near future:

- Energy consumption (energy harvesting, clean energies);
- Security guarantees (robust to attacks, privacy guarantees);
- Machine learning;
- Physical Layer improvements (antennas, modulation, and coding schemes) to reduce latency and increase transmission rates;
- IoT and social media services.

As such, the IoT is only one of many areas where communication systems must propose, analyze, and design new protocols, schemes, and systems on the path to the Sixth Generation (6G) of mobile communication systems. There are still many challenges that must be overcome to introduce the 6G seamlessly in the coming years, as detailed in one paper on this Special Issue, contribution 1. For instance, providing high-precision placement and high-speed transmission is crucial for various applications, such as autonomous vehicles, Drone-aided solutions in wireless networks, and event detection in Wireless Sensor Networks (WSNs), among others, where the use of Artificial Intelligence (AI) algorithms surely will play a prominent role. To this end, the use of Cloud and Edge resources will be required, as well as intelligent offloading schemes that allow for reducing energy consumption and increasing data processing efficiency, which is still an open research topic. We invite the reader to study the points provided in contribution 1 for an in-depth analysis of the challenges and open questions regarding the introduction of 6G mobile communications, as well as identifying new applications and services that are not yet implemented in our day-to-day digital life, such as digital twins. Additionally, regarding the physical layer, another paper presents invaluable details on the design of rectenas, antennas, and rectifiers for reduced and sustainable energy usage in communication systems, with a focus on Energy Harvesting (EH) techniques.

Not only are the physical layer challenges discussed in this issue, but many works also focus on providing a comprehensive view of the main challenges and technological issues that need to be solved before the introduction of 6G mobile communication systems. Issues such as security services, efficient resource allocation and utilization, including the use of Integrated Sensing and Communication Systems (ISACs), Unmanned Aerial Vehicles (UAVs), Device-to-Device (D2D) communications, Edge and Cloud computing, among others. From this, we believe that after a thorough read of this issue, the reader will have a clear idea of what has been done in recent years in these different areas and what is still required to investigate, and what are some of the issues that have not been tackled but will be required on the road to the development of the 6G. Hence, for a graduate student initiating their research or for a senior researcher with a clear and solid research trajectory seeking new opportunities to apply their expertise, this issue will be useful in providing a general vision of where telecommunication systems are headed.

We now discuss in detail the main works presented in this issue, which range from details in the physical layer, where coding, modulation, and waveforms are studied and analyzed. We then delve into MIMO systems, which, although primarily belonging to the physical layer, require special attention since they will be the fundamental technology that enables increased transmission rates and bandwidth required in 6G systems. We then present works on D2D and UAV-assisted systems, which significantly extend the coverage and capacity of classical communication systems, and also provide new services and applications that were previously unavailable. Then, we present works on advancements in sensing (the primary function of IoT devices). Next, we present advances in Edge and Cloud computing. We end this brief editorial with a discussion on energy harvesting schemes that will allow battery-reduced or even battery-free communication systems in the future. Some relevant conclusions are presented at the end.

2. Physical Layer: Antennas, Modulation, Coding, and Waveforms

In this special issue, a paper detailing the waveforms (shape, structure, and frequency) used in cellular systems from 0G to 6G and beyond provides an interesting overview of how wireless systems have evolved from purely analog systems to modern and future digital systems as stated in contribution 2. Aside from the parameters, the waveforms have a significant impact on key performance issues, including capacity, coverage, power consumption, battery life, spectral efficiency, bandwidth utilization, and resistance to interference and noise.

Waveforms in cellular networks are a major component in the operation and service provision in modern communication systems. By studying the evolution of waveforms that have been used, we will provide invaluable information for designing and developing new and improved waveforms in the future, considering that modulation schemes are designed in part by the waveform of the signals. Additionally, it is essential to note that as generations in mobile communications systems evolve, the primary services provided by these systems also evolve accordingly. As such, in future systems, it is highly relevant to consider the primary services and applications when designing and proposing well-suited waveforms to achieve optimal network performance.

https://doi.org/10.3390/telecom701(

The insights and lessons learned from this path in cellular networks can also be applied to other systems, such as the IoT, P2P networks, and many others, by considering parameters including the communication radio, transmission rate, and packet size, among others. As such, this paper is relevant not only in cellular networks but in all wireless communication systems. Of particular interest in this context is the use of terahertz communications, Visible Light Communications, ultramassive MIMO, and the use of Artificial Intelligence as a core technology at the core of 6G. Additionally, the use of sound is of great interest for providing additional security guarantees as detailed in contribution 3. These new transmission techniques would require drastically different waveforms and modulation schemes to implement practical/operational systems. Additionally, the rise of Integrated Sensing and Communication systems (which is discussed in another paper of this issue) is of major interest to modern applications, such as autonomous driving and the navigation of swarms of drones. This paper also describes emerging waveforms, such as orthogonal chirp division multiplexing, affine frequency division multiplexing, orthogonal time frequency space, and index modulation, and provides the fundamentals of each of these techniques to put them in context for beyond 6G systems.

Another issue addressed in this Special Issue is the use of clean energy to power up nodes in the IoT context. As such, the work presented in contribution 4 shows opportunities in the Energy Harvesting field, focusing on materials, AI techniques, hybrid EH, and novel architectures to obtain energy from RF sources, such that researchers working in this area can optimize the use of battery-free (and battery-based pollution-free) future communication systems.

In the context of Smart Cities and IoT, an interesting area is underwater communications. To monitor pollution in rivers, lakes, or dams, or to prevent flooding, nodes can be placed in strategic locations submerged in water, forcing communications in a denser medium than air, where most wireless communications are usually designed. In contribution 5, the authors propose efficient coding schemes for wireless underwater communications. To this end, they first characterized the noise in underwater communications and then proposed the use of polar coding to reduce the Bit Error Rate (BER). Conventional coding schemes are no longer adequate, as underwater channels suffer from severe multipath interference, high Doppler shifts, strong fading, low data rates, and limited bandwidth.

These papers open some opportunities to design WSNs located underwater, or the use of aquatic robots and swarms of aquatic robots to perform multiple tasks by having a reliable communication system in these conditions. For instance, consider a scenario where rivers are monitored with these sensor networks, such that in case of abnormal conditions (a rise in the water level or an unexpected change in the flow direction or velocity), nodes communicate among other nodes in the water or even with nodes on land to alert of possible emergency events. In such cases, nodes can be placed in remote areas, such as the middle of forests or mountains, where communication with central disaster management offices is difficult or even impossible. As such, nodes can perform long-range communications underwater through multiple multi-hop transmissions within the river, allowing them to communicate with a node on land that can then transmit to surrounding UAVs or vehicles passing by. Although this scenario may seem futuristic or impractical today, many of the works presented in this special issue tackle critical technical details to make it possible in the near future.

3. MIMO Systems

Although MIMO systems are part of the physical layer, they are of special relevance to future communication systems, so we present the works in this area in a separate section. MIMO systems are required for high data rate systems, and also, in the field of Energy

https://doi.org/10.3390/telecom7010012

Harvesting, for simultaneously transmitting information and recharging nodes in the IoT context, as mentioned and detailed in contribution 4.

Specifically, MIMO systems use multiple antennas for both transmission and reception. The authors in contribution 6 propose an antenna design for the 28 GHz band, featuring gains of 13 dBi and a bandwidth of 1 GHz. The development of MIMO systems is necessary to increase data rates, particularly in IoT, autonomous driving, and Smart City applications. Antenna design at 28 GHz is particularly challenging due to high path loss and atmospheric attenuation. In this regard, the authors design a four-port MIMO structure with reduced size by carefully selecting the element spacing and employing a decoupling strategy that reduces interference among the individual antennas.

Signal detection in MIMO systems is not straightforward due to the complexity of the channels. In this regard, the authors in contribution 7 present a Deep Neural Network (DNN) for detecting signals in MIMO systems, comparing their proposal to traditional methods such as maximum likelihood, minimum mean square error, and zero-forcing methods. They prove that the use of DNNs achieves better results in terms of computational complexity and accuracy. However, they also identify some issues that have to be studied in future designs, not only for DNN-based algorithms, but for the use of Machine Learning in MIMO system design in general, such as the use of activation functions for low-power communication systems, and the use of hybrid schemes integrating other AI methods with traditional methods.

Another work in MIMO systems is presented in contribution 8, where the authors propose a transceiver design for a multiuser MIMO full-duplex amplify-and-forward relay downlink communication system. This is performed using an imperfect loop interference cancellation. Also, an iterative method is developed to solve the optimization problem. This work focuses on relay systems where data is sent from the transmitter to the relay and then from the relay to the receiver, with the objective of amplifying the signal. This is of particular interest in long-range distance communications or in environments with high noise and interference, typical in big cities.

4. Device-to-Device Communicartions

With the exponential growth of IoT infrastructure, the need for connected devices has also become increasingly relevant as the road to 6G implementation progresses. Even though 5G has been an important issue to study, to this date, many devices are still connected to a central unit, which poses some security risks and loss of efficiency (increased latency and energy consumption). Additionally, some applications rely absolutely on efficient D2D communications, such as connected vehicles and UAV swarms. Hence, the relevance of proposing novel schemes to enable direct connections among neighboring devices, such as the one proposed in contribution 9, where a low-energy consumption scheme is proposed. Their proposal allows nodes to determine whether direct communication or communication through the cellular network is advised, given the current system conditions. Specifically, the authors investigate a bandwidth-efficient scheme that enables nodes to communicate using licensed or unlicensed spectrum in either an underlay or overlay architecture. In this regard, the authors are concerned with power control and interference mitigation in the resource allocation scheme, utilizing machine learning algorithms to predict future scenarios in complex systems. In particular, they propose the use of a Reinforcement Learning algorithm to achieve adequate power management, selecting the instantaneous power transmission of each device and thereby reducing interference in the cellular system.

https://doi.org/10.3390/telecom70100

5. UAV-Assited Communications

Another relevant open research area is the use of UAVs to aid wireless systems, such as WSNs and Cellular Networks, as UAVs can provide temporal bandwidth to increase capacity and coverage, while also reducing energy consumption. To this end, the work presented in contribution 10 proposes a channel model for ground-to-UAV applications in the metropolitan context in Indian cities. The proposed model considers both LoS (Line-of-Sight) and NLoS (Non-Line-of-Sight) conditions, which are commonly found in these UAV-aided systems. The need for such models becomes increasingly relevant in modern and future communication systems, providing high-bandwidth and low-latency services in the context of smart cities, including autonomous driving and intelligent sensing services. Such models must consider the elevation angle, shadowing, and path loss phenomena, as well as horizontal and vertical polarization, among other parameters. Of particular interest is the work presented in contribution 10, where the model is compared to real propagation conditions, demonstrating an adequate match. The use of UAVs in this context represents a cost- and energy-efficient alternative to implementing fixed Base Stations, where traffic is not sufficiently high to justify the use of mobile channels that are continuously available, or where traffic only increases at specific events, such as zones close to stadiums or public places. Additionally, the use of swarms of UAVs, either for public figure displays (which will replace the use of fireworks in the near future) or to aid wireless systems, will definitely require communication models such as the one presented in this paper.

Although this work is primarily focused on Indian cities, where suburban, urban, dense urban, and high-rise urban environments coexist, it is clear that many other cities worldwide could benefit from such propagation models by adjusting for their specific conditions of shadowing and path loss. Especially considering that communication models for low-altitude devices have been largely overlooked.

6. Sensing

A relatively new research area in modern communication systems is Integrated Sensing and Communication Systems (ISAC), which enhances efficiency by combining two of the most energy-intensive and resource-consuming tasks into a single procedure: sensing and data transmission. To this end, the work developed in contribution 11 proposes the use of Physical Layer Abstraction (PLA) to simplify the system-level simulations, replacing link-level simulations. By using the PLA method, it is possible to determine the Signal-to-Noise Ratio (SNR) in the wireless channel, thereby enabling the acquisition of distance, positioning, and detection information. The paper details these algorithms under two different conditions: low and high noise, which in turn produce low and high errors. The authors study the system under an OFDM technique, clearly illustrating the number of symbols required to achieve an adequate performance.

In this regard, the development of ISAC mechanisms and the work presented in contribution 11, in particular, will be of great relevance for many services and applications in the near future, such as autonomous vehicles, swarms of UAVs to aid wireless systems, and energy reduction in WSNs, among others. All of these, in the context of Smart Cities, will require the implementation of thousands or millions of nodes that must perform both sensing and communication tasks, which have been performed separately to date.

In this sense, the paper in contribution 3 presents a work where sensing is performed by nodes in certain regions where cyber-attacks are expected. For instance, to monitor restricted natural areas where logging or hunting is prohibited, but illegal trespassers are equipped with jamming devices and have previously disabled the WSN by interfering with the alarm transmission of nodes. Since most nodes communicate using the RF spectrum,

https://doi.org/10.3390/telecom7010012

it is expected that the jamming attacks would occur in this part of the electromagnetic spectrum. To prevent the disabling of event reporting, the authors propose a technique that utilizes the sounds of native bird species as a means of communication. Specifically, the authors propose inserting the monitoring data into the prerecorded sounds of birds commonly found in the region in a way that transmits information using sound that is not distorted, allowing it to pass undetected to people on the ground, even people familiar with the specific bird's sounds of the region. Specifically, the work presented in contribution 3 develops a covert channel utilizing steganography techniques to conceal information within sound waves. To this end, a robust system is analyzed and designed, where nodes must be turned on and off according to the specific nature of the natural sounds of birds. A teletraffic analysis is developed to carefully select the system's parameters to achieve a high throughput WSN. This proposal focuses on utilizing the sound of birds, but it could be extended to provide robust WSNs against jamming attacks using other types of sounds. For instance, in big cities, the sounds of vehicles or construction equipment can be used to conceal data transmissions.

7. EDGE and Cloud Computing

EDGE and Cloud computing are a major area that was primarily viewed as a computational field. However, in recent years, telecommunication systems have benefited from the advantages posed by having higher processing and storage capabilities. In particular, the use of EDGE computing provides low-latency and increased bandwidth services close to users, which are invaluable for many modern and future systems. However, integrating EDGE services into 5G and 6G systems requires detailed design. Specifically, in 6G systems, where increased data rates and the number of connected devices are expected, EDGE capabilities would enable hyper-connected and autonomous operation in a scalable manner. In contribution 12, the authors outline the benefits of utilizing EDGE computing in mobile networks, including enhanced Quality of Service (QoS), reduced latency, and decreased computational load. First, the work in contribution 12 provides details on the integration of 5G mobile networks with Multi-Access Edge Computing (MEC), analyzing the software components and their architecture, and then it performs an integration between MEC and 5G technology, which is validated in a virtualized environment with a 5G core using a Kubernetes cluster.

8. Reduced and Sustainable Energy Consumption and Harvesting

With the exponential growth of autonomous nodes in the IoT context, the use of alternative energy sources, other than batteries that could leak hazardous and toxic substances into natural zones or in residential homes, is of extreme importance. It is crucial to study the use of alternative and clean sources to power millions of devices for various applications. As such, Energy Harvesting is a key technology that can deliver a solution in present and future communication systems, avoiding frequent battery changes, potential leaks, and service interruptions due to energy shortages in the nodes. Building on this, in this Special Issue, some techniques and methods are studied and identified in contribution 4. Energy Harvesting is of special interest, given the possibility of wirelessly recharging from already present signals, such as those from TV, WiFi, and cellular systems' waves, which can be found pervasively in modern cities. As mentioned in this work, EH requires an antenna, a rectifier circuit that converts RF signals into DC voltage, and a storage unit. As such, in this work, the reader will find recent innovations in EH techniques and architectures, highlighting the main challenges and issues hindering the massive adoption of this technology in IoT applications, such as rectenna design, power management circuits, the use of AI algorithms, and materials' properties that have to be

https://doi.org/10.3390/telecom7010

improved for an efficient transformation of RF signals, making an emphasis on smart cities and biomedial applications, where certain wearables that report critical hrlath-related data must continuously operate.

The authors of this work first describe the main components of an EH system, including techniques for efficiently storing energy from RF signals. They then provide a general overview of recent advancements and applications, placing the reader in context and enabling a clear understanding of the direction of this research topic. From this, it is then easy to understand the main challenges and limitations of this technology, which are discussed in detail in the Section 9 of the paper.

9. Conclusions and Discussion

After a detailed review of this special issue, the reader will gain a clear understanding of how communication systems have evolved to reach the already implemented 5G mobile communication systems in various areas, including waveforms and energy harvesting techniques. Then, the reader will have a clear understanding of a specific proposal to enhance system performance, such as advances in the areas of MEC, UAVs, ISAC, and MIMO. Throughout this special issue, the reader will also identify the use of Machine Learning algorithms in the communication system context, which will be increasingly used in the years to come to optimize various operational and design issues.

We would like to extend our sincere gratitude to all the authors and the editorial team who made this special issue possible, as reflected in the publication of these papers.

Author Contributions: Conceptualization, M.E.R.-Á.; methodology, M.E.R.-Á., and I.Y.O.-F.; investigation, M.E.R.-Á., and I.Y.O.-F.; writing—original draft preparation, M.E.R.-Á.; writing—review and editing, I.Y.O.-F.; visualization, I.Y.O.-F.; supervision, I.Y.O.-F.; funding acquisition, M.E.R.-Á., and I.Y.O.-F. All authors have read and agreed to the published version of the manuscript.

Conflicts of Interest: The authors declare no conflicts of interest.

List of Contributions:

1. Mohammed, S.A.; Murad, S.S.; Albeyboni, H.J.; Soltani, M.D.; Ahmed, R.A.; Badeel, R.; Chen, P. Supporting Global Communications of 6G Networks Using AI, Digital Twin, Hybrid and Integrated Networks, and Cloud: Features, Challenges, and Recommendations. *Telecom* **2025**, *6*, 35. https://doi.org/10.3390/telecom6020035.
2. Arabian, F.; Shoushtari, M. A Comparative Study of Waveforms Across Mobile Cellular Generations: From 0G to 6G and Beyond. *Telecom* **2025**, *6*, 67. https://doi.org/10.3390/telecom6030067.
3. Rodriguez-Gomez, A.I.; Rivero-Angeles, M.E.; Orea Flores, I.Y.; Gallegos-García, G. Performance Analysis of a Sound-Based Steganography Wireless Sensor Network to Provide Covert Communications. *Telecom* **2024**, *5*, 652–679. https://doi.org/10.3390/telecom5030033.
4. Arinze, S.N.; Obi, E.R.; Ebenuwa, S.H.; Nwajana, A.O. RF Energy-Harvesting Techniques: Applications, Recent Developments, Challenges, and Future Opportunities. *Telecom* **2025**, *6*, 45 https://doi.org/10.3390/telecom6030045.
5. Al-Askery, A.J.; Hasan, F.S.; Yassin, Y.A. Polar-Coded Differential/Quadrature Chaos Shift Keying Communication Systems for Underwater Acoustic Channels. *Telecom* **2024**, *5*, 476–486. https://doi.org/10.3390/telecom5020024.
6. Shaban, M. Development and Implementation of High-Gain, and High-Isolation Multi-Input Multi-Output Antenna for 5G mmWave Communications. *Telecom* **2025**, *6*, 14. https://doi.org/10.3390/telecom6010014.
7. Shoukat, H.; Khurshid, A.A.; Daha, M.Y.; Shahid, K.; Hadi, M.U. A Comparative Analysis of DNN and Conventional Signal Detection Techniques in SISO and MIMO Communication Systems. *Telecom* **2024**, *5*, 487–507. https://doi.org/10.3390/telecom5020025.

8. Shao, Y.; Gulliver, T.A. Transceiver Optimization for Multiuser Multiple-Input Multiple-Output Full-Duplex Amplify-and-Forward Relay Downlink Communications. *Telecom* **2024**, *5*, 216–227. https://doi.org/10.3390/telecom5010011.
9. Attar, I.S.; Mahyuddin, N.M.; Hindia, M.H.D.N. A Semi-Distributed Scheme for Mode Selection and Resource Allocation in Device-to-Device-Enabled Cellular Networks Using Matching Game and Reinforcement Learning. *Telecom* **2025**, *6*, 12. https://doi.org/10.3390/telecom6010012.
10. Patel, A.K.; Joshi, R.D. Fifth-Generation (5G) Communication in Urban Environments: A Comprehensive Unmanned Aerial Vehicle Channel Model for Low-Altitude Operations in Indian Cities. *Telecom* **2025**, *6*, 9. https://doi.org/10.3390/telecom6010009.
11. Ramos, A.; Inca, S.; Ferrer, M.; Calabuig, D.; Roger, S.; Monserrat, J.F. Simulation Framework for Detection and Localization in Integrated Sensing and Communication Systems. *Telecom* **2025**, *6*, 4. https://doi.org/10.3390/telecom6010004.
12. Xavier, R.; Silva, R.S.; Ribeiro, M.; Moreira, W.; Freitas, L.; Oliveira-Jr, A. Integrating Multi-Access Edge Computing (MEC) into Open 5G Core. *Telecom* **2024**, *5*, 433–450. https://doi.org/10.3390/telecom5020022.

https://doi.org/10.3390/telecom70100

telecom

Article

Development and Implementation of High-Gain, and High-Isolation Multi-Input Multi-Output Antenna for 5G mmWave Communications

Mahmoud Shaban

Department of Electrical Engineering, College of Engineering, Qassim University, Buraydah 52571, Saudi Arabia; s.mahmoud@qu.edu.sa

Abstract: This work introduces a high-performance multi-input multi-output (MIMO) antenna design to operate at the 28 GHz band. The proposed four-port MIMO antenna, in which each port comprises a 1 × 8 series-fed array, achieves peak gains of 13 dBi along with bandwidths of 1 GHz. Enhanced antenna performance is achieved through the optimal spacing of antenna elements and a decoupling methodology comprising a well-designed metamaterial unit cell, leading to reduced interference between antenna arrays. The design shows significantly suppressed mutual coupling to be less than −40 dB, a diversity gain that is very close to 10 dB, an envelope correlation coefficient of 0.00012, and a channel capacity loss of 0.147 bit/s/Hz, at 28 GHz. The experimental assessments confirmed these developments, endorsing the suggested design as a robust contender for 5G mmWave communications.

Keywords: antennas; MIMO systems; 5G wireless communications; mmWave; series-fed array

1. Introduction

The surging demand for data driven by the proliferation of electronic devices has placed significant pressure on current communication systems. This escalating demand necessitates higher data rates, demanding significant advancements in wireless technology and broadband integration. Each successive generation of wireless technology introduces new features and surpasses the performance of its predecessor. Fifth-generation (5G) wireless technology influenced a diverse spectrum of frequencies, each with distinct characteristics. Sub-1 GHz bands offer extensive coverage for outdoor users, and mid-band frequencies including 2.5, 3.5, 3.7–4.2, and 4.8–5 GHz offer a balance between coverage and data capacity. Conversely, millimeter-wave (mmWave) bands operating above 24 GHz promise exceptionally high data rates but currently suffer from limited coverage.

The Federal Communications Commission (FCC) has designated certain frequency ranges for 5G deployment, such as the 24 GHz, 28 GHz, 37, 40, and 47 GHz bands for licensed operations [1–5]. Additionally, the 64 to 71 GHz spectrum has been reserved for unlicensed applications, enabling technologies such as Wi-Gig and various ultra-fast wireless communication systems [6,7].

Multiple-input multiple-output (MIMO) antenna systems, which utilize multiple antennas at both the transmitting and receiving ends, substantially improve wireless communication performance in several key areas. By enabling the concurrent exchange of multiple data streams, MIMO technology boosts data rates, strengthens signal reliability, and elevates overall network efficiency. This has driven significant interest in advancing

MIMO antenna designs, especially for operation in the mmWave frequency spectrum, making it a prominent and dynamic research focus in modern telecommunications [8–10].

Antennas designed for the 28 GHz band (N261, 27.5–28.35 GHz) must address the challenges of mmWave propagation, including high path loss, atmospheric attenuation, and the need for dense network deployments. A critical requirement is a wide bandwidth to cover the allocated spectrum, enabling support for high data rates and channel aggregation in 5G new radio (NR). High gain (10–15 dBi or higher) and directional beamforming capabilities are essential to compensate for signal attenuation and enable precise user tracking in dynamic environments.

For MIMO configurations, antennas must achieve low mutual coupling (isolation > 30 dB) and ultra-low envelope correlation coefficients (ECC < 0.1) to ensure independent operation and maximize spatial multiplexing gains. Compact physical dimensions are equally vital as mmWave systems often require dense antenna arrays (e.g., 4×4 or 8×8 MIMO) integrated into small devices like smartphones or IoT modules.

Critical requirements for successful MIMO antenna design in 5G involve achieving high spectral efficiency, minimizing interference, and dynamically adapting the bandwidth—key factors in advancing 5G capabilities. Meeting these demands relies on spectrum-efficient infrastructure, resilient interference mitigation strategies, and flexible bandwidth allocation to address shifting network requirements. MIMO technology has been widely acknowledged for its ability to enhance spectral efficiency via methods like spatial multiplexing and beamforming. These approaches leverage adaptive radiation patterns, polarization diversity, and directional signal transmission to suppress interference and optimize performance in dense communication environments [11–15].

MIMO antenna design faces several critical challenges, including minimizing interference between antenna elements (mutual coupling), reducing size and complexity, maximizing signal strength (gain), broadening the usable frequency range (bandwidth), and ensuring efficient operation. Previous research has explored various innovative antenna designs to overcome these challenges and advance MIMO technology.

One notable example is the 5G-compatible MIMO antenna presented in reference [16]. This design operates at 28 GHz and features a unique "infinity-shaped" structure incorporating three circular rings. This design aimed to enhance performance at high frequencies and demonstrated an antenna radiation efficiency of 92% and a peak gain of 6.1 dBi.

A notable improvement in this field is the MIMO antenna with a "tree-like" configuration proposed in [17]. Tailored for 5G systems, this design integrates four curved radiating elements to achieve broad bandwidth coverage. Its radiation patterns are characterized at operational frequencies of 28 GHz, 33 GHz, and 38 GHz, demonstrating peak gains of 10.58, 8.87, and 11.45 dBi, respectively.

The work in reference [18] presents an MIMO antenna optimized for dual-band 28/38 GHz operation, delivering high-gain performance. This configuration achieves peak realized gains of 4.15 dBi and 7.73 dBi at the respective frequencies, with total efficiencies reaching 80.13% and 85.44%.

Furthermore, an MIMO antenna comprising a crescent-shaped slot elliptical operating at dual frequencies (28/38 GHz) was introduced in reference [19]. Utilizing a linear array of 1×4 elements, this design achieved a bandwidth of 7.8 GHz and peak gain of 10.4 dBi.

Reference [20] presents an implementation that achieves high isolation (up to 23 dB) between MIMO antenna elements by employing spatial and polarization diversity. This antenna operates across the 27.25–29 GHz and 34.5–41 GHz frequency bands, achieving gains of 9.5 and 11.7 dBi at 28 and 38 GHz, respectively.

Reference [21] introduced an MIMO antenna with four-element series slot-fed array for circular polarization diversity, utilizing 45° inclined slots and a straight microstrip line feeder. The realizable antenna footprint is 20 × 27.7 mm^2. The measured results confirmed that the suggested design features a bandwidth of 3.8 GHz and a peak gain of 11.86 dBi.

Metamaterials offer a powerful tool for optimizing the performance of MIMO systems. By carefully integrating these engineered materials, it is possible to achieve significant improvements in gain, isolation, and overall system efficiency. The proposed design, with its unique combination of features, demonstrates the potential of metamaterials in revolutionizing future wireless communication technologies. Research has explored various approaches to enhance isolation in dense antenna arrays. These include utilizing periodic metamaterial photonic bandgap structures for applications like MIMO and synthetic aperture radar (SAR) [22–26], investigating the impact of surface wave effects caused by metasurface in MIMO/SAR systems [26], and utilizing metamaterials as electromagnetic bandgap (EM-BG) layers to minimize mutual coupling and improve performance in challenging communication environments [27–31].

The proposed design presented here offers several key advantages:

- It achieves a high peak gain of 13 dBi operating at 28.8 GHz that enhances signal directivity and extends the communication range.
- It realizes a bandwidth of 1.0 GHz, around 28.8 GHz, measuring a reduced reflection coefficient of −27 dB, enabling the low loss operation.
- It demonstrates isolation levels of 34.5 dB across its ports, effectively rejects interference and ensures strong, clear signal transmission.

These results were experimentally validated to test the design's reliability, creating a promising candidate for future 5G mmWave. The organization of this paper is as follows: Section 2 discusses the methodologies and materials, Section 3 presents the results and discussion, and Section 4 provides the conclusions.

2. Materials and Methods

The design of a rectangular antenna starts with determining the length and width of the patch forming the array. The effective relative permittivity for the substrate dielectric material is computed by [32]:

$$\varepsilon_{reff} = \frac{\varepsilon_r + 1}{2} + \frac{\varepsilon_r - 1}{2}\left[1 + 12\frac{h}{W}\right]^{\frac{-1}{2}} \tag{1}$$

The effective length of the patch, L_{eff}, is calculated by

$$\Delta L = 0.412h\frac{\left(\varepsilon_{reff} + 0.3\right)\left(\frac{W}{h} + 0.264\right)}{\left(\varepsilon_{reff} - 0.254\right)\left(\frac{W}{h} + 0.8\right)} \tag{2}$$

A suitable patch width that ensures low loss and therefore higher radiation efficiency is determined by

$$W = \frac{c}{2f_r}\sqrt{\frac{2}{\varepsilon_r + 1}} \tag{3}$$

The fringing fields at the edges of a microstrip patch antenna effectively extend its physical length (L) by an additional length (ΔL) on each end. The total effective length (L_{eff}) is therefore expressed as follows:

$$L_{eff} = L + 2\Delta L \tag{4}$$

Therefore, the length of the patch is calculated as follows:

$$L = \frac{c}{2f_r\sqrt{\varepsilon_{reff}}} - 2\Delta L \tag{5}$$

The input resistance (R_{in}) of the antenna at resonance frequency and at the patch edges (at y = 0) is determined by the parallel combination of the conductances of the two radiating slots. This is expressed as follows:

$$R_{in}(0) = \frac{1}{2(G_1 + G_{12})} \tag{6}$$

Here, G_1 denotes conductance of single radiating slot in the patch, G_{12} represents the conductance arising from the mutual coupling between the two radiating slots of the patch. These terms can be computed using the methodology outlined in [32].

$$G_1 = \frac{1}{120\pi^2}\int_0^{\pi}\left[\frac{sin\left(\frac{k_0w}{2}cos\theta\right)}{cos\theta}\right]^2 sin^3\theta d\theta \tag{7}$$

$$G_{12} = \frac{1}{120\pi^2}\int_0^{\pi}\left[\frac{sin\left(\frac{k_0w}{2}cos\theta\right)}{cos\theta}\right]^2 J_0(k_0 L\sin\theta)sin^3\theta d\theta \tag{8}$$

Here, J_0 represents the first kind of Bessel function of the order zero, and $k_0 = 2\pi/\lambda_0$ is the free-space phase constant for λ_0, the free-space wavelength. Input impedance at distance y_0 from the edge along the patch length can be expressed as follows [32]:

$$R_{in}(y_0) = R_{in}(0)cos^2\left(\frac{\pi}{L}y_0\right) \tag{9}$$

Antenna directivity (D) quantifies the effectiveness of the array in concentrating radiation in a specific direction. It can be calculated using the array factor (AF) and element pattern, providing insights into the array's ability to focus radiation in the desired direction. The antenna gain is directly proportional to its directivity, which quantifies how efficiently the antenna radiates energy compared to an ideal isotropic radiator, and is proportional to its directivity. In other words, higher directivity leads to higher gain, reflecting the antenna's ability to concentrate radiation in a particular direction. The directivity of the patch antenna is calculated as follows [32]:

$$D = \left(\frac{2\pi w}{\lambda_0}\right)^2 \frac{\pi}{I_1} \tag{10}$$

where

$$I_1 = \iint_0^{\pi}\left[\frac{sin\left(\frac{K_0w}{2}cos\theta\right)}{cos\theta}\right]^2 sin^3\theta\, cos^2\left(\frac{K_0L_{eff}}{2}sin\theta sin\phi\right)d\theta d\phi \tag{11}$$

When designing a series-fed antenna array, several key design considerations and equations come into play. The spacing between antenna elements is crucial for achieving the desired radiation pattern. The spacing distance (d) between adjacent elements in a uniform linear array is typically set to half the guided wavelength (i.e., $\lambda_g/2$) to promote constructive interference in the major lobe direction and enable good radiation efficiency.

The array factor (AF) of a uniform linear array with N elements quantifies the combined contribution of all elements to the array's overall radiation pattern and it is given by [32]

$$AF(\theta) = \sum_{n=1}^{N} e^{j(n-1)kdcos(\theta)} = \frac{sin\left(\frac{N}{2}kd\ cos\theta\right)}{\frac{1}{2}kd\ cos\theta} \tag{12}$$

This equation describes the combined contribution of each element to the overall radiation pattern. Designing an antenna array requires the careful selection of both N and d to achieve optimal radiation characteristics, including gain, directivity, and beamwidth. Increasing N proportionally enhances gain and directivity and narrows the major lobe beamwidth, thereby improving angular resolution, which is required for specific wireless applications such as radar. However, a larger N can lead to increased sidelobe levels, which may cause interference, necessitating amplitude tapering techniques to mitigate these effects. Practical constraints such as physical size, cost, and feeding network losses further limit the feasible number of elements in compact array designs. Element spacing is equally critical, with an optimal spacing of d = $\lambda/2$ minimizing mutual coupling, preventing grating lobes, and ensuring uniform phase progression across the array. Using smaller spacing (d < $\lambda/2$) yields a compact array with a slightly enhanced major lobe gain due to constructive interference, but it also increases mutual coupling and the risk of pattern distortion. Conversely, larger spacing (d > $\lambda/2$) can reduce coupling issues but often introduces undesirable grating lobes, which degrade overall array performance. In special cases, such as d = λ or greater, the formation of significant grating lobes makes the design impractical for high-performance applications. Practical applications in 5G/MIMO systems typically require a balanced approach, with a spacing of d = $\lambda/2$ (or slightly less) and a moderate number of elements to achieve a compact design. In this study, the simulation results indicate that an array with N = 8 elements and spacing slightly less than half of the guided wavelength ($\lambda_g/2$) can achieve a high-gain, high-efficiency MIMO antenna.

An important step for an efficient antenna array is to attain good impedance matching that is critical for maximizing power transfer efficiency within the array. Ensuring that each antenna element and the feed network are impedance-matched to the system impedance (often 50 Ω) minimizes reflections and enhances overall performance.

MIMO antennas are assessed using several parameters that measure their performance. Among these metrics, the envelope correlation coefficient (ECC) stands out as a key measure of the correlation between signal envelopes received from different antennas.

In this regards, an ECC calculation helps in evaluating the correlation among signals and their impact on MIMO performance, which can be determined using

$$ECC = \frac{\left|S_{ii}^{*}S_{ij} + S_{ji}^{*}S_{jj}\right|^{2}}{\left(1 - \left(|S_{ii}|^{2} + |S_{ji}|^{2}\right)\right) + \left(1 - \left(|S_{jj}|^{2} + |S_{ij}|^{2}\right)\right)} \tag{13}$$

where the terms S_{ii}, S_{jj}, S_{ii}^*, and S_{ji}^* denote to the complex and conjugate complex components of the scattering matrix, representing the received signals by antennas i and j, respectively. A lower ECC value indicates strong diversity performance, suggesting an improved reception of uncorrelated signals and enhanced diversity gain.

Another important parameter is the diversity gain (DG), which quantifies the received signal quality achieved by employing multiple numbers of antennas to reduce signal fading through spatial diversity. It is measured in dB and is related to the ECC as follows:

$$DG = 10\sqrt{1 - (ECC)^{2}} \tag{14}$$

The channel capacity loss (CCL) metric measures the reduction in the maximum achievable data rate of the MIMO system. It is calculated using the following expression:

$$CCL = -\log_2\left|\beta^R\right|$$
where
$$\beta^R = \begin{pmatrix} \beta_{11} & \beta_{12} & \cdots & \beta_{1N} \\ \beta_{21} & \beta_{22} & \cdots & \beta_{2N} \\ \vdots & \vdots & \vdots & \vdots \\ \beta_{N1} & \beta_{N2} & \cdots & \beta_{NN} \end{pmatrix} \quad (15)$$
where
$$\beta_{ii} = 1 - \sum_{n=1}^{N} \left|S_{in}^{*} S_{ni}\right|$$
and
$$\beta_{ij} = -\sum_{n=1}^{N} \left|S_{in}^{*} S_{nj}\right|$$

The matrix β^R represents the received signal coefficients in an MIMO system. The diagonal elements β_{ii} correspond to the power received by antenna ii, accounting for the influence of signals from all other antennas. The off-diagonal elements β_{ij} describe the correlation between the received signals by antennas i and j, in the MIMO array.

3. Results and Discussion

3.1. Single-Array Antenna

Figure 1 shows the design of a single-port antenna array built on a substrate material of Rogers RT-5880, which has a dielectric constant (ε_r) of 2.2, a thickness of 0.787 mm, and a loss tangent of 0.0009. The array consists of 1 × 8 rectangular elements, and each patch is fed by a microstrip feedline. Every element features a main radiating patch with dimensions $W_p \times L_p$. The antenna impedance was calibrated using inset feeders, each sized $W_{inset} \times L_{inset}$, to improve impedance matching. To enhance the antenna's bandwidth, two rectangular slots with dimensions $W_{slot} \times L_{slot}$ were incorporated, positioned at a distance of Wp/4 from the right and left patch edges. The patches were fed using a 50 Ω feedline, with dimensions of $W_{mf} \times L_{mf}$. The substrate and ground plane were designed with dimensions $W_s \times L_s$ and $W_g \times L_g$, respectively. The precise optimization of the slot dimensions was performed to achieve optimal impedance matching at the target resonant frequency. Through parametric sweep and optimization, it was identified that the length of the inset feeders, L_{inset}, was the most effective dimension that greatly influenced the antenna array input impedance.

Table 1 provides a comprehensive summary of all the design parameters. Figure 1 displays details of the proposed design. In the antenna's reflection coefficient described by the scattering parameter, S_{11}, depicted in Figure 2, resonance is observed at a frequency of 28.8 GHz, reaching −28.5 dB reflection coefficients.

The bandwidths surrounding this resonant frequency is determined to be ~1 GHz (from 28.3 to 29.3 GHz). These simulation results are essential for predicting and assessing the antenna's performance before fabricating a prototype of the proposed design.

The parametric examination highlights that, among all the design dimensions, the lengths of insets (L_{inset} or y_0) constitute the most key parameter that directly influences the input impedance of the antenna. Varying L_{inset} from 0.1 to 0.6 mm caused substantial resonance frequency shifts and additionally altered bandwidth and reflection coefficient values, as demonstrated in Figure 3. These changes arise from the complex interchange between the antenna's input impedance and the two inset lengths. Modifying these lengths significantly alters the impedance characteristics, which in turn affects the resonance, band-width, and the antenna reflection properties. The optimized performance was observed when the inset length was set to 0.3 mm.

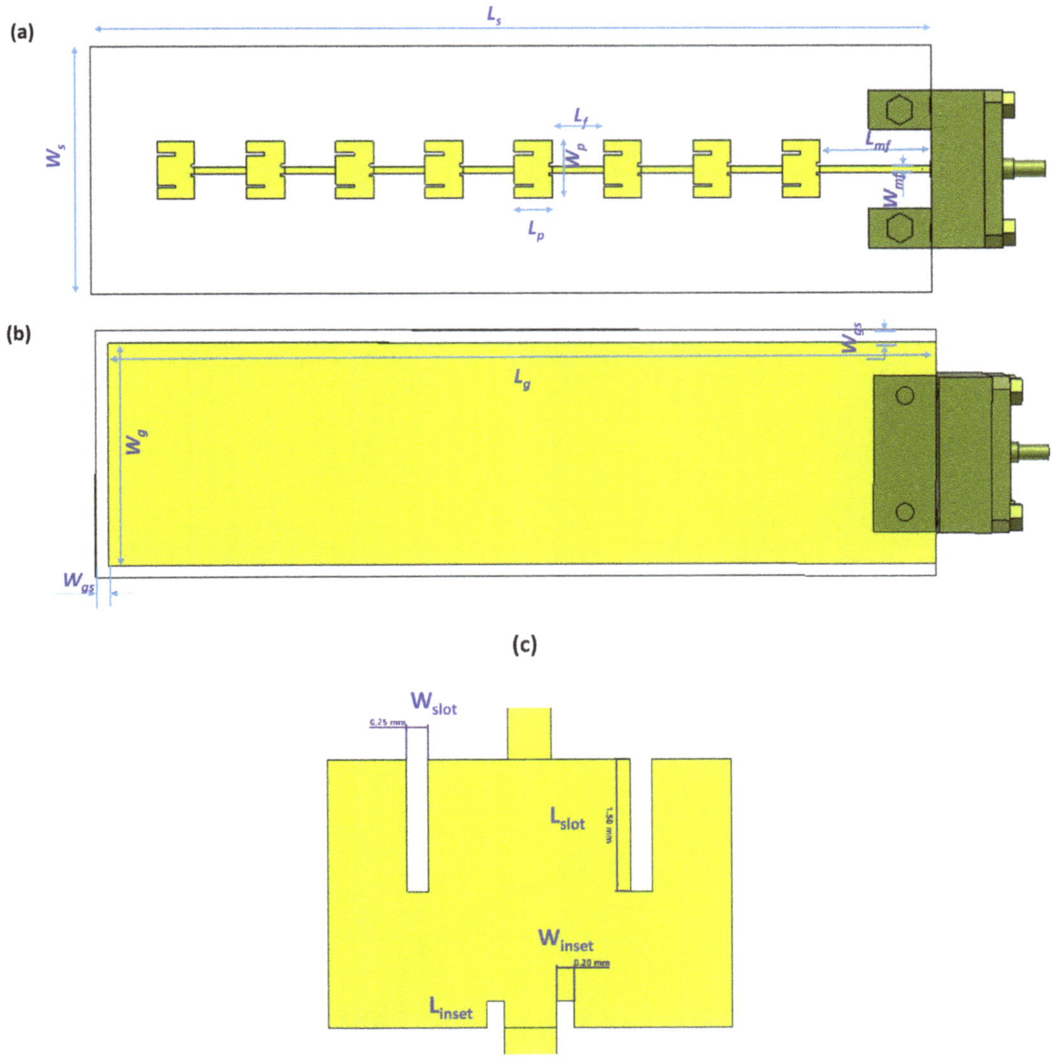

Figure 1. (**a**) Front, (**b**) back view of single-port antenna; and (**c**) dimensions of slots created in each single-patch.

Table 1. Optimized dimensions of single-array antenna.

Parameter	W_p	L_p	W_s	L_s	W_f	L_f	W_{mf}	L_{mf}	W_g	L_g	W_{gs}	W_{slot}	L_{slot}	W_{inset}	L_{inset}
Value (mm)	4.65	3.06	20	68	0.5	4.16	0.6	9.02	18	67	1	0.25	1.5	0.2	0.3

Figure 4 presents the computed peak gain of the proposed antenna as a function of frequency. The series-fed antenna array comprising 1×8 elements achieves a peak gain of 13 dBi at the resonance frequency of 28.8 GHz. This demonstrates the antenna's capability to focus electromagnetic (EM) energy in a specific direction within the investigated band, highlighting its fitness for the N261 band operation. The peak gain reflects the antenna's efficiency in transmitting and receiving signals with enhanced precision and effectiveness across the designed frequency range. Regarding radiation efficiency, the simulation results confirm that this gain corresponds to a radiation efficiency of 91.85% and a total efficiency of 82.9%.

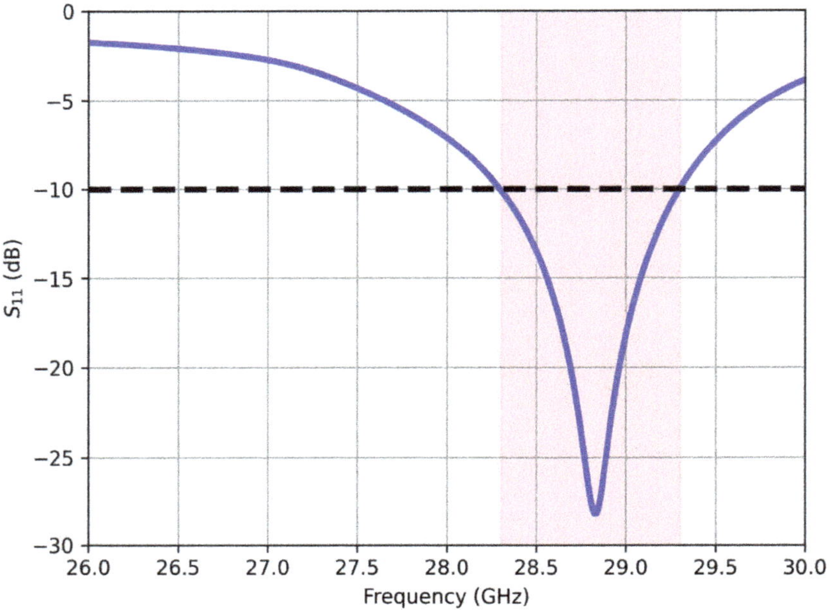

Figure 2. Simulated reflection coefficient of the single-port antenna as a function of frequency. The highlighted area shows bandwidth range.

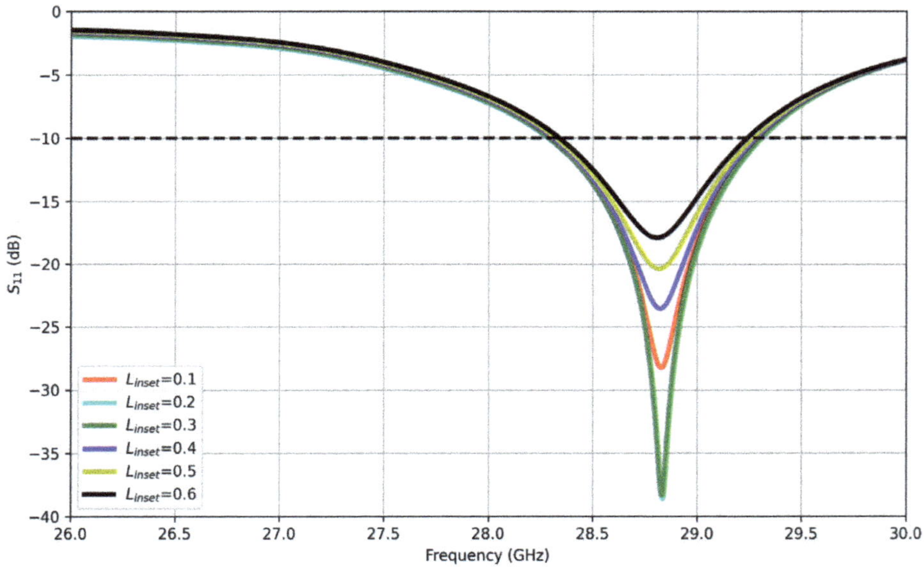

Figure 3. A parametric analysis of the inset length for the proposed antenna design.

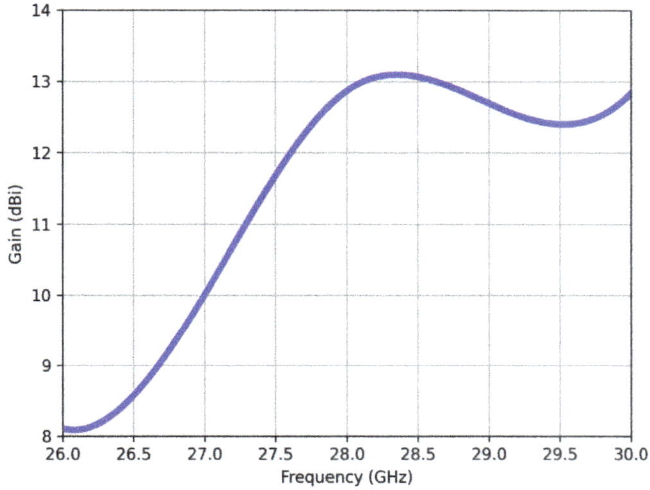

Figure 4. Frequency dependence of peak gain of single-array antenna.

To ascertain the primary direction(s) in which the proposed antenna predominantly concentrates EM radiation, an analysis of the radiation pattern is conducted. The investigation aims to determine the main lobes and nulls in the radiation pattern, which provide crucial information about the antenna's directional characteristics. This analysis helps in identifying the specific azimuth and elevation angles where the antenna exhibits maximum radiation concentration, thereby aiding in optimizing the antenna's orientation and performance for targeted communication or sensing applications. The radiation patterns displayed in Figure 5 offer a detailed insight into the performance of a single 1 × 8 series-fed array antenna across various frequencies in both the electric field plane (E-plane) and magnetic field plane (H-plane). At 28.8 GHz in the E-plane, the antenna realizes a gain of 11.6 dBi, with the main lobe pointed at θ = 3° and a beamwidth of 64.2°. In the H-plane, the major lobe displays a gain of 12.7 dBi, directed at θ = 5°, with a beamwidth of 19.3°.

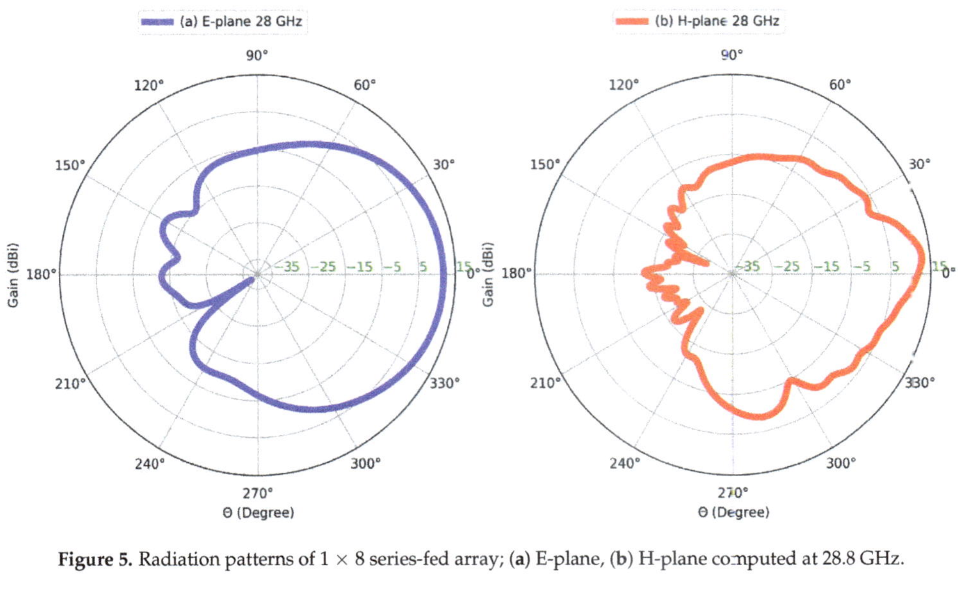

Figure 5. Radiation patterns of 1 × 8 series-fed array; (**a**) E-plane, (**b**) H-plane computed at 28.8 GHz.

3.2. Four-Port MIMO Antenna

The MIMO arrangement in this design is tailored for a four-port antenna operating at 28.8 GHz resonance frequency. Particular emphasis is placed on optimizing the spacing distance between elements to improve isolation, minimize mutual coupling, and enhance the overall performance of the MIMO system. The structure comprises four individual ports compactly integrated within an area of 68 × 80 mm^2, employing the same substrate as that of the single-array antenna. Figure 6a,b display the MIMO antenna structure in top and bottom views, showing SMAs (Sub-Miniature version A) RF connectors for each port. The distance between each array antenna and the MTM cells is 5.68 mm, which is very close to $\lambda/2$, which is 5.36 mm. The simulation results in Figure 6c depict the S-parameters, specifically the mutual coupling coefficients (S_{12}, S_{32}, S_{42}) within the MIMO system. Port 2, chosen as the central port, is notably influenced by its adjacent ports. The results indicate that mutual coupling remains below −41 dB across the entire operational bandwidth, underscoring the effectiveness of the proposed decoupling technique in mitigating interference among the MIMO ports. This high level of isolation demonstrates the success of the decoupling strategy, ensuring improved performance and reduced interference in the suggested MIMO configuration.

The significant mutual coupling reduction is primarily attributed to the design of the metamaterial (MTM) cell, as illustrated in Figure 7a,b. The extracted parameters—relative permittivity (ε_r), relative permeability (μ_r), and refractive index (n)—demonstrate positive and negative values that change with frequency. This unique behavior of the MTM cell plays a critical role in enhancing isolation and minimizing interference within the MIMO system. Specifically, the design output features negative ε_r values through the investigated frequency band and demonstrates resonant behavior. This underlines the MTM cell behavior. Furthermore, it enhances the gain as the MTM enables focusing EM energy in specific directions. Moreover, regarding isolation, it absorbs or prevents surface wave propagation in certain directions. The split-ring-like cells operate similar to bandstop filters, filtering out unwanted interference and thus improving the attenuation of mutual coupling.

Figure 8 presents the computed frequency-dependent ECC and DG of the proposed MIMO construction. Across the frequency range, the ECC values remain consistently below 0.00012, significantly lower than the conventional threshold of 0.5. This indicates excellent isolation and minimal signal correlation between antenna ports. At the same time, the DG achieved very close level to the 10 dB, reflecting strong signal reception in the desired direction and the effective mitigation of fading effects. Additionally, the CCL values exhibit strong performance. At 28.8 GHz, the CCL is calculated to be 0.147 bit/s/Hz. This value is well below the standard reference threshold of 0.4 bit/s/Hz, demonstrating that the proposed antenna design effectively preserves channel capacity. Such a low CCL reflects strong signal integrity, reduced interference, and the efficient utilization of the available bandwidth, which are critical for high-performance wireless communications. This result underscores the robustness and reliability of the antenna design, making it highly suitable for applications requiring high data rates and reliable connectivity.

The simulated radiation pattern of the MIMO structure is depicted in Figure 9. At 28.8 GHz, the antenna demonstrates a gain of 12 dBi in the E-plane, with the major lobe oriented at $\theta = 1°$ and a beamwidth of 61.6°. In the H-plane, the major lobe achieves a gain of 12.9 dBi, directed at $\theta = 5°$, with a beamwidth of 18.9°. These results underscore the directional characteristics of the MIMO structure in the E- and H-plane, highlighting its ability to focus energy efficiently in specific directions.

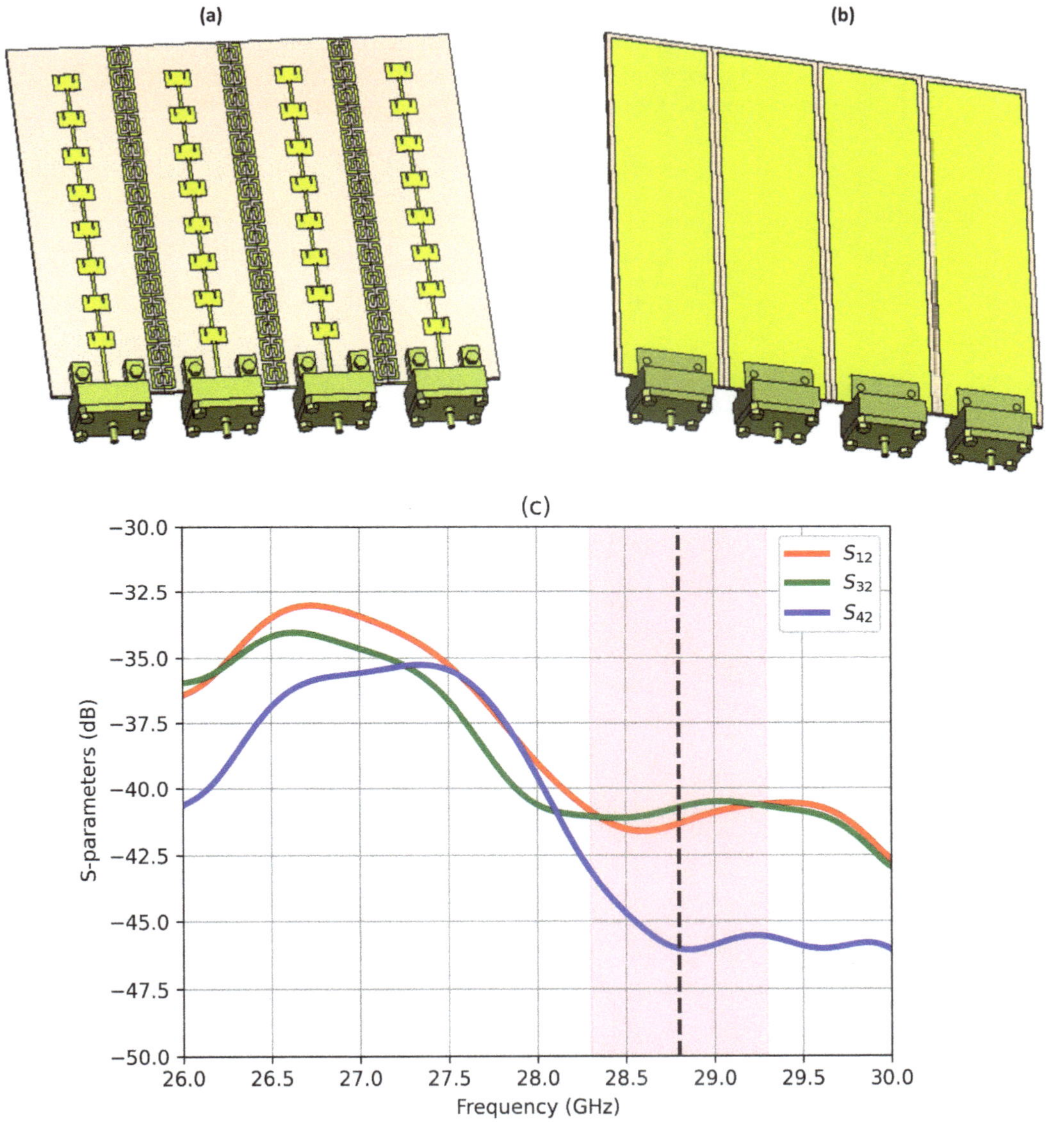

Figure 6. (**a**) Top, (**b**) bottom view of four-port MIMO antenna; and (**c**) simulated s-parameters versus frequency.

3.3. Experimental Results

A Vector Network Analyzer (VNA) with a frequency range of up to 67 GHz was employed to evaluate the scattering parameters of the fabricated structure. The antennas being tested were attached to the VNA via end-launch RF connectors, as shown in Figure 10. The depicted measurement setup serves multiple purposes: (a) gauging the reflection coefficients of individual ports, (b) assessing the mutual coupling coefficients among the ports, (c) presenting a sample result from the measurement of S_{11}, and (d) a comparison between the simulation and experimental results. The measured and simulated results demonstrate good alignment in all of the measured frequency band. However, minor discrepancies between the simulated and measured results can be accredited to aspects such as impedance mismatch, the influence of the SMA connector, environmental sceneries, and/or modeling assumptions that may affect the experimental setup parameters. Nonetheless, the measured bandwidths do not significantly deviate from the simulated values, confirming the consistent performance of the realized antenna prototype across the entire measured frequency band.

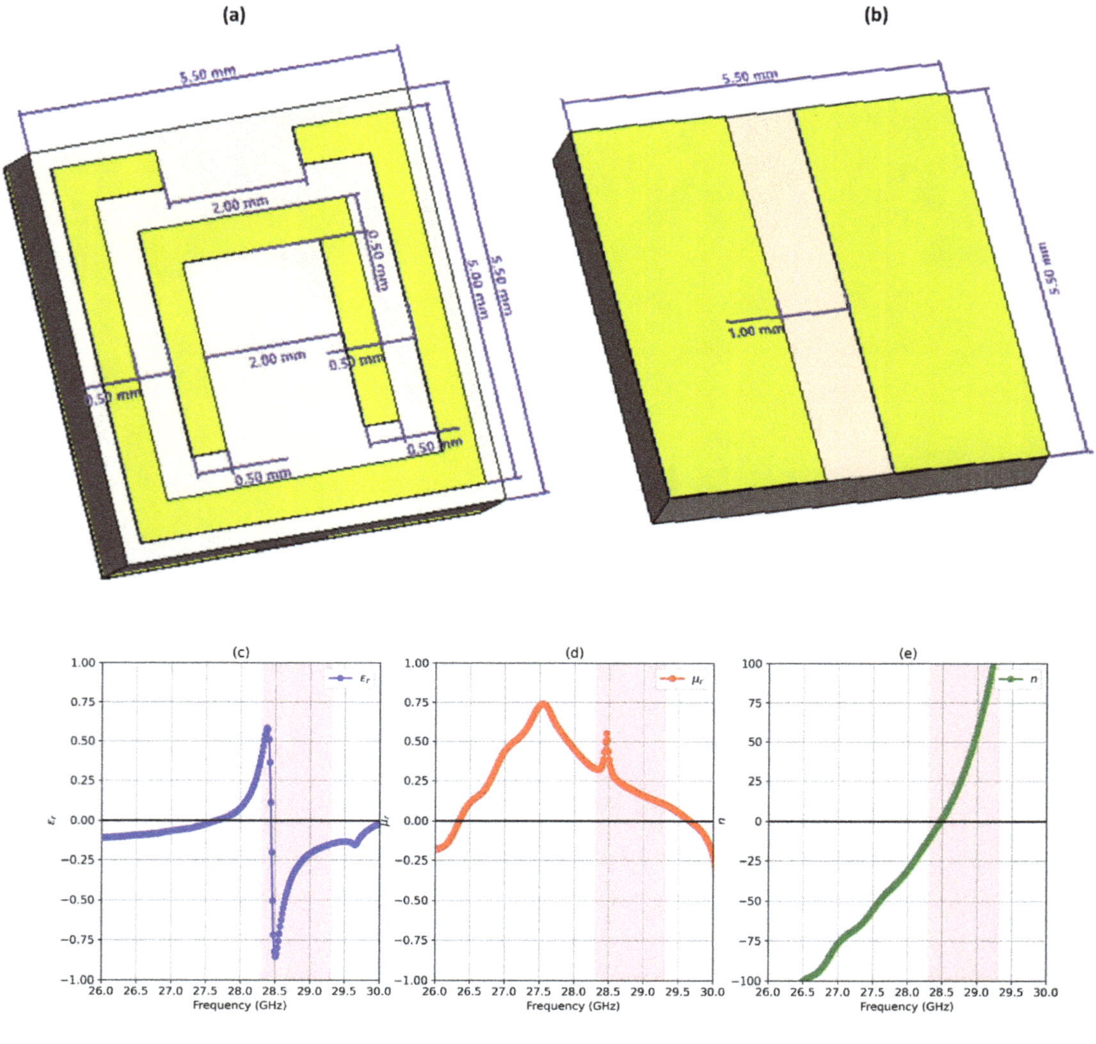

Figure 7. (**a**) Top, (**b**) bottom views of MTM unit cell and its computed (**c**) ε_r, (**d**) μ_r, and (**e**) n parameters.

Table 2 provides a comparative analysis of the proposed MIMO antenna design alongside the latest reported works. The results validate that the proposed design achieves highly competitive performance across key metrics, particularly in bandwidth, gain, and isolation. The comparison highlights the superiority of the proposed MIMO antenna over existing designs, showcasing improved bandwidth efficiency, gain, and isolation characteristics. By contextualizing the proposed design within recent advancements, this assessment underscores its effectiveness for advancing MIMO technology in next-generation 5G communication systems.

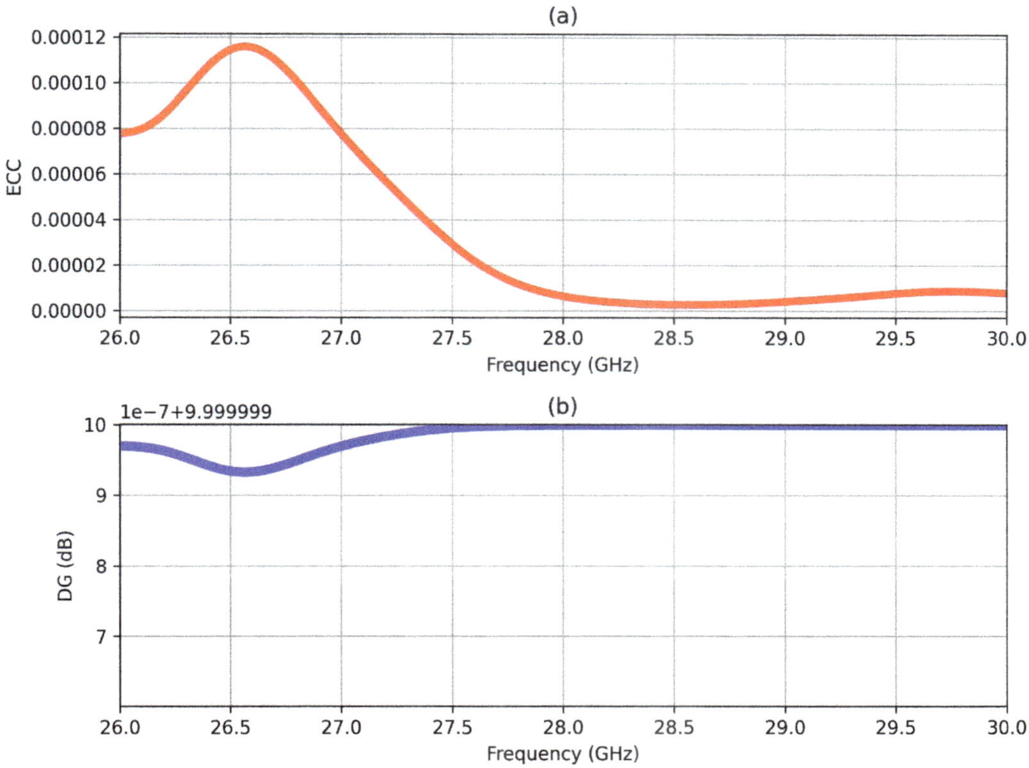

Figure 8. Computed frequency dependence of (**a**) ECC and (**b**) DG of four-port MIMO antenna.

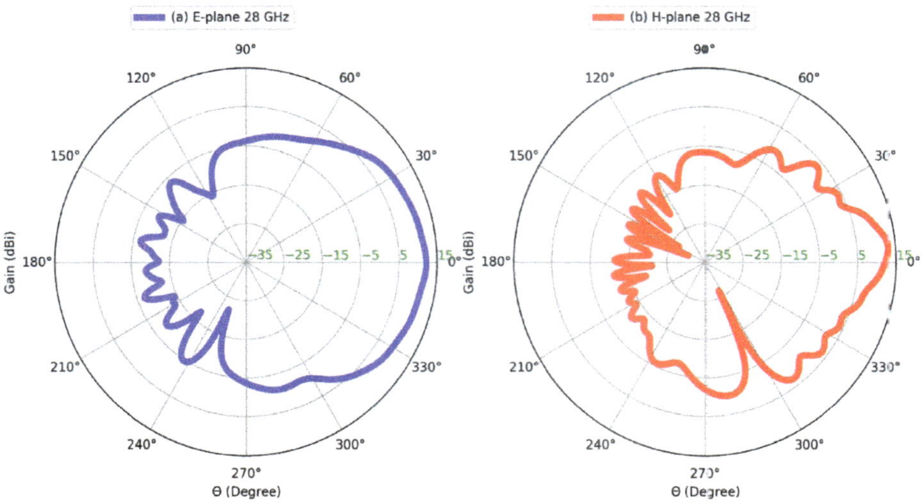

Figure 9. Radiation patterns of the 4-port MIMO antenna at 28.8 GHz: (**a**) E-plane and (**b**) H-plane.

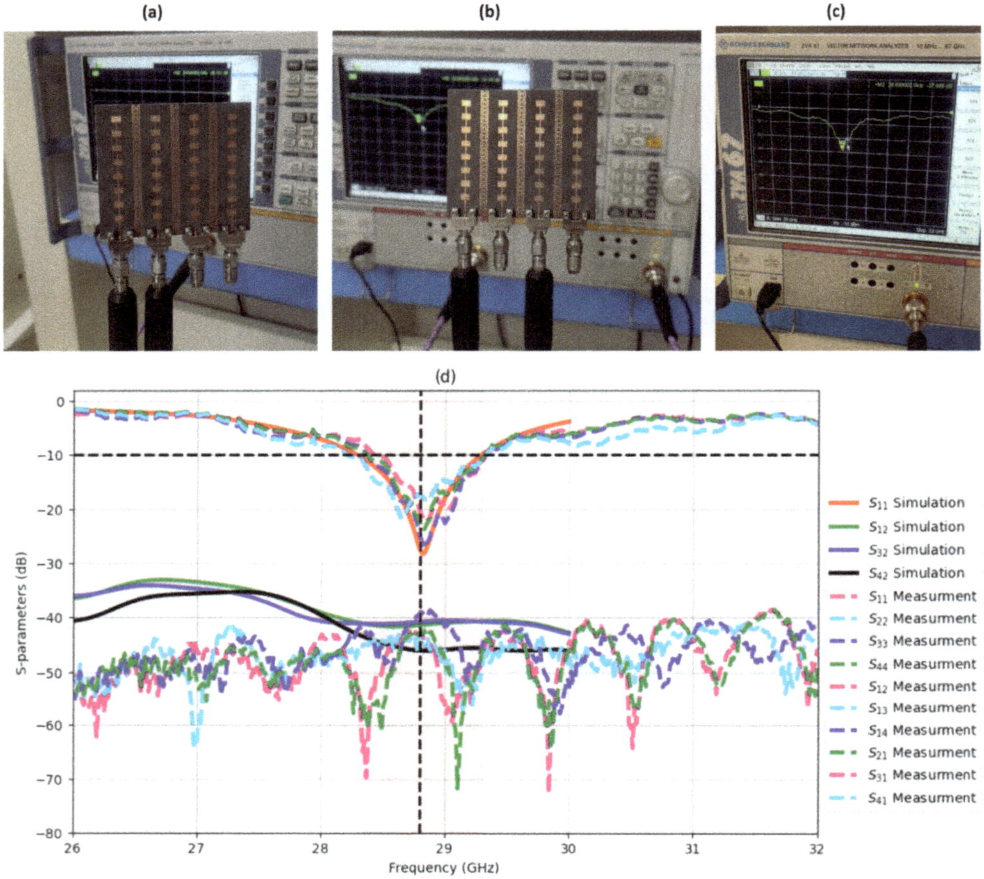

Figure 10. Measurement setup of (**a**) evaluating the reflection coefficients of individual ports, (**b**,**c**) assessing mutual coupling coefficients among the ports, and (**d**) displaying a sample result from the measurement of S_{11}.

Table 2. Comparing attained results of this work with those of recent designs in terms of many assessment metrics.

Reference No.	Frequency (GHz)	No. of Ports	Dimensions (mm^2)	Bandwidth (GHz)	Gain (dBi)	Isolation (dB)
[21]	28	2	20 × 27	3.6	11.9	37
[23]	28/39/48.7	2	10 × 16	1.8 (28 GHz) 6.2 (39 GHz) 3.8 (48.7 GHz)	9.8 (28 GHz) 9.5 (39 GHz) 11.5 (48.7 GHz)	17
[33]	28	4	30 × 35	4.1	8.3	22
[34]	28	4	32 × 8	2.6	7.5	25
[35]	28/38	2	55 × 110	1.07 (28 GHz) 1.43 (38 GHz)	7 (28 GHz) 8 (28 GHz)	27
[36]	28/38	2	7.5 × 8.8	1.23 (28 GHz) 1.06 (28 GHz)	6.60 (28 GHz) 5.86 (38 GHz)	20
This work	28.8	4	68 × 80	1.0	13	41

The proposed design showed high performance regarding its gain and isolation. The latter is owed to the precisely designed MTM, which played a crucial role in achieving high isolation and low mutual coupling, as clearly demonstrated by the experimental measurements. Additionally, the proposed design, with its achieved results, is well suited for low-power devices, including those powered by batteries [37]. Regarding its applicability in real-world scenarios, several additional factors should be considered when deploying the proposed MIMO antenna system. These include environmental conditions, integration with existing communication infrastructure, potential interference from other devices, and the impact of physical obstructions. Furthermore, practical aspects such as manufacturing tolerances, cost-effectiveness, and scalability for mass production must also be addressed to ensure seamless its implementation in real-world applications.

4. Conclusions

This study presents a high-performance MIMO antenna design tailored for a 28 GHz band. The compact four-port MIMO structure, with dimensions of 68×80 mm^2, incorporates a 1×8 series-fed array for each port, achieving a peak gain of 13 dBi and a bandwidth of 1 GHz at 28.8 GHz. Enhanced performance is achieved through calibrated antenna element spacing and a decoupling strategy utilizing carefully designed MTM unit cells, which successfully minimize cross-talk interaction among antenna elements. The design demonstrates exceptional performance metrics, including mutual coupling below −41 dB, an ECC of 0.00012, a CCL of 0.147 bit/Hz/s and DG approaching 10 dB. Experimental validation confirms these advancements, positioning the proposed design as a strong contender for mmWave communication systems operating in the 28 GHz band.

Funding: This research received no external funding.

Institutional Review Board Statement: Not applicable.

Informed Consent Statement: Not applicable.

Data Availability Statement: All data supporting the findings of this study are already included within the article.

Acknowledgments: The researcher would like to thank the Deanship of Graduate Studies and Scientific Research at Qassim University for financial support (QU-APC-2025).

Conflicts of Interest: The author declares no conflicts of interest.

References

1. Federal Communications Commission. *Report and Order and Further Notice of Proposed Rulemaking. The Matter of Revision of the Commission's Rules to Ensure Compatibility with Enhanced*; Federal Communications Commission: Washington, DC, USA, 2016; Volume 911, pp. 94–102.
2. Attiah, M.L.; Isa, A.A.; Zakaria, Z.; Abdulhameed, M.K.; Mohsen, M.K.; Ali, I. A survey of mmWave user association mechanisms and spectrum sharing approaches: An overview, open issues and challenges, future research trends. *Wirel. Netw.* **2020**, *26*, 2487–2514. [CrossRef]
3. Mandloi, M.; Gurjar, D.; Pattanayak, P.; Nguyen, H. *5G and Beyond Wireless Systems*; Springer: Berlin, Germany, 2021.
4. Aboagye, S.; Saeidi, M.A.; Tabassum, H.; Tayyar, Y.; Hossain, E.; Yang, H.C.; Alouini, M.S. Multi-band wireless communication networks: Fundamentals, challenges, and resource allocation. *IEEE Trans. Commun.* **2024**, *72*, 4333–4383. [CrossRef]
5. Hong, W.; Jiang, Z.H.; Yu, C.; Hou, D.; Wang, H.; Guo, C.; Hu, Y.; Kuai, L.; Yu, Y.; Jiang, Z.; et al. The role of millimeter-wave technologies in 5G/6G wireless communications. *IEEE J. Microw.* **2021**, *1*, 101–122. [CrossRef]
6. Daraseliya, A.; Korshykov, M.; Sopin, E.; Moltchanov, D.; Andreev, S.; Samouylov, K. Coexistence analysis of 5G NR unlicensed and WiGig in millimeter-wave spectrum. *IEEE Trans. Veh. Technol.* **2021**, *70*, 11721–11735. [CrossRef]
7. Xu, X.; Chen, Q.; Jiang, H.; Huang, J. Millimeter-wave NR-U and WiGig coexistence: Joint user grouping, beam coordination, and power control. *IEEE Trans. Wirel. Commun.* **2021**, *21*, 2352–2367. [CrossRef]

8. Alnemr, F.; Ahmed, M.F.; Shaalan, A.A. A compact 28/38 GHz MIMO circularly polarized antenna for 5 G applications. *J. Infrared Millim. Terahertz Waves* **2021**, *42*, 338–355. [CrossRef]
9. Sabek, A.R.; Ali, W.A.; Ibrahim, A.A. Minimally coupled two-element MIMO antenna with dual band (28/38 GHz) for 5G wireless communications. *J. Infrared Millim. Terahertz Waves* **2022**, *43*, 335–348. [CrossRef]
10. Cuneray, K.; Akcam, N.; Okan, T.; Arican, G.O. 28/38 GHz dual-band MIMO antenna with wideband and high gain properties for 5G applications. *AEU-Int. J. Electron. Commun.* **2023**, *162*, 154553. [CrossRef]
11. Ali, S.A.; Wajid, M.; Kumar, A.; Alam, M.S. Design challenges and possible solutions for 5G SIW MIMO and phased array antennas: A review. *IEEE Access* **2022**, *10*, 88567–88594. [CrossRef]
12. Tiwari, P.; Gahlaut, V.; Kaushik, M.; Rani, P.; Shastri, A.; Singh, B. Advancing 5G connectivity: A comprehensive review of MIMO antennas for 5G applications. *Int. J. Antennas Propag.* **2023**, *2023*, 5906721. [CrossRef]
13. Wang, Z.; Zhang, J.; Du, H.; Wei, E.I.; Ai, B.; Niyato, D.; Debbah, M. Extremely large-scale MIMO: Fundamentals, challenges, solutions, and future directions. *IEEE Wirel. Commun.* **2023**, *31*, 117–124. [CrossRef]
14. Raj, T.; Mishra, R.; Kumar, P.; Kapoor, A. Advances in MIMO antenna design for 5G: A comprehensive review. *Sensors* **2023**, *23*, 6329. [CrossRef] [PubMed]
15. Jemaludin, N.H.; Al-Gburi, A.J.; Elabd, R.H.; Saeidi, T.; Akbar, M.F.; Ibrahim, I.M.; Zakaria, Z. A comprehensive review on MIMO antennas for 5G smartphones: Mutual coupling techniques, comparative studies, SAR analysis, and future directions. *Results Eng.* **2024**, *10*, 102712. [CrossRef]
16. Kamal, M.M.; Yang, S.; Ren, X.-c.; Altaf, A.; Kiani, S.H.; Anjum, M.R.; Iqbal, A.; Asif, M.; Saeed, S.I. Infinity Shell Shaped MIMO Antenna Array for mm-Wave 5G Applications. *Electronics* **2021**, *10*, 165. [CrossRef]
17. Sehrai, D.A.; Abdullah, M.; Altaf, A.; Kiani, S.H.; Muhammad, F.; Tufail, M.; Irfan, M.; Glowacz, A.; Rahman, S. A Novel High Gain Wideband MIMO Antenna for 5G Millimeter Wave Applications. *Electronics* **2020**, *9*, 1031. [CrossRef]
18. Tiwari, R.N.; Sharma, D.; Singh, P.; Kumar, P. A flexible dual-band 4× 4 MIMO antenna for 5G mm-wave 28/38 GHz wearable applications. *Sci. Rep.* **2024**, *14*, 14324. [CrossRef]
19. Wang, L.; Li, H.; Shi, Y.; Chen, M.; Li, C.; Wang, H.; Zheng, L. An elliptical dual-band antenna with crescent slot for 5G mmWave applications. *J. Phys. Conf. Ser.* **2024**, *2807*, 012033. [CrossRef]
20. Sghaier, N.; Belkadi, A.; Hassine, I.B.; Latrach, L.; Gharsallah, A. Millimeter-wave dual-band MIMO antennas for 5G wireless applications. *J. Infrared Millim. Terahertz Waves* **2023**, *44*, 297–312. [CrossRef]
21. Ullah, U.; Al-Hasan, M.; Koziel, S.; Mabrouk, I.B. Series-slot-fed circularly polarized multiple-input–multiple-output antenna array enabling circular polarization diversity for 5G 28 GHz indoor applications. *IEEE Trans. Antennas Propag.* **2021**, *69*, 5607–5616. [CrossRef]
22. Sharma, P.; Tiwari, R.N.; Singh, P.; Kumar, P.; Kanaujia, B.K. MIMO antennas: Design approaches, techniques and applications. *Sensors* **2022**, *22*, 7813. [CrossRef]
23. Kumar, P.; Ali, T.; Pai, M.M. Electromagnetic metamaterials: A new paradigm of antenna design. *IEEE Access* **2021**, *22*, 18722–18751. [CrossRef]
24. Ketzaki, D.A.; Yioultsis, T.V. Metamaterial-based design of planar compact MIMO monopoles. *IEEE Trans. Antennas Propag.* **2013**, *61*, 2758–2766. [CrossRef]
25. Alibakhshikenari, M.; Virdee, B.S.; Shukla, P.; See, C.H.; Abd-Alhameed, R.A.; Falcone, F.; Quazzane, K.; Limiti, E. Isolation enhancement of densely packed array antennas with periodic MTM-photonic bandgap for SAR and MIMO systems. *IET Microw. Antennas Propag.* **2020**, *14*, 183–188. [CrossRef]
26. Alibakhshikenari, M.; Virdee, B.S.; See, C.H.; Abd-Alhameed, R.A.; Falcone, F.; Limiti, E. Surface wave reduction in antenna arrays using metasurface inclusion for MIMO and SAR systems. *Radio Sci.* **2019**, *54*, 1067–1075. [CrossRef]
27. Alibakhshikenari, M.; Khalily, M.; Virdee, B.S.; See, C.H.; Abd-Alhameed, R.A.; Limiti, E. Mutual-coupling isolation using embedded metamaterial EM bandgap decoupling slab for densely packed array antennas. *IEEE Access* **2019**, *9*, 51827–5184[illegible] [CrossRef]
28. Ramanathan, S.; Maria, A.B. A Compact Four-Element MIMO Antenna for 5 G Millimeter-Wave (37–39 GHz) Applications. *J. Infrared Millim. Terahertz Waves* **2024**, *23*, 604–620. [CrossRef]
29. Alsalman, A.; Alhumaid, A.; Alnogithan, A.; Hamad, E.K.; Shaban, M. Reconfigurable 28/38 GHz wideband and high isolation MIMO antenna for advanced mmWave applications. *J. Electr. Eng.* **2024**, *75*, 467–483. [CrossRef]
30. Mahmoud, S. Design and Modeling of a Reconfigurable Multiple Input, Multiple Output Antenna for 24 GHz Radar Sensor *Modelling* **2025**, *6*, 2. [CrossRef]
31. Alsaab, N.; Alhassoon, K.; Alsaleem, F.; Alsunaydih, F.N.; Madbouly, S.O.; Khaleel, S.A.; Ameen, A.M.; Shaban, M. High Performance Series-Fed Array Multiple-Input Multiple-Output Antenna for Millimeter-Wave 5G Networks. *Sensors* **2025**, *2*[illegible] 1036. [CrossRef]
32. Balanis, C.A. *Antenna Theory: Analysis and Design*; John Wiley & Sons: Hoboken, NJ, USA, 2005.

33. Khalid, M.; Iffat Naqvi, S.; Hussain, N.; Rahman, M.; Fawad Mirjavadi, S.S.; Khan, M.J ; Amin, Y. 4-Port MIMO antenna with defected ground structure for 5G millimeter wave applications. *Electronics* **2020**, *9*, 71. [CrossRef]
34. Abbasi, N.A.; Virdee, B.; Din, I.U.; Ullah, S.; Althuwayb, A.A.; Rashid, N.; Soruri, M.; See, C.H.; Alibakhshikenari, M. High-Isolation Array Antenna Design for 5G mm-Wave MIMO Applications. *J. Infrared Millim. Terahertz Waves* **2025**, *46*, 12. [CrossRef]
35. Marzouk, H.M.; Ahmed, M.I.; Shaalan, A.H. Novel dual-band 28/38 GHz MIMO antennas for 5G mobile applications. *Prog. Electromagn. Res. C* **2019**, *93*, 103–117. [CrossRef]
36. Farahat, A.E.; Hussein, K.F. Hussein. Dual-band (28/38 GHz) wideband MIMO antenna for 5G mobile applications. *IEEE Access* **2022**, *10*, 32213–32223. [CrossRef]
37. Citroni, R.; Mangini, F.; Frezza, F. Efficient integration of ultra-low power techniques and energy harvesting in self-sufficient devices: A comprehensive overview of current progress and future directions. *Sensors* **2024**, *24*, 4471. [CrossRef]

telecom

MDPI

Article

A Semi-Distributed Scheme for Mode Selection and Resource Allocation in Device-to-Device-Enabled Cellular Networks Using Matching Game and Reinforcement Learning

Ibrahim Sami Attar [1], Nor Muzlifah Mahyuddin [1,*] and M. H. D. Nour Hindia [2]

[1] School of Electrical and Electronic Engineering, Universiti Sains Malaysia, Nibong Tebal 14300, Malaysia; ibrahim_sami@student.usm.my

[2] Department of Electrical Engineering, Faculty of Engineering, University of Malaya, Kuala Lumpur 50603, Malaysia; nourhindia@um.edu.my

* Correspondence: eemnmuzlifah@usm.my

Abstract: Device-to-Device (D2D) communication is a promising technological innovation that is significantly considered to have a substantial impact on the next generation of wireless communication systems. Modern wireless networks of the fifth generation (5G) and beyond (B5G) handle an increasing number of connected devices that require greater data rates while utilizing relatively low power consumption. In this study, we present joint mode selection, channel assignment, and power allocation issues in a semi-distributed D2D scheme (SD-scheme) that underlays cellular networks. The objective of this study is to enhance the data rate, Spectrum Efficiency (SE), and Energy Efficiency (EE) of the network while maintaining the performance of cellular users (CUs) by creating a threshold of data rate for each CU in the network. Practically, we propose a centralized approach to address the mode selection and channel assignment problems, employing greedy and matching algorithms, respectively. Moreover, we employed a State-Action-Reward-State-Action (SARSA)-based reinforcement learning (RL) algorithm for a distributed power allocation scheme. Furthermore, we suggest that the sub-channel of the CU is shared among several D2D pairs, and the optimum power is determined for each D2D pair sharing the same sub-channel, taking into consideration all types of interferences in the network. The simulation findings illustrate the enhancement in the performance of the proposed scheme in comparison to the benchmark schemes in terms of data rate, SE, and EE.

Keywords: device-to-device (D2D); resource allocation; greedy algorithm; matching theory; power allocation; reinforcement learning (RL); SARSA

1. Introduction

The growing demand for improved broadband services for mobile devices and the rapid growth of different fields of application, such as vehicle-to-vehicle communication, automating factories, cellular healthcare services, and augmented and virtual reality solutions, require the development of new (B5G) network architectures. These architectures need to be capable of supporting lower energy consumption and higher area capacity compared to current networks [1]. Considering these demands, the problem of limited network capacity is an important challenge for the evolution of emerging wireless networks. In addition to the limited availability of the spectrum, the development of the cellular network is leading to concerning levels of energy consumption [2]. The importance of EE has been growing as a consequence of economic, operational, and environmental considerations [3]. Furthermore, the rapid increase in data rate demands and power-intensive mobile systems and applications, in combination with the scarcity of available spectrum resources, necessitates the exploration of innovative networking solutions.

D2D communication is a promising technology among the emerging B5G solutions, providing potential solutions for the aforementioned demands. D2D communication enables direct transmission of data traffic instantaneously via the D2D transmitter toward a D2D receiver without passing via a cellular gNB [4]. Moreover, it requires low power transmission that enhances EE and allows for the spectrum sharing of the available resources, ultimately improving spectral efficiency. D2D communications utilizing the resources of wireless cellular networks will be crucial in enhancing the capacity of incoming B5G systems [5]. The advantages of implementing D2D communication over wireless cellular networks are becoming widely recognized, particularly for data offloading, content sharing, EE, coverage expansion, and enhanced utilization of the spectrum. Furthermore, an additional demand in B5G networks is the capacity to deal with extensive communications resulting from the rapid increase in connected devices in conventional cellular networks [6].

Mode selection and resource allocation are crucial issues for creating and maintaining direct connections between D2D users within cellular networks. Furthermore, the distribution of network resources among D2D pairs and cellular users can be achieved efficiently, which could improve SE and EE by controlling interferences in the network.

Mode selection is an essential challenge in D2D communications that enhances EE and average data rate and minimizes interference in the network. It decides whether D2D users can perform in direct mode or cellular mode. The mode selection procedure is flexible, leading to decreased latency and increased spectrum resource utilization. The gNB can assign three D2D communication modes to each D2D pair, including direct D2D mode (DM), relay-assisted D2D mode (RM), and local route D2D mode (LM) [7].

D2D communication can utilize either a licensed or unlicensed spectrum for channel assignment to establish direct connections, assigned as in-band and out-band D2D communications, respectively [8]. According to out-band D2D communication, users communicate via direct communication utilizing an unlicensed spectrum, separated from that utilized by cellular users in the network. On the other hand, in-band D2D communication may be categorized into two classifications: underlay and overlay. According to the framework of underlaying in-band communication, the spectrum is shared and assigned to D2D pairs and CUs. Moreover, in-band communication overlay identifies specific parts of the whole spectrum for D2D communications and the other portion for CUs. This study investigates the idea of underlaying in-band D2D communication, focusing on improving the system's performance. However, D2D pairs experience interference when implemented with the underlay in-band architecture, caused by the utilization of shared sub-channels [9]. Consequently, the control of interference becomes a significant research concern within the field of D2D communication networks.

Despite the improvements in cellular networks enabled by D2D communication, many concerns should be addressed. Conventional cellular networks are significantly affected by the interferences imposed by D2D communications due to the sharing of cellular user resources. These concerns have a substantial effect on the performance of cellular user communications and could impact the future development of D2D technology. Therefore, it is essential to figure out efficient strategies for the purpose of allocating resources to D2D communication pairs without maximizing the complexity of the network's topology. Specifically, effective allocation of resources for D2D communications is an important challenge in the cellular network. Following the mode selection procedure, optimal sub-channel resource allocation is necessary to achieve the goal of D2D technology [10].

Power control and interference mitigation should be considered through the resource allocation process. Additionally, to enhance EE and guarantee QoS for both D2D pairs and

CUs, appropriate power allocation approaches need to be designed. Blindly implementing power control to D2D communications in a cellular network could reduce system efficiency. D2D transmitters should optimize their transmission power to enhance EE, SINR demands, and system performance.

Machine Learning (ML) has been demonstrated to be highly beneficial in multiple areas due to its ability to precisely predict future scenarios and address complex problems with huge datasets [11]. Artificial intelligence approaches have been indicated to be an excellent technique for tackling complicated non-convex optimization problems in communication networks. In the field of wireless communication networks, reinforcement learning (RL) based on ML techniques has been utilized to solve the issue of power allocation [12].

2. Related Work

Various current investigations have tackled the problem of mode selection, channel assignment, and power allocation for D2D communication underlay cellular networks. For instance, in [13], the resource allocation problem has been formulated based on the investigation of the system model in many-to-many matching D2D communication underlying cellular networks. An overlapping coalition game method that utilizes a candidate sequence is introduced to improve the D2D transmission data rate. The simulation findings showed that the efficiency of the proposed method, which incorporates an optimization technique for candidate sequences, outperforms that of the overlapping approach with random optimization.

In [14], a genetic algorithm-based joined power and channel allocation was investigated in D2D-underlayed cellular networks. Efficient allocation of transmitted power and shared channels for every user is achieved by a genetic algorithm optimization, that assists in minimizing interference. The simulation outcomes illustrated that the proposed approach outperformed the fixed, random, and particle swarm optimization approaches in terms of maximizing the total resource utilization. In [15], the authors addressed the issues of mode selection, spectrum usage, and power management in D2D underlay cellular communications by the use of a hierarchical game approach. For mode selection and channel assignment issues, the optimal solution is achieved by employing the hedonic coalition game. Moreover, the non-cooperative game is employed to tackle the issue of power management in the proposed scheme. The algorithm effectively addressed the issue of fair allocation of resources among users, while simultaneously improving the system throughput.

In [16], the interference management issue was investigated in order to effectively enable the coexistence of two technologies: a massive MIMO and a D2D communication system that shares uplink network resources under cellular networks. To achieve a distributed solution, they formulated the problem via a matching theory and presented a resource optimization technique utilizing the principles of many-to-many matching to enhance the performance of the system. The numerical outcomes illustrated that the suggested method effectively improved the network performance by leveraging the diversification benefits of massive MIMO and matching users based on their preferences. Moreover, in [17], the paper presented a dynamic resource allocation strategy for D2D communications under cellular networks. The suggested resource allocation technique incorporates a Q-learning-based power management algorithm to allocate optimum powers for D2D pairs, with the goal of optimizing the throughput. The introduced algorithm employs a Q-learning technique to enhance the transmission power for the D2D pairs that utilize the same resource block at the establishment of a unique D2D pair. The model findings showed that the suggested technique can achieve quick convergence and outperform the random power allocation algorithm in terms of overall throughput. In [18], the authors examined techniques for allocating resources in D2D communications underlaying cellular networks. A unique DC programming method for the difference of convex functions is provided to successfully address this complex resource allocation problem. The simulation findings showed that the introduced system reaches the maximum weighted sum-data rate in comparison to the benchmark methods, while simultaneously guaranteeing the satisfaction of QoS demands for each D2D pair and CUs.

Furthermore, in [19], the quality of cell-edge user coverage was enhanced with the development of a D2D-relay communication system for underlay cellular networks. Subsequently, the Lagrange dual approach-based power allocation technique was developed to effectively allocate power levels to the users in the network. The suggested technique converges in time, and an optimal closed-form solution is derived. The simulation findings demonstrated that the suggested strategy highly improved the network coverage, the data rates of users, and the SE of the system. In [20], a multi-agent Q-learning approach is presented to enhance the throughput of D2D communications underlying cellular networks. First, the multi-agent Q-learning technique-based channel resource allocation is implemented. Furthermore, to tackle the issue of slow convergence in the Q-learning algorithm for the D2D communications system, the authors included a Fuzzy C-Means algorithm into a multi-agent Q-learning framework. This integration aimed to enhance the utilization of power management by employing a Q-learning approach. The results showed that the implementation of the multi-agent Q-learning approach enhances the system's throughput. In [21], the authors proposed an auction technique based on a D2D relay selection algorithm underlaying cellular networks. The cooperative willingness of user relay devices was evaluated from a social level. As social relationships become stronger, the willingness to cooperate increases and the transmission power of the relay also increases. Subsequently, they determined a relationship between the outage probability and the transmission power. Next, an auction approach was employed to stimulate relays to enhance the transmitted power via monetary incentives. The results demonstrated the technique's superiority over previous relay methods, as it not only enhanced the system's data rate but also decreased the probability of a communications outage.

Additionally, the researchers in [22] suggested a graph coloring approach to address interference issues in D2D communications-based cellular networks. The primary objective is to exploit the weighted prioritization of spectrum resources, allowing several D2D pairs to use the same resources as cellular users inside the network. Once the spectrum allocation has been achieved, the power management process is employed to reduce the transmission power of D2D pairs which leads to minimizing the interferences and enhancing the energy consumption of the entire cell. The simulation outcomes illustrated that the suggested method successfully mitigated co-channel interference, enhanced system throughput, and minimized power consumption in comparison to the traditional techniques. The authors in [23] suggested a method based on the D2D multicast clusters approach along with a Q-Learning-aided approach to tackle the issue of joint sub-channel assignment and power management for D2D communication under cellular networks. An agglomerative hierarchical clustering approach using unsupervised ML was suggested to establish clusters of D2D pairs, considering user preferences and ensuring reliable D2D multicast communications across D2D users. The numerical simulation outcomes indicated that the suggested approach provided substantial benefits for the throughput and EE of the introduced system compared to existing techniques. In [24], the paper examined the issues of joint channel assignment and power management in D2D communications underlaying cellular networks based on the concept of the NOMA technique. The suggested approach divided the issue into two sub-issues: channel assignment and power management. The assignment of channels to the pairs is shown using a matching game theory, providing a stable solution while considering the interferences imposed by users in the network. The power allocation problem is addressed in both phases utilizing difference-of-convex programming that is optimized iteratively by the Frank–Wolfe method. The performance investigations have shown that the suggested method enhanced SE, fairness, and network connectivity. In [25], the authors examined the issue of resource allocation for D2D communications in cellular networks. A one-to-many matching game was implemented to enhance spectrum utilization and tackle frequency interference problems. They introduced a resource allocation algorithm based on the relationship between the distance of the users and the interference levels to assign channel resources efficiently. This technique enables D2D pairs to efficiently reuse channel resources utilized by CUs in close proximity while minimizing interference between D2D pairs and CUs. Moreover, the particle swarm optimization technique was employed to tackle the optimum power allocation problem, aiming to obtain the highest transmission rate of the network. The simulation findings demonstrated that this approach enhanced system data rate and performance, while simultaneously minimizing the computational complexity.

In [26], a resource allocation technique is suggested, based on the QL algorithm, for D2D communication in the unlicensed spectrum. This approach allows for dynamic allocation of transmitted power to D2D users based on updating network traffic conditions, leading to improved performance of the coexisting system. The agent's states are determined by different factors such as fairness, SNR, and data rate of CUs. The agent's actions can be determined by the different duty cycles and transmission power levels. The agent can learn the duty cycles and optimum power transmission via iterative interaction responses to the network. The comprehensive simulation findings indicated that the proposed technique outperforms the compared approaches in resource allocation fairness and throughput. The authors in [27] tackled a joint optimization issue of resource allocation, optimum power management, and relay-selection in a two-way relaying approach underlaying cellular networks. The power management problem is solved using particle swarm optimization, while the relay selection problem is addressed via the one-to-one stable matching technique. The numerical outcomes indicated that employing stable matching in the relay selection problem significantly increased the performance of the network, D2D data rate, and EE. In [28], the authors investigated the enhancement of system fairness and throughput in an underlying D2D communications scheme by employing joint channel allocation and power management methodologies. They suggested an iterative resource allocation approach based on RL by considering the channel parameters. The authors proposed an enhanced reinforcement learning-based SARSA algorithm. The simulation findings illustrated the enhancement of the system's throughput, EE, and SE. In [29], the authors examined EE optimization for D2D communications underlaying cellular networks. The primary objective of the study is to improve the performance of the scheme as well as reduce the power conception of the equipment while maintaining the QoS for every user. The Q-learning algorithm is applied to achieve optimum communication between the users and gNB in the proposed network. The simulation findings demonstrated that the proposed system enhanced the transmission performance as compared with the traditional systems.

In [30], the authors proposed a joint channel allocation and power allocation issue for D2D connections within cellular networks. The main goal of this research is to optimize the EE of the proposed scheme while satisfying the QoS of CUs and D2D communications. The suggested problem is NP-hard and complex to solve; therefore, an iterative solution is introduced. The authors introduced Dinkelbach's method to address the problem and obtain the optimal solution. The simulation findings showed the advantage of the suggested technique relative to conventional schemes.

3. Research Contributions

This paper introduces joint mode selection, channel assignment, and power allocation issues of D2D communications underlaying cellular networks in an uplink scenario, where a single sub-channel can be shared via a CU and several D2D pairs. The main goal of this study is to enhance the data rate, SE, and EE of the proposed network, while simultaneously satisfying the minimal QoS demands for CUs and D2D pairs. The suggested approach is a semi-distributed architecture, in which mode selection and channel assignment issues are centralized, while the power management issue is solved with a distributed technique. The complexity of this technique is lower in comparison to centralized solutions. Firstly, the optimal D2D mode can be achieved by utilizing a greedy algorithm that chooses the best mode among DM, RM, and LM based on maximum SINR. Furthermore, the provided channel assignment method is based on a matching algorithm that takes into consideration the priorities of both the D2D pairs and the CUs, which differs from most of the existing research that mainly investigates the preferences of D2D pairs only. The channel assignment technique based on two-sided preference achieves stable matching with minimal complexity. Moreover, the channel assignment method enables the reuse of a single CU sub-channel throughout several D2D pairs, resulting in higher SE. A higher number of D2D pairs may be served with limited spectrum resources through the implementation of this type of resource-sharing scheme.

In addition, power management is achieved for each D2D pair by applying the SARSA-based RL algorithm. This low-complexity distributed RL algorithm has the ability to calculate the optimum power for each D2D pair that enhances the EE of the network. While several studies have attempted to address the issue of resource allocation throughout D2D pairs, they have either insufficiently accounted for the potential that D2D users might interfere with each other or have assigned resources based on the assumption that D2D pairs have a constant power transmission. The proposed approach differs from previous studies that either assumed a fixed D2D power or neglected the interferences between the D2D pairs. The main contributions of the introduced scheme can be illustrated as described below:

1. The problem of joint mode selection, channel assignment, and power allocation is formulated for D2D communications underlaying cellular networks by utilizing the uplink resources of CUs. The optimization problem is formulated to enhance the data rate, SE, and EE of the network while considering QoS characteristics related to D2D pairs and CUs simultaneously.
2. By employing a greedy algorithm, a mode selection technique is introduced to choose the optimum mode throughout DM, RM, and LM across every D2D pair in the network. The computational representation is formulated based on the highest SINR.
3. channel assignment method is introduced, using a matching algorithm to assign the optimal sub-channel to the D2D pairs in the network. The channel assignment approach based on a two-sided preferences list provides stable matching with low complexity. The first preference list consists of the data rates of D2D pairs arranged in descending order according to their highest value. The second preference list consists of the interference effect of CUs on D2D pairs when they share the same sub-channel, arranged in ascending order according to their lowest impact value.
4. low-complexity distributed SARSA-based RL algorithm is implemented to address the issue of power control and allocate the optimum power level for each D2D pair to enhance the EE of the network.
5. The effectiveness of the suggested method has been shown by simulations, particularly in terms of the data rate, SE, and EE of the network in comparison to conventional systems.

The remainder of this paper is organized as follows: Section 4 provides a comprehensive clarification about the system model. Section 5 defines the mode selection, channel assignment, and power allocation problem. Section 6 discusses the challenge and illustrates the methodologies utilized for joint mode selection, channel assignment, and the power allocation scheme. Section 7 presents and discusses the simulation parameters as well as the simulation results. Section 8 concludes the paper.

4. System Model

In this study, we explore the principle of the D2D communication scheme underlaying cellular networks, which involves spectrum sharing across multiple D2D users. For every D2D pair, we analyze three different modes, which include DM, RM, and LM, as illustrated in Figure 1. The DM enables direct transmission of data from the transmitter to the receiver of each D2D pair. Moreover, the second D2D mode is RM which aims to establish relay communication between two distant devices. An idle user is used as a relay node in this mode to facilitate the creation of a connection between the transmitter and receiver of the D2D pairs. Regarding the LM, it is suggested that auxiliary antennas be installed on the gNB to enhance the connections of the LM [31]. The data are transferred via the gNB instead of entering the core, which means that the gNB acts as a node to support the connections of faraway users to establish a D2D pair.

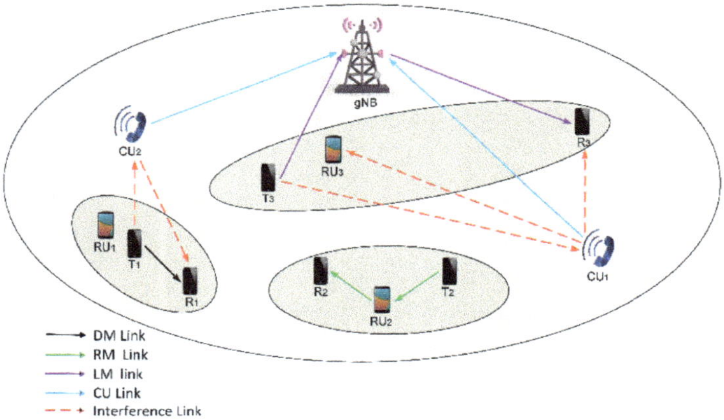

Figure 1. System model.

In this study, we examine an environment where N represents the D2D pairs' number. Furthermore, T_n, R_n, and RU_n represent the D2D transmitter, receiver, and relay of the pair n, respectively. Let us suppose W is the total bandwidth in the network which is partitioned to a number of sub-channels indicated as K. Table 1 illustrates the scheme symbols.

Table 1. Scheme symbols.

Symbol	Description
N	D2D pairs' number
M	CUs number
T_n	D2D transmitter
R_n	D2D receiver

Table 1. *Cont.*

Symbol	Description
RU_n	Relay user equipment
W	System Bandwidth
K	Sub-channels' number
G_n, G_{RUm}	Channel gain for *nth* D2D pair, R_n to *mth* CU, respectively
G_{nm}, G_{mn}	Channel gain across *nth* D2D pair and *mth* CU, *mth* CU and *nth* D2D pair, respectively
P_n, P_C, P_Z	Power transmission of the D2D pair, CU, and gNB, respectively
G_{TR}, G_{TRU}, G_{TZ}	Channel gain between T_n to R_n, RU_n, and gNB, respectively
G_{RUR}, G_{ZR}	Channel gain across RU_n, and gNB to R_n
σ	Noise power
$\Upsilon_{n,k}$, $\Upsilon_{m,k}$	Received SINR for D2D pair n and cellular user m utilizing sub-sub-channel k, respectively
$\mathcal{I}_D$, $\mathcal{I}_R$, $\mathcal{I}_L$	Interference impact of CUs, other D2D transmitters, and D2D relays on the *nth* DM, RM, and LM, respectively
$G_{T'R}$, $G_{T'RU}$, $G_{T'Z}$	Channel gain between other D2D transmitters to R_n, RU_n, and gNB, respectively
$G_{RU'R}$, $G_{RU'RU}$, $G_{RU'Z}$	Channel gain across other D2D relays to R_n, RU_n, and gNB, respectively
$\Pi^1_{n,k}$, $\Pi^2_{n,k}$, $\Pi^3_{n,k}$	Binary mode sub-channel assignment indicator of DM, RM, and LM respectively
$R_{n,k}$, $R_{m,k}$	Data rate for *nth* D2D pair and *mth* CU, respectively
$\mathcal{M}_D$, $\mathcal{M}_R$, $\mathcal{M}_L$	Mode selection sets of DM, RM, and LM respectively
$\mathcal{Q}^D$, $\mathcal{Q}^R$, $\mathcal{Q}^L$	DM, RM, and LM, each has respective quotas
ℓ^p, ℓ^s	Primary and secondary preference lists, respectively
s_t, a_t, $\mathcal{R}_t$	State, action, and reword of the SARSA algorithm
s_{t+1}, a_{t+1}	Next state and next action of the SARSA algorithm
α	The learning rate
$\mathcal{Y}$	The discount factor
$\mathcal{V}^*(s)$	The value function

Let $P_n = \{p_1, p_2, \ldots, p_n, \ldots, p_N\}$, P_C, P_Z denote the transmission power of the T_n, CU, gNB, respectively. The transmitted power for every D2D pair P_n is determined from the available power levels set, ranging from p_{min} to p_{max}, while the transmission power of CUs is supposed to be fixed. In the proposed scheme, Π is the binary mode sub-channel assignment indicator matrix with $\Pi \in \{0,1\}$, where $\Pi^1_{n,k}$, $\Pi^2_{n,k}$, and $\Pi^3_{n,k}$ represent the binary mode sub-channel assignment indicators for DM, RM, and LM, respectively. If the *nth* D2D pair is utilizing the sub-channel k of the CU m, then $\Pi = 1$, otherwise, $\Pi = 0$. The SINR of the *nth* D2D pair utilizing *kth* shared sub-channel of the *mth* CU at time slot t is given as follows:

$$\Upsilon_{n,k}(t) = \frac{P_n G_{TR}(t)\Pi^1_{n,k} + (P_n G_{TRU}(t) + P_n(t)G_{RUR}(t))\Pi^2_{n,k} + (P_n G_{TZ}(t) + P_Z G_{ZR}(t))\Pi^3_{n.k}}{\mathcal{I}_{D2D} + P_C(t)G_{mn} + \sigma} \quad (1)$$

where G_{TR}, G_{TRU}, G_{TZ}, represent the channel gain between T_n to R_n, RU_n, and gNB, respectively. G_{RUR} and G_{ZR} refer to the channel gain between RU_n and gNB to R_n. Furthermore, G_{mn} refers to the channel gain of the D2D pair *nth* and the *mth* CU. σ denotes the noise power. The expression $\mathcal{I}_{D2D}$ can be illustrated as follows:

$$\mathcal{I}_{D2D} = \mathcal{I}_D + \mathcal{I}_R + \mathcal{I}_L, \quad (2)$$

where $\mathcal{J}_D$, $\mathcal{J}_R$, and $\mathcal{J}_L$ demonstrate the impact of the interferences on the nth D2D pair in DM, RM, and LM caused by gNB, other D2D T_n, and other RU_n. The mathematical representation of $\mathcal{J}_D$, $\mathcal{J}_R$, and $\mathcal{J}_L$ can be shown as follows:

$$\mathcal{J}_D = P_Z G_{ZR}(t) \Pi^3_{n,k} + \sum_{n=1}^{N} P_n G_{T'R} \Pi^1_{n,k} + \sum_{n=1}^{N} P_n G_{RUR}(t) \Pi^2_{n,k}, \tag{3}$$

$$\begin{aligned} \mathcal{J}_R = (P_Z G_{ZRU}(t) + P_Z G_{ZR}(t))\, \Pi^3_{n,k} + \sum_{n=1}^{N} (P_n G_{T'RU}(t) + P_n G_{T'R}(t))\, \Pi^1_{n,k} \\ + \sum_{n=1}^{N} (P_n G_{RU'RU}(t) + P_n G_{RU'R}(t)) \Pi^2_{n,k}, \end{aligned} \tag{4}$$

$$\mathcal{J}_L = \sum_{n=1}^{N} (P_n G_{T'Z}(t) + P_n G_{T'R}(t)) \Pi^1_{n,k} + \sum_{n=1}^{N} (P_n G_{RU'Z}(t) + P_n G_{RU'R}(t)) \Pi^2_{n,k}, \tag{5}$$

where $G_{T'R}$, $G_{T'RU}$, and $G_{T'Z}$ are the channel gain between other D2D transmitters to R_n, RU_n, and gNB, respectively. Moreover, $G_{RU'R}$, $G_{RU'RU}$, and $G_{RU'Z}$ represent the channel gain between other D2D relays to R_n, RU_n, and gNB, respectively. The SINR of the mth CU utilizing the kth sub-channel at time slot t may be expressed as follows:

$$Y_{m,k}(t) = \frac{P_C(t) G_{mn}}{\sum_{n=1}^{N} p_n(t) G_n + \sum_{n=1}^{N} p_n(t) G_{RUm} + \sigma}, \tag{6}$$

The data rate of D2D pair n utilizing the uplink kth sub-channel can be determined at time slot t as follows:

$$R_{n,k}(t) = W \log_2(1 + Y_{n,k}(t)), \tag{7}$$

Furthermore, the data rate of CU m utilizing the kth channel can be determined at time slot t as follows:

$$R_{m,k}(t) = W \log_2(1 + Y_{m,k}(t)), \tag{8}$$

SE shows the effectiveness of using the available spectrum in terms of the data rate obtained regarding a given bandwidth. Thus, the SE for the nth D2D communication pair can be given as follows:

$$SE_{n,k}(t) = \frac{\sum_{n=1}^{N} \sum_{k=1}^{K} R_{n,k}(t)}{W}, \tag{9}$$

Based on the obtained data rate and energy consumption, the EE of the D2D communications scheme at time slot t is given as follows:

$$EE_{n,k}(t) = \frac{\sum_{n=1}^{N} \sum_{k=1}^{K} R_{n,k}(t)}{\sum_{n=1}^{N} p_n(t) + p_{cir}}, \tag{10}$$

where p_{cir} denotes the D2D pair circuit power consumption.

5. Problem Formulation

In this study, an optimization problem in D2D networks is investigated, specifically concentrating on joint mode selection, channel assignment, and power allocation optimization issues. In our proposed system, D2D users can choose among the available three D2D modes, including DM, RM, or LM based on maximum SINR. Moreover, a network that has been completely loaded is regarded as having no dedicated channels for D2D pairs to utilize. Moreover, the optimum power level can be obtained from the range P_{min} to P_{max}. This paper aims to optimize the sum data rate, SE, and EE of the proposed D2D communications scheme while guaranteeing the QoS demands for both D2D pairs and CUs. The following is the formulation of the optimization problem:

$$\max_{\mathcal{M},\Pi,P} \sum_{n=1}^{N} \sum_{k=1}^{K} R_{n,k}(t), SE_{n,k}(t), \text{ and } EE_{n,k}(t), \tag{11}$$

s.t.

$$\sum_{n=1}^{N} \sum_{k=1}^{K} R_{n,k}(t) \geq R_{min}^{th}(t), \qquad \forall k \in 1 \ldots K, \tag{11a}$$

$$\sum_{n} \mathcal{M} = 1 \qquad n \in N, \tag{11b}$$

$$\sum_{n=1}^{N} \sum_{k=1}^{K} \Pi_{n,k}^{1}, \Pi_{n,k}^{2}, \text{ or } \Pi_{n,k}^{3} \leq K \qquad \forall k \in 1 \ldots K, \tag{11c}$$

$$\sum_{m} \Pi_{m} = 1 \qquad m \in UEs, \tag{11d}$$

$$\sum_{n=1}^{N} \Pi_{k,n} \leq 3, \qquad \forall n \in N, \tag{11e}$$

$$P_{min} \leq P_n \leq P_{max}, \qquad \forall N, \quad RU, \tag{11f}$$

$$P_C, P_Z = P_{max}, \tag{11g}$$

Constraint (11a) specifies the minimal data rate for the nth D2D pair in shared sub-channel k. Constraint (11b) denotes that each D2D pair n chooses one mode among the D2D modes including DM, RM, or LM. Constraint (11c) indicates that the binary mode sub-channel indicator matrix for each D2D including DM, RM, and LM is equivalent to or less than the total number of sub-channels k. Moreover, the constraint (11d) indicates that every cellular user m utilizes a distinct sub-channel k. The constraint (11e) indicates that each sub-channel may be utilized a maximum of three times. The constraints (11f) and (11g) denote that T_n, CUs, and gNB utilize specific transmission power.

To sufficiently address the optimization problem expressed in (11), it should be divided into two sub-issues: joint mode selection and channel assignment, as well as power management. Since it is an MINLP problem, the optimization problem is NP-hard and involves computational difficulties.

6. Proposed Joint Mode Selection and Resource Allocation Scheme (SD-Scheme)

In the present part, we introduce an SD-scheme underlaying cellular networks. Joint mode selection, channel assignment, and power allocation are considered with the aim of optimizing the sum data rate, SE, and EE. First, the mode selection issue is tackled by employing a greedy algorithm based on maximum SINR to select the optimum mode among DM, RM, and LM for every D2D pair. After that, the matching algorithm is implemented to tackle the problem of sub-channel assignment by exploiting the two-sided preference lists to optimize the utilization of spectrum resources. Finally, the power allocation issue is solved by introducing SARSA-based RL to obtain the optimum power for each D2D pair in the proposed scheme.

Several important factors inspired the decision to choose the SARSA algorithm for this study. Firstly, the modeling of the network in the complicated and proposed scenario of resource allocation for D2D communication is simply unpracticable. Therefore, the model-free feature of this system is particularly advantageous in this particular scenario. Moreover, the SARSA algorithm is highly applicable to decentralized decision-making and enables agents to self-sufficiently learn the most optimal policies, which aligns effectively with the architecture of the proposed network. The proposed approach utilizes an SARSA algorithm to train and update the power level states by considering the environment feedback information.

6.1. Mode Selection and Channel Assignment Scheme (C-Scheme)

Firstly, this study examines a mode selection technique for D2D communication by employing a greedy algorithm. The method focuses on direct, relay, and local route D2D modes to enhance the performance of the D2D scheme. The mode selection issue is tackled based on the maximum SINR to calculate the best mode for each D2D pair, while considering the distance between the transmitter and the receiver. This will guarantee the chosen mode optimizes signal quality while taking into account the physical relationship of the communication devices, thereby maximizing overall performance. However, a threshold level is considered for the distance between the transmitter and the receiver for each D2D communication in the proposed network.

Let us suppose that $\mathscr{M}$ is a set of 0 and 1 elements that are applied to represent which mode is chosen. The following formula is applied to select the best mode:

$$\mathscr{M}_{D2Dn} = \mathscr{M}_D + \mathscr{M}_R + \mathscr{M}_L, \tag{12}$$

where $\mathscr{M}_D$, $\mathscr{M}_R$, and $\mathscr{M}_L$ represent the mode selection sets of DM, RM, and LM, respectively. If the data rate in a DM is greater than in an RM and an LM for each given n D2D communication pair, then $\mathscr{M}_D = 1$ and $\mathscr{M}_R = \mathscr{M}_L = 0$, and similarly for other cases. While a value in the set of the mode selection $\mathscr{M}_{D2Dn}$ is 1, the associated mode is selected ($\mathscr{M}_D$, $\mathscr{M}_R$, or $\mathscr{M}_L$), subsequently adding that particular data rate of that D2D pair n to the sum data rate of the network. Conversely, when a value in the mode selection set $\mathscr{M}_{D2Dn}$ is valued as 0, it leads to a missing contribution to the sum data rate of that particular D2D pair n.

Once the optimal D2D mode is determined in a scenario including DM, RM, and LM, the gNB employs the matching method to assign optimum reused sub-channels for the D2D pairs, which increases the spectrum utilization in the proposed network. This part introduces the model of a channel assignment issue to optimize the sum data rate and SE accordingly. We define the channel assignment formula in which D2D pair n shares the sub-channel k with CU m at time slot t as follows:

$$\Pi = \left(\Pi^1_{n,k}, \Pi^2_{n,k}, \Pi^3_{n,k}\right)_{N\times 3K}, \tag{13}$$

The vectors $\Pi^1_{n,k}$, $\Pi^2_{n,k}$, and $\Pi^3_{n,k}$ indicate the possibility that the D2D pair is assigned to DM, RM, or LM, using the shared sub-channel k with CUs. For each D2D mode in the proposed scheme, the quota ($\mathcal{Q}$) can be determined. $\mathcal{Q}$ represents the threshold of mode-channel assignment for every D2D pair within the framework. The $\mathcal{Q}$ features can be illustrated as follows:

- $\sum_{k=1}^{K}\sum_{n=1}^{N}\Pi^{1}_{n,k} \le \mathcal{Q}^D, \quad \forall n \in \mathscr{M}^D$
- $\sum_{k=1}^{K}\sum_{n=1}^{N}\Pi^{2}_{n,k} \le \mathcal{Q}^R, \quad \forall n \in \mathscr{M}^R$
- $\sum_{k=1}^{K}\sum_{n=1}^{N}\Pi^{3}_{n,k} \le \mathcal{Q}^L, \quad \forall n \in \mathscr{M}^L$

Based on the above criteria, it is crucial that the overall set of D2D pairs across the modes, that are assigned to sub-channel k, should not exceed $\mathcal{Q}$. In the matching game, the establishment of the two-sided preference lists can be represented as follows:

$$\ell^p = \left\{R^D_{n,k}(t), R^R_{n,k}(t), R^L_{n,k}(t)\right\}, \tag{14}$$

$$\ell^s = \{P_C(t)G_{nm}\}, \tag{15}$$

where ℓ^p and ℓ^s denote preference lists that consist of D2D pairs arranged in descending sequence according to their highest data rate and CUs organized in ascending sequence according to their minimal interference effect, respectively.

While performing the matching game theory, every D2D pair within the cell is proposed to earn the sub-channels with its higher priority. Consequently, the gNB admits the D2D pairs with the highest priority while refusing the remaining pairs. For further clarification, if the D2D pair n gives a proposal to pick sub-channel k based on its greatest utility in ℓ^p, then sub-channel k is subsequently assigned to the exact D2D pair n according to the least interference impact utility function in ℓ^s. Moreover, the matching process continues till all devices in the network are paired to enhance the system performance.

To provide a more detailed explanation of the resource allocation process, we now describe the matching game-based sub-channel assignment in greater depth.

Matching Process Execution:

- Each D2D pair initially proposes to the sub-channel that provides the highest data rate based on its preference list.
- The gNB evaluates all proposals and initially assigns sub-channels to D2D pairs while ensuring that the total number of assigned pairs does not exceed the predefined threshold $\mathcal{Q}$.
- If a sub-channel receives multiple proposals, the gNB selects the D2D pairs that maximize SE and rejects lower priority requests.
- Rejected D2D pairs then propose to their next preferred sub-channel, and this process iterates until a stable matching is achieved, meaning no further changes can improve the overall network performance.

This iterative matching ensures an efficient and interference-aware sub-channel allocation strategy that enhances both spectral efficiency and system stability. The complete mode selection and channel assignment approach is detailed in Algorithm 1.

Algorithm 1. C-scheme algorithm

Input: M, N, K
Output: $\Pi_{n,k}$
Initialization $\Pi_{n,k} = \text{zeros}(N, 3\text{K})$
for 1 to M
determine $\Upsilon_{m,k}(t)$
determine $R_{m,k}(t)$
end for
for 1 to N
calculate $\Upsilon_{n,k}(t)$
calculate $R_{n,k}(t)$, $SE_{n,k}(t)$, $EE_{n,k}(t)$
end for
find $\mathscr{M}_{D2Dn}$ based on maximum $\Upsilon_{n,k}(t)$
find Q from the matrix $\mathscr{M}_{D2Dn}$
for 1 to N
for 1 to K
calculate $R^D_{n,k}(t), R^R_{n,k}(t), R^L_{n,k}(t)$ with regards to $\mathscr{M}_{D2Dn}$
calculate the interference impact of CUs
end for

Algorithm 1. *Cont.*

```
end for
sort ℓ^p in descending order
sort ℓ^s in ascending order
the most preferred sub-channel k is matched by D2D pair n based on ℓ^p and ℓ^s
if R_{n,k} ≥ R_th
set Π_{n,k} = 1
else
set Π_{n,k} = 0
end if
```

6.2. Proposed RL-Based Power Allocation Scheme

The power allocation optimization can be achieved using a dynamic distribution scheme. A wireless network with real-time communications demands immediate training and learning of the D2D pairs to provide distributed power allocation without imposing a significant load over the gNB. Thus, ML provides a potential solution with a wide range of applications in the execution of dynamic resource allocation and tackles many challenges associated with prospective communications networks. The transmitter of the D2D pair, performing as the intelligent agent in this scenario, has the ability to learn and make the most suitable decision to enhance the network performance.

One of the most advanced ML techniques is RL. RL utilizes an approach based on trial and error to determine the best resource allocation decisions. Moreover, RL works efficiently without any previous information about the system environment, in contrast to the traditional techniques. RL may enhance performance by facilitating rapid detection of optimum solutions or decisions in comparison with conventional centralized techniques.

SARSA is a reinforcement learning approach utilized to identify the best action in a dynamic resource allocation system. This research presents an SARSA-based approach to solving the issue of power distribution, including the following elements:

Agent: the agent is the D2D transmitter and serves as a crucial element in the power allocation issue.

State: The state of the SARSA algorithm includes essential network information, such as interference levels, channel conditions, and user location. These factors describe the present state of the environment, operating as inputs to an agent's decision-making in power distribution. In this case, the agent indicates the connection of D2D pairs.

Action: The action is an activity performed via the agent. The power distribution levels established by the D2D pairs constitute the action, which comprises a range of powers from P_{min} to P_{max}.

Reward: The reward function in the SARSA-based power management approach is defined as the EE of each D2D pair within the system. The agent's interactions with the environment in SARSA are illustrated in Figure 2.

In the specific state s_t, the action a_t is selected, and the reward $\mathcal{R}_t$ is allocated to the D2D pair (agent) for each action performed. The agent subsequently transitions to the newly created state s_{t+1} and executes another action, a_{t+1} for its present state s_{t+1}. Moreover, the pattern s_t-a_t-$\mathcal{R}_t$-s_{t+1}-a_{t+1} defines the sequence of procedures for the suggested SARSA algorithm. The Q-value is firstly set to zero value, then the proposed algorithm modifies the Q-table in accordance with the current policy.

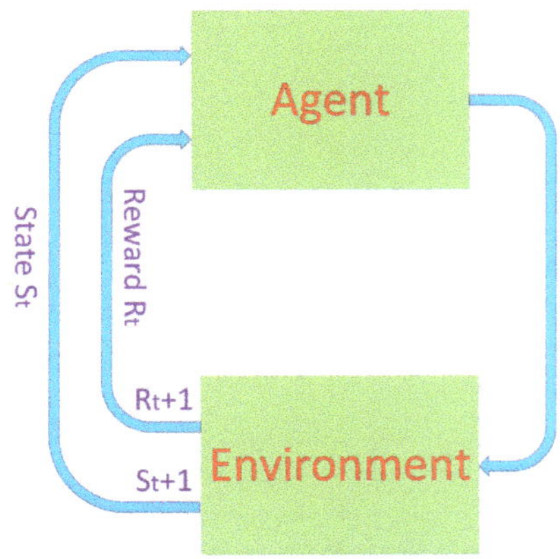

Figure 2. Agent–environment interactions in SARSA.

The algorithm shows the agent inside the SARSA framework has to accomplish several episodes. At each time step t, the agent in the state $\mathscr{s}_t$ chooses an action $\mathscr{a}_t$ based on greedy strategy. Afterwards, the agent receives the reward $\mathscr{R}_t$ and proceeds to the following state $\mathscr{s}_{t+1}$, where they choose the action $\mathscr{a}_{t+1}$ in dependence on the Q-table. The state–action equation can be represented as:

$$\mathcal{Q}(\mathscr{s}_t, \mathscr{a}) = (1-\mathscr{a})\mathcal{Q}(\mathscr{s}_t, \mathscr{a}) + \alpha[\mathscr{R}_{t+1} + \mathscr{Y}\mathcal{Q}(\mathscr{s}_{t+1}, \mathscr{a}_{t+1})], \quad (16)$$

Here, α represents the learning rate of the agent, $\mathscr{R}_{t+1}$ denotes the reward function of the next state, and $\mathscr{Y}$ signifies the discount factor. In the framework of RL, the agent aims to optimize the reward by adopting an optimal policy. The optimum policy can be calculated through the Bellman formula:

$$\mathscr{V}^*(\mathscr{s}) = \underset{\mathscr{a} \in \mathscr{A}}{max}\mathcal{Q}^*(\mathscr{s}, \mathscr{a}), \quad (17)$$

Subsequently, the value function is identified by the following equation:

$$\mathscr{V}(\mathscr{s}) = \underset{\mathscr{a} \in \mathscr{A}}{max}\mathcal{Q}(\mathscr{s}, \mathscr{a}), \quad (18)$$

The given equation is utilized to identify the optimal action value in order to optimize $\mathcal{Q}(\mathscr{s}_t, \mathscr{a}_t)$ for each state involved.

$$\mathscr{a} = \underset{\mathscr{a} \in \mathscr{A}}{argmax}\mathcal{Q}(\mathscr{s}, \mathscr{a}), \quad (19)$$

To decide what action $\mathscr{a}$ is going to be chosen during a particular time t, the Exploration and Exploitation Policy (EEP) feature is employed as follows:

$$a_t = \begin{cases} \underset{\mathscr{a} \in \mathscr{A}}{argmax}\mathcal{Q}(\mathscr{s}, \mathscr{a}) & exploitation \\ \underset{\mathscr{a} \in \mathscr{A}}{rand}(\mathscr{a}) & exploration \end{cases}, \quad (20)$$

In Equation (20), the 'ε − greedy' strategy is used while performing EEP, indicating the fact that the probabilities of exploitation and exploration are ε and $1-\varepsilon$, accordingly. Furthermore, a Markov Decision Process model of D2D communications underlaying cellular networks is provided to allocate power for each D2D pair using SARSA-based RL. Moreover, the actions of the agents, which are represented as $\mathscr{a}$, include a set of transmission powers, are indicated by P_n, and are allocated to the D2D transmitters T_n.

The reward function is formulated based on EE for each D2D pair n employing sub-channel k as follows:

$$R_n = \begin{cases} EE_{n,k}\left(\sqrt{R_{n,k}(t) - R^{th}_{min}}\right) & if\ R_{n,k}(t) \geq R^{th}_{min} \\ EE_{n,k}\left(-e^{\left(R_{n,k}(t) - R^{th}_{min}\right)}\ \right) & if\ R_{n,k}(t) < R^{th}_{min} \end{cases}, \quad (21)$$

Exploration involves a comprehensive examination of the network, gathering information, and randomly selecting actions to evaluate their efficiency. However, exploitation exploits the advantage of previous decisions according to the Q-table. Exploration and exploitation have trade-off features in power distribution techniques that utilize RL.

SARSA is more appropriate for our case than Q-learning because of its on-policy nature, which updates action values according to the current policy, hence offering a more adaptive response to network changes. This functionality is especially beneficial in D2D communication systems, where interference levels and network topology may vary. The Q-learning method, while efficient in static situations, fails to update according to real-time rules, potentially limiting its flexibility in dynamic contexts. Our investigation demonstrates that SARSA surpasses Q-learning in EE and power consumption, particularly under changing interference levels, highlighting its enhanced capability to handle real-time interference in high-density networks. Algorithm 2 defines the extensive structure of the SARSA strategy.

Algorithm 2. SARSA algorithm for power allocation issue

```
Initialize: N, M, P_max, P_min, Π_{n,k}, Q(s, a) table, Y, ε, α
For episode ∈ {1,...,EP} do
Reset s, t = 0
Select the level of power between P_max and P_min, utilizing policy derived Q (ε-greedy)
For t ∈ {0,...,T − 1} do
Every agent performs an action a ∈ A as well as observes R and s_{t+1}
Check (11a), (11f), and (11g)
If the conditions are satisfied, then
Establish action a ∈ A, R, s_{t+1}
End if
Each agent takes an action a_{t+1} ∈ A and observes R and s_{t+1}
Update the Q-table
S ← s_{t+1}, A ← a_{t+1}
End until all D2D pairs connect or the total iteration numbers is reached
End for
Output: optimum power for each D2D pair
```

7. Simulation Results

This part demonstrates the performance evaluation of the proposed SD-scheme regarding sum data rate, EE, SE, outage probability, and power saving. The SD-scheme that effectively joins the centralized mode selection and channel assignment scheme (C-scheme) is introduced based on greedy and matching algorithms, respectively, with the distributed power allocation scheme based on the SARSA algorithm. The suggested approach is compared with multiple traditional schemes, including a channel allocation scheme based on matching theory with the goal of optimizing EE [24], a channel allocation scheme based on the greedy algorithm [30], and power allocation schemes based on Q-learning in [17,23,29]. Table 2 contains the parameters employed in the simulation.

Table 2. Simulation parameters.

Parameter	Value
Entry 1	data
Cell radius	500 m
CUs number	20
D2D pairs number	2–38
Transmitter to receiver distance	10–100 m
Total Bandwidth	5 MHz
Noise power	−174 dBm//Hz
Sub-channel bandwidth	240 KHz
D2D transmitter power	0–23 dBm
gNB transmitting power	35 dBm
SINR threshold	3 dBm
Shadowing standard deviation	8 dB
Circuit power	0.1 watt
CU transmitting power	23 dBm
Discount factor	0.9
Learning rate	0.2
Epsilon	0.2

Figure 3 shows the EE comparison of the proposed SD-scheme with the suggested C-scheme, a channel allocation scheme based on matching theory with the goal of optimizing EE [24], a traditional channel allocation scheme based on the greedy algorithm [30], and power allocation schemes based on Q-learning-RL in [17,23,29] versus the number of D2D pairs. The EE rises with the incremental in the number of D2D pairs. As illustrated in Figure 3, the introduced approach demonstrates performance superiority and outperforms the benchmark schemes. The substantial boost in EE of the suggested technique demonstrates its efficacy in managing resources with the increase in D2D pairs, resulting in allocating optimal transmission power to each pair based on its requirements. Conversely, conventional systems provide limited enhancements, stabilizing at lower EE values as the number of D2D pairs escalates. The papers [24,30] confirm the worst EE since these schemes utilize conventional centralized approaches which lead to high control overhead, delayed or inflexible decisions, and insufficient optimum power allocation. In contrast, the proposed SARSA-based RL outperformed the Q-learning algorithm schemes [17,23,29], especially, where the pair number equals 10 and higher. The reason is that Q-learning often exhibits excessive exploration in some scenarios, strongly seeking optimum behaviors that enhance throughput, perhaps resulting in increased energy consumption. This conduct may adversely affect EE, particularly in D2D networks where sustaining low consumption of energy is essential. Furthermore, SARSA offers a more appropriate balance between exploration (engaging in new actions) and exploitation (utilizing the currently optimal action). This balanced approach in EE guarantees that power consumption is low during long exploratory stages since the SARSA strategy is continually adjusted depending on actual performance.

Figure 4 demonstrates the effectiveness of the proposed SD-scheme on total power saving versus the D2D pair numbers in the network. It is obvious that the power saving of the suggested scheme and the benchmark schemes increase as the number of D2D pairs increases. When the number of D2D pairs increases, the total power saving obtained by the suggested technique shows a constant and substantial enhancement, especially at the highest D2D pairs number. The introduced scheme outperforms the conventional systems as shown in the figure. The reason is that the SARSA algorithm is on-policy learning that responds to varying network situations instantaneously. Moreover, the introduced SARSA algorithm optimized the power transmission for every D2D pair continually in response to the channel conditions and interferences established by the CUs or other D2D pairs inside the cell.

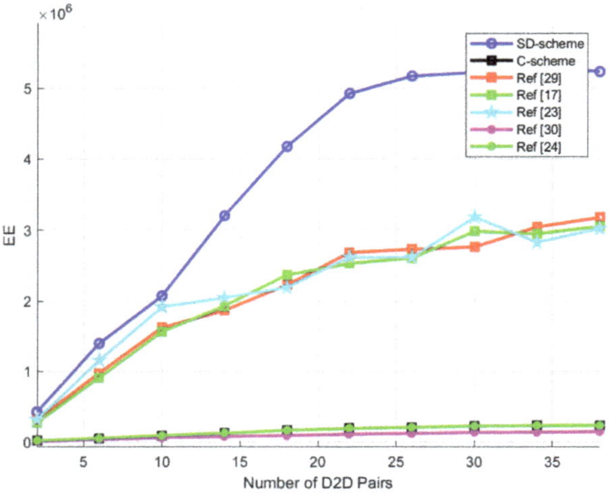

Figure 3. EE versus number of D2D pairs [17,23,24,29,30].

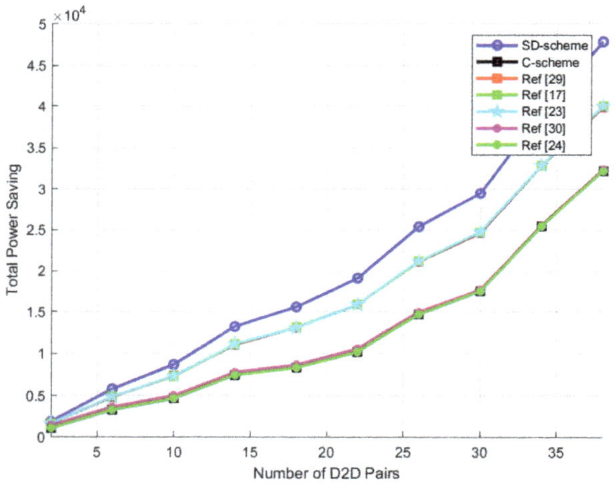

Figure 4. Total power saving versus number of D2D pairs [17,23,24,29,30].

Figure 5 compares the sum data rate of the SD-scheme to benchmark algorithms with various numbers of D2D pairs. The increment in the sum data rate is proportional to the D2D pair increase. It is clear that the [24,30] approaches outperform the proposed scheme in terms of sum data rate since these approaches employ maximum transmission power which results in reducing EE in the system. Moreover, in comparison to the benchmark schemes, the SD-scheme provides a higher sum data rate. The reason is that optimizing mode selection and channel assignment mitigates the effects of co-tier interference resulting from spectrum sharing between CUs and other D2D pairs. Consequently, the sum data rate of the network is enhanced accordingly. Clearly, the sum data rates increase slightly when the number of D2D pairs is between 25 and 40, due to the second reuse of the CUs' spectrum that increases the interference in the network. Furthermore, the suggested strategy enhances network performance by effectively balancing the utilization of resources and interference mitigation.

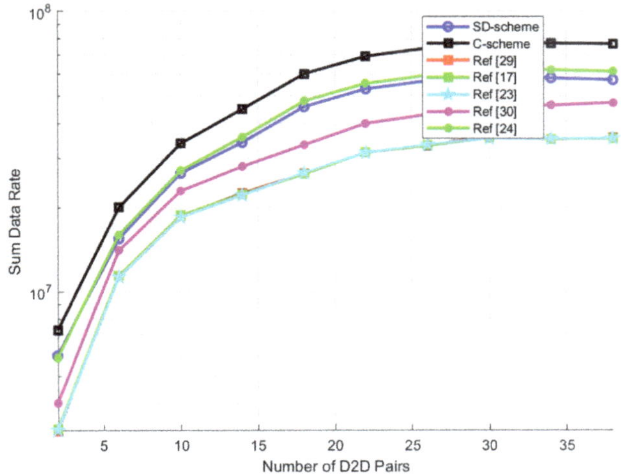

Figure 5. Sum data rate versus the number of D2D pairs [17,23,24,29,30].

Table 3 provides an overview of performance values, particularly EE, total power savings, and sum data rate, to elucidate the benefits of the SD-scheme over the C-scheme. The findings clearly indicate that the SD-scheme achieves better performance based on elevated EE, total power savings, and sum data rate, demonstrating its efficacy in resource allocation for D2D communications within cellular networks.

Table 3. Performance values.

No. of Pairs	EE (Bit/Sec/Watt)		Total Power Saving (Watt)		Sum Data Rate (Bit/Sec)	
	SD-Scheme	C-Scheme	SD-Scheme	C-Scheme	SD-Scheme	C-Scheme
5	1.12	0.71	0.49	0.27	1.3×10^7	2.0×10^7
10	2.06	0.95	0.91	0.49	2.6×10^7	3.4×10^7
15	3.45	1.54	1.40	0.77	3.8×10^7	4.9×10^7
20	4.52	2.12	1.74	0.93	4.9×10^7	6.4×10^7
25	5.14	2.93	2.43	1.37	5.5×10^7	7.2×10^7
30	5.21	3.15	2.93	1.74	5.9×10^7	7.8×10^7
35	5.25	3.40	4.22	2.72	5.8×10^7	7.8×10^7

Figure 6 shows the SE evaluation of the SD-scheme with the benchmark schemes in relation to the D2D pair numbers. The term SE rises incrementally with the number of D2D pairs since the sub-channel reuse indicator improves correspondingly with the increase in D2D pairs. Consequently, the co-tier interferences among shared channels of CUs increase. The SD-scheme demonstrates superior efficiency relative to other approaches. This approach demonstrated better SE when the number of D2D pairs ranged from 2 to 25, related to the minimal interference across the shared channels during the single reuse of the CUs' uplink channel. Moreover, the SE decreases in cases where there are 25–40 D2D pairs due to multiple times reusing the uplink spectrum of CU resources. When the transmitted power of the D2D pair is set to the maximum value, an increased data rate and SE are obtained due to a strong signal which results in increasing the interference and decreasing EE accordingly. The significant efficacy of the proposed strategy is attributed to its adaptive resource allocation mechanism, which responds flexibly to changing network situations. This flexibility ensures potential improvements in SE, which is especially important in heavily loaded D2D environments where interference management is critical.

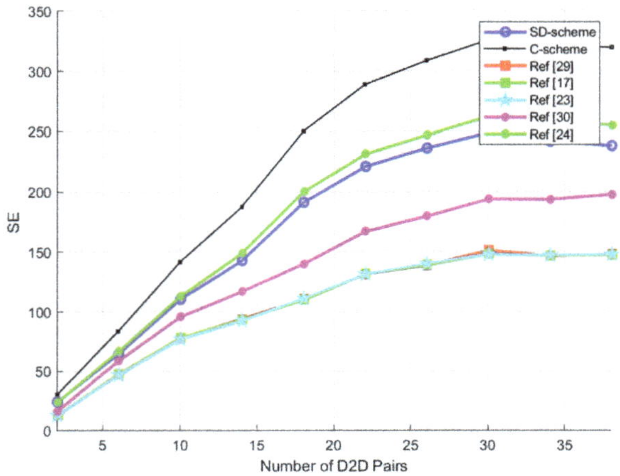

Figure 6. SE versus the number of D2D pairs [17,23,24,29,30].

Figure 7 shows the comparison of the outage probability with various numbers of D2D pairs. The outage probability of the SD-scheme increases in accordance with the number of D2D pairs because of the increased network interferences. Nonetheless, the suggested method and other channel-optimized algorithms reduce the probability of outages by ensuring the most efficient utilization of the resources in order to meet the minimal QoS demands for each D2D pair. By optimizing reused channels, the suggested system efficiently allows multiple D2D pairs using one particular CU's sub-channel. Despite the number of D2D pairs growing, this approach maintains low outage probability and reduced interference. Because of the unmanaged interferences of the resources that are shared among D2D pairs and between D2D pairs and CUs from the other side in benchmark methods, the outage probability significantly increases.

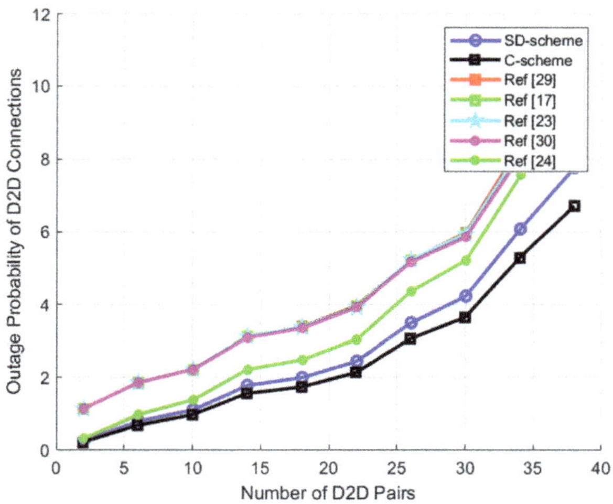

Figure 7. Outage probability versus the number of D2D pairs [17,23,24,29,30].

The EE analysis for varying transmission power of CUs is illustrated in Figure 8. It is clear that there are slight decreases in EE with the incremental increase in CUs transmitting power due to high interference created by CUs and the higher transmission power of D2D pairs required to satisfy QoS demands. The EE of the suggested SD-scheme demonstrates better performance in comparison to traditional schemes. The reason is that a lack of scalability and flexibility with a higher power level can be observed in traditional systems like [24,30], which show better establishing EE but rapidly decrease as CU power rises. The suggested approach, on the other hand, demonstrates its better energy management and adaptation to varied network conditions by maintaining better EE despite changes in power levels.

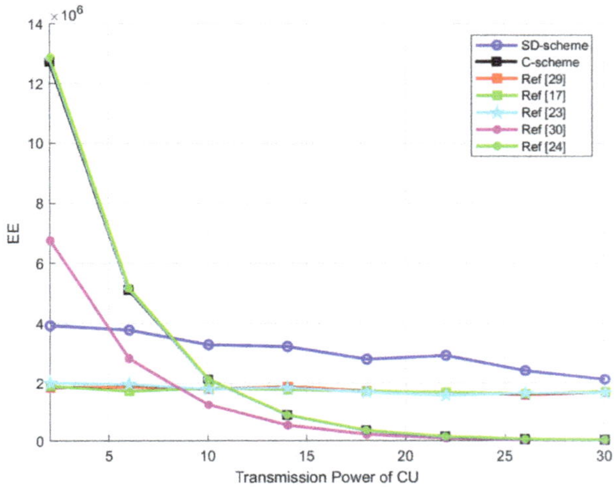

Figure 8. EE versus transmission power of CUs [17,23,24,29,30].

Figure 9 shows the sum data rate comparison of the suggested SD-scheme with traditional schemes versus the transmission power levels of CUs. The sum data rate demonstrates a slight decrease with the increase in CU transmission power. The first reason is due to the increased interference of D2D pairs that share the same spectrum of CUs. Secondly, the D2D pairs experience higher competition for limited resources in the network. The conventional approaches fail to efficiently allocate resources which results in low SINR and low sum data rate accordingly.

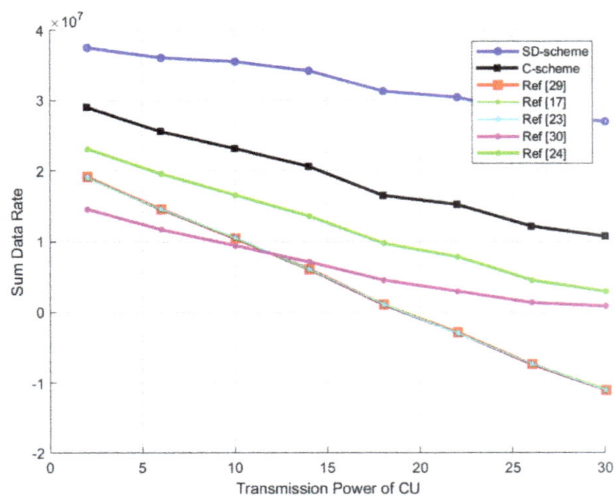

Figure 9. Sum data rate versus transmission power of CUs [17,23,24,29,30].

Figure 10 demonstrates the total power saving of the proposed SD-scheme over varying transmission power levels of CUs. It is obvious that the total power saving remains stable up to 15 dBm, after that, the total power saving increases sharply due to several reasons. Firstly, the pairs can barely perform with the same power level without any requirement to enhance the transmission power strategies. Furthermore, the interference from CUs substantially affects the D2D links; consequently, the D2D pairs tend to overcome the interference threshold. As illustrated in Figure 10, the proposed SD-scheme outperforms the benchmark approaches. The reason is that the proposed SARSA-based RL allocates optimum power for each D2D pair according to the immediate demands of users and ensures that the D2D pairs are utilizing the lowest possible power to prevent co-tier interference and guarantee communication between the transmitter and receiver for the D2D link.

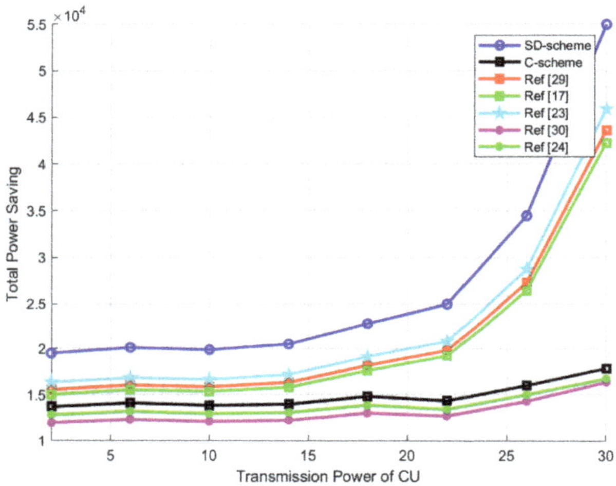

Figure 10. Power saving versus transmission power of CUs [17,23,24,29,30].

Figure 11 illustrates the SE different levels in relation to different available techniques versus the CU transmission power. When the transmission power of CUs escalates, the SE decreases to specific thresholds, except for the [24,30] methods, since these approaches employ maximum transmission power. The reason behind this decrease is the conflict of sub-channel reuse since the transmission power of CUs increases. Hence, the D2D pairs experience higher interference due to shared resources between them.

Figure 12 demonstrates the probability of connections for D2D modes including DM, RM, and LM versus the maximum distance of relays to D2D pairs. Regarding DM, the figure indicates a consistent probability with the variety of relay distances since the distance between relays and D2D pairs has no impact on direct D2D mode. Furthermore, the probability of connection of RM decreases with the incremental distance, highlighting the dependence of D2D pairs on the proximity of relays to establish efficient communication links. When the distance of relays increases, the D2D pairs tend to communicate with each other through LM. Consequently, LM increases with the increase in the distance between relays to D2D pairs.

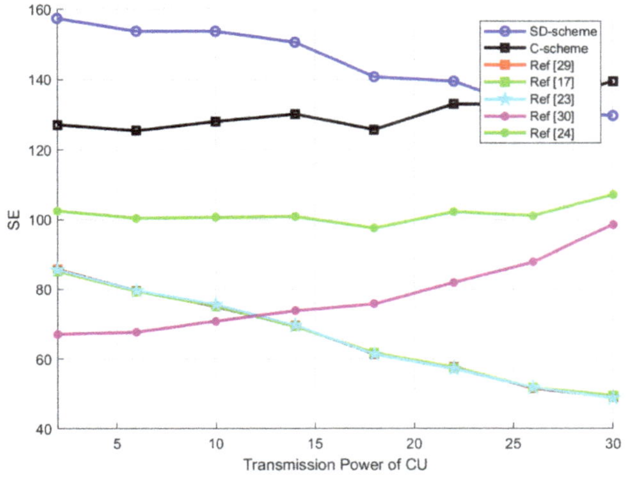

Figure 11. SE versus transmission power of CUs [17,23,24,29,30].

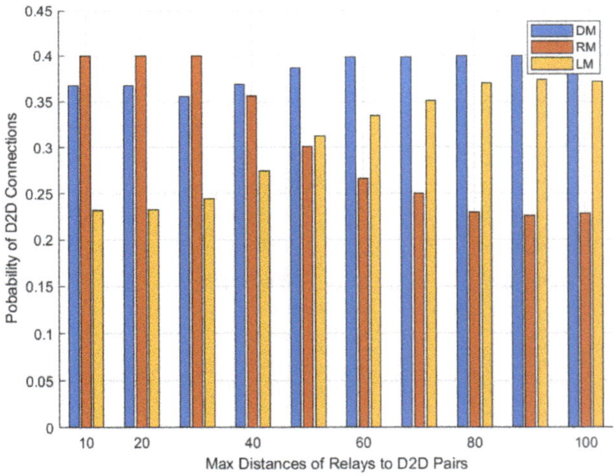

Figure 12. Probability of D2D connections versus maximum distance of relays to D2D pairs (m).

Figure 13 demonstrates the probability of connections of D2D pairs including DM, RM, and LM in comparison to the highest distance between the D2D pair. It is obvious that the DM reaches the highest probability of connections when the maximum distance across the D2D pair is limited. Furthermore, the probability of connections for DM decreases with the increase in distance since the DM mode depends on the proximity between transmitter and receiver, which allows the D2D pair to operate efficiently. On the other hand, the probability of connections for both RM and LM increases with the increase in the maximum distance between the D2D pair. The reason is that when the distance is medium, the pair tend to choose RM to maintain the QoS threshold. Moreover, when the distance is high, the pair prefers to choose the LM for the same previous reason since the power of the gNB is greater than that of the D2D transmitter. Consequently, the D2D pair chooses either DM, RM, or LM depending on the maximum SINR which is impacted by the distance.

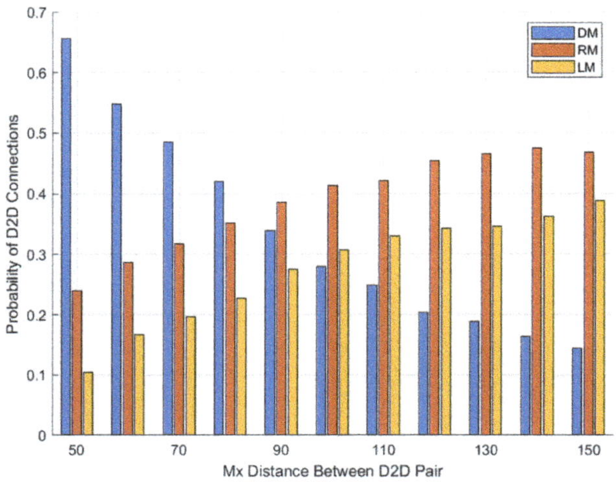

Figure 13. Probability of D2D connections versus max distance between D2D pairs.

Figure 14 demonstrates the number of disconnected pairs versus the number of D2D pairs in the network with the goal of comparing the performance of the proposed SD-scheme with the C-scheme. When the number of D2D pairs increases, the proposed schemes increase in terms of the number of disconnected D2D pairs because of the increased interference between D2D pairs and CUs. It is obvious that the proposed SD-scheme achieves slightly better performance than the C-scheme because of its better ability to mitigate interference and utilize resources effectively. The performance divergence across both schemes increases with the number of D2D pairs, indicating that the SD-scheme exhibits more adaptability and stability under higher network loads. This enhancement is due to the SD-scheme's optimized resource allocation technique, which reduces connection failures and guarantees superior connectivity for D2D users.

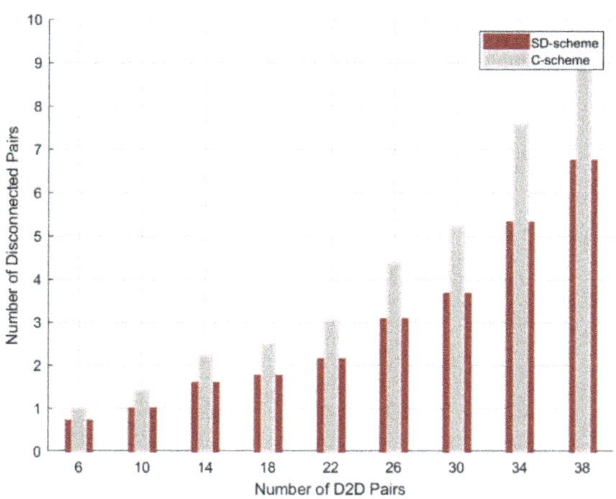

Figure 14. Number of disconnected pairs versus number of D2D pairs.

8. Conclusions

This paper presents an SD-scheme for mode selection, channel assignment, and power allocation in D2D communications underlaying cellular networks. The primary goal of this research is to enhance D2D performance while maintaining the QoS demands of CUs.

Furthermore, the sub-channel of each single CU can be reused across several D2D pairs, and the detrimental effects of the interference that occurs across D2D pairs are considered during the process of resource allocation. The joint problems of mode selection, channel assignment, and power allocation are MINLP and NP and difficult to resolve. Therefore, we designed a hybrid scheme: a centralized mode selection and channel assignment approach, followed by a distributed power management approach. The initial process involves the concept of a centralized greedy-based mode selection and two-sided preference lists channel assignment approach, followed by the implementation of a distributed power control approach in the subsequent phase. Moreover, an SARSA-based RL power control method has been proposed to iteratively update the transmission power for each D2D pair utilizing the same assigned sub-channel of individual CUs with the goal of improving the EE of the D2D communications in the network. The simulation findings demonstrated that the introduced scheme yields better performance with low complexity and outperforms traditional and Q-learning schemes in terms of data rate, SE, and EE. Future research may include examining the influence of other networking factors such as users' mobility as well as including modern equipment like unmanned aerial vehicles and satellites to the network.

Author Contributions: Conceptualization, N.M.M.; Data curation, N.M.M.; Formal analysis, I.S.A.; Investigation, I.S.A.; Methodology, I.S.A. and M.H.D.N.H.; Project administration, N.M.M.; Software, I.S.A.; Supervision, N.M.M.; Validation, M.H.D.N.H.; Writing—original draft, I.S.A.; Writing—review and editing, M.H.D.N.H. All authors have read and agreed to the published version of the manuscript.

Funding: This research received no external funding.

Data Availability Statement: No new data were created in this study. All relevant data are included within the paper.

Conflicts of Interest: The authors declare no conflicts of interest.

References

1. Siddiqui, M.U.A.; Qamar, F.; Tayyab, M.; Hindia, M.N.; Nguyen, Q.N.; Hassan, R. Mobility Management Issues and Solutions in 5G-and-Beyond Networks: A Comprehensive Review. *Electronics* **2022**, *11*, 1366. [CrossRef]
2. Alhashimi, H.F.; Hindia, M.N.; Dimyati, K.; Hanafi, E.B.; Safie, N.; Qamar, F.; Azrin, K.; Nguyen, Q.N. A Survey on Resource Management for 6G Heterogeneous Networks: Current Research, Future Trends, and Challenges. *Electronics* **2023**, *12*, 647. [CrossRef]
3. Gismalla, M.S.M.; Azmi, A.I.; Bin Salim, M.R.; Abdullah, M.F.L.; Iqbal, F.; Mabrouk, W.A.; Othman, M.B.; Ashyap, A.Y.I.; Supa'At, A.S.M. Survey on Device to Device (D2D) Communication for 5GB/6G Networks: Concept, Applications, Challenges, and Future Directions. *IEEE Access* **2022**, *10*, 30792–30821. [CrossRef]
4. Alibraheemi, A.M.H.; Hindia, M.N.; Dimyati, K.; Izam, T.F.T.M.N.; Yahaya, J.; Qamar, F.; Abdullah, Z.H. A Survey of Resource Management in D2D Communication for B5G Networks. *IEEE Access* **2023**, *11*, 7892–7923. [CrossRef]
5. Alibraheemi, A.M.H.; Hindia, M.N.; Izam, T.F.T.M.N.; Dimyati, K. Spectrum Efficient Mode Selection and Resource Allocation Optimization for D2D Communication in HetNet: A Multi-Agent Q-Learning Approach. *IEEE Access* **2024**, *12*, 131217–131229. [CrossRef]
6. Alhashimi, H.F.; Hindia, M.N.; Dimyati, K.; Hanafi, E.B.; Izam, T.F.T.M.N. Joint Optimization Scheme of User Association and Channel Allocation in 6G HetNets. *Symmetry* **2023**, *15*, 1673. [CrossRef]
7. Chen, C.-Y.; Sung, C.-A.; Chen, H.-H. Capacity maximization based on optimal mode selection in multi-mode and multi-pair D2D communications. *IEEE Trans. Veh. Technol.* **2019**, *68*, 6524–6534. [CrossRef]
8. Attar, I.S.; Mahyuddin, N.M.; Hindia, M.H.D.N. Joint mode selection and resource allocation for underlaying D2D communications: Matching theory. *Telecommun. Syst.* **2024**, *87*, 663–678. [CrossRef]
9. Jayakumar, S.; Nandakumar, S. A review on resource allocation techniques in D2D communication for 5G and B5G technology. *Peer-to-Peer Netw. Appl.* **2021**, *14*, 243–269. [CrossRef]
10. Gu, W.; Zhu, Q. Social-aware-based resource allocation for NOMA-Enhanced D2D communications. *Appl. Sci.* **2020**, *10*, 2446. [CrossRef]

11. Hassan, A.N.; Al-Chlaihawi, S.; Khekan, A.R. Artificial intelligence techniques over the fifth generation mobile networks. *Indones. J. Electr. Eng. Comput. Sci.* **2021**, *24*, 317–328. [CrossRef]
12. Alhashimi, H.F.; Hindia, M.N.; Dimyati, K.; Hanafi, E.B.; Tengku Mohmed Noor Izam, T.F. Reinforcement Learning Based Power Allocation for 6G Heterogenous Networks. In *Lecture Notes in Computer Science (Including Subseries Lecture Notes in Artificial Intelligence and Lecture Notes in Bioinformatics)*; Springer Science and Business Media Deutschland GmbH: Berlin, Germany, 2024; pp. 128–141. [CrossRef]
13. Sheng, J.; Liu, S.; Huang, T.; Wu, Y. Overlapping Coalition Game for Resource Allocation in Many-to-Many D2D Communication. *Wirel. Commun. Mob. Comput.* **2022**, *2022*, 1738530. [CrossRef]
14. Rosas, A.A.; Shokair, M.; Dessouky, M.I. Genetic Based Approach for Optimal Power and Channel Allocation to Enhance D2D Underlaied Cellular Network Capacity in 5G. *Comput. Mater. Contin.* **2022**, *72*, 3751–3762. [CrossRef]
15. Sun, Y.; Miao, M.; Wang, Z.; Liu, Z. Resource Allocation Based on Hierarchical Game for D2D Underlaying Communication Cellular Networks. *Wirel. Pers. Commun.* **2021**, *117*, 281–291. [CrossRef]
16. Dejen, A.A.; Wondie, Y.; Forster, A. Distributed Throughput and Energy Efficient Resource Optimization When D2D and Massive MIMO Coexist. *J. Commun. Inf. Netw.* **2022**, *7*, 278–295. [CrossRef]
17. Jiang, S.; Zheng, J. A Q-learning based Dynamic Power Control Algorithm for D2D Communication Underlaying Cellular Networks. In Proceedings of the 13th International Conference on Wireless Communications and Signal Processing, WCSP 2021, Changsha, China, 20–21 October 2021; IEEE: Piscataway, NJ, USA, 2021. [CrossRef]
18. Chang, H.-H.; Liu, L.; Bai, J.; Pidwerbetsky, A.; Berlinsky, A.; Huang, J.; Ashdown, J.D.; Turck, K.; Yi, Y. Resource Allocation for D2D Cellular Networks with QoS Constraints: A DC Programming-Based Approach. *IEEE Access* **2022**, *10*, 16424–16438. [CrossRef]
19. Xing, X.; Cao, J.; Zhou, H. Improving Quality of Service for Cell-Edge Users in D2D-Relay Networks. *Wirel. Pers. Commun.* **2022**, *126*, 1789–1804. [CrossRef]
20. Wei, Y.; Qu, Y.; Zhao, M.; Zhang, L.; Yu, F.R. Resource allocation and power control policy for device-to-device communication using multi-agent reinforcement learning. *Comput. Mater. Contin.* **2020**, *63*, 1515–1532. [CrossRef]
21. Wang, H.; Wang, Y.; Tang, L.; Xia, Y. D2D Social Selection Relay Algorithm Combined with Auction Principle. *Sensors* **2022**, *22*, 9265. [CrossRef]
22. Hamid, A.K.; Al-Wesabi, F.N.; Nemri, N.; Zahary, A.; Khan, I. An optimized algorithm for resource allocation for D2D in heterogeneous networks. *Comput. Mater. Contin.* **2022**, *70*, 2923–2936. [CrossRef]
23. Jiang, F.; Zhang, L.; Sun, C.; Yuan, Z. Clustering and resource allocation strategy for D2D multicast networks with machine learning approaches. *China Commun.* **2021**, *18*, 196–211. [CrossRef]
24. Awad, M.K.; Baidas, M.W.; El-Amine, A.A.; Al-Mubarak, N. A matching-theoretic approach to resource allocation in D2D-enabled downlink NOMA cellular networks. *Phys. Commun.* **2022**, *54*, 101837. [CrossRef]
25. Gao, J.; Meng, X.; Yang, C.; Zhang, B.; Yi, X. Resource Allocation for D2D Communication Underlaying Cellular Networks: A Distance-Based Grouping Strategy. *Wirel. Commun. Mob. Comput.* **2023**, *2023*, 8594323. [CrossRef]
26. Pei, E.; Zhu, B.; Li, Y. A Q-learning based Resource Allocation Algorithm for D2D-Unlicensed communications. In Proceedings of the IEEE Vehicular Technology Conference, Helsinki, Finland, 25–28 April 2021. [CrossRef]
27. El-Nakhla, O.M.; Obayya, M.I.; Kishk, S.E. Stable Matching Relay Selection (SMRS) for TWR D2D Network With RF/RE EH Capabilities. *IEEE Access* **2022**, *10*, 22381–22391. [CrossRef]
28. Jayakumar, S.; Nandakumar, S. Reinforcement learning based distributed resource allocation technique in device-to-device (D2D) communication. *Wirel. Netw.* **2023**, *29*, 1843–1858. [CrossRef]
29. Lee, S.-H.; Shi, X.-P.; Tan, T.-H.; Lee, Y.-L.; Huang, Y.-F. Performance of Q-learning based resource allocation for D2D communications in heterogeneous networks. *ICT Express* **2023**, *9*, 1032–1039. [CrossRef]
30. Gour, R.; Tyagi, A. Joint uplink–downlink resource allocation for energy efficient D2D underlaying cellular networks with many-to-one matching. *Phys. Commun.* **2023**, *58*, 102016. [CrossRef]
31. Wang, H.; Xiao, P.; Li, X. Channel Parameter Estimation of mmWave MIMO System in Urban Traffic Scene: A Training Channel-Based Method. *IEEE Trans. Intell. Transp. Syst.* **2024**, *25*, 754–762. [CrossRef]

telecom

MDPI

Article

Fifth-Generation (5G) Communication in Urban Environments: A Comprehensive Unmanned Aerial Vehicle Channel Model for Low-Altitude Operations in Indian Cities

Ankita K. Patel * and Radhika D. Joshi

Department of Electronics and Telecommunication Engineering, College of Engineering Pune Technical University, Pune 411005, India; rdj.extc@coeptech.ac.in
* Correspondence: ankitaketan@gmail.com; Tel.: +91-940-4239-178

Abstract: Unmanned aerial vehicles (UAVs) significantly shape the evolution of 5G and 6G technologies in India, particularly in reconfiguring communication networks. Through their deployment as base stations or relays, these aerial vehicles substantially enhance communication performance and extend network coverage in areas characterized by high demand and challenging topographies. Accurate modelling of the UAV-to-ground channel is imperative for gaining valuable insights into UAV-assisted communication systems, particularly within India's rapidly expanding metropolitan cities and their diverse topographical complexities. This study proposes an approach to model low-altitude channels in urban areas, offering specific scenarios and tailored solutions to facilitate radio frequency (RF) planning for Indian metropolitan cities. The proposed model leverages the International Telecommunication Union recommendation (ITU-R) for city mapping and utilizes frequency ranges from 1.8 to 6 GHz and altitudes up to 500 m to comprehensively model both line-of-sight (LoS) and non-line-of-sight (NLoS) communications. It employs the uniform theory of diffraction to calculate the additional path loss for non-line-of-sight (NLoS) communication for both vertical and horizontal polarizations. The normal distribution for additional shadowing loss is discerned from simulation results. This study outlined the approach to derive a comprehensive statistical channel model based on the elevation angle and evaluate model parameters at various frequencies and altitudes for both vertical and horizontal polarization. The model was subsequently compared with existing models for validation, showing close alignment. The ease of implementation and practical application of this proposed model render it an invaluable tool for planning and simulating mobile networks in urban areas, thus facilitating the seamless integration of advanced communication technologies in India.

Keywords: UAV; low altitude; ITU-R Model; 5G communication; uniform theory of diffraction

1. Introduction

The advent of the fifth generation (5G) of communications can revolutionize services by providing enhanced end user experiences, seamless coverage, high data rates, and low latency, thus significantly improving performance and reliable communications [1]. The rollout of 5G services in India began in October 2022 and will expand across 28 states and eight union territories by 2024. According to Ericsson's projections, the number of 5G users in India is expected to reach 700 million by the end of 2028 [2]. This technology is anticipated to play a pivotal role in achieving the goals of the Digital India Program established by the Government of India. The launch of 5G services in India faces several challenges, primarily the need to fundamentally redesign the communication system's core architecture. Unlike 4G, 5G struggles with long-distance data transmission, necessitating a denser network with small cell architecture—consisting of numerous low-power base stations for better coverage. This shift demands significant infrastructure investments, including more cell towers and improved supporting technologies to fully leverage 5G's capabilities and ensure seamless connectivity nationwide. Implementing this advanced infrastructure demands substantial investments in additional cell towers and supporting technologies to maximize 5G's potential.

To overcome these obstacles, unmanned aerial vehicles (UAVs) present a promising solution, particularly within India's rapidly evolving metropolitan landscapes. As cities expand geographically, UAVs can function as dynamic aerial base stations, relays, or aerial servers [3–5]. We can significantly enhance communication performance and coverage by integrating these flying machines with existing base stations. This synergy facilitates improved data transmission and extends the reach of 5G networks, ultimately paving the way for a robust and interconnected future. UAV base stations are distinct from UAVs primarily due to their advanced communication infrastructure. Unlike UAVs, which may lack a dedicated communication system, UAV base stations—often referred to as drone base stations or flying base stations—are specifically designed to enhance connectivity. These stations are equipped with essential components such as antennas, transceivers, and power supplies, enabling them to establish reliable communication with ground users effectively. This robust setup allows for seamless data transfer, real-time monitoring, and improved operational efficiency in various applications, including surveillance, mapping, and emergency response.

Furthermore, incorporating UAVs with wireless sensor networks can support various applications, such as agriculture, surveys, package and food delivery, aerial photography, etc. [6,7]. UAVs will be invaluable in emergency and public safety scenarios in which the terrestrial network is compromised. By deploying UAVs with communication systems, authorities can establish temporary networks to coordinate rescue efforts and relay vital information [8].

The deployment of UAVs as base stations or communication relays necessitates a thorough assessment of the communication infrastructure requirements for effective service establishment. This consideration is particularly crucial given the limited energy capacity of UAVs. To achieve a successful network deployment, it is vital to determine several key factors: the optimal number of base stations required, the operational altitude for the UAVs, and the specific level of service expected by users. When UAVs are utilized for air-to-ground communication at low altitudes—essentially mirroring traditional cellular network principles—efficient deployment becomes increasingly paramount. A robust and precise channel model is essential in this context, as it provides the foundation for accurately calculating path loss and power consumption. By carefully analyzing these technical variables, operators can ensure that their network will provide adequate coverage and service quality across targeted geographic regions, ultimately enhancing user experience and operational reliability.

The impetus behind this proposal arises from two crucial factors that significantly impact the development of communication models in Indian metropolitan cities: (i) the current lack of a comprehensive generalized model that encompasses all urban environments within these cities; (ii) the necessity for a singular model that effectively captures the interplay between essential elements of UAV communication such as polarization, altitude, elevation, and frequency. In light of the absence of established models that address low-altitude UAV links—particularly those influenced by shadowing loss components—this proposal aims to introduce a generalized channel model for UAV communication. This initiative is especially pertinent for Indian metropolitan cities, which need advanced communication infrastructures to support the rollout of 5G technology.

The urban fabric of Indian metropolitan cities can be distinctly categorized into four environmental scenarios: (i) the suburban environment, which encompasses the extended city peripheries that have characteristics akin to rural areas; (ii) the urban environment, representing any typical area within the city; (iii) the dense urban environment, characterized by closely clustered buildings and limited open spaces; and (iv) the high-rise urban environment, highlighting newly developed regions featuring skyscraper-style architecture. These varied urban settings necessitate the development of a channel model to accurately assess path loss and shadowing variations at different heights. Presently, established generalized models relying on statistical parameters from ITU-R P.1410-5 fall short of addressing low altitudes below 500 m, requiring many applications associated with 5G communication. Deploying low-altitude systems ensures that devices stay within close range, leading to more reliable data connections. This proximity enhances connection quality, allowing for stable transmissions and seamless connectivity, even in densely populated areas where device saturation is an issue. A robust and comprehensive channel model is essential for integrating UAVs as aerial base stations into the existing ground station infrastructure. This model should consider diverse scenarios, frequency ranges, and altitudes pertinent to urban communication. This paper focuses on constructing a model that reflects the four identified scenarios within Indian metropolitan cities while aligning with the frequency and altitude requirements specific to 5G technology. The key contributions of this paper are multifaceted:

1. A statistical model is proposed to predict the link between the UAV (transmitter) and the end users (receivers).
2. The proposed model is based on ITU-R statistical parameters for city layout and the uniform theory of diffraction to account for additional path loss.
3. The additional path loss is mapped with a normal distribution, and distribution parameters are assessed.
4. The model's formulation depends on the elevation angle, and its parameters are computed for different frequencies and altitudes, considering both vertical and horizontal polarization.
5. The proposed model is compared with existing models for validation.
6. The impact of variation in frequency and UAV altitude on path loss is discussed.

Through this comprehensive approach, we aim to provide a significant contribution to the field of urban communication infrastructure, enabling better integration of UAV technology within the context of 5G networks in India.

This paper's propagation model is established on simulations conducted in a randomly generated urban environment. This paper is structured as follows: Section 2 overviews air-to-ground channel modelling. Section 3 presents a simulation of a city featuring four environments, which represent Indian metropolitan cities. In Sections 4 and 5, a new model is proposed and subjected to analysis, followed by validation of the model in Section 6. Finally, Section 7 offers concluding remarks on this study.

2. Air to Ground Channel Modelling

Two primary methods are utilized for categorizing the air-to-ground channel: an empirical channel based on measurement data and a deterministic or geometry-based channel model derived from simulated data [9].

The empirical channel model used measurement data for C-band- and L-band-using aircraft in scenarios like suburban, urban, over-water, hilly, and mountainous environments [10–13]. The data provided by measurement gives insight into the statistical properties of different terrains. In [14], the authors investigated the impact of airframe shadowing on the channel. They also provided a solution to mitigate this effect by effectively deploying multiple antennas. The statistical properties of large-scale and small-scale fading were analyzed using measurements for suburban areas, rural areas, and open areas of a city in the context of air-to-ground UAV communication, as detailed in [15,16]. Determination of path loss for low-altitude UAV channels at frequencies 2.4 and 5.9 GHz was achieved by using a multilink measurement campaign [17]. The authors proposed path loss based on elevation angle and distance and further analyzed the shadow fading across different frequency bands and evaluated spatial correlations. A medium-sized UAV conducted low-altitude measurements in a suburban area to investigate multipath effects [18]. In Reference [19], the authors used a fixed-wing UAV for an air-to-ground channel measurement campaign conducted at low altitudes in a rural area, at 2.7 GHz. The paper offers details about the measurement system and analysis of large-scale fading, including path loss, shadow fading, and small-scale fading. The UAV provided a path loss model and identified the impact of altitude on the channel. While measurement data offer firsthand insights into channel characteristics, it is essential to note that such data are site-specific and may not be broadly applicable to general scenarios. Furthermore, specific measurement environments are complex and involve high costs.

The deterministic channel modelling approach using ray tracing can demonstrate the channel's performance when considering environmental conditions. A ray tracing method was used to characterize a propagation channel model for rural, suburban, and dense urban scenarios at 28 GHz to find the effect of multipath components at the receiver [20]; suburban open-environment channel characteristics are analyzed in [21]. A ray tracing-based Monte Carlo method was used to analyze and evaluate UAV air-to-ground channel model performance [22]. Geometry-based stochastic models are widely used to evaluate spatial and temporal variation and characterize air-to-ground channels. A regular-shaped, geometry-based stochastic model for a UAV MIMO channel is proposed in [23]. The statistical characteristics of the channel were ascertained by considering parameters such as altitude and flight movement. A model with mixed bouncing was proposed to capture AG channel characteristics [24]. Taking into consideration the changing moving direction and speed of the MS, a channel model of 3D wideband MIMO was proposed [25], and the statistical properties were verified by simulation data and theoretical data. The models are tailored to specific geometries and unsuitable for different scenarios. A multi-UAV OFDMA communication system that utilizes geographical information develops a realistic channel model incorporating blockage-aware parameters [26].

In the existing body of literature, generic models are available to determine statistical parameters for modelling the UAV air-to-ground channel [27–33]. In [27], the authors introduce a generic statistical model tailored for low-altitude platforms, where the channel model parameters are derived via 3D ray-tracing at frequency bands of 700, 2000, and 5800 MHz for altitudes up to 2 km. Furthermore, the probability of the line-of-sight (LoS) channel of a UAV is modelled utilizing a sigmoid function of the vertical angle between the UAV and the user, as presented in [28]. This model is leveraged to characterize the coverage radius for an aerial base station of the UAV as a function of path loss, alongside assessments of the optimal UAV altitude in [29,30]. The WIRELESS Insite software version 2.5, program is employed to simulate a dense urban environment at varying altitudes and frequencies, further refined through analytical formulation using the knife edge refraction model [31]. A statistical model specific to high altitude, based on elevation, is delineated in [32]. The methodology for estimating coverage using the generalized UAV air-to-ground channel model for altitudes up to 100 m is explained in [33]. In comparison, for altitudes up to 500 m, it is delineated in [34]. Authors in [35] furnish a comprehensive channel model for UAV linkages based on the alpha-beta model framework.

The existing body of literature concerning the generic channel model does not sufficiently investigate low-altitude communication links for unmanned aerial vehicles (UAVs) within the complex and varied urban landscapes of India. Previous research underscores the critical need for an elevation-based model, which can improve the precision of path loss calculations and account for shadowing variations that occur at different altitudes. The objective of this study is to introduce a comprehensive air-to-ground channel model that takes elevation angles into account. This model will utilize statistical parameters recommended by the International Telecommunication Union (ITU-R), specifically tailored to suit various environments, including suburban, urban, dense urban, and high-rise urban areas. By doing so, this study aims to support the effective expansion and implementation of 5G communication systems across the diverse urban settings in India. The proposed system aims to enhance the coverage area while managing the energy requirements of UAVs operating as aerial base stations (ABSs). It incorporates power reduction techniques such as trajectory optimization through efficient flight planning, flexible user assignment to nearby base stations, and dynamic adjustment of the coverage area based on demand. These strategies are designed to address the limited energy capacity of UAVs.

3. Modeling of City Environment

The building's layout and characteristics significantly impact the radio frequency (RF) model in an urban setting. These factors include building heights, materials, spacing between buildings, and obstacles such as trees and other structures. Understanding these conditions and constraints is essential for accurately developing an RF model to account for signal propagation and coverage in urban areas effectively. The statistical ITU-R recommendation P.1410-5 [36] model allows for us to create the build-up area for an urban environment without specific information about building shapes and distribution. As per [34], only three parameters are needed to model the area: α, which is the ratio of build-up area to the total land area (dimensionless); β, the mean number of buildings per unit area (buildings/km^2); γ, a scale parameter that describes the distribution of building heights according to a Rayleigh probability density function.

$$p(h_b) = \frac{h_b}{\gamma^2}\exp(\frac{-{h_b}^2}{\gamma^2}) \tag{1}$$

where $p(h_b)$ is the probability distribution of the building with height h_b.

The statistical ITU-R model parameters are utilized to create four different environments that can be used to map the area of an Indian metropolitan city. Table 1 summarizes the above-mentioned statistical parameters based on ITU-R recommendation P.1410-5 [36] These parameters are utilized to create four distinct environments that can be correlated with urban areas. The distribution of buildings is generated using the Rayleigh distribution to ensure a realistic approach when analyzing shadowing effects.

Table 1. ITU-R P.1410-5 model parameters.

Environment	α	β	Γ
Suburban	0.1	750	8
Urban	0.3	500	15
Dense urban	0.5	300	20
High-rise urban	0.5	300	50

Designing a geometric model that meets these criteria to represent an acceptable city layout is a significant challenge. This paper utilizes a mathematical approximation proposed by Al Hurani [27] to shape the city environment, as shown in Figure 1. The simulation considers a square plot with an area of one square kilometer, where the total number of buildings is Ns, the side length is denoted as Ds, the building width is Ws, and the interbuilding space is Ss, all measured in meters. As per [27], the ITU-R statistical parameters can be linked with D_S, W_S, and S_S with the relationship $W_s = 1000\sqrt{\frac{\alpha}{\beta}}$, with interbuilding spacing $S_s = \frac{1000}{\sqrt{\beta}} - W_s$ and side length $D_s = \frac{W_S+S_S}{1000}\sqrt{N_S}$.

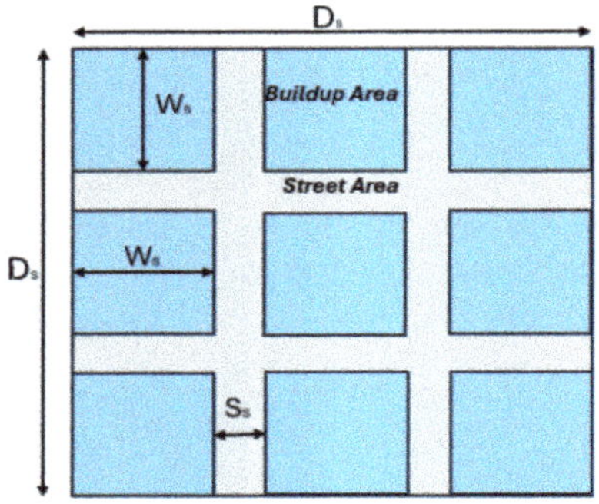

Figure 1. Selected layout for city areas.

The city layout can be easily generated using MATLAB 2023. In the real-world environment, buildings do not have a regular structure. To mitigate that in simulation, the UAV's position is determined by the elevation and azimuthal angles at each point along the street. Calculations are carried out for specific elevation angles using the median value obtained from the azimuthal angle. This approach ensures that the results are not dependent on the azimuthal angle, accounting for the uneven spacing of buildings in the real world.

Simulation Method

The simulation is performed using MATLAB. It is divided into two parts: The first is that LoS probability in the street is studied as a function of elevation angle for four types of build areas. Second, the additional path loss resulting from the shadowing effects of buildings is investigated using the uniform theory of diffraction (UTD) method [35,36]. Our analysis considers the building's walls constructed of concrete and brick with relative permittivity ε of 15 and 4, respectively and conductivity σ (0.015 S/m). Diffraction loss for vertical and horizontal polarization is computed separately. Figure 2 shows the geometry that encompasses both LoS and NLoS communication.

The simulation involves creating four virtual environments within a 1 km $\times$ 1 km area to mimic different scenarios in MATLAB. The UAV's position is determined along the street. The azimuthal angle ranges from 0 to 360 degrees in increments of 9 degrees. Calculations are carried out for specific elevation angles using the median value obtained from the azimuthal angle. This approach ensures that the results are not dependent on the azimuthal angle, accounting for the uneven spacing of buildings in the real world. The simulation covers elevation angles from 1 to 89 degrees, assuming a maximum UAV altitude of 500 m. This setup reflects the use of the UAV as a base station or relay to meet the demand of 5G and beyond.

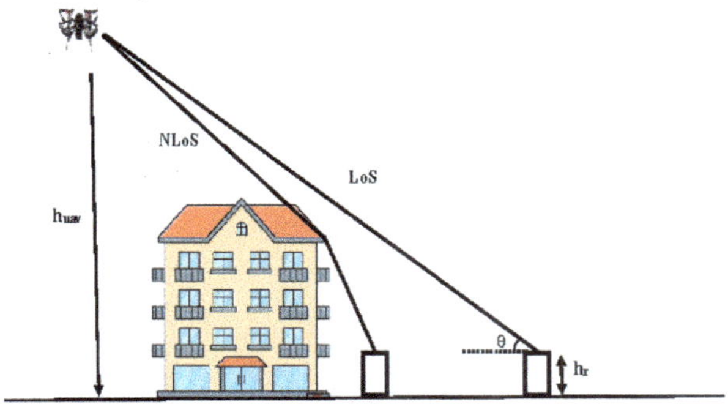

Figure 2. Geometry of LoS and NLoS scenario.

4. New Propagation Model

The simulation results are categorized into two distinct components: the first pertains to the LoS probability on the street; the FSPL model is applicable for calculating the average propagation loss and the second is additional path loss due to the shadowing effect of propagation buildings for NLoS conditions. Our previous investigation [34] shows that an elevation angle of 60 to 80 degrees provides the most realistic coverage area. However, it is also important to consider other elevation angles, as they may be significant for studying interference and other related factors.

4.1. Modelling Line of Sight Probability

The LoS probability on the street is obtained based on elevation angle for four environments with UAV altitudes up to 500 m and azimuthal angles between 0 and 360 degrees. The LoS probability proposed in our previous work [34] is given by

$$P_{LoS} = \frac{1}{a_3 + e^{-(-a_1 + a_2(\theta - a_4))}} \tag{2}$$

where a_1, a_2, a_3, and a_4 are the empirical parameters. The empirical parameter values for the suburban environment are (2.1778, 0.3557, 1, 0), the urban environment (3.0734, 0.1565, 0.9989, 0.158), the dense urban environment (3.4912, 0.1304, 1.007, 0.3344), the high-rise urban environment below elevation angle 45 degrees (4.2234, 0.8815, 1.5747, 0.114), and the high-rise urban environment above elevation angle 45 degrees (4.7313, 0.1209, 0.9801, 13.144).

4.2. Shadowing Loss

The analysis of additional path loss resulting from the shadowing effect of buildings using the uniform theory of diffraction (UTD) [37–39] and the method outlined in ITU-R P.526-14 for diffraction-based propagation using UTD [40]. This methodology applies to both vertical and horizontal polarizations to predict diffraction loss induced by a finitely conducting wedge. It is pertinent in diffraction scenarios around building corners, across terrain featuring a wedge-shaped hill, or over roof ridges. The method requires the conductivity and relative dielectric constant of the obstructing wedge and assumes that no transmission occurs through the wedge material. Furthermore, the method accounts for diffraction in both shadow and line-of-sight regions, facilitating a smooth transition between these zones. Figure 3 visualizes the geometry associated with wedge diffraction as delineated by UTD.

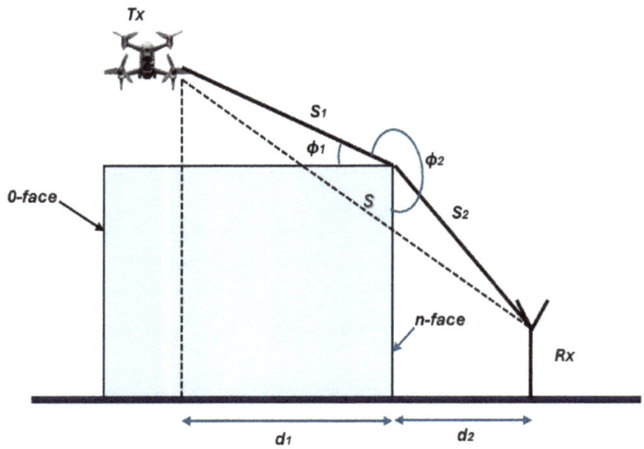

Figure 3. Geometry of wedge diffraction.

The UTD formulation for the electric field at the field point:

$$E_{UTD} = E_0 \frac{e^{-jk(s_1+s_2)}}{s_1} D \sqrt{\frac{s_1}{s_2(s_1+s_2)}} \tag{3}$$

where E_{UTD} is the electric field at the field point; E_0 is relative source amplitude, s_1 and s_2 are distanced from the source point to the diffracting edge and the diffracting edge to the field point, respectively; k is the wave numberz; D is the diffraction coefficient depending on the polarization (parallel or perpendicular to the plane of incidence) of the incident field on edge; (refer to [39] for formulas of D for vertical and horizontal polarization); and s is the distance between transmitter and receiver with obstruction. The relative field strength at the field point to the field in the absence of the obstruction is

$$E_{UTD}(dB) = 20\log\left(\left|\frac{sE_{UTD}}{e^{-jks}}\right|\right) \tag{4}$$

The relative diffraction loss based on UTD model is

$$L_{UTD} = 20\log\left(\frac{\sqrt{s_1 s_2(s_1+s_2)}}{s|D|}\right) \tag{5}$$

This simulation study explores the diffraction loss experienced in various environmental contexts by examining vertical and horizontal polarization. The calculations are conducted at an elevation angle of 70 degrees, a frequency of 2.1 GHz, and an altitude of 200 m across four distinct environments: suburban, urban, dense urban, and high-rise.

The normalized histogram of the diffraction loss is presented in Figure 4a–d. The histogram indicates that the distribution of loss values closely resembles a normal distribution. It was observed that the histogram aligns well with this pattern. To further support this finding, the probability density function (pdf) of the normal distribution is fitted to the simulated data, as shown in Figure 4. This fitting highlights the consistency of the diffraction loss across different environments, thereby enhancing the credibility of the simulation results. The pdf of the normal distribution is given by

$$p(z) = \frac{1}{\sigma\sqrt{2\pi}} e^{-\left(\frac{(x-\mu)^2}{2\sigma^2}\right)} \tag{6}$$

where p is the normalized probability, μ is the mean value in dB, and σ is the standard deviation. Table 2 gives the normal distribution's mean and standard deviation.

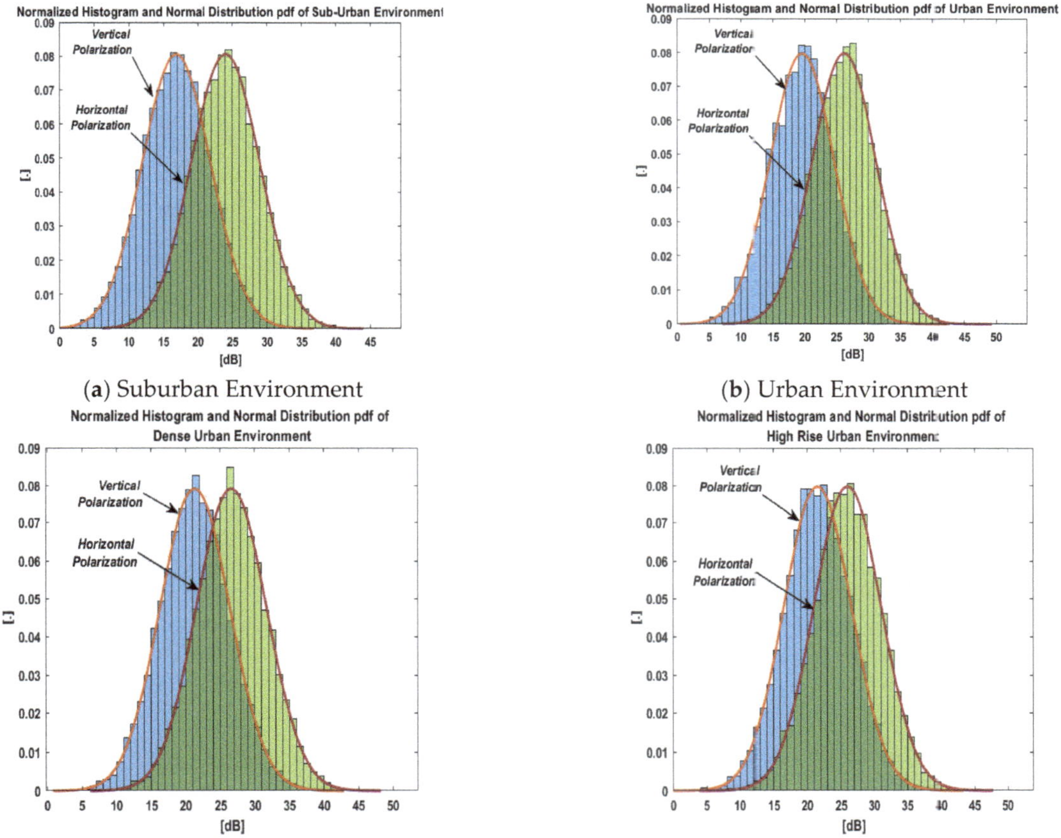

(**a**) Suburban Environment (**b**) Urban Environment (**c**) Dense-urban Environment (**d**) High-rise-urban Environment

Figure 4. (**a**–**d**) Normalized histogram of shadowing loss at 2.1 GHz for elevation angle 70°.

Table 2. Mean and variance of normal distribution at 70 degrees and frequency 2.1 GHz.

Environment	Vertical		Horizontal	
	Mean(dB)	**Standard Deviation (dB)**	**Mean(dB)**	**Standard Deviation (dB)**
Suburban	16.8603	5.0019	23.9999	5.0019
Urban	19.5104	5.0119	26.1119	5.0119
Dense Urban	21.3564	5.0124	26.6517	5.0124
High-Rise urban	16.5303	5.0136	21.0535	5.0136

Further, the cumulative distribution function (CDF) of vertical and horizontal polarization across all environments is simulated for elevation angles ranging from 10 to 80 degrees, as illustrated in Figure 5. The dashed curve in the figure denotes the CDF of the normal distribution of the simulated data. The close correspondence between the simulated data and the normally distributed data is evident from the figure. The graph demonstrates that the CDF for elevation angles between 10° and 40° is nearly identical for vertical polarization, whereas this is not the case for horizontal polarization. Here, altitudes range from 100 to 500 m.

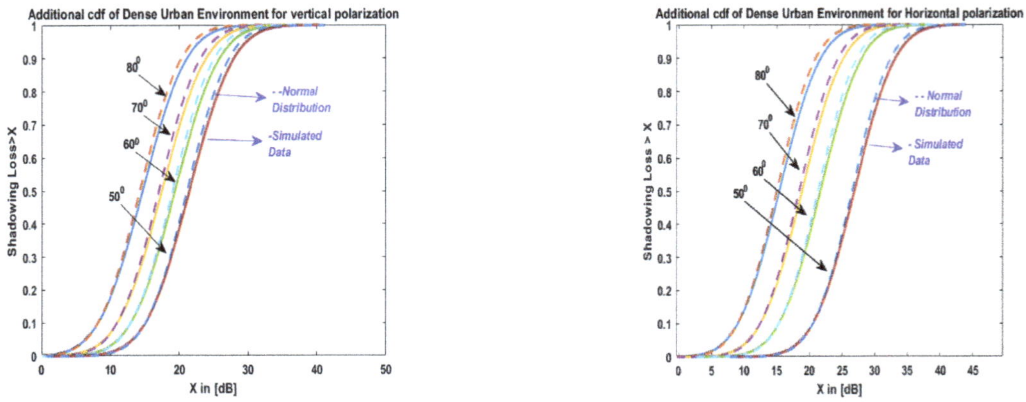

Figure 5. CDF of shadowing loss for horizontal and vertical polarization at 2.1 GHz for dense urban environment.

These revealed a significant increase in shadowing loss at these higher frequencies and altitudes. Figures 6 and 7 show the parameters of a normal distribution mean and standard deviation for three different carrier frequencies and elevation angles ranging from 1 to 89 degrees across all four environmental scenarios, considering a UAV altitude of 200 m. A rational function was developed to effectively model the mean of a normal distribution, while a linear function was used to represent the standard deviation based on the simulation results depicted in Figures 4 and 5. Equation (7) approximates the normal distribution's mean value, and Equation (8) approximates the standard deviation.

$$\mu = \frac{p_1\theta + p_2}{\theta + p_3} \tag{7}$$

$$\sigma = s_1\theta + s_2 \tag{8}$$

where θ is the elevation angle in degrees; p_1, p_2, and p_3 are empirical parameters for the mean; and s_1, s_2 are for standard deviation.

An extensive series of experiments was conducted to analyze communication dynamics at various altitudes, ranging from 100 m to 500 m, reflecting the height of base stations in an Indian city. The frequencies examined during this research are 1.8 GHz, 2.1 GHz, and 5.8 GHz, frequencies used by many operators in India. The objective was to accurately model the mean and standard deviation of a normal distribution associated with these parameters. This study focused on both vertical and horizontal polarizations, and it was carried out under a wide range of environmental conditions, ensuring that the findings would be robust and applicable in real-world scenarios.

The results of our extensive experimentation provide valuable insights into the empirical parameters of the means and standard deviations of the normal distribution as per Equations (7) and (8), which are systematically presented in Tables 3–5. Within Tables 3 and 4, we identify empirical parameters labeled p_1, p_2, and p_3. These parameters reflect the mean values obtained from our rigorous experimental processes, showcasing the central tendencies across different conditions.

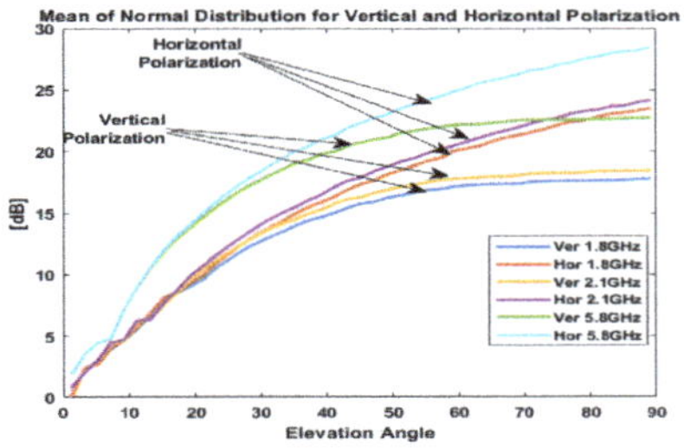

(**a**) Suburban

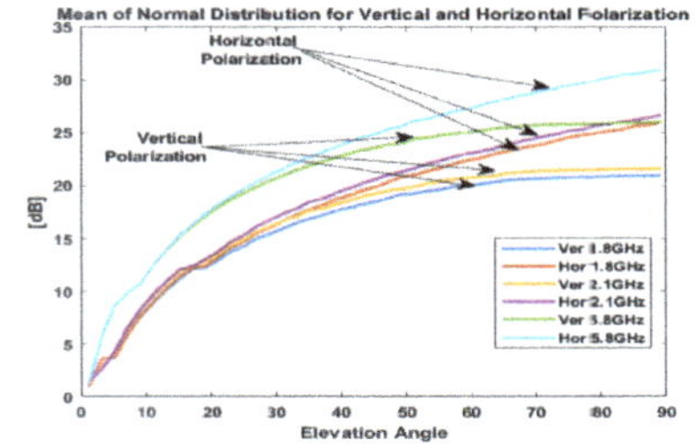

(**b**) urban

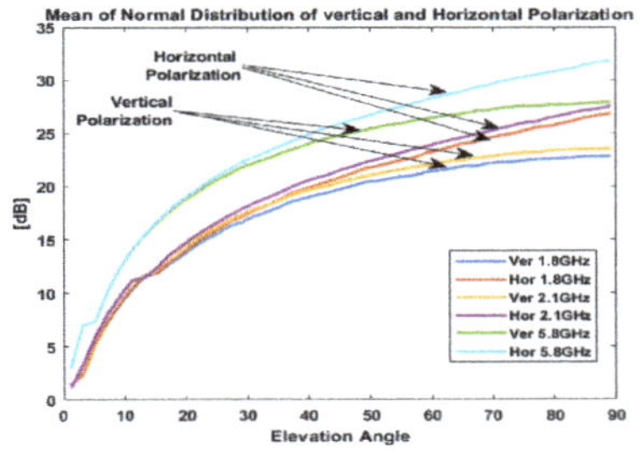

(**c**) Dense Urban

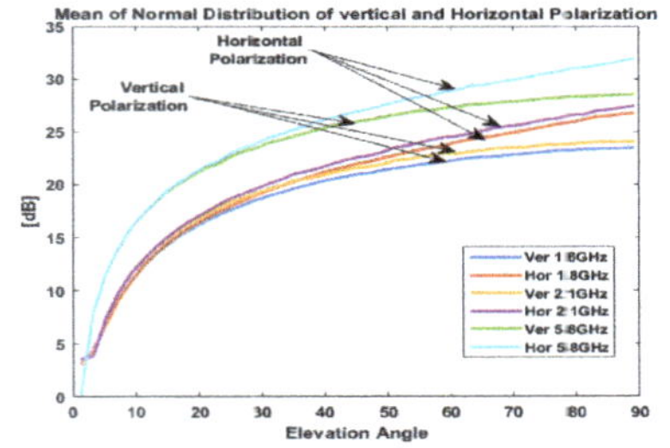

(**d**) High-rise Urban

Figure 6. (**a**–**d**) mean of normal distribution for horizontal and vertical polarization at 1.8 GHz, 2.1 GHz, and 5.8 GHz for different environments for a range of elevation angles.

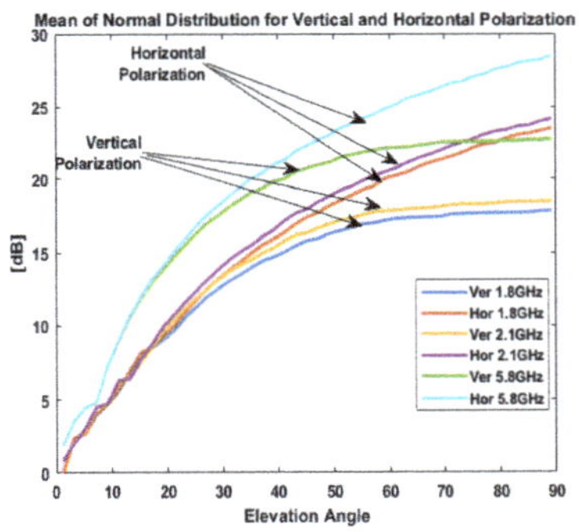

(**a**) Suburban

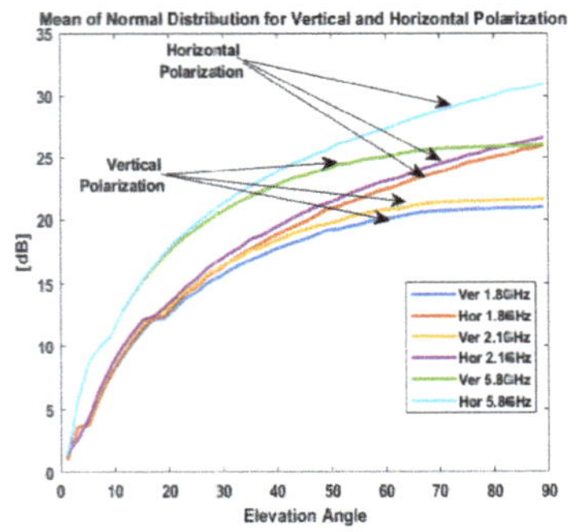

(**b**) urban

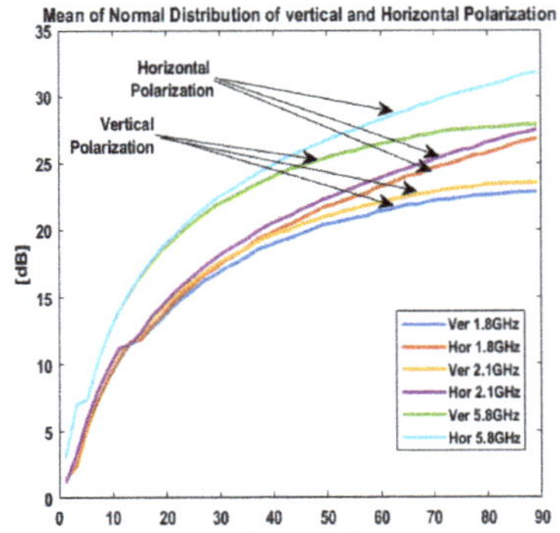

(**c**) Dense Urban

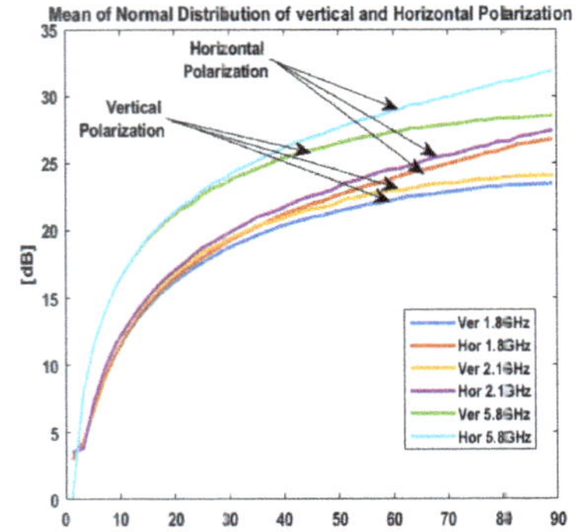

(**d**) High-rise Urban

Figure 7. (**a**–**d**) Standard deviation of normal distribution for horizontal and vertical polarization at 1.8 GHz, 2.1 GHz, and 5.8 GHz for different environments for a range of elevation angles.

Table 3. Mean for vertical polarization at different frequencies.

Suburban		**100 m**	**200 m**	**300 m**	**400 m**	**500 m**
1.8 GHz	p_1	23.7216	24.0883	23.8048	23.7678	23.5960
	p_2	−29.9015	−36.7989	−36.9605	−31.5423	−34.1584
	p_3	22.2064	22.2152	22.2862	22.2410	21.3403
2.1 GHz	p_1	25.1339	25.2131	24.9089	24.6789	24.4703
	p_2	−32.2446	−32.6191	−31.6761	−29.4293	−28.6999
	p_3	29.2789	25.1215	23.2878	22.3367	21.3743
5.8 GHz	p_1	28.8010	28.9738	28.7539	28.6436	28.5876
	p_2	−30.8454	−29.2240	−28.5110	−27.0202	−26.3835
	p_3	20.9991	18.5201	17.2056	16.7119	16.4857
Urban						
1.8 GHz	p_1	25.4228	26.5718	26.3567	26.2121	26.0853
	p_2	−2.9948	−15.0461	−21.0732	−26.0114	−26.2173
	p_3	23.9305	19.5814	17.3439	16.1217	15.5793
2.1 GHz	p_1	25.9287	27.0221	27.0222	26.8835	26.7900
	p_2	−3.3181	−15.4761	−17.9286	−19.9051	−23.0595
	p_3	22.5635	18.3154	16.8960	15.9480	15.3296
5.8 GHz	p_1	29.6679	31.0097	31.2253	31.1271	30.9798
	p_2	−19.9921	−3.8644	−12.0476	−12.5775	−13.5733
	p_3	17.2139	14.6677	13.7547	13.1230	12.6010
Dense Urban						
1.8 GHz	p_1	25.2118	28.2947	28.7122	28.5990	28.6188
	p_2	10.1286	−9.2078	−11.0606	−19.5715	−19.9987
	p_3	22.4716	19.1733	17.9622	16.4188	16.0471
2.1 GHz	p_1	25.4447	28.9292	29.3238	29.3915	29.3611
	p_2	1.5633	−5.1620	−14.5809	−18.3496	−19.6060
	p_3	19.8798	18.6637	17.0602	16.3199	15.8336
5.8 GHz	p_1	29.7586	33.1945	33.5504	33.5410	33.4350
	p_2	36.7097	16.3986	4.3173	0.6839	−4.1420
	p_3	17.5222	16.0942	14.6104	13.8227	13.1878
High-Rise Urban						
1.8 GHz	p_1	20.3463	27.4242	28.4428	28.7052	28.8546
	p_2	17.5012	4.0327	0.3840	−8.1527	−7.1619
	p_3	14.0465	14.1650	13.1850	12.2438	12.1136
2.1 GHz	p_1	21.0427	27.9894	29.0438	29.3998	29.4792
	p_2	22.0963	4.9665	2.1906	−2.9504	−5.6475
	p_3	13.8743	13.5637	12.7458	12.1928	11.7493
5.8 GHz	p_1	24.1529	31.0331	32.2939	32.7668	32.9584
	p_2	−14.2152	−19.4081	−22.4160	−21.4092	−22.5993
	p_3	6.2512	7.7838	7.6857	7.7134	7.5998

Table 4. Means for horizontal polarization at different frequencies.

Suburban		**100 m**	**200 m**	**300 m**	**400 m**	**500 m**
1.8 GHz	p_1	35.6745	38.3942	38.8261	39.2372	39.1544
	p_2	−18.1666	−23.6267	−28.3137	−20.8376	−30.5913
	p_3	57.6867	54.1762	51.9193	52.0133	50.4158
2.1 GHz	p_1	31.0113	36.4189	37.9059	38.3037	38.4629
	p_2	15.1345	17.0104	19.3257	14.9433	8.1047
	p_3	32.2843	34.7100	35.0109	34.1919	33.3575
5.8 GHz	p_1	36.2385	39.8708	40.4364	40.7366	41.0820
	p_2	−19.4166	−6.3016	−13.2586	−14.0001	−8.0419
	p_3	32.9455	35.0245	33.8670	33.5594	34.004
Urban						
1.8 GHz	p_1	30.7810	30.9439	37.3921	37.6546	37.9306
	p_2	17.4879	17.4879	10.6488	1.9634	−2.0582
	p_3	34.8856	35.3342	35.9805	34.6766	34.1562
2.1 GHz	p_1	31.0113	36.4189	37.9059	38.3037	38.4629
	p_2	15.1345	17.0104	19.3257	14.9433	34.1919
	p_3	32.2843	34.7100	35.0109	8.1047	33.3575
5.8 GHz	p_1	34.0326	39.2117	40.5478	41.0711	41.3309
	p_2	35.0653	39.9744	25.1888	26.9476	27.1345
	p_3	24.0709	26.5295	26.0679	25.9490	25.8098
Dense Urban						
1.8 GHz	p_1	28.0617	35.4410	38.0706	38.9074	39.5604
	p_2	23.2720	20.3020	26.1260	14.1716	13.6524
	p_3	27.9972	31.4166	33.1805	32.4072	32.5977
2.1 GHz	p_1	25.4447	28.9292	29.3238	29.3915	29.3611
	p_2	1.5633	−5.1620	−14.5809	−18.3496	−19.6060
	p_3	19.8798	18.6637	17.0602	16.3199	15.8336
5.8 GHz	p_1	32.2079	39.3991	41.5958	42.5604	42.8968
	p_2	53.9679	55.2319	48.7093	48.0354	40.7040
	p_3	21.2534	24.7038	25.1447	25.2794	24.7796
High-Rise Urban						
1.8 GHz	p_1	21.3318	32.2853	35.5201	36.7368	37.5098
	p_2	23.1789	29.8386	37.4605	26.3565	27.7044
	p_3	15.8443	21.6186	23.4886	23.1616	23.4518
2.1 GHz	p_1	22.0077	32.7059	35.9666	37.4077	38.0281
	p_2	27.8509	30.2808	39.9798	36.0752	32.6578
	p_3	15.5492	20.4536	22.5132	22.9589	22.8287
5.8 GHz	p_1	24.9204	34.9199	38.0406	39.4709	40.2425
	p_2	−12.0049	−2.3964	2.7190	7.4398	7.3907
	p_3	7.05918	12.0028	13.7630	14.6158	14.9451

Table 5. Standard deviations for vertical/horizontal polarization in different.

Environment	s_1	s_2
Suburban	0.00112	4.944
Urban	0.00136	4.936
Dense urban	0.00141	4.931
High-rise urban	0.001591	4.927

Table 5, meanwhile, focuses on the parameters s_1 and s_2, which signify the standard deviations associated with our findings. A noteworthy outcome of our analysis is the discovery that the s_1 and s_2 parameters exhibit remarkable consistency, showing no significant variation across a spectrum of conditions, including different altitudes, polarizations, and frequencies tested. This consistency allows us to derive the standard deviation from a single, unified equation, greatly simplifying the overall calculation process. For additional frequencies within the range of 2 and 6 GHz, we can obtain values through interpolation, utilizing the parameters presented in Tables 3–5. It is crucial to apply these parameters with the specified precision in decimal points in all calculations to ensure accuracy.

The implications of these findings are expected to be significant, particularly in the context of evaluating additional shadowing path loss. Our work takes into account the complex interactions associated with various frequencies, altitudes, polarizations, and environmental contexts. Ultimately, this extensive experimentation has culminated in the development of a comprehensive and refined solution for the UAV channel model, specifically attuned for the unique challenges of an Indian metropolitan area. This model is particularly focused on radio frequency (RF) planning at low altitudes, effectively addressing the intricate nuances introduced by differing frequencies and polarizations, thereby enhancing the planning capabilities vital for urban environments.

4.3. Shadowing Path Loss

As discussed earlier, the shadowing path loss for the environment was given by normal distribution with mean μ and standard deviation σ (Equations (7) and (8)). The path loss is given by

$$Ls = normrand\left(\frac{p_1\theta + p_2}{\theta + p_3}, s_1\theta + s_2\right) \tag{9}$$

5. Proposed Channel Model

As discussed in previous sections, the path loss for LoS scenario is calculated using free space path loss (FSPL), and for NLoS scenarios, additional shadowing loss is included, formulated in terms of the normal distribution. The total path loss for the proposed channel model is

$$L = L_{FSPL} + L_A + L_{ex} \tag{10}$$

where L_{FSPL} is the average free space path loss, which is determined by the distance and frequency. L_A stands for the antenna loss. L_{ex} stands for the additional shadowing loss, which accounts for obstruction between transmitter and receiver as well as additional loss due to environmental effects.

5.1. Free Space Path Loss

The free space path loss (FSPL), described as a function of the elevation angle between UAV and receiver, is given by

$$L_{FSPL} = 20log\frac{\Delta h}{\sin\theta} + 20logf \; - \; 27.55 \quad (11)$$

where Δh = $h_{UAV} - h_r$ is the height difference between transmitter (UAV) (h_{UAV}) and receiver (hr), measured in kilometers, and f is the frequency, measured in MHz.

5.2. Antenna Loss

As per ITU-R P.1336-1 [41], the relative gain G(θ) of the antenna with respect to the isotropic antenna (dBi) can be given as

$$G(\theta) = \begin{matrix} G_0 - 12\left(\frac{\theta}{\theta_3}\right)^2 & 0 \leq \theta < \theta_3 \\ G_0 - 12 - \; 10log\frac{\theta}{\theta_3} & \theta_3 \leq \theta < 90 \end{matrix} \quad (12)$$

where θ_3 = 107.6 x$10^{-0.1 \times G_0}$; G_0 is the maximum value of gain in or near the horizontal plane in dBi; θ is the absolute elevation angles relative to the maximum gain angle in degrees; θ_3 is the 3 dB beam width at the vertical plane in degrees. The path loss due to transmitting and receiving antennas is calculated by

$$L_A = -2 * G(\theta) \quad (13)$$

5.3. Excess Path Loss (L_{ex})

The excessive path loss for NLoS communication can be obtained from the probability of line-of-sight communication.

$$L_{ex} = L_s + 20\log(1 - P_{LoS}) \quad (14)$$

where L_s is the additional path loss or shadowing loss due to shadowing when the wedges obstruct the line-of-sight ray.

5.4. Path Loss

The path loss represented by Equation (6) is

$$L = L_{FSPL} + L_A + L_{ex}$$

By substituting values from Equations (9), (11)–(14), the path loss can be represented by

$$L = 20log\frac{(h_{uav} - \; h_r)}{sin\theta} + A - 20log\frac{\theta}{\theta_3} + Ls + 20\log(1 - P_{LoS}) \quad (15)$$

where

$$A = 20logf - 51.55 - 2G_0 \quad (16)$$

5.5. Proposed Model Analysis

The analysis includes the simulation of proposed path loss in four different environments at varying frequencies and altitudes, considering elevation angles as depicted in Figures 8 and 9. The findings provide a detailed analysis of path loss across various environments at a frequency of 5.8 GHz, specifically at a UAV altitude of 200 m. As shown in Figure 8a, the results reveal that the suburban environment experiences the lowest level of path loss, indicative of more favorable communication conditions. In contrast, the dense urban environment is characterized by significantly higher path loss, attributed to the greater density of buildings and obstacles that contribute to increased signal attenuation and shadowing.

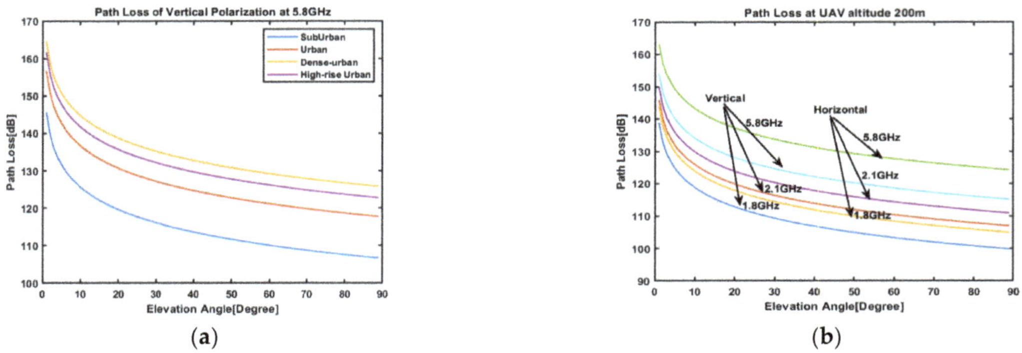

Figure 8. Proposed model path loss for (**a**) different environments at frequency 5.8 GHz and altitude 200 m and (**b**) dense urban environments at different frequencies and polarization at altitude 200 m.

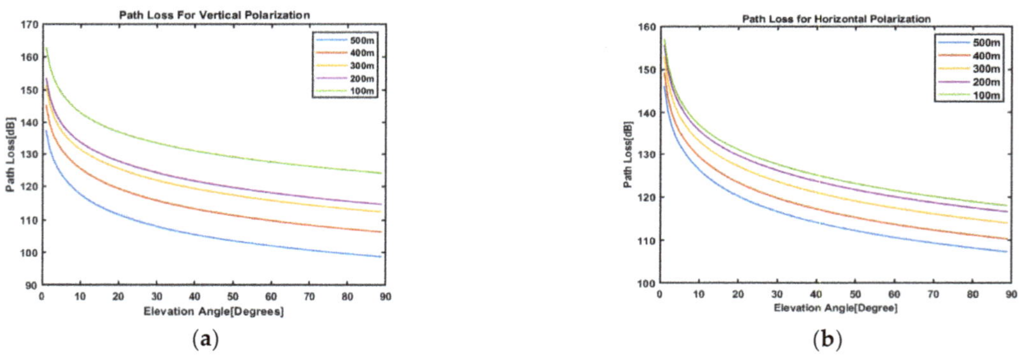

Figure 9. (**a**,**b**) Proposed model path loss for dense urban environment at UAV altitude 100–500 m at frequency 5.8 GHz for vertical and horizontal polarization, respectively.

This trend is consistent across additional frequencies of 1.8 GHz and 2.1 GHz, observed at varying altitudes. To further understand the behavior of path loss in the dense urban setting, simulations were conducted for 1.8 GHz, 2.1 GHz, and 5.8 GHz, focusing on both vertical and horizontal polarization of the signals. Figure 8b illustrates that as frequency increases, there is a corresponding rise in path loss for both types of polarization, highlighting the challenges posed by higher frequency signals in urban landscapes.

In addition, this study examines the relationship between the altitude of the UAV and its impact on the path loss. Figure 9a,b depict a strong correlation: as the UAV's altitude increases, resulting in a notable decrease in overall path loss due to increase in line-of-sight communication between transmitter and receiver. This finding provides valuable insight into how variations in altitude can significantly influence the effectiveness of communication systems, particularly in urban environments where obstructions are prevalent. Overall, this comprehensive analysis deepens our understanding of the dynamics affecting path loss in various scenarios.

6. Proposed Model Validation

The proposed model is designed to enhance the simulation of coverage estimation for deploying 5G services in the bustling metropolitan cities of India. This model seeks to streamline the computational process involved in simulations, making it quicker and more efficient. The accuracy of any channel model is fundamentally tied to the mathematical assumptions it employs. For instance, if the actual layout of buildings in the target area diverges significantly from the assumptions made by the model, the predictions regarding signal behavior and coverage may suffer. To provide a robust validation of the proposed channel model presented in this paper, path loss comparisons were conducted against established models: Model 1 [27] and Model 2 [31] are specifically designed for dense urban environments. The simulation parameters detailed in Table 6 are utilized for validation.

Table 6. Model parameters.

Model Parameter	Value
Frequency	5.8 GHz
h_{UAV}	200 m
h_r	2 m
G_o	2.15
θ_3	67

The results shown in Figure 10 demonstrate a close alignment between the existing model and the proposed model. Notably, the existing models do not account for the effect of polarization on path loss. Furthermore, other models' scenarios are validated against the estimates found in [27], where a similar close correlation is also observed. The close alignment of the proposed model with the existing model ensures that the mathematical framework of the model effectively reproduces key aspects of path loss and the propagation characteristics essential for successful 5G deployment.

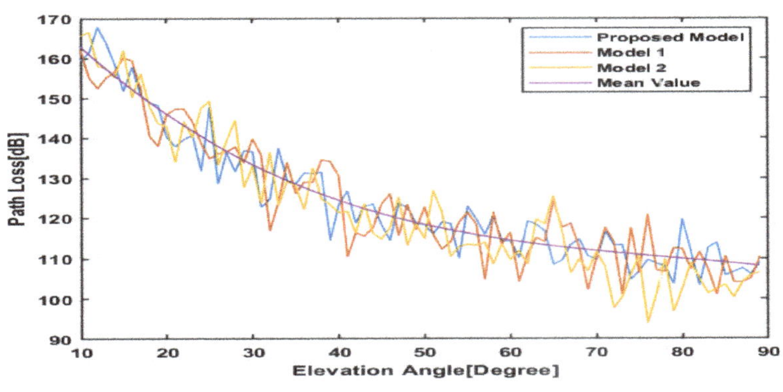

Figure 10. Proposed model vs other models.

7. Conclusions

An elevation-based shadowing statistical model was developed for low-altitude UAV (unmanned aerial vehicle) channels in various environments, including suburban, urban, dense urban, and high-rise urban settings. This model, formulated as a function of elevation angle, covers a frequency range from 1.8 GHz to 6 GHz and altitudes up to 500 m, considering both vertical and horizontal polarization. The shadowing path loss for non-line-of-sight (NLoS) conditions is presented with elevation angle. It is a valuable tool for radio network planning to support the expansion of 5G communication in Indian metropolitan cities. Based on extensive simulations, the model provides an efficient formula that minimizes computational time while maintaining a strong correlation with existing air-to-ground models. Additionally, it analyzes the impact of frequency and UAV height adjustments, offering a structured framework for determining optimal UAV altitudes, ultimately maximizing ground coverage and enhancing communication performance. Derived from extensive simulations, the model provides a simplified formula for implementation, effectively reducing computational time. Notably, it demonstrates a strong correlation with other air-to-ground models. Furthermore, the model analyzes the effects of frequency and UAV height adjustments, providing a straightforward framework for determining the optimal UAV altitude to maximize ground coverage.

Author Contributions: Conceptualization, A.K.P. and R.D.J.; methodology, A.K.P.; software, A.K.P.; validation, A.K.P.; formal analysis, A.K.P. and R.D.J.; investigation, A.K.P. and R.D.J.; resources, A.K.P.; data curation, A.K.P.; writing—original draft, A.K.P.; writing—review and editing, A.K.P. and R.D.J.; visualization, A.K.P. and R.D.J. All authors have read and agreed to the published version of the manuscript.

Funding: This research received no external funding.

Data Availability Statement: The contributions made in this study are documented within the article. For any additional inquiries, please direct your questions to the corresponding author.

Acknowledgments: This work was supported by the College of Engineering, Pune.

Conflicts of Interest: Authors declare that there are no known conflicts of interest related to this study.

References

1. Tyokighir, S.S.; Mom, J.; Ukhurebor, K.E.; Igwue, G. New developments and trends in 5G technologies: Applications and concepts. *Bull. Electr. Eng. Inform.* **2024**, *13*, 254–263. [CrossRef]
2. What is 5G? How will it transform our world? Available online: https://www.ericsson.com/en/5g (accessed on 30 November 2023).
3. Amponis, G.; Lagkas, T.; Zevgara, M.; Katsikas, G.; Xirofotos, T.; Moscholios, I.; Sarigiannidis, P. Drones in B5G/6G Networks as Flying Base Stations. *Drones* **2022**, *6*, 39. [CrossRef]
4. Rolly, R.M.; Malarvezhi, P.; Lagkas, T.D. Unmanned aerial vehicles: Applications, techniques, and challenges as aerial base stations. *Int. J. Distrib. Sens. Netw.* **2022**, *18*, 1–9. [CrossRef]
5. Masaoka, R.; Tran, G.K.; Nakazato, J.; Sakaguchi, K. The Future of Flying Base Stations: Empirical and Numerical Investigations of mmWave-Enabled UAVs. *Future Internet* **2024**, *16*, 5. [CrossRef]
6. Mohsan, S.A.H.; Othman, N.Q.H.; Li, Y.; Alsharif, M.H. Unmanned aerial vehicles (UAVs): Practical aspects, applications, open challenges, security issues, and future trends. *Intell. Serv. Robot.* **2023**, *16*, 109–137. [CrossRef]
7. Ghamari, M.; Rangel, P.; Mehrubeoglu, M.; Tewolde, G.S.; Sherratt, R.S. Unmanned Aerial Vehicle Communications for Civil Applications: A Review. *IEEE Access* **2022**, *10*, 102492–102531. [CrossRef]
8. Shah, A.F.M.S. Architecture of Emergency Communication Systems in Disasters through UAVs in 5G and Beyond. *Drones* **2023**, *7*, 25. [CrossRef]
9. Khawaja, W.; Guvenc, I.; Matolak, D.W.; Fiebig, U.-C.; Schneckenburger, N. Survey of air-to-ground propagation channel modelling for unmanned aerial vehicles. *IEEE Commun. Surv. Tutor.* **2019**, *21*, 2361–2391. [CrossRef]
10. Matolak, D.W.; Sun, R. Air-ground channel characterization for unmanned aircraft systems-part i: Methods, measurements, and models for over-water settings. *IEEE Trans. Veh. Technol.* **2016**, *66*, 26–44. [CrossRef]
11. Dhillon, H.S. Downlink coverage analysis for a finite 3-d wireless network of unmanned aerial vehicles. *IEEE Trans. Commun.* **2017**, *65*, 4543–4558.
12. Sun, R.; Matolak, D.W. Air-ground channel characterization for unmanned aircraft systems-part ii: Hilly and mountainous settings. *IEEE Trans. Veh. Technol.* **2017**, *66*, 1913–1925. [CrossRef]
13. Matolak, D.W.; Sun, R. Air-ground channel characterization for unmanned aircraft systems-part iii: The suburban and near urban environments. *IEEE Trans. Veh. Technol.* **2017**, *66*, 6607–6618. [CrossRef]
14. Sun, R.; Matolak, D.W.; Rayess, W. Air-ground channel characterization for unmanned aircraft systems—Part IV: Airframe shadowing. *IEEE Trans. Veh. Technol.* **2017**, *66*, 7643–7652. [CrossRef]

15. Schneckenburger, N.; Jost, T.; Shutin, D.; Walter, M.; Thiasiriphet, T.; Schnell, M.; Fiebig, U.C. Measurement of the L-band air-to-ground channel for positioning applications. *IEEE Trans. Aerosp. Electron. Syst.* **2016**, *52*, 2281–2297. [CrossRef]
16. Takizawa, K.; Kagawa, T.; Lin, S.; Ono, F.; Tsuji, H.; Miura, R. C-band aircraft-to-ground (A2G) radio channel measurement for unmanned aircraft systems. In Proceedings of the International Symposium on Wireless Personal Multimedia Communications (WPMC), Sydney, Australia, 7–10 September 2014; pp. 754–758.
17. Lyu, Y.; Wang, W.; Sun, Y.; Yue, H.; Chai, J. Low-Altitude UAV Air-to-Ground Multilink Channel Modeling and Analysis at 2.4 and 5.9 GHz. *IEEE Antennas Wirel. Propag. Lett.* **2023**, *22*, 2135–2139. [CrossRef]
18. Qiu, Z.; Chu, X.; Calvo-Ramirez, C.; Briso, C.; Yin, X. Low Altitude UAV Air-to-Ground Channel Measurement and Modeling in Semiurban Environments. *Wirel. Commun. Mob. Comput.* **2017**, *2017*, 1587412. [CrossRef]
19. Lyu, Y.; Wang, W.; Chen, P. Fixed-Wing UAV Based Air-to-Ground Channel Measurement and Modeling at 2.7 GHz in Rural Environment. *IEEE Trans. Antennas Propag.* **2024**, *1*. [CrossRef]
20. Cui, Z.; Briso-Rodríguez, C.; Guan, K.; Zhong, Z.; Quitin, F. Multi-frequency air-to-ground channel measurements and analysis for UAV communication systems. *IEEE Access* **2020**, *8*, 110565–110574. [CrossRef]
21. Zhang, X.; He, R.; Yang, M.; Ai, B.; Wang, S.; Li, W.; Sun, W.; Li, L.; Huang, P.; Xue, Y. Measurements and modeling of large-scale channel characteristics in subway tunnels at 1.8 and 5.8 GHz. *IEEE Antennas Wirel. Propag. Lett.* **2023**, *22*, 561–565. [CrossRef]
22. Yang, M.; Ai, B.; He, R.; Ma, Z.; Zhong, Z. Vehicle-to-vehicle channel characteristics in intersection environment. In Proceedings of the 2022 IEEE International Conference on Communications Workshops (ICC Workshops), Seoul, Republic of Korea, 16–20 May 2022; pp. 1124–1129.
23. Li, Y.; Cheng, X. Modelling and simulation for UAV-aided vehicular MIMO communication channels. *IET Commun.* **2019**, *13*, 3044–3051. [CrossRef]
24. Jiang, H.; Zhang, Z.; Wu, L.; Dang, J. Three-dimensional geometry-based UAV MIMO channel modeling for A2G communication environments. *IEEE Commun. Lett.* **2018**, *22*, 1438–1441. [CrossRef]
25. Cheng, X.; Li, Y. A 3-D geometry-based stochastic model for UAV-MIMO wideband nonstationary channels. *IEEE Internet Things J.* **2018**, *6*, 1654–1662. [CrossRef]
26. Yi, P.; Zhu, L.; Xiao, Z.; Zhang, R.; Han, Z.; Xia, X.G. 3-D Positioning and Resource Allocation for Multi-UAV Base Stations Under Blockage-Aware Channel Model. *IEEE Trans. Wirel. Commun.* **2024**, *23*, 2453–2468. [CrossRef]
27. Al-Hourani, A.; Kandeepan, S.; Jamalipour, A. Modeling air-to-ground path loss for low altitude platforms in urban Environments. In Proceedings of the IEEE Global Communications Conference, Austin, TX, USA, 8–12 December 2014; pp. 2898–2904.
28. Al-Hourani, A. On the Probability of Line-of-Sight in Urban Environments. *IEEE Wirel. Commun. Lett.* **2020**, *9*, 1178–1181. [CrossRef]
29. Akram, A.L.; Chandrasekharan, S.; Kaandorp, G.; Glenn, W.; Jamalipour, A.; Kandeepan, S. Coverage and rate analysis of Aerial Base Station. *IEEE Trans. Aerosp. Electron. Syst.* **2016**, *52*, 3077–3081.
30. Al-Hourani, A.; Kandeepan, S.; Lardner, S. Optimal LAP Altitude for Maximum Coverage. *IEEE Wirel. Commun. Lett.* **2014**, *3*, 569–572. [CrossRef]
31. Safwat, N.E.; Newagy, F.; Hafez, I.M. Air-to-ground channel model for UAVs in dense urban environments. *IET Commun.* **2020**, *14*, 1016–1021. [CrossRef]
32. Holis, J.; Pechac, P. Elevation Dependent Shadowing Model for Mobile Communications via High Altitude Platforms in Built-Up Areas. *IEEE Trans Antennas Propag.* **2008**, *56*, 1078–1084. [CrossRef]
33. Patel, A.K.; Joshi, R.D. Area Coverage Analysis of Low Altitude UAV Bases Station using Statistical Channel Model. In Proceedings of the International Conference on Signal and Information Processing (IConSIP), Pune, India, 26–27 August 2022; pp. 1–6.
34. Patel, A.K.; Joshi, R.D. Coverage Estimation Using Probabilistic Line-of-Sight Model for Unmanned Aerial Vehicle Communication. In Proceedings of the UASG 2021: Wings 4 Sustainability UASG Lecture Notes in Civil Engineering, International Conference on Unmanned Aerial System in Geomatics, Roorkee, India, 2–4 April 2021; Jain, K., Mishra, V., Pradhan, B., Eds.; Springer: Berlin/Heidelberg, Germany, 2023; Volume 304, pp. 533–544.
35. Kumar, M.; Arti, M.K. A simple characterization of UAV communication systems. *Sādhanā* **2022**, *47*, 210. [CrossRef]
36. ITU-R. *Recommendation P.1410-5 Propagation Data and Prediction Methods for The Design of Terrestrial Broadband Millimetric Radio Access Systems*; P Series; Radiowave Propagation; International Telecommunication Union: Geneva, Switzerland, 2005.
37. Kouyoumjian, R.G.; Pathak, P.H. A uniform geometrical theory of diffraction for an edge in a perfectly conducting surface. *Proc. IEEE* **1974**, *62*, 1448–1461. [CrossRef]
38. Luebbers, R. Finite conductivity uniform GTD versus knife edge diffraction in prediction of propagation path loss. *IEEE Trans. Antennas Propag.* **1984**, *32*, 70–76. [CrossRef]
39. Brugarolas-Ortiz, E.; Rodríguez-Rodríguez, I.; Rodríguez, J.-V.; Juan-Llácer, L.; Pardo-Quiles, D. Using the Uniform Theory of Diffraction to Analyze Radio Wave Propagation along Urban Street Canyons for Device-to-Device Communication. *Electronics* **2023**, *12*, 593. [CrossRef]

40. ITU-R. *Recommendation P.526-14 Propagation by Diffraction, P Series, Radiowave Propagation*; International Telecommunication Union: Geneva, Switzerland, 2018.
41. ITU-R. *Recommendation F.1336-1 Reference Radiation Patterns of Omnidirectional, Sectoral and other Antennas in Point-To-Multipoint Systems for Use in Sharing Studies in the Frequency Range from 1 GHz to About 70 GHz*; International Telecommunication Union: Geneva, Switzerland, 2000.

telecom

MDPI

Article

Simulation Framework for Detection and Localization in Integrated Sensing and Communication Systems

Andrea Ramos [1], Saúl Inca [1], Mireia Ferrer [1], Daniel Calabuig [1], Sandra Roger [2] and Jose F. Monserrat [1,*]

[1] iTEAM Research Institute, Universitat Politècnica de València, 46022 Valencia, Spain; anrapil@iteam.upv.es (A.R.); sauin@iteam.upv.es (S.I.); miferma@upv.edu.es (M.F.); dacaso@iteam.upv.es (D.C.)
[2] Computer Science Department, Universitat de València, 46100 Burjassot, Spain; sandra.roger@uv.es
* Correspondence: jomondel@iteam.upv.es

Abstract: Integrated Sensing and Communication (ISAC) systems have emerged as a key component for Sixth Generation (6G) networks, enhancing resource efficiency and enabling diverse applications. Currently, ISAC systems have been recognized as a leading trend for future standardization, i.e., International Mobile Telecommunications (IMT)-2030. As in the previous IMT-2020 standardization, the emphasis has been on developing a methodology for assessing network conditions, with one of the crucial approaches incorporating system-level simulations. However, within this framework, there has been a notable absence of proposed abstractions for the physical layer of ISAC systems, which are valuable for system-level simulators. The physical abstraction process helps reduce computational simulation costs, enabling efficient and rapid evaluation of system conditions. Therefore, this paper aims to fill this gap by outlining the key aspects and metrics recommended for a physical layer abstraction in sensing applications within ISAC frameworks. Applying physical abstraction in the context of target localization and detection algorithms may enable an initial understanding and evaluation of ISAC system performance. These algorithms are proposed as an example of simulating the sensing functionalities to be abstracted, which are based on a stochastic geometric channel model. Orthogonal Frequency Division Multiplexing (OFDM) symbols play a crucial role in target position estimation. The findings show that doubling OFDM symbols improves the detection probability by 3 dB in terms of Signal to Noise Ratio (SNR). Finally, the proposed Physical Layer Abstraction (PLA) method produces performance metrics as figures and lookup tables tailored for system-level simulators.

Keywords: Integrated Sensing and Communication; Sixth Generation; physical abstraction; target detection; target localization

1. Introduction

The transition from Fifth Generation (5G) to next generation communication systems (NGCS) marks the evolution towards a fully connected, intelligent society. This evolution is distinguished by the integration of artificial intelligence (AI) and sensing technologies into Sixth Generation (6G) networks [1,2], aiming to address the growing demands for higher speeds, lower latency, and greater reliability in mobile communications. A key innovation in this context is the development of Integrated Sensing and Communication (ISAC) systems. By merging sensing and communication functionalities within a single framework, ISAC systems promise to efficiently share spectrum and hardware resources. ISAC leverages millimeter Wave (mmW) and Super High Frequency (SHF) bands, which

 https://doi.org/10.3390/telecom6010004

offer wide bandwidths and enhanced spatial resolution, enabling precise sensing and high-speed communication [3]. This integration paves the way for diverse and novel applications anticipated in 6G networks.

In recent years, within the ISAC framework, recent technical advances in ISAC include signal processing methods [4], ISAC channel modeling [5–7], channel estimation techniques built on Machine Learning (ML)-based compressed sensing [8], or even novel protocols that involve Quantium-ISAC [9]. In addition, the International Telecommunications Union (ITU) has declared it as one of the main trends in International Mobile Telecommunications (IMT)-2030 [10]. Given all the attention, the research community has primarily focused on advancements, often overlooking the need for a standardized framework to evaluate ISAC. In this context, [11] proposes a methodology to evaluate ISAC, highlighting the main factors that must be considered in channel modeling, i.e., channel correlation between communication and sensing and spatial consistency. Even though the mentioned contribution describes one of the main elements of the physical layer, the authors do not mention any computational reduction solution for the simulation process. As we progress toward emulating a more comprehensive system, we anticipate increased complexity and higher computational costs, just like in predecessor technologies of Fourth Generation (4G) or 5G.

Evaluation through simulations has been segmented into link-level and system-level simulations in the literature and previous standardization [12]. According to the ITU, the link level evaluates aspects of the physical layer, and the system level assesses overall network conditions. In the context of system-level simulations, the physical layer abstraction has been instrumental in simplifying the detailed and exhaustive process of physical aspects like modulation or channel effects. Therefore, this abstraction process has been introduced to reduce the computational cost of system-level simulations in IMT-2020, enabling comprehensive evaluations of system performance and optimization strategies [13,14].

Given this background, it is reasonable to expect a similar trajectory for ISAC, as it remains a wireless technology. The abstraction of the physical layer should now be included in the ISAC framework. Moreover, ISAC is not only a technology that encompasses communication capabilities, but also sensing ones. Therefore, considering this concept, evaluating sensing services must be included in the evaluation process.

In sensing services, positioning and detection are highlighted in the literature [1] and in the Third Generation Partnership Project (3GPP) Technical Specification (TS) 22.137 [15] as the most prominent ISAC use cases. These services can serve as a starting point and be integrated into Physical Layer Abstraction (PLA). Nevertheless, the aforementioned intentions have not been found in the literature. PLA might have been neglected within the ISAC framework. New efforts and novel types of simulations are essential to provide future system simulators with abstractions that facilitate the analysis of ISAC systems.

Since there is still a lack of contributions addressing the abstraction of the physical layer in the ISAC framework, this paper aims to fill this gap with the following contributions:

- A background on PLA, emphasizing the rationale behind the approach and revitalizing key elements from the existing literature.
- A description of the proposed PLA method's essential elements, focusing on those utilized in previous technologies, including channel generation and Signal to Noise Ratio (SNR).
- Two sensing services are considered, target location and detection, whose algorithms are formulated as examples of the PLA method.
- An evaluation of the sensing algorithms' performance, offering quantitative insights.
- The mentioned evaluation is used to abstract the performance metrics of the algorithms.

As a result of this abstraction process, a set of lookup tables is generated, which can be incorporated into system-level simulations. The search tables are submitted to a repository available in [16].

Notation: Matrices are denoted by bold uppercase letters, vectors are defined by bold lowercase letters, and the standard font represents scalars. The transpose conjugate is represented as $(^*)$.

2. ISAC Physical Layer Abstraction

Wireless network simulations have been divided into link-level and system-level simulations. On the one hand, link-level simulation refers to obtaining the performance of individual links. However, since the network can grow with many devices, it can increase the complexity of emulating other interferences or environmental features. Here, system-level simulations are dedicated to evaluating the entire system's efficiency. Due to these characteristics, simulation costs can be very high in terms of time and computational costs

Instead of modeling all of the complex and detailed physical layer operations, its abstraction can simplify the process during system-level simulations. The PLA method replaces simulations at the link level and provides a computationally efficient way to test the desired network. Typically, the information from PLA is delivered in lookup tables containing the pre-computed process and indexing operations stored inside an array.

Lookup tables have been widely used for PLA methods like Exponential Effective SNR Mapping (EESM), Effective SNR Mapping (ESM) [14,17–19], or Mutual Information (MI)-ESM [20]. These methods have appeared in the predecessor technologies of 4G and 5G. We can rescue from these methods the essential elements considered to perform the link abstraction process: a channel model and an established SNR. These elements form the foundation of our abstraction method. Figure 1 illustrates a flowchart of the proposed approach. In the channel generation component, we have incorporated the creation of communication and sensing channels for the ISAC system. In addition, we have developed a prior knowledge channel, a sensing channel without the presence of the target. Details of this channel generation process will be elaborated on in the following sections.

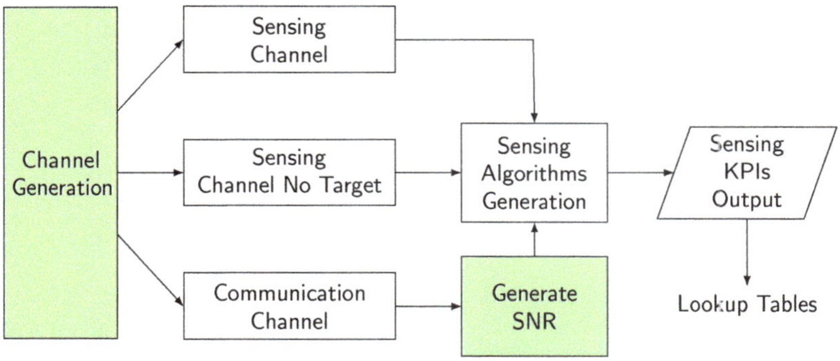

Figure 1. General scheme of the proposed ISAC physical layer abstraction for sensing services.

Following the flowchart, we determine the communication SNR, the primary input for generating the sensing algorithms. Distance, positioning, and detection algorithms show the abstraction's applicability, although other techniques or sensing services can be used. Then, this process generates lookup tables of the following Key Performance Indicators (KPIs) [15] to obtain the estimated distance, position, and detection information:

- Position accuracy or error: This refers to the closeness of the measured position of the target or user to its true position value. In other words, the mean positioning error in the horizontal plane or 2D reference plane can also be considered the error on the vertical axis (i.e., the altitude). This KPI ensures that the data and results are trustworthy and can be used confidently.

- False alarm probability: This describes the conditional probability of falsely detecting the presence of a target object or environment when the target object or environment is not present.
- Detection probability: Conversely, the detection probability is the conditional probability of correctly detecting the presence of a target when the target is indeed present.

3. Main Elements of Abstraction: Channel and SNR Generation

The channel generation leverages the Stochastic Cluttered Environment (SCE) model from [7], where scatterers are distributed randomly throughout the scenario and multiple signal hops are considered when generating the desired channels. Once the scatterers are located, the model facilitates the computation of channel matrices between various points within the system. For example, it models a Base Station (BS) functioning simultaneously as a transmitter and receiver, facilitating monostatic sensing applications. Additionally, it allows for the previous BS to serve as the transmitter and a distinct point to act as the receiver, thereby obtaining the communication channel between the two points.

In this paper, we consider a monostatic Orthogonal Frequency Division Multiplexing (OFDM) ISAC system, which is composed of a BS, a target, and a set of scatterers generated as in [7]. Figure 2 shows a representation of this system for the two possible channels: (a) the sensing channel and (b) the communication channel. The Multiple Path Components (MPCs) of both channels are represented as arrows. Round-trip red arrows represent target-involved MPCs; black arrows depict other MPCs. The black round-trip and one-way arrows in Figure 2b enclose the inherent MPCs for the communication-only channel.

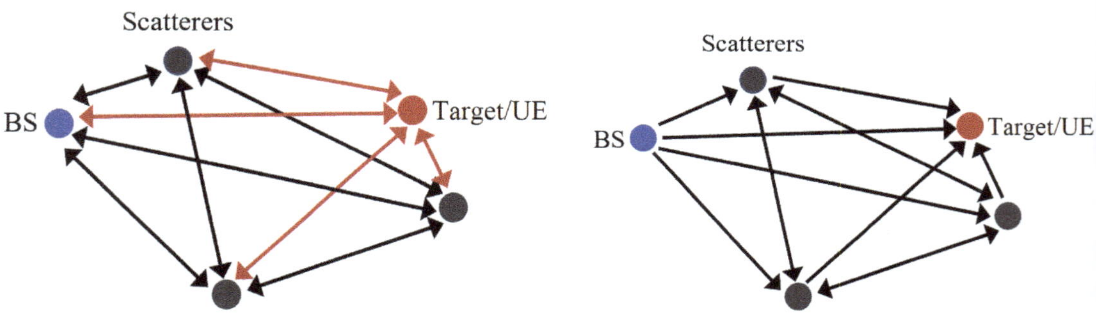

(**a**) Sensing channel (**b**) Communication channel

Figure 2. System model for (**a**) the monostatic sensing channel from the BS, and (**b**) the BS–target communication channel if the target acts as a receiver. Arrows represent direct links. The red arrows in the sensing channel represent the links involving the target.

We assume that the BS uses L transmitted OFDM symbols over K sub-carriers to sense the same scenario as the one represented in Figure 2a. Let M be the number of antennas at the BS. The received sensing signal at the BS after sensing the target and scatterers, for the l-th OFDM symbol and the k-th subcarrier, is

$$\mathrm{y}_l^{\mathrm{sen}}(k) = \mathrm{H}_{\mathrm{sen}}(k)\mathrm{s}_l(k) + \mathrm{z}_l(k), \tag{1}$$

where $\mathrm{H}_{\mathrm{sen}}(k) \in \mathbb{C}^{M\times M}$ is the sensing channel matrix, $\mathrm{s}_l(k) \in \mathbb{C}^M$ is the transmitted symbol vector, and $\mathrm{z}_l(k) \in \mathbb{C}^M$ is the noise vector.

To abstract the performance of the sensing applications, we need to use some measure of the channel quality that serves as a reliable indicator of this performance. In conventional communication systems, such measures are typically the SNR of the received signal. As depicted in Figure 2a, where the received signal of interest in (1) results from the sensing system, it is affected by all MPCs, even those not involved with the target (black arrows). Therefore, considering an SNR of the received sensing signal requires that the signal term account for the effect of all those MPCs. The primary challenge with this type of SNR is that certain MPCs devoid of interactions with the target may substantially contribute to the total signal power. For instance, scatterers near the BS can reflect a significant portion of the transmitted power. Consequently, the power of the sensing signal in (1) is not a good indicator of the performance of sensing applications.

Alternative measures of channel quality can be explored to overcome this problem. For example, by assuming that the target behaves like a conventional User Equipment (UE) in a communication system, it becomes feasible to focus on the communication channel between the BS and UE (target). In this scenario, the system model depicting this channel is illustrated in Figure 2b. In the context of the communication channel, all MPCs terminate at the target, thereby ensuring interaction with the target for all components. Assuming that the target behaving as a UE has a single antenna, the received signal at this UE is

$$\mathrm{y}_l^{\mathrm{com}}(k) = \mathrm{H}_{\mathrm{com}}(k)\mathrm{s}_l(k) + \mathrm{n}_l(k), \tag{2}$$

where $\mathrm{H}_{\mathrm{com}}(k) \in \mathbb{C}^{1\times M}$ is the communication channel matrix, and $\mathrm{n}_l(k) \in \mathbb{C}$ is the noise vector for the l-th OFDM symbol and the k-th sub-carrier.

The received signal in (2) is considered to compute the SNR for the PLA method. Since all the MPCs associated with the received communication signal interact with the target, the resulting SNR serves as a reliable indicator for sensing applications. Assuming that the elements of $\mathrm{s}_l(k)$ are independent and identically distributed (i.i.d.) from a zero-mean and unit-variance normal distribution (i.e., the transmitted signal has a covariance matrix equal to the identity), the power of the signal in $\mathrm{y}_l^{\mathrm{com}}(k)$ can be obtained as $\mathrm{tr}(\mathrm{H}_{\mathrm{com}}(k)\mathrm{H}^*_{\mathrm{com}}(k)) = \|\mathrm{H}_{\mathrm{com}}(k)\|^2$. Moreover, assuming that the noise is i.i.d. from a zero-mean normal distribution with variance σ^2, the noise power in $\mathrm{y}_l^{\mathrm{com}}(k)$ is directly σ^2. Therefore, the communications SNR denoted for all the K sub-carriers of an OFDM symbol can be calculated as

$$\mathrm{SNR}_{\mathrm{com}} = \frac{\sum_{k=1}^{K} \|\mathrm{H}_{\mathrm{com}}(k)\|^2}{K\sigma^2}. \tag{3}$$

4. Sensing Algorithms Generation: Target Detection and Localization

This section presents two algorithms that will be used to illustrate the simulation framework: one for target localization and another for target detection. As mentioned in the previous section, the received sensing signal in (1) includes all MPCs, even those that do not involve the target. In this case, it is essential to differentiate the part of the received sensing signal that contains relevant information about the target from the rest of the received signal. However, this is not an easy task. One way to address this problem is to perform background subtraction, as in [21]. To allow for this, the sensing BS needs to acquire previous sensing knowledge of the scenario without the target, i.e., to estimate the sensing channel corresponding to MPCs of only scatterers that are not the target of interest. This sensing channel comprises only the black arrows in Figure 2a and can be denoted as $\mathrm{H}_{\mathrm{pre}}(k) \in \mathbb{C}^{M\times M}$ for the k-th sub-carrier. This channel can be obtained through environment reconstruction and channel estimation techniques, as proposed in [22,23]. Then, the effective sensing channel matrix can be defined as

$$\mathrm{H}_{\mathrm{eff}}(k) = \mathrm{H}_{\mathrm{sen}}(k) - \mathrm{H}_{\mathrm{pre}}(k), \tag{4}$$

where $\mathrm{H}_{\mathrm{eff}}(k) \in \mathbb{C}^{M\times M}$ contains only those MPCs involving the target, i.e., only the red arrows in Figure 2a. The following section presents a method to estimate $\mathrm{H}_{\mathrm{eff}}(k)$.

4.1. Effective Sensing Channel Estimation

Although we assume that $\mathrm{H}_{\mathrm{pre}}(k)$ is known, note that $\mathrm{H}_{\mathrm{sen}}(k)$ is unknown to the BS, thus precluding the direct computation of (4). The BS can only obtain knowledge about $\mathrm{H}_{\mathrm{sen}}(k)$ from the received signal $\mathrm{y}_l^{\mathrm{sen}}(k)$. However, as evident from (1), we must first eliminate the transmitted signal's influence. To achieve this, we exploit the fact that the covariance matrix of the transmitted signal is the identity. In particular, we obtain

$$\hat{\mathrm{H}}_{\mathrm{eff}}(k) = \frac{1}{L}\sum_{l=1}^{L}(\mathrm{y}_l^{\mathrm{sen}}(k) - \mathrm{H}_{\mathrm{pre}}(k)\mathrm{s}_l(k))\mathrm{s}_l(k)^*, \tag{5}$$

which, from (1), is

$$\hat{\mathrm{H}}_{\mathrm{eff}}(k) = \frac{1}{L}\sum_{l=1}^{L}(\mathrm{H}_{\mathrm{eff}}(k)\mathrm{s}_l(k) + \mathrm{z}_l(k))\mathrm{s}_l(k)^*. \tag{6}$$

Now, we operate in (6) to obtain

$$\hat{\mathrm{H}}_{\mathrm{eff}}(k) = \mathrm{H}_{\mathrm{eff}}(k)\frac{1}{L}\sum_{l=1}^{L}\mathrm{s}_l(k)\mathrm{s}_l(k)^* + \frac{1}{L}\sum_{l=1}^{L}\mathrm{z}_l(k)\mathrm{s}_l(k)^*. \tag{7}$$

The term $\frac{1}{L}\sum_{l=1}^{L}\mathrm{s}_{l,k}\mathrm{s}_{l,k}^*$ is an estimator of the covariance matrix of the transmitted signal, which is, as mentioned before, the identity. Therefore, $\hat{\mathrm{H}}_{\mathrm{eff}}(k) \in \mathbb{C}^{M\times M}$ can be treated as an estimator of $\mathrm{H}_{\mathrm{eff}}(k)$.

4.2. Distance and Angular Direction Estimation Algorithm

In this section, a localization algorithm based on the estimation of the sensing channel is presented. To extract the embedded information of $\hat{\mathrm{H}}_{\mathrm{eff}}(k)$, we can first acquire the taps of the channel using an Inverse Discrete Fourier Transform (IDFT), which allows us to identify the temporal sample where the target is located. This can be expressed as

$$\hat{\mathrm{h}}_{\mathrm{eff}}(n) = \mathrm{IDFT}\{\hat{\mathrm{H}}_{\mathrm{eff}}(n)\}(t), \tag{8}$$

where $\hat{\mathrm{h}}_{\mathrm{eff}} \in \mathbb{C}^{M\times M}$ represents the impulse response of the channel $\hat{\mathrm{H}}_{\mathrm{eff}}(k)$ at the n-th temporal sample. Given the expression in (8), we can evaluate the channel response when the BS emits power towards a specific Angle of Departure (AoD). This analytical process involves the utilization of steering vectors, $\mathrm{a}(\theta)$, which denotes the antenna array's configuration, where θ is the AoD. Then, the estimated energy of $\hat{\mathrm{h}}_{\mathrm{eff}}(n)$ in a certain θ and at a certain temporal sample n is

$$P(\theta, n) = ||\hat{\mathrm{h}}_{\mathrm{eff}}(n)\mathrm{a}(\theta)||^2 = \mathrm{a}^*(\theta)\hat{\mathrm{h}}_{\mathrm{eff}}^*(n)\hat{\mathrm{h}}_{\mathrm{eff}}(n)\boldsymbol{a}(\theta). \tag{9}$$

The maximum of the energy is where the target is supposed to be, so the final step of this procedure is to estimate the values of θ and n by maximizing the energy as

$$[\hat{\theta}, \hat{n}] = \arg\max_{[\theta,n]}(P(\theta, n)). \tag{10}$$

The estimated sample $\hat{n}$ is used to estimate the distance as

$$\hat{d} = \frac{\hat{n}c}{2f_s}, \tag{11}$$

where c is the speed of light and f_s is the sampling frequency. The next subsection will use the estimated angle $\hat{\theta}$.

A summary of the above steps is provided in Algorithm 1 to benefit the reader's understanding.

Algorithm 1 Distance and angle estimation algorithm

Input: $\mathrm{H}_{\mathrm{pre}}(k)$, s, y, L

Output: $\hat{\theta}$, $\hat{d}$, $\hat{\mathrm{h}}_{\mathrm{eff}}(n)$

1. **for** $k = 1 : K$ **do**
   - (a) $\hat{\mathrm{H}}_{\mathrm{eff}}(k) = \frac{1}{L}\sum_{l=1}^{L}(\mathrm{y}_{l,k} - \mathrm{H}_{\mathrm{pre}}(k)\mathrm{s}_{l,k})\mathrm{s}_{l,k}^*$
   - (b) Compute $\hat{\mathrm{h}}_{\mathrm{eff}}(n) = \mathrm{IDFT}\{\hat{\mathrm{H}}_{\mathrm{eff}}(k)\}(n)$

   end for
2. Compute $P(\theta, n) = ||\hat{\mathrm{h}}_{\mathrm{eff}}(n)\mathrm{a}(\theta)||^2$ for each angle θ and each sample n
3. Find $\hat{\theta}$ and $\hat{n}$ for the maximum received power:
   $[\hat{\theta}, \hat{n}] = \arg\max_{[\theta,n]}(P(\theta, n))$
4. Calculate $\hat{d}$ for the corresponding $\hat{n}$: $\hat{d} = \frac{\hat{n}c}{2f_s}$

4.3. Position Estimation Algorithm

After obtaining the estimated angle and distance of the target from Algorithm 1, we can estimate the position by applying basic geometry. Specifically, if the height of the target (i.e., z^{UE}) and the coordinates of the BS, $(x^{\mathrm{BS}}, y^{\mathrm{BS}}, z^{\mathrm{BS}})$, are assumed to be known, the estimated UE coordinates can be calculated by solving for $\hat{x}^{\mathrm{UE}}$ and $\hat{y}^{\mathrm{UE}}$ from the following equations:

$$\hat{\theta} = \mathrm{acos}\left(\frac{x^{\mathrm{BS}} - \hat{x}^{\mathrm{UE}}}{\sqrt{|x^{\mathrm{BS}} - \hat{x}^{\mathrm{UE}}|^2 + |y^{\mathrm{BS}} - \hat{y}^{\mathrm{UE}}|^2}}\right), \tag{12}$$

$$\hat{d} = \sqrt{(x^{\mathrm{BS}} - \hat{x}^{\mathrm{UE}})^2 + (y^{\mathrm{BS}} - \hat{y}^{\mathrm{UE}})^2 + (z^{\mathrm{BS}} - z^{\mathrm{UE}})^2}. \tag{13}$$

Thus, the estimated coordinates can be given as

$$\hat{x}^{\mathrm{UE}} = x^{\mathrm{BS}} + \cos(\hat{\theta})\sqrt{\hat{d}^2 - (z^{\mathrm{BS}} - z^{\mathrm{UE}})^2}, \tag{14}$$

$$\hat{y}^{\mathrm{UE}} = y^{\mathrm{BS}} + \sin(\hat{\theta})\sqrt{\hat{d}^2 - (z^{\mathrm{BS}} - z^{\mathrm{UE}})^2}. \tag{15}$$

4.4. Target Detection Algorithm

A target or user is detected when the energy of the effective channel $\hat{\mathrm{h}}_{\mathrm{eff}}(n)$ at any given temporal sample n exceeds a pre-established threshold, denoted by ξ. A Boolean variable "withTarget" is set to true in the algorithm when a target is detected and is false otherwise.

The determination of the threshold is of paramount importance and requires meticulous calibration. In this work, the threshold has been chosen as $\xi = \Delta_\xi \hat{E}$, where $\hat{E} = \frac{1}{K}\sum_{n=1}^{K}||\hat{\mathrm{h}}_{\mathrm{eff}}(n)||^2$ represents the mean energy of $\hat{\mathrm{h}}_{\mathrm{eff}}(n)$ over time and Δ_ξ corresponds to a selected constant factor. This formulation dynamically adjusts the threshold ξ to match the scenario's noise power. For the sake of clarity, a summary of the above steps is provided in Algorithm 2.

Algorithm 2 Target detection algorithm

Input: $\hat{\mathrm{h}}_{\text{eff}}(n)$, Δ_{ξ}
Output: withTarget

1. withTarget = false
2. $\hat{E} = \frac{1}{K}\sum_{n=1}^{K} ||\hat{\mathrm{h}}_{\text{eff}}(n)||^2$
3. $\xi = \Delta_{\xi}\hat{E}$
4. **for** $n = 1$ **to** N **do**

        if $||\hat{\mathrm{h}}_{\text{eff}}(n)||^2 > \xi$ **then**

            withTarget = true

            break

    end for

5. Evaluation and Abstraction

The proposed algorithms are evaluated in a stochastic scenario based on [7], aligned with 3GPP Indoor Hotspot (InH) specifications [24]. This procedure involves an exhaustive series of simulations in which input parameters are detailed in Table 1. We used the carrier frequency recommended by 3GPP TR 38.901 in calibration process. Since our scenario is InH, a carrier frequency of 30 GHz was selected to obtain valuable information about the potential of the mmW band in ISAC system.

Table 1. Simulation assumptions considered to evaluate KPIs and develop the lookup tables.

Simulation parameters	
Number of simulation seeds (J)	30,000
Number of scatterers	30
Number of OFDM symbols (L)	$2^n, n = 1,2,\ldots 10$
Baseline evaluation configuration parameters	
Carrier frequency for evaluation	30 GHz
Sampling frequency	30.44 MHz
Number of sub-carriers (K)	2048
BS antenna height	3 m
UE antenna height	1.5 m
Total transmit power per TR $\times$ P	SNR dependent
SNR	$\{-20, -19, \ldots 20\}$ dB
Additional parameters for system-level simulation	
Number of BS antenna elements (M)	16
Number of UE antenna elements	1
Hall size	120×50 m
BS–UE distances (d)	32 and 64 m
ISAC Channel model	Stochastic [7]

The BS location remains constant across all simulations to streamline the simulation process. In a 120×50 m hall, the UE is positioned at pre-determined distances 32 and 64 m from the BS. These distances represent the typical link lengths within this spatial configuration but in a random direction, ensuring a Line of Sight (LoS) condition.

A unique set of input parameters characterizes each simulation seed. This includes variations in the number of OFDM symbols (L) used for channel estimation, different SNR levels, and two possible distances between the BS and the UE. Outputs, such as distance and position target estimations, are given for these variations.

For the results analysis, we aim to define the distribution of the estimation errors; for this objective, we have conducted a post-processing methodology that will be described below.

5.1. Distance and Positioning Abstraction

Distance and positioning algorithms are evaluated using KPIs, including mean distance error (E_d) and mean positioning error (E_p), as described in Section 2. These metrics can be obtained as

$$E_d = \frac{1}{J}\sum_{j=1}^{J}(d_j - \hat{d}_j), \tag{16}$$

$$E_p = \frac{1}{J}\sum_{j=1}^{J}\sqrt{(x_j^{UE} - \hat{x}_j^{UE})^2 + (y_j^{UE} - \hat{y}_j^{UE})^2}, \tag{17}$$

where J is the total number of simulation seeds, d_j refers to the true distance from the BS to the target in the j-th seed, and x_j^{UE} and y_j^{UE} denote the true position coordinates of the UE.

Figure 3a,b illustrate the Cumulative Distribution Functions (CDFs) for E_d and E_p under varying OFDM symbols at 0 dB SNR, with the target positioned at 64 m. The scale shift is applied at a distance error equal to 5 m to zoom in and better visualize the values of the CDFs corresponding to errors around 0 m. Fitting these CDFs to known distributions is challenging because their extensive tails extend toward the maximum estimation error corresponding to the last OFDM symbol. This implies a non-negligible probability of the employed algorithms pinpointing the target in the last OFDM symbol.

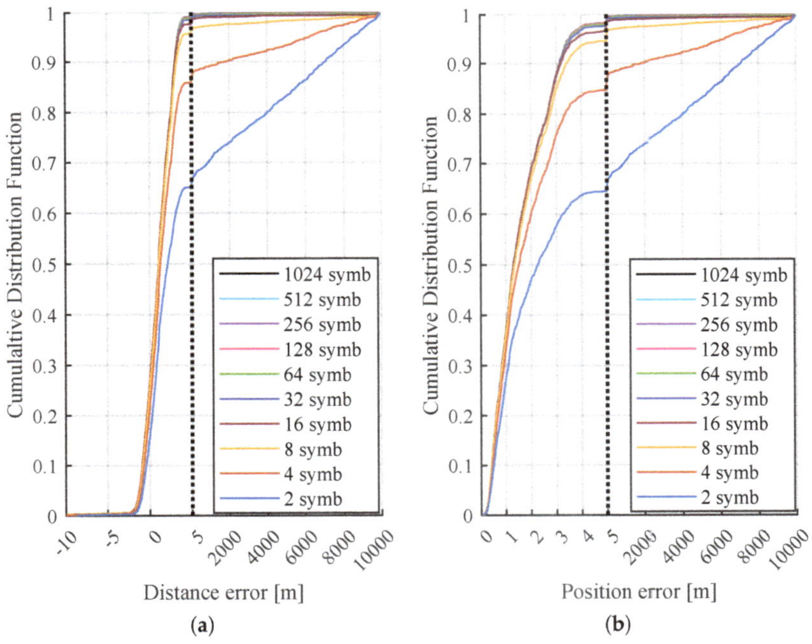

Figure 3. CDFs of the distance errors (**a**) and position errors (**b**) for 2–1024 OFDM symbols and SNR= 0 dB. Here, the target is positioned at $d = 64$ m.

Two different estimation conditions are valid for the abstraction proposal in the CDFs. On the left-hand side, we can find the results obtained under target-dominated conditions, where the distance and position errors are observed to be low. On the right, it reveals that for extremely large estimation errors (exceeding 10 m), their behavior exhibits a linear trend, aligning with the characteristics of a uniform distribution. Thus, the CDFs are a mix of two different distributions, one of which is the uniform distribution. Under noise-dominated conditions, distance and position estimates become uniformly distributed, indicating equal probability across all possible target locations.

Considering all mapping levels of SNR (ranging from −20 dB to 20 dB) and the full range of OFDM symbol values, the distribution curves exhibit minimal variation in distance and position error cases under target-dominated conditions. This description is illustrated in Figure 4a,b, where legends are omitted since multiple curves represent results across all possible values. This observation indicates that errors consistently follow the same distribution regardless of these two parameters.

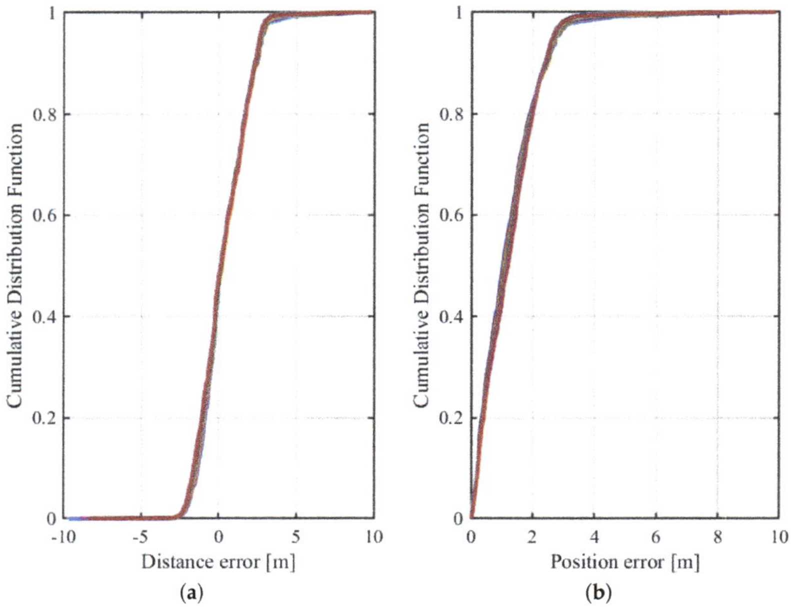

Figure 4. Approximated CDFs of the distance and positioning errors under target-dominated conditions for 2–1024 OFDM symbols applied for all SNR levels defined in Table 1. Here, the target is positioned at $d = 64$ m.

Results of the probability of estimation under target-dominated conditions versus SNR and the number of OFDM symbols are calculated and provided as lookup tables, which are available in [16].

5.2. Target Detection Abstraction

Target detection abstraction is evaluated using false alarm probability (P_{FA}) and the detection probability (P_D), as detailed in Section 2.

A set of thresholds has been defined by selecting threshold values Δ_ξ between 0 dB and 3 dB. The Boolean variable "withTarget" is stored to analyze P_{FA} and P_D later for each simulation seed. Figure 5 shows the curves of the false alarm probability versus the selected threshold factor for two examples corresponding to extreme cases. First, the case where the SNR is −20 dB and the algorithm uses only 2 OFDM symbols and, second, the case where the SNR is 20 dB, and the algorithm uses 1024 OFDM symbols. These curves show a notable decay in the false alarm probability as the threshold factor Δ_ξ increases.

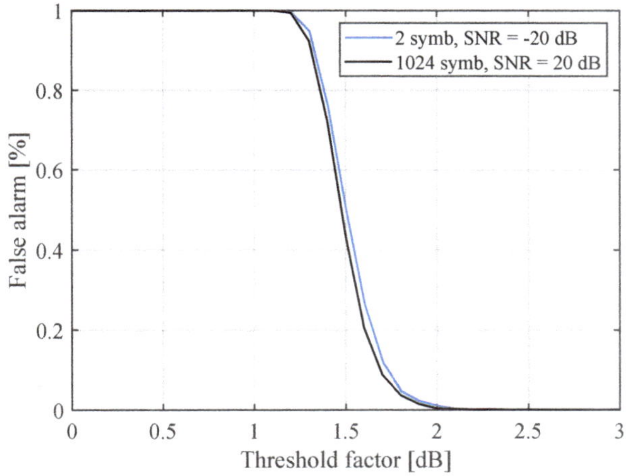

Figure 5. False alarm vs. threshold factor, with the target at $d = 64$ m.

To examine the effect of Δ_ξ on the target detection probability, Figure 6 depicts this probability against various selected factors Δ_ξ ranging from 0 to 9 dB. A correlation between the false alarm and target detection probabilities can be identified. For example, when $\Delta_\xi = 0$ dB, the probability of detection and false alarm reach 100%, meaning that the algorithm continuously detects targets even when the target is not present in the scenario. Nevertheless, If the threshold exceeds 3 dB, as shown by the light blue ($\Delta_\xi = 6$ dB) and red lines ($\Delta_\xi = 9$ dB), we are over-dimensioning the detection. In other words, we are being too strict, which likely increases the likelihood of incorrect detections.

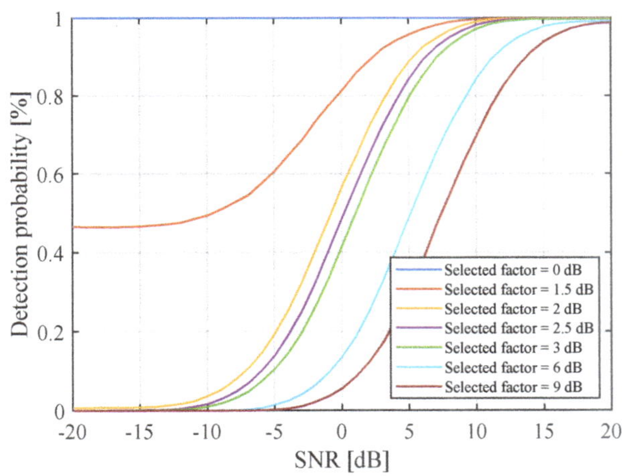

Figure 6. Detection probability vs. SNR for 8 OFDM symbols and different values of the selected factor, with the target positioned at $d = 64$ m.

Conversely, we observe that with $\Delta_\xi = 3$ dB, the false alarm probability is close to 0%, according to Figure 5. This behavior yields a sufficiently reasonable detection probability concerning the SNR. In the case of over-dimensioning Δ_ξ, the detection capability of the algorithm decreases since the curve is shifted towards increasing SNR values.

For the abstraction of the target detection algorithm, Δ_ξ has been fixed at 3 dB, ensuring a null probability of false alarm. Figure 7 shows a clear relationship between the detection probability and the number of OFDM symbols for a given SNR. Doubling the number of OFDM symbols effectively achieves 3 dB improvement in SNR. Lookup tables illustrating the detection probability across various SNR values and OFDM symbol counts for these thresholds are available in [16].

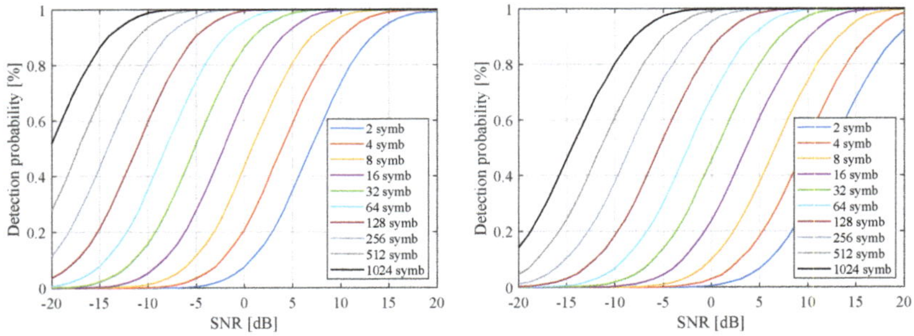

(**a**) BS-UE distance = 32 m (**b**) BS-UE distance = 64 m

Figure 7. Detection probability vs. SNR for different OFDM symbols.

6. Conclusions

This study addresses a critical gap in the abstraction of the physical layer for sensing applications within ISAC systems. By leveraging existing channel models and developing algorithms for target detection and localization services, we demonstrate the feasibility of abstracting performance metrics essential for future system-level simulations.

The findings highlight two distinct conditions for estimation: target-dominated, characterized by low errors, and noise-dominated, where high noise levels result in a uniform distribution of errors across all possible outcomes. We focused on the target-dominated condition to simplify the PLA process, which offers the most actionable insights for enhancing system accuracy and reliability. Furthermore, doubling the number of OFDM symbols was observed to achieve a 3 dB improvement in SNR, providing a straightforward method to enhance estimation and detection performance. Results are provided in lookup tables, publicly available in [16], enabling comprehensive evaluations of sensing services within the ISAC framework.

Future work should expand this abstraction to encompass complex environments, such as Non Line of Sight (NLoS) scenarios, to enhance the robustness of ISAC systems and widen their applicability. Additionally, outdoor scenarios should be investigated to evaluate sensing services over extended distances. Furthermore, future studies should integrate additional sensing services beyond target detection and localization to create a more versatile and comprehensive abstraction framework.

Author Contributions: Conceptualization, A.R., S.I., D.C. and S.R.; methodology, A.R., S.I., D.C. and S.R.; software, A.R., S.I. and M.F.; validation, A.R., S.I., M.F., D.C., S.R. and J.F.M.; formal analysis, D.C., S.R. and J.F.M.; investigation, A.R., S.I. and M.F.; resources, A.R., S.I., M.F., D.C., S.R. and J.F.M.; data curation, S.I. and D.C.; writing—original draft preparation, A.R., S.I., M.F., D.C. and S.R.; writing—review and editing, A.R., D.C., S.R. and J.F.M.; visualization, J.F.M.; supervision, J.F.M.; project administration, J.F.M.; funding acquisition, J.F.M. All authors have read and agreed to the published version of the manuscript.

Funding: This research received no external funding.

Data Availability Statement: The research data has been provided in [16].

Conflicts of Interest: The authors declare no conflicts of interest.

References

1. Pin Tan, D.K.; He, J.; Li, Y.; Bayesteh, A.; Chen, Y.; Zhu, P.; Tong, W. Integrated Sensing and Communication in 6G: Motivations, Use Cases, Requirements, Challenges and Future Directions. In Proceedings of the 2021 1st IEEE International Online Symposium on Joint Communications and Sensing, Dresden, Germany, 23–24 February 2021; pp. 1–6. [CrossRef]
2. Jiang, W.; Han, B.; Habibi, M.A.; Schotten, H.D. The road towards 6G: A comprehensive survey. *IEEE Open J. Commun. Soc.* **2021**, *2*, 334–366. [CrossRef]
3. Ali, A.; Gonzalez-Prelcic, N.; Heath, R.W.; Ghosh, A. Leveraging Sensing at the Infrastructure for mmWave Communication. *IEEE Commun. Mag.* **2020**, *58*, 84–89. [CrossRef]
4. Wei, Z.; Qu, H.; Jiang, W.; Han, K.; Wu, H.; Feng, Z. Iterative Signal Processing for Integrated Sensing and Communication Systems. *IEEE Trans. Green Commun. Netw.* **2023**, *7*, 401–412. [CrossRef]
5. Rahman, M.L.; Zhang, J.A.; Huang, X.; Guo, Y.J.; Lu, Z. Joint communication and radar sensing in 5G mobile network by compressive sensing. *IET Commun.* **2020**, *14*, 3977–3988. [CrossRef]
6. Zhang, Z.; He, R.; Ai, B.; Yang, M.; Li, C.; Mi, H.; Zhang, Z. A General Channel Model for Integrated Sensing and Communication Scenarios. *IEEE Commun. Mag.* **2023**, *61*, 68–74. [CrossRef]
7. Calabuig, D.; Mrozowski, A.; Prado-Alvarez, D.; Monserrat, J.F.; Zhang, Y.; Yang, W.; Chen, Y.; Chen, Y. Stochastic Cluttered Environments for the Channel Modeling of ISAC Systems. In Proceedings of the 2023 IEEE Globecom Workshops (GC Wkshps), Lumpur, Malaysia, 4–8 December 2023; pp. 1195–1200. [CrossRef]
8. Wang, H.; Hussain, F.; Wang, X.; Li, X.; Zhao, N.; Dev, K. Machine learning-enabled MIMO-FBMC communication channel parameter estimation in IIoT: A distributed CS approach. *Digit. Commun. Netw.* **2022**, *9*, 306–312. [CrossRef]
9. Liu, Y.C.; Cheng, Y.B.; Pan, X.B.; Sun, Z.Z.; Pan, D.; Long, G.L. Quantum integrated sensing and communication via entanglement. *Phys. Rev. Appl.* **2024**, *22*, 034051. [CrossRef]
10. ITU-R. *Report ITU-R M.2516-0: Future Technology Trends of Terrestrial International Mobile Telecommunications Systems Towards 2030 and Beyond*; Technical Report; ITU-R: Geneva, Switzerland, 2022.
11. Ramos, A.; Mrozowski, A.; Prado-Alvarez, D.; Monserrat, J.F.; Zhang, Y.; Yu, Z.; Chen, Y. Evaluation Methodology for 6G Sensing-Assisted Communication System Performance. *IEEE Access* **2024**, *12*, 7841–7852. [CrossRef]
12. ITU-R. *Report ITU-R M.2412-0: Guidelines for Evaluation of Radio Interface Technologies for IMT-2020*; Technical Report; ITU: Geneva, Switzerland, 2017.
13. Chenand, L.; Chen, W.; Wang, B.; Zhang, X.; Chen, H.; Yang, D. System-level simulation methodology and platform for mobile cellular systems. *IEEE Commun. Mag.* **2011**, *49*, 148–155. [CrossRef]
14. Lagen, S.; Wanuga, K.; Elkotby, H.; Goyal, S.; Patriciello, N.; Giupponi, L. New Radio Physical Layer Abstraction for System-Level Simulations of 5G Networks. In Proceedings of the ICC 2020—2020 IEEE International Conference on Communications (ICC), Dublin, Ireland, 7–11 June 2020; pp. 1–7. [CrossRef]
15. 3GPP. *TS 22.137 (V1.0.0): Service Requirements for Integrated Sensing and Communication; Stage 1 (Release 19)*; Technical Report; 3GPP: Nice, France, 2023.
16. Ramos, A.; Inca, S.; Ferrer, M.; Calabuig, D.; Roger, S.; Monserrat, J.F. Lookup Tables: "Physical Layer Abstraction of Detection and Localization Algorithms in ISAC Systems". Repository Link. 2024. Available online: https://rb.gy/e9my7u (accessed on 1 October 2024).
17. Olmos Bonafé, J.J.; Ruiz Boqué, S.; García Lozano, M.; Martín Sacristán, D. Link abstraction models based on mutual information for LTE downlink. In Proceedings of the 11th MCM of COST 2100, Athens, Greece, 31 May–1 June 2010.
18. Hanzaz, Z.; Schotten, H.D. Performance evaluation of Link to system interface for Long Term Evolution system. In Proceedings of the 2011 7th International Wireless Communications and Mobile Computing Conference, Istanbul, Turkey, 4–8 July 2011; pp. 2168–2173. [CrossRef]
19. Mezzavilla, M.; Miozzo, M.; Rossi, M.; Baldo, N.; Zorzi, M. A lightweight and accurate link abstraction model for the simulation of LTE networks in ns-3. In Proceedings of the 15th ACM International Conference on Modeling, Analysis and Simulation of Wireless and Mobile Systems, New York, NY, USA, 21–22 October 2012; MSWiM'12, pp. 55–60. [CrossRef]
20. Brueninghaus, K.; Astely, D.; Salzer, T.; Visuri, S.; Alexiou, A.; Karger, S.; Seraji, G.A. Link performance models for system level simulations of broadband radio access systems. In Proceedings of the 2005 IEEE 16th International Symposium on Personal, Indoor and Mobile Radio Communications, Berlin, Germany, 11–14 September 2005; Volume 4, pp. 2306–2311 [CrossRef]
21. Ramos, A.; Keskin, M.F.; Wymeersch, H.; Inca, S.; Monserrat, J.F. Enhancing Sensing-Assisted Communications in Cluttered Indoor Environments Through Background Subtraction. In Proceedings of the 2024 IEEE Wireless Communications and Networking Conference (WCNC), Dubai, United Arab Emirates, 21–24 April 2024; pp. 1–6. [CrossRef]
22. Huawei. *6G Integrated Sensing and Communication—Sensing-assisted Environment Reconstruction and Channel Reconstruction*; Technical Report 5; Communications of HUAWEI RESEARCH: Shenzhen, China, 2023.

23. Huawei. *Environment Reconstruction Based on Hybrid Aperture Sensing: Technologies and Applications;* Technical Report 5; Communications of HUAWEI RESEARCH: Shenzhen, China, 2023.
24. 3GPP. *TR 38.901: Study* on Channel Model for Frequencies from 0.5 to 100 GHz (Release 17); Technical Report; 3GPP: Nice, France, 2022.

telecom

MDPI

Article

Performance Analysis of a Sound-Based Steganography Wireless Sensor Network to Provide Covert Communications

Ariadna I. Rodriguez-Gomez *, Mario E. Rivero-Angeles *, Izlian Y. Orea Flores * and Gina Gallegos-García *

Centro de Investigación en Computación, Instituto Politécnico Nacional, Mexico City 07738, Mexico
* Correspondence: arodriguezg2020@cic.ipn.mx (A.I.R.-G.); erivero@cic.ipn.mx (M.E.R.-A.); iorea@ipn.mx (I.Y.O.F.); ggallegos@cic.ipn.mx (G.G.-G.)

Abstract: Given the existence of techniques that disrupt conventional RF communication channels, the demand for innovative alternatives to electromagnetic-based communications is clear. Covert communication, which claims to conceals the communication channel, has been explored using bio-inspired sounds in aquatic environments, but its application in terrestrial areas is largely unexplored. This work develops a mathematical analysis of a wireless sensor network that operates stealthily in outdoor environments by using birdsong audio signals from local birds for covert communication. Stored bird sounds are modified to insert sensor data while altering the sound minimally, both in characteristics and random silence/song patterns. This paper introduces a technique that modifies a fourth-level coefficient detail with a wavelet transform, then applies an inverse transform to achieve imperceptible audio modifications. The mathematical analysis includes a statistical study of the *On/Off* periods of different birds' songs and a Markov chain capturing the system's main dynamics. We derive the system throughput to highlight the potential of using birdsong as a covert communication medium in terrestrial environments. Additionally, we compare the performance of the sound-based network to the RF-based network to identify the proposed system's capabilities.

Keywords: covert communications; sound-based steganography; performance analysis

1. Introduction

Wireless Sensor Networks (WSNs) are composed of a set of sensors with the capability to collect and transmit data in a determinate area to sense a specific phenomenon. WSNs can be applied to a great number of fields, to name a few: e-health, e-logistics, e-agriculture, etc. [1]. These networks are crucial for tracking environmental parameters, enabling early detection of changes and efficient resource management. Given this, the use of WSNs stands out as a critical technical development, providing real-time remote data collection and analysis to improve our comprehension of dynamic ecological systems and assist preventative conservation measures [2–4].

The primary goal of this study is to develop a covert communication system for signaling alerts about environmental threats to protected ecosystems, such as illegal logging and species extraction. This system transmits acoustic signals from various nodes within the network to a central base station, which decodes the messages and relays them to the designated environmental protector. It ensures regular monitoring while providing immediate alerts during unauthorized activities.

Most WSNs communicate via electromagnetic signals, such as radio frequency (RF) or infrared [5], using commercial protocols like Bluetooth and IEEE802. However, attackers can disrupt RF communications to avoid detection. Therefore, using alternative communication methods, such as sound, provides an additional security layer, making the system more robust against attacks. There is a need to explore and analyze new alternatives to electromagnetic-based communication to ensure covert communications. Covert communication aims to hide the communication channel to prevent eavesdropping and jamming attacks [6,7].

Covert communication plays a critical role in ensuring the security and reliability of sensitive data transmission. The primary significance of covert communications lies in their ability to conceal the existence of the communication itself, thereby preventing detection by unauthorized parties. This is particularly crucial in scenarios involving protecting critical infrastructure, military operations, and environmental monitoring, where the interception of communications can lead to severe consequences. One of the main challenges in covert communications is maintaining a balance between data hiding and communication efficiency. The communication system must ensure that the hidden messages are not detectable by adversaries while also guaranteeing timely and reliable data transmission. This involves sophisticated techniques to embed data within carrier signals in an imperceptible yet robust manner against various types of attacks. Another significant challenge is the potential for increased complexity and energy consumption. Covert communication systems often require additional processing to encode and decode hidden messages, which can strain the resources of sensor nodes in WSNs. Therefore, designing energy-efficient methods that minimize the impact on node performance while maintaining effective covert communication is essential.

This study focuses on implementing an undetected communication system to prevent attackers from disrupting critical environmental alerts. By maintaining the element of secrecy, the system aims to provide an uninterrupted line of defense against activities that could harm the ecosystem, allowing timely responses from guardians.

In this work, we focus on bird sounds as the main communication element because they have an omnipresent quality in natural environmental conditions during daylight time, and in the majority of cases, the size of the birds and their ability to fly lets them hide in their surroundings. To effectively conceal information in the bird's sounds, we first characterize the periods where birds remain silent (*Off* period) and when they make sounds (*On* periods) in such a way as for nodes to generate the previously recorded sounds closely following the statistical properties of these *On/Off* periods and transmitting data packets only during the *On* periods.

At present, between the different proposals for creating a covert communication channel, there exists an approach that is focused on using bio-inspired sounds [8–18], in which they present the efficiency of the proposed communication schemes through computer simulations or physical prototypes. In these works, the authors often only consider a single sender and receiver without delving into how such a scheme would behave in a networked setting. Conversely, we develop a mathematical analysis of a communication system that not only considers a single sender and receiver but also provides network metrics such as the impact of the number of nodes in the system. Also, in our work, we consider the statistical main characteristics of the bird's specific sounds to hide the information better, i.e., to have the prerecorded sounds have the same statistical behavior as the birds. Specifically, we model these songs as an *On/Off* process using phase-type distributions to vary the Coefficient of Variation accordingly, and we consider that the information from nodes is transmitted only during the *On* periods while nodes remain silent during the *Off* periods. The use of the phase-type distributions, such as the Erlang and Hyper-Exponential distributions, allows the use of Markov chains to model the system. This produces a covert channel that closely follows the specific patterns of the bird's songs, making the communication system imperceptible to potential attackers. Besides, this natural *On/Off* process entails an important energy reduction in the system that we can measure using the derived mathematical model.

In addition to closely matching the sound/silent (*On/Off*) periods to conceal the data transmission effectively, we also avoid distortion in the bird's sounds by means of steganographic techniques, which are a set of processes to hide a message in a container signal, such as text, images, audio, video, or in a specific protocol, and is employed to protect the exchanged data from malicious attacks [19]. In the majority of works mentioned before, they make use of synthetic sounds that closely resemble the original sounds to use as a carrier medium. However, in our case, we are using the original audio as a container of the message hidden on it. Within the techniques used to perform steganographic methods, there is a modification over the frequency domain using the discrete wavelet transform (DWT), which provides a resolution similar to the human ear time-frequency perception [19,20]. The DWT produces 2 sets of coefficients: the approximation coefficients (cA) and the detail coefficients (cD), which are related to low and high frequencies, respectively.

To hide the information sent by the nodes in the WSN, which in our case comprises a small number of bits to indicate the presence of unusual activity through the sound of shrill sounds like an electric saw or human detection sensors, we consider the time when birds are emitting their characteristic sounds (which we call songs throughout this work) inserting data in such a way as to modify them as less as possible. To this end, we develop a teletraffic analysis that captures the main dynamics of the system, such as system throughput, idle time probability, collision probability, and an energy consumption estimation. In this sense, a network model is developed below based on Continuous Time Markov chains where pre-recorded bird songs are used as carrier signals, which are modified with steganographic techniques for sharing messages using sound waves as the transmitting medium. In this sense, using a novel technique by applying the wavelet transform, the audio signal is decomposed into coefficients [21], in which a modification is made over the detail coefficients. Then, the signal is reconstructed by using the inverse wavelet transform. After these, by using the statistical variables of the song and silence times of the previously modified audios, we associate them to a phase distribution, which are based on the exponential distribution, hence they have memoryless properties, which means that the current state depends uniquely on the previous state. This distribution accurately model the behavior of the nodes in the sensor network system determined to reproduce a bio-inspired behavior.

Random medium access protocols involve N nodes transmitting data on a single channel, with potential signal distortions from simultaneous transmissions. Timing can be continuous or slotted. Detection of ongoing transmissions may or may not be possible. The simplest protocols are Pure ALOHA, allowing transmission at any time, and its variant, Slotted ALOHA, which aligns transmissions with timed slots for increased efficiency. More complex protocols like Carrier Sense are not here. The S-ALOHA protocol is simple and widely used in Ad hoc networks and sensor networks. It is also used in competing 802.11 hotspots where they cannot sense each other's presence. This protocol is well-studied, with the purpose of identifying critical processes such as collision rates, free slots, throughput, and estimation of energy consumption. It is used for system analysis and to consider new alternatives for communication.

Building on this, the performance of the system is evaluated considering that the nodes transmit their data packets using the S-ALOHA protocol, which functions as a MAC protocol where the shared channel is slotted, i.e., nodes can only transmit at the beginning of the slot [22]. In our case, it provides communication capabilities over a hidden channel, providing a different form of communication than radio frequency. The system evaluation transmission rates above the rates reported in previous works that make use of bio-inspired audios; additionally, the proposed analysis allows the system administrator to predict the system's behavior under different conditions, such as different number of nodes and transmission rates.

Summarizing, this work introduces an innovative approach to wireless sensor network (WSN) communications by utilizing bird song audio for covert communication through audio steganography techniques, thus offering a unique alternative to traditional radio communication methods. Diverging from existing methods that focus on specific insertion points in natural sounds, our project stands out by altering the entire spectrum of bird song audios for embedding secret messages. This strategy not only diversifies the communication mediums used in WSNs but also provides a solution for combating electromagnetic jamming attacks, a prevalent issue in wireless networks. Additionally, the project proposes a method based on statistical behavior for energy efficiency in the Medium Access Control (MAC) layer. This method, inspired by the sound and silence patterns in bird songs, aligns the sleep/awake cycles of network nodes with these natural patterns. By integrating the unique properties of natural sounds into technological applications, our approach marks a distinct departure from conventional techniques in covert communication and WSN functionality. Building on this, the main contributions of this work are as follows:

- We conduct a comprehensive statistical analysis of bird sounds to determine the Coefficient of Variation (CoV) of active and inactive periods. This allows us to accurately model and account for effective data transmission within a covert channel communication system.
- We perform a teletraffic analysis to calculate the optimal number of nodes that can transmit simultaneously. We align our model closely with the natural behavior of bird sounds by employing phase-type distributions, such as Erlang and Hyper-Exponential distributions.
- We develop a novel steganography technique that inserts information into bird sounds with minimal distortion. This technique ensures that the covert nature of the communication channel is maintained, avoiding detection by potential attackers.
- We investigate the Slotted ALOHA protocol to derive key performance metrics for a WSN utilizing a covert channel based on sound. This analysis provides insights into the proposed system's throughput, transmission rate, and energy consumption.

The rest of the paper is organized as follows. Section 2 presents a survey of related work. In Section 3, we review a reference framework for our study on information insertion and wireless communication. Then, in Section 4, we introduce a mathematical model based on Markov chains. We also discuss the insertion of information in this section. In Sections 7 and 8, we present our test and results, where we analyze different metrics that were mentioned earlier. Finally, Section 9 concludes this paper and outlines future research directions.

2. Related Work

Our project is intended to create an alternative to radio communication in WSN by using covert communication using audio steganography techniques. In the literature, various studies have focused on concealing information in audio for covert communication. For instance, Kasetty et al. [23] used Discrete Wavelet Transform (DWT) and Singular Value Decomposition (SVD) to hide dialogue within musical audio. They applied SVD to the approximation coefficients, effectively embedding secret audio signals. Similarly, Alsabhany et al. [24] proposed Adaptive Multi-Level Phase Coding (AMPC) to address performance and security issues, achieving 24 Kbps at 32 dB SNR using Interval-Centered Quantization (ICQ). Bharti et al. [25] developed a robust method against LSB removal and resampling attacks, demonstrating that their approach maintains audio quality and withstands Gaussian white noise.

In terrestrial environments, researchers have explored two natural sounds for covert communication. Jiang et al. [8] used cricket sounds, achieving a rate of 46.9 bit/s by encoding information into pulse sequences. Additionally, Jiang et al. [16] proposed a method for covert communication by mimicking the songs of the Verditer Flycatcher, achieving a transmission rate of about 58 bps with high covertness and low bit error rates in forest environments. However, they did not consider specific *On/Off* periods. Bird songs, particularly from Passeriform birds, offer a higher frequency range (2–7 kHz) compared to cricket sounds (around 4.3 kHz), making them suitable for our purposes [26–29].

In underwater scenarios, researchers have used aquatic animal sounds for covert communication. Liu et al. [10] mimicked dolphin whistles and clicks for synchronization and information encoding, achieving 37 bit/s. They also used whale sounds in a Direct Sequence Spread Spectrum (DSS) signal for covert communication [12]. Jia et al. [11] used sea lion sounds, achieving a baud rate of 20 bit/s. Jiang et al. [9] and Li et al. [17] used sperm-whale call pulses, achieving transmission rates of 37.5 bit/s and 27 bit/s. These works focused on synchronization and information codes, while our approach modifies the entire spectrum of bird songs during *On* periods. Xie et al. [18] also proposed a method to enhance underwater communication security by mimicking dolphin whistles, achieving a rate of 45.28 bps.

Ahn et al. [13] used machine learning to classify dolphin whistles for covert communication, achieving low Bit Error Rates (BER). They later modeled Right Humpback Whales' whistles for information hiding [15]. Xie et al. [18] used Time-Frequency Contour techniques to improve signal-to-noise ratios in dolphin whistles. Unlike these works, we use real bird sounds and modify original recordings to insert information.

The jamming attack disrupts communication channels by introducing external electromagnetic energy. Mihajlov et al. [30] showed how jamming attacks affect WSNs using MAC protocols. Pirayesh et al. [31] highlighted the vulnerability of wireless networks to jamming attacks due to open wireless channels. Our work proposes an alternative communication method to avoid electromagnetic jamming attacks.

Kafetzoglou et al. [32] proposed an energy-saving method in the MAC layer using sleep/awake schedules, synchronizing nodes only with their parent and child nodes. Our approach proposes a statistical behavior of sleep/awake periods based on the natural behavior of bird sounds to better conceal the covert channel.

In summary, while significant progress has been made in the field of covert communication using various natural sounds across different environments, our approach stands out by leveraging real bird songs for terrestrial wireless sensor networks. Unlike in previous works, that have focused on synthetic signals or water environments, our method changes original bird recordings by inserting information into them, thus enhancing the covert nature of the communication channel. By overcoming weaknesses such as those associated with electromagnetic jamming attacks and minimizing energy consumption by making use of natural behavior patterns, our technique presents itself as a viable solution for security in WSNs.

3. Wireless Sound-Based Channel

For this work, we considered the birds that are usually found in the Meztitlan desert region or the Oak Pine Forest, namely: *Pheucticus melanocephalus, Myadestes occidentalis, Turdus migratorius, Toxostoma curvirostre, Campylorhynchus brunneicapillus, Cardinalis cardinalis*. In this region, the aforementioned species of birds are omnipresent, also according to the species taxonomy, we consider the passeriform order was considered because the birds that belong to it are known to be birds who sing. We collected many recordings of these birds and analyzed both the statistical properties of the active/silent (*On/Off*) periods and the frequency components of the sounds they emit to use them as carrier signals for the communication method. The birds songs' are structured as follows: The simplest sound made by a bird can be called a "note", a series of them are called "syllables" and the set of "syllables" are named as "phrases" [33]. In our work, the period corresponds to the syllables. To this end, we obtained the histograms of both the *On* and *Off* periods where the *On* period starts when the first note is audible. We then approximate the probability density function of these periods considering an exponential distribution when the Coefficient of Variation (CoV) is equal to 1, to an Erlang distribution when the $CoV < 1$ and to an Hyper-Exponential distribution when $CoV > 1$. The reason for the use of these phase-type distributions is to propose a teletraffic analysis using Continuous Time Markov Chains to obtain the probability of having different sounds simultaneously as explained in a subsequent section.

3.1. Audio Selection

The features of the sounds that we considered have different characteristics of the duration of the song and silence periods, frequencies, and amplitudes. Hence, our work considers a two-level steganography system that hides information in an efficient manner both in terms of the dynamics of the songs (active/inactive periods) and the signal characteristics. This leads to producing a model that is not focused on specific characteristics of a single bird, and it englobes the species considered. The recordings of the bird's sounds were extracted from the xenocanto.org webpage, which is a collaborative project, dedicated to uploading recordings of different bird species, where the labels of the birds are verified. However, these recordings do not always provide high audio quality and the majority of them include environmental noise.

3.2. Wavelet Transform

Unlike the Fourier transform, which decomposes a signal solely into its frequency components to analyze the spectral content, the wavelet transform breaks down a signal into various scales, reflecting the level of resolution, and time locations, which represent the temporal location of the wavelet function along the time-axis. This approach provides a comprehensive analysis that encompasses both the frequency and time domains [34]. Wavelet transforms achieve high temporal resolution at the cost of frequency resolution at higher frequencies. Conversely, at lower frequencies, they offer high-frequency resolution but with reduced temporal accuracy, a behavior similar to that of the human ear [20].

This analysis is enabled by the wavelet transform's inherent scaling and translation properties. Specifically, the scaling function facilitates a detailed examination in terms of frequency, while the translation function permits temporal analysis. Following this logic, the wavelet series can be expressed as shown in Equation (1):

$$f(x) = \sum_{j}\sum_{k} c_{jk}\psi_{jk}(t) \tag{1}$$

where c_{jk} are the wavelet coefficients at each of the scales and temporal positions and ψ_{jk} are the wavelet function scaled and translated, as given by Equation (2):

$$\psi_{jk}(t) = 2^{j/2}\psi(2^{j}t - k) \tag{2}$$

To calculate the wavelet coefficients c_{jk}, it is essential to select a wavelet function along with the corresponding low-pass and high-pass filters. For our analysis, we have chosen the Haar wavelet due to its simplicity and capacity to detect transients or sudden changes in an audio signal. The associated filters for the Haar wavelet, which belongs to the orthogonal wavelets family, are as follows in Equations (3) and (4):

$$H_{Low} = \frac{1}{\sqrt{2}}[1,1] \tag{3}$$

$$H_{High} = \frac{1}{\sqrt{2}}[1,-1] \tag{4}$$

Here, the low-pass filter computes the sum of adjacent elements, while the high-pass filter calculates their difference. The resultant of the high-pass filter produces the detail coefficients (cD), and the post-filtered signal contains residual noise. Passing the signal through a reconstruction high-pass filter produces the signal details (D). On the other hand, the result of the low-pass filter is the approximation coefficients (cA). In these coefficients, the smoothed signal is stored and when passing it through a low-pass reconstruction filter, the approximation of the signal (A) is obtained [21].

And the Equation (1) can be rewritten as Equation (5):

$$f(x) = \sum_{j}\sum_{k}\left(cA_{jk}\phi_{jk}(t) + cD_{jk}\psi_{jk}(t)\right) \tag{5}$$

where $\phi_{jk}(t)$ and $\psi_{jk}(t)$ are the scaling and translation function that corresponds to the cA and the cD, respectively.

3.3. Information Coverage

In this section, we explain in detail how the audio is processed as a stereo signal with two channels, both of which were used for the coverage signal. For the wavelet analysis, we employed the Haar wavelet function over each of the arrays of each audio channel, as previously mentioned, which generates the $cA1$ and $cD1$ coefficients. Subsequently, we used the resulting $cA1$ coefficient for further analysis. This process involved applying another wavelet analysis to the cA's coefficients, continuing up to four levels of analysis. Figure 1 illustrates this multilevel analysis. The final $cA4$ coefficients serve as the carrier signal in which the information is concealed.

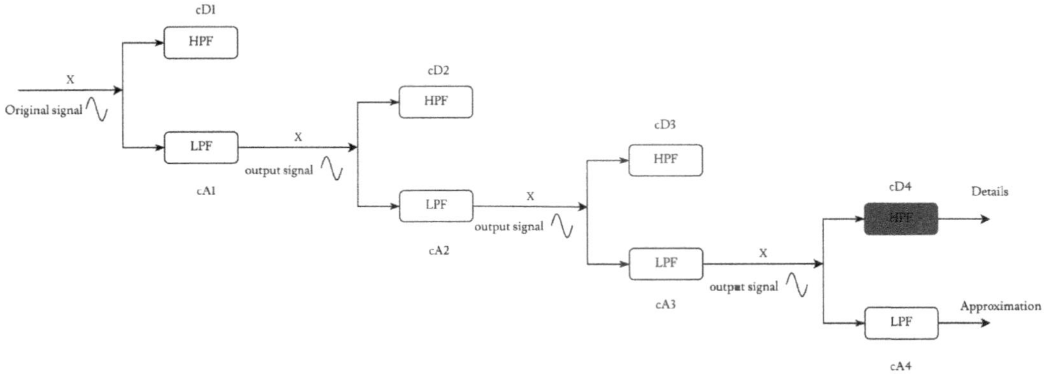

Figure 1. Signal decomposition in four levels after applying a wavelet transformation. The coefficient in gray is the Detail coefficient that hosts the modifications.

The hiding procedure for the information of the nodes starts by selecting a percentage of modification in each of the audios, this percentage varies for each audio, Table 1 presents a set of the percentages selected for some of the audios in the dataset. The audio signal gets distorted and provides less coverage at a higher modification percentage, none of the audios could support more than 2% of modification over the detail coefficient at the fourth level. The percentage selected is transformed into the number of bits to modify (C) at the fourth detail coefficient. The coefficient is segmented into C chunks. The last value of each of the chunks is modified to the result of $x_{new} = (x_{original} + m) \times \alpha$ (see Figure 2), the probe message to be covered was a message of "1's" of length = C, so $m = 1$. The α value is set to achieve greater discretion to the modification and it was obtained by empirical methods to achieve a coefficient that allows a modification without being noticed, for this probe $\alpha = 0.009$. The inverse transform is made by replacing the original 4th detail coefficient with the modified one (Figure 3). The results of the modified audios are reported in Table 1. Finally in Figure 4, we present the reconstructed signal separated by channel. It is important to note that the sounds used in this study are real and have not been simulated or synthesized.

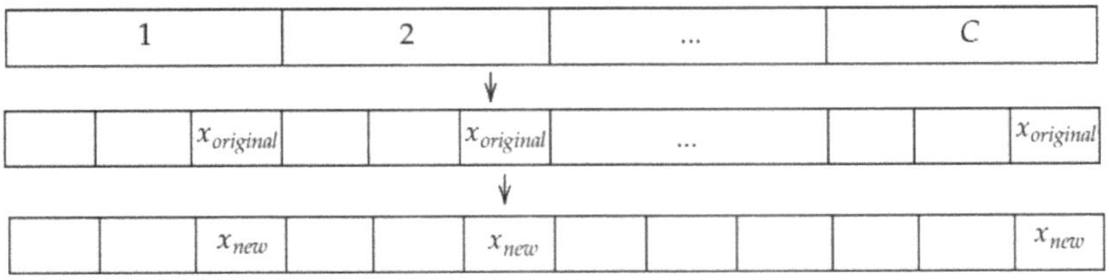

Figure 2. The cA coeffient is divided into C chunks, then we select the last value of each of the chunks, and the insertion is made by calculating $x_{new} = (x_{original} + m) \times \alpha$.

Table 1. Report of Modifications.

Audio Name	Duration of Audio	Coefficients Modified	% Modified	bps
XC3989Cuitlacoche	10 s	644	1	2060
XC196554Cuitlacoche	13 s	1617	2	3980
XC5127Solitario	36 s	2190	1	1946
XC5145Solitario	28 s	856	0.5	978
XC11493Cucarachero	27 s	1666	1	1974
XC220232Cucarachero	55 s	99	0.03	57
XC158418Cardenal	10 s	627	1	2006
XC32571Zorzal	15 s	438	1	934

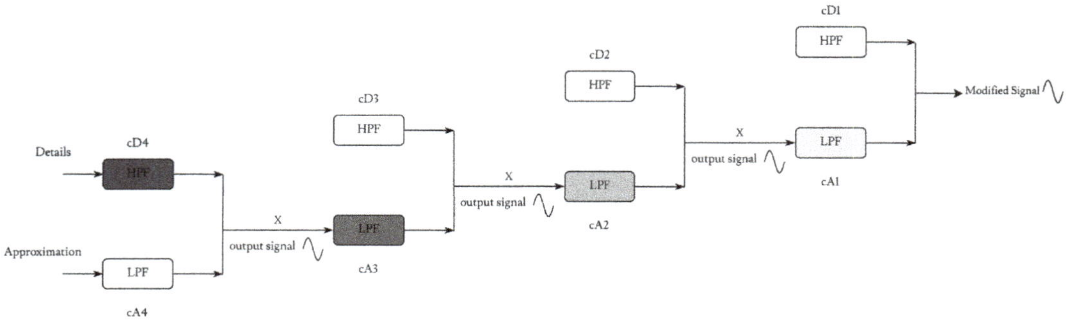

Figure 3. Signal reconstruction after modifying the Detail coefficient at a fourth transformation. The spread of the modification is signalized with the degradation of the modification.

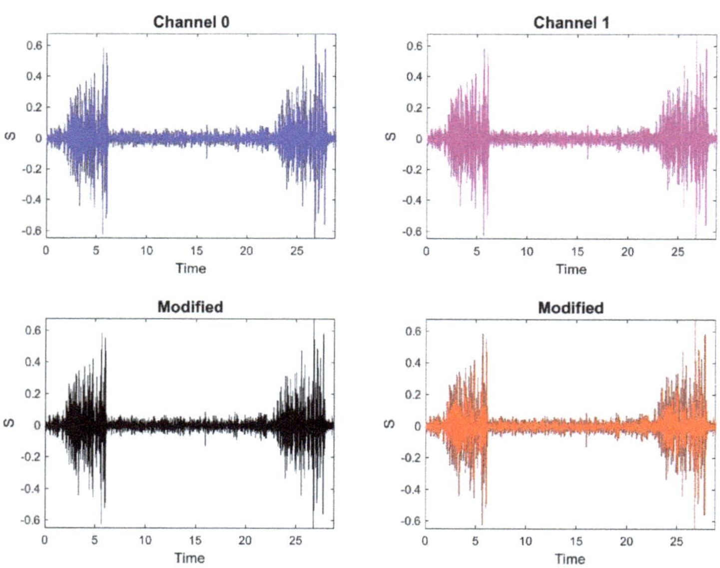

Figure 4. Signal reconstructed of the XC3951Solitario audio recording is shown before modification (**top part** of the figure) and after modification (**bottom part** of the figure), separated by channel.

4. Network Description

The system comprises N nodes distributed in the region of interest to detect certain events. These nodes transmit data packets regarding the environmental conditions (movement, presence of people, etc.) to a sink node in a single hop using the Slotted ALOHA protocol and sound as the communication channel instead of the usually used RF signals. Building on this, nodes have preregistered bird sounds (preferably birds from that same region where the WSN is installed to avoid potential suspicions). In those sounds, the nodes will insert their data using the abovementioned procedure to avoid any perceptible signal distortion. Additionally, nodes will emit the bird's sounds (and, thus, send the packet data) and remain silent according to the statistical characteristics of these active/inactive periods, effectively implementing an *On/Off* scheme.

As such, we have N_{On} active nodes and $N_{Off} = N - N_{On}$ inactive nodes at any instant of the system operation, where $0 \leq N_{On}, N_{Off} \leq N$. These nodes perform their state change (*On/Off*) that follows poisson process behavior represented by λ_{On} to activate the node, and λ_{Off} to deactivate it. This variables are calculated from analyzing the expected value ($E[x]$), variance ($\sigma_x^2(x)$) and the variation coefficient (CoV).

The transmission of data packets is measured using the following teletraffic metrics: throughput ($\bar{S}$), the probability of having a free slot ($\bar{FS}$), and the probability of packet collisions ($P_{collided}$). These metrics depend on the probability τ, which represents the likelihood of a transmission occurring in the current slot, the probability that only one node transmits during the current slot (P_S), and the probability of an idle slot (P_F).

To calculate the system's lifespan, we determine the total energy available to the system ($Total\_Energy$), the energy consumption of transmitting nodes ($Energy\_tx$ and tx respectively), the energy consumption of nodes that are turned on but not transmitting ($Energy\_On$), and the energy consumption of nodes in an idle state ($Energy\_Off$).

Finally, we assess the system's capability to transmit bits per second (bps) by considering the bps capacity that each audio file can hide and the system's throughput.

5. Characterization of the *On/Off* Periods

In this section, we first characterize the active and inactive periods of the selected birds' sounds.

We first obtain the absolute value of the amplitude of the spectrogram as shown in Figure 5. Observing this bird song, we can see that even in the silent period, the sound is present, different from 0, although clearly, the bird is not singing during this time. As such, we have to establish a threshold to digitalize the signal to obtain the lapses of time when the bird is singing or in silence as we show in Figure 6a where we show the analog signal magnified and in Figure 6b we show the digital representation where the On/Off periods are now clearly identified. Note that the digital representation is only used to characterize active and inactive periods, not to digitalize the bird's song which will be done in a subsequent section.

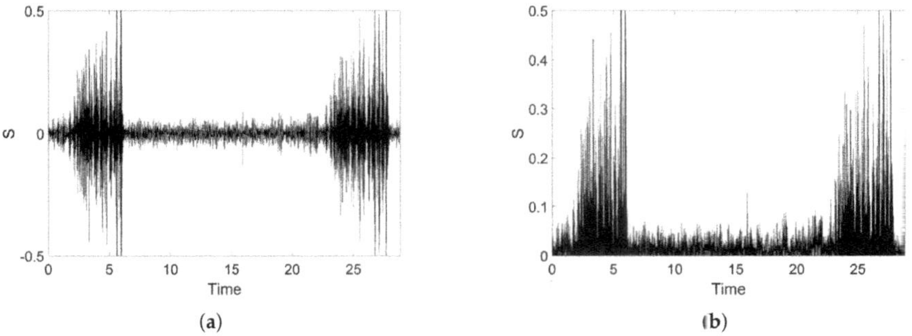

Figure 5. Spectrum analysis and absolute value of the amplitude of the spectrum of the *Myadestes occidentalis* birdsong. (**a**) Audio spectrum of *Myadestes occidentalis* birdsong. (**b**) Absolute value of spectrum of *Myadestes occidentalis* birdsong.

We measured each of these active (silent) times, which to us, they represent the times when sensor nodes will be able to transmit data packets (remain silent), effectively following a close behavior of the natural bird song, concealing the system to potential intruders. From these measurements, we obtain the histograms, and from the histograms we obtain the probability distribution function (pdf).

From this, we can characterize the random variables of both the active/inactive times by measuring all the *On* and *Off* times of the bird's sounds and then we calculate the mean ((6) and (7)), variance ((8) and (9)), standard deviation (10), and Coefficient of Variation (11) obtained respectively as:

$$E(x)_{On} = \sum_{-\infty}^{\infty} xP(X = siging) = \lambda_{On} \tag{6}$$

$$E(x)_{Off} = \sum_{-\infty}^{\infty} xP(X = silence) = \lambda_{Off} \tag{7}$$

$$\sigma_{On}^2 = \sum_{-\infty}^{\infty} (x - \lambda_{On})^2 P(X = singing) \tag{8}$$

$$\sigma_{Off}^2 = \sum_{-\infty}^{\infty} (x - \lambda_{Off})^2 P(X = silence) \tag{9}$$

$$\sigma_{On(Off)} = \sqrt{\sigma_{On(Off)}^2(x)} = \sqrt{\int_{-\infty}^{\infty} (x - \lambda_{On(Off)})^2 P(X = singing(silence)} \tag{10}$$

$$CoV = \frac{\sigma_x}{E(x)} \tag{11}$$

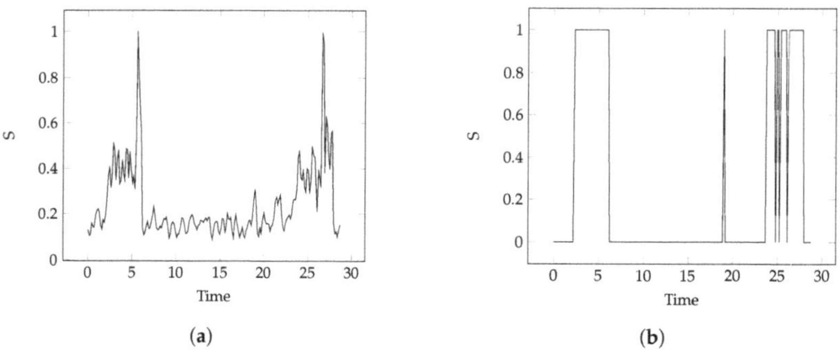

Figure 6. Sampled spectrum of birdsong and spectrum digitalized signal of the *Myadestes occidentalis* birdsong. (**a**) Sampled spectrum of the *Myadestes occidentalis* birdsong. (**b**) Spectre digitalized of the *Myadestes occidentalis* birdsong.

By identifying the statistical characteristics of these periods, we now propose different phase-type distributions to approximate to these empirical distributions. Namely, we use the Exponential distribution, given by (12), to approximate distributions that have a $CoV = 1$; the Hyper-Exponential distribution, depicted in (15) to approximate probability distributions with a $CoV > 1$; and an Erlang distribution, given by (20), to approximate distributions that have a $CoV < 1$. To characterize these distributions we match the first moment and the second central moment of these distributions to the observed measurements.

$$f_x(x) = \lambda e^{-\lambda x} \rightarrow E[X] = \sum_{0}^{\infty} x\lambda e^{-\lambda x} = \frac{1}{\lambda} \rightarrow \lambda = \frac{1}{E[x]} \tag{12}$$

And the second central moment (variance) is calculated as $\sigma_x^2(x) = E[x^2] - (E[x])^2$, then we have Equation (13):

$$E\left[x^2\right] = 2\frac{1}{\lambda^2} \rightarrow \sigma_x^2(x) = E[x^2] - (E[x])^2 = \frac{1}{\lambda^2} \rightarrow \sigma_x(x) = \frac{1}{\lambda} \tag{13}$$

Hence, the CoV is given in Equation (14):

$$CoV = \frac{\frac{1}{\lambda}}{\frac{1}{\lambda}} = 1 \tag{14}$$

The Hyper-Exponential is composed by two exponential distribution, with parameters λ_1 and λ_2 which are selected with probability p_1 and p_2 respectively where $p_1 = p$, and $p_2 = 1 - p$.

As such, the pdf of the Hyper-Exponential distribution can be expressed as Equation (15):

$$f_x(x) = \sum_{i=1}^{2} p_i\lambda_i e^{-\lambda_i x} = p_1\lambda_1 e^{-\lambda_1 x} + p_2\lambda_2 e^{-\lambda_2 x} \tag{15}$$

and the mean is obtained as in Equation (16):

$$\begin{aligned} E[x] &= \sum_0^\infty x\left(p_1\lambda_1 e^{-\lambda_1 x} + p_2\lambda_2 e^{-\lambda_2 x}\right) = \sum_0^\infty x\left(p\lambda_1 e^{-\lambda_1 x} + (1-p)\lambda_2 e^{-\lambda_2 x}\right) \\ &= p\left(\frac{1}{\lambda_1}\right) + (1-p)\left(\frac{1}{\lambda_2}\right) \end{aligned} \tag{16}$$

To calculate the variance we first obtain $E[x^2]$ and then we substitute this value to find σ_x^2 and then we obtain σ_x, as demonstrated by Equation (17).

$$\begin{aligned} E[x^2] &= \sum_0^\infty x^2(p_1\lambda_1 e^{-\lambda_1 x} + p_2\lambda_2 e^{-\lambda_2 x})dx = \frac{2p_2}{\lambda_2^2} + \frac{2p_1}{\lambda_1^2} \\ \sigma_x^2 &= E[x^2] - (E[x])^2 = \frac{2\lambda_1^2(1-p) + 2\lambda_2^2 p - (\lambda_1(1-p) - \lambda_2 p)^2}{\lambda_1^2\lambda_2^2} \\ \sigma_x &= \sqrt{\frac{2\lambda_1^2(1-p) + 2\lambda_2^2 p - (\lambda_1(1-p) - \lambda_2 p)^2}{\lambda_1^2\lambda_2^2}} \end{aligned} \tag{17}$$

Finally, the CoV can be expressed as in Equation (18):

$$\begin{aligned} CoV &= \frac{\sqrt{\frac{2\lambda_1^2(1-p)+2\lambda_2^2 p-(\lambda_1(1-p)-\lambda_2 p)^2}{\lambda_1^2\lambda_2^2}}}{p\left(\frac{1}{\lambda_1}\right) + (1-p)\left(\frac{1}{\lambda_2}\right)} \\ &= -\frac{\lambda_1\lambda_2\sqrt{-(\lambda_1^2(p-1) - \lambda_2^2 p + (-\lambda_1(p-1) + \lambda_2 p)^2)}}{\lambda_1^2\lambda_2^2(\lambda_1(p-1) - \lambda_2 p)} \end{aligned} \tag{18}$$

Therefore, solving p from Equation (16) we get Equation (19):

$$\begin{aligned} E[x] &= p\left(\frac{1}{\lambda_1}\right) + (1-p)\left(\frac{1}{\lambda_2}\right) = \frac{p(\lambda_2 - \lambda_1) + \lambda_1}{\lambda_1\lambda_2} \\ p(\lambda_2 - \lambda_1) &= E[x]\lambda_1\lambda_2 - \lambda_1 \rightarrow p = \frac{E[x]\lambda_1\lambda_2 - \lambda_1}{\lambda_2 - \lambda_1} \end{aligned} \tag{19}$$

And by $p = \frac{E[x]\lambda_1\lambda_2 - \lambda_1}{\lambda_2 - \lambda_1}$, we know that $\lambda_2 - \lambda_1$ should be greater than 0 by its causal property, so $\lambda_2 - \lambda_1 > 0$. Hence, $\lambda_2 > \lambda_1$. The theoretical analysis indicates that the condition $\lambda_2 > \lambda_1$ with $0 < p < 1$ generally leads to a CoV greater than 1. This outcome is particularly pronounced when λ_2 is significantly larger than λ_1.

For the case of the Erlang distribution corresponds to the sum of k exponential distributions. Then the pdf is obtained as the $(k-1)$th convolution of exponential distributions with parameter λ. Thus, the Erlang distribution is obtained as Equation (20).

$$f_x(x) = \lambda e^{-\lambda x} \frac{(\lambda x)^{k-1}}{(k-1)!} \quad (20)$$

The mean of an Erlang random variable is calculated as k times the mean of the exponential distribution, $1/\lambda$, presented in Equation (21)

$$E[x] = \sum_{0}^{\infty} x f_x(x) = k \times \frac{1}{\lambda} \rightarrow \lambda = \frac{k}{E[x]} \quad (21)$$

and $E[x^2]$ is k times $E[x^2]$ of the exponential distribution, then we get Equation (22):

$$E[x^2] = k \times \frac{1}{\lambda^2}; \sigma_x^2(x) = \frac{k}{\lambda^2} \quad (22)$$

From this, we can obtain the CoV for the Erlang distribution as presented in Equation (23):

$$CoV = \frac{\sqrt{\frac{k}{\lambda^2}}}{\frac{k}{\lambda}} = k^{-\frac{1}{2}} \quad (23)$$

Note that we already have the values of the mean, variance and CoV of the empirical data, directly obtained form the recordings of the bird's sounds. Then, we can obtain the parameters of these distributions by solving for λ when the $CoV = 1$, for λ_1, λ_2, and p when the $CoV \geq 1$ and for k and λ when the $CoV \leq 1$. Once we have found the parameters of the proposed distributions to approximate the *On/Off* distributions, we calculate the accuracy of the approximation using the Chi-square method as shown in Equation (24):

$$\chi^2 = \sum \frac{(ObsevedValue_i - ExpectedDistribution_i)^2}{ExpectedDistribution_i} \quad (24)$$

The *Chi-squared* analysis, is evaluated by a p_{value}, which is dependent of the χ^2 and the freedom degrees, which are the number of independent variables minus one $(IV-1)$. Then, the accuracy of the fit with the proposed phase-type distributions is obtained as follows: if the p_{value} is less than the critical value, the hypothesis should be rejected. In our project, we selected a $p_{value} = 0.05$.

We analyzed the audio recordings by categorizing them into statistical subgroups, which we refer to as Stratum. The 'General Stratum' encompasses all analyzed audios. The 'Pine-Oak Forest Stratum' includes recordings of the bird species *Pheucticus melanocephalus*, *Myadestes occidentalis*, and *Turdus migratorius*. The 'Meztitlan Stratum' comprises audios featuring *Toxostoma curvirostre*, *Campylorhynchus brunneicapillus*, *Campylorhynchus brunneicapillus*, and the *Cardinalis cardinalis*.

The parameters obtained for different birds and the distributions used to approximate to both the *On* and *Off* periods were calculated and for all the birds considered, both active and inactive times have a $CoV > 1$. Hence, we propose the Hyper-Exponential distribution to characterize both *On* and *Off* times. Where the minimum value for the CoV reported was for the *Pheucticus melanocephalus* in the Silence times with $CoV = 1.065180$; and the major CoV reported was for the General stratum in singing times with $CoV = 2.264716$. Additional data have been included in Appendix A.

6. Teletraffic Analysis of the WSN

In this section, we develop a mathematical analysis based *On* Continuous Time Markov Chains (CTMC) that captures the main dynamics of the system, such as the nodes that become active (inactive). according to the statistical characteristics of the particular birds. As we described in the previous section, the CoV of all the selected bird's sounds are higher than 1. Hence, we use a Hyper-Exponential distribution to approximate to the *On/Off* periods of the nodes in the WSN. However, we first develop the mathematical analysis considering the exponential case since the Hyper-Exponential case is an extension of this exponential model.

From this, the Markov Chain corresponds to an irreductible chain with valid states $\{\Omega_{N_{On},N_{Off}} : 0 \le N_{On} \le N, 0 \le N_{Off} \le (N - N_{On})\}$ where N is the total number of nodes in the system, N_{On} (N_{Off}) is the number of active (inactive) nodes, i.e., nodes in the *On* (*Off*) mode. Note that $N = N_{On} + N_{Off}$. From state (N_{On}, N_{Off}) the valid transitions are as follows (See Figure 7, and Table 2):

- To state $(N_{On}+1, N_{Off}-1)$ when an inactive node becomes active and $N_{Off} > 0$. This occurs with rate $N_{Off} \times \lambda_{On}$. As such, the mean time of the node in the *Off* state is considered to be $1/\lambda_{On}$.
- To state $(N_{On}-1, N_{Off}+1)$ when an active node becomes inactive and $N_{On} > 0$. This occurs with rate $N_{On} \times \lambda_{Off}$. As such, the mean time of the node in the *On* state is considered to be $1/\lambda_{Off}$.

This Markov chain is numerically solved to find the probability that there are N_{On} and N_{Off} nodes in the WSN, i.e., the steady-state probabilities, $\pi_{N_{On},N_{Off}}$.

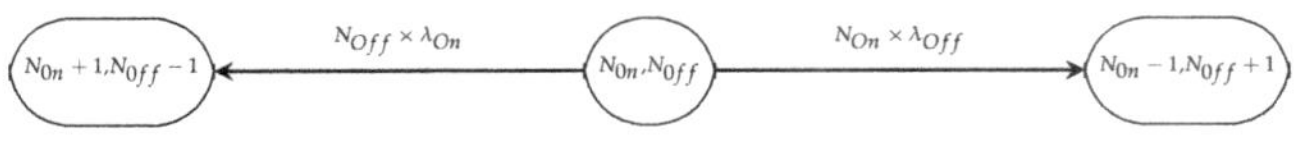

Figure 7. Markov chain for the case when the *On/Off* periods are exponentially distributed.

Table 2. Description of the possible transitions in the proposed network model.

Description of Node State Transitions				
Next State	**Transition Rate**	**A Node in State**	**Turns**	**Turns to a State of**
$(N_{On}+1, N_{Off}-1)$	$N_{Off} \times \lambda_{On}$	Off	On	On
$(N_{On}-1, N_{Off}+1)$	$N_{On} \times \lambda_{Off}$	On	Off	Off

For the Hyper-exponential case, when the $CoV > 1$, recall that this distribution is formed by two exponential distributions, with parameters λ_{On} and λ_{Off}, with probabilities P_{On} and P_{Off} respectively. Then, a node in phase 1 (2) of the *On* (*Off*) state can transit to phase 1 (2) of the *Off* (*On*) state and vice versa. Building on this, the Markov Chain also corresponds to an irreducible chain with valid states $\{\Omega_{N^{(1)}_{On},N^{(2)}_{On},N^{(1)}_{Off},N^{(2)}_{Off}} : 0 \le N^{(1)}_{On} + N^{(2)}_{On} \le N, 0 \le N^{(1)}_{Off} + N^{(2)}_{Off} \le (N - (N^{(1)}_{On} + N^{(2)}_{On}))\}$ where N is the total number of nodes in the system, $N_{On} = N^{(1)}_{On} + N^{(2)}_{On}$ ($N_{Off} = N^{(1)}_{Off} + N^{(2)}_{Off}$), and is the number of active (inactive) nodes, i.e., nodes in the *On* (*Off*) mode. Note that $N = N^{(1)}_{On} + N^{(2)}_{On} + N^{(1)}_{Off} + N^{(2)}_{Off}$. From state $(N^{(1)}_{On}, N^{(2)}_{On}, N^{(1)}_{Off}, N^{(2)}_{Off})$ the valid transitions are as follows (See Figure 8, and Table 3):

- To state $(N^{(1)}_{On}-1, N^{(2)}_{On}, N^{(1)}_{Off}+1, N^{(2)}_{Off})$ when a node is active in phase 1 and turns off in phase 1 of the *Off* mode. This occurs with rate of $N^{(1)}_{On} \times \lambda^{(1)}_{Off} \times P_{Off}$, this is the reason why there is one node less in state *On* with phase 1 and one more node in the *Off* mode with phase 1.

- To state $(N_{On}^{(1)}-1, N_{On}^{(2)}, N_{Off}^{(1)}, N_{Off}^{(2)}+1)$ when a node is active in phase 1 and turns off in phase 2 of the *Off* mode. This occurs with rate of $N_{On}^{(1)} \times \lambda_{Off}^{(1)} \times (1-P_{Off})$, this is the reason why there is one node less in state *On* with phase 1 and one more node in the *Off* mode with phase 2.
- To state $(N_{On}^{(1)}, N_{On}^{(2)}-1, N_{Off}^{(1)}+1, N_{Off}^{(2)})$ when a node is active in phase 2 and turns off in phase 1 of the *Off* mode. This occurs with rate of $N_{On}^{(2)} \times \lambda_{Off}^{(2)} \times P_{Off}$, this is the reason why there is one node less in state *On* with phase 2 and one more node in the *Off* mode with phase 1.
- To state $(N_{On}^{(1)}, N_{On}^{(2)}-1, N_{Off}^{(1)}, N_{Off}^{(2)}+1)$ when a node is active in phase 2 and turns off in phase 2 of the *Off* mode. This occurs with rate of $N_{On}^{(2)} \times \lambda_{Off}^{(2)} \times (1-P_{Off})$, this is the reason why there is one node less in state *On* with phase 2 and one more node in the *Off* mode with phase 2.
- To state $(N_{On}^{(1)}+1, N_{On}^{(2)}, N_{Off}^{(1)}-1, N_{Off}^{(2)})$ when a node is turned off in phase 1 and turns on in phase 1 of the *On* mode. This occurs with rate of $N_{Off}^{(1)} \times \lambda_{On}^{(1)} \times P_{On}$, this is the reason why there is one node less in state *Off* with phase 1 and one more node in the *On* mode with phase 1.
- To state $(N_{On}^{(1)}+1, N_{On}^{(2)}, N_{Off}^{(1)}, N_{Off}^{(2)}-1)$ when a node is turned off in phase 2 and turns on in phase 1 of the *On* mode. This occurs with rate of $N_{Off}^{(2)} \times \lambda_{On}^{(2)} \times P_{On}$, this is the reason why there is one node less in state *Off* with phase 2 and one more node in the *On* mode with phase 1.
- To state $(N_{On}^{(1)}, N_{On}^{(2)}+1, N_{Off}^{(1)}-1, N_{Off}^{(2)})$ when a node is turned off in phase 1 and turns on in phase 2 of the *On* mode. This occurs with rate of $N_{Off}^{(1)} \times \lambda_{On}^{(1)} \times (1-P_{On})$, this is the reason why there is one node less in state *Off* with phase 1 and one more node in the *On* mode with phase 2.
- To state $(N_{On}^{(1)}, N_{On}^{(2)}+1, N_{Off}^{(1)}, N_{Off}^{(2)}-1)$ when a node is turned off in phase 2 and turns *On* in phase 2 of the *On* mode. This occurs with rate of $N_{Off}^{(2)} \times \lambda_{On}^{(2)} \times (1-P_{On})$, this is the reason why there is one node less in state *Off* with phase 2 and one more node in the *On* mode with phase 2.

Table 3. Description of the possible transitions on the chain proposed.

Next State	Transition Rate	A Node in State	Turns	To a State of
$(N_{On}^{(1)}-1, N_{On}^{(2)}, N_{Off}^{(1)}+1, N_{Off}^{(2)})$	$N_{On}^{(1)}\lambda_{Off}^{(1)}P_{Off}$	$On^{(1)}$	*Off*	$Off^{(1)}$
$(N_{On}^{(1)}-1, N_{On}^{(2)}, N_{Off}^{(1)}, N_{Off}^{(2)}+1)$	$N_{On}^{(1)}\lambda_{Off}^{(1)}(1-P_{Off})$	$On^{(1)}$	*Off*	$Off^{(2)}$
$(N_{On}^{(1)}, N_{On}^{(2)}-1, N_{Off}^{(1)}+1, N_{Off}^{(2)})$	$N_{On}^{(2)}\lambda_{Off}^{(2)}P_{Off}$	$On^{(2)}$	*Off*	$Off^{(1)}$
$(N_{On}^{(1)}, N_{On}^{(2)}-1, N_{Off}^{(1)}, N_{Off}^{(2)}+1)$	$N_{On}^{(2)}\lambda_{Off}^{(2)}(1-P_{Off})$	$On^{(2)}$	*Off*	$Off^{(2)}$
$(N_{On}^{(1)}+1, N_{On}^{(2)}, N_{Off}^{(1)}-1, N_{Off}^{(2)})$	$N_{Off}^{(1)}\lambda_{On}^{(1)}P_{On}$	$Off^{(1)}$	*On*	$On^{(1)}$
$(N_{On}^{(1)}, N_{On}^{(2)}+1, N_{Off}^{(1)}-1, N_{Off}^{(2)})$	$N_{Off}^{(1)}\lambda_{On}^{(1)}(1-P_{On})$	$Off^{(1)}$	*On*	$On^{(2)}$
$(N_{On}^{(1)}+1, N_{On}^{(2)}, N_{Off}^{(1)}, N_{Off}^{(2)}-1)$	$N_{Off}^{(2)}\lambda_{On}^{(2)}P_{On}$	$Off^{(2)}$	*On*	$On^{(1)}$
$(N_{On}^{(1)}, N_{On}^{(2)}+1, N_{Off}^{(1)}, N_{Off}^{(2)}-1)$	$N_{Off}^{(2)}\lambda_{On}^{(2)}(1-P_{On})$	$Off^{(2)}$	*On*	$On^{(2)}$

The Markov chain is numerically solved to find the steady-state probabilities, $\pi_{N_{On}^{(1)}, N_{On}^{(2)}, N_{Off}^{(1)}, N_{Off}^{(2)}}$.

The total nodes in the system are distributed over four possible existing states as they follow a Hyper-exponential distribution in the Singing and Silence times. To calculate the number of possible states of this Markov Chain, consider for instance the case when $N = 2$.

Then, the possible states in the system are {(0,0,0,2), (0,0,2,0), (0,2,0,0), (2,0,0,0), (0,0,1,1), (0,1,0,1), (1,0,0,1), (0,1,1,0), (1,0,1,0), (1,1,0,0)} which corresponds to all valid configurations. It is important to mention that configurations such as (2,−1,1,0) are not possible as they are considering negative nodes.

And for the general case, the valid configurations follow Equation (25):

$$|valid| = 4 + 6(N-1) + 2(N-1)(N-2) + \frac{1}{6}(N-1)(N-2)(N-3), N \in \mathbb{Z}^{+} \quad (25)$$

where 4 belongs to the base case of all the nodes are in the same state with the same phase, $6(N-1)$ when there are only 2 states and phases that hold all the nodes in the system, $2(N-1)(N-2)$ when all the nodes are distributed in 3 of the states and phases, and $\frac{1}{6}(N-1)(N-2)(N-3)$ when all the nodes are distributed in the all the four possible states and phases.

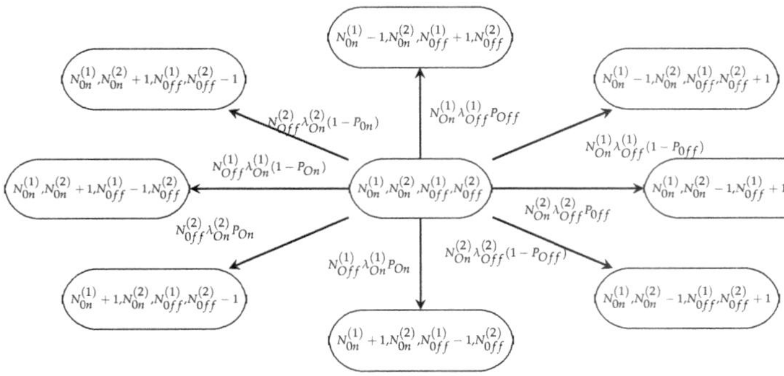

Figure 8. Markov chain of the transition process of a node which depends on the birdsongs behavior, Hyper-exponentional case.

And finally, for completeness reasons, since no bird song has either *On* or *Off* modes determined by a $CoV < 1$, we also develop the Erlang (Hypo-exponential) model considering that there may be bird songs that follow this Erlang distribution for the *On/Off* periods that can be used in future works. This model also corresponds to an irreducible, Markov Chain with valid states $\{\Omega_{N_{On}^{(1)}, N_{On}^{(2)}, \ldots, N_{On}^{(k)}, N_{Off}^{(1)}, N_{Off}^{(2)}, \ldots, N_{Off}^{(k)}} : 0 \leq N_{On}^{(1)} + N_{On}^{(2)} + \ldots + N_{On}^{(k)} \leq N, 0 \leq N_{Off}^{(1)} + N_{Off}^{(2)} + \ldots + N_{Off}^{(k)} \leq (N - (N_{On}^{(1)} + N_{On}^{(2)} + \ldots + N_{On}^{(k)}))\}$ where N is the total number of nodes in the system, $N_{On} = N_{On}^{(1)} + \ldots + N_{On}^{(k)}$ ($N_{Off} = N_{Off}^{(1)} + \ldots + N_{Off}^{(k)}$), and is the number of active (inactive) nodes, i.e., nodes in the *On (Off)* mode. Note that $N = N_{On}^{(1)} + \ldots + N_{On}^{(k)} + N_{Off}^{(1)} + \ldots + N_{Off}^{(k)}$. Note that the Erlang model corresponds to the sum of k_{On} (k_{Off}) exponential random variables with parameter λ_{On} (λ_{Off}). As such, in the *On (Off)* state, nodes have to transit k_{On} (k_{Off}) exponentially distributed random times before going to the *Off (On)* state. Hence, valid transitions from state $((N_{On}^{(1)}, \ldots, N_{On}^{(k)}, N_{Off}^{(1)}, \ldots, N_{Off}^{k}))$ are as follows (See Figure 9, and Table 4):

- To state $(N_{On}^{(1)} - 1, N_{On}^{(2)} + 1, \ldots, N_{On}^{(k)}, N_{Off}^{(1)}, \ldots, N_{Off}^{(k)})$ when a node is active in phase 1 and becomes active in phase 2 of the *On* mode. This occurs with rate of $N_{On}^{(1)} \times \lambda_{On}$, this is the reason why there is one node less in state *On* with phase 1 and one more node in the *On* mode with phase 2.
- To state $(N_{On}^{(1)}, \ldots, N_{On}^{(k-1)} - 1, N_{On}^{(k)} + 1, N_{Off}^{(1)}, \ldots, N_{Off}^{(k)})$ when a node is active in phase $k-1$ and becomes active in phase k of the *On* mode. This occurs with rate of $N_{On}^{(k-1)} \times \lambda_{On}$, this is the reason why there is one node less in state *On* with phase $k-1$ and one more node in the *On* mode with phase k.

- To state $(N_{On}^{(1)}, \ldots, N_{On}^{(k)} - 1, N_{Off}^{(1)} + 1, \ldots, N_{Off}^{(k)})$ when a node is active in phase k and becomes inactive in phase 1 of the *Off* mode. This occurs with rate of $N_{On}^{(k)} \times \lambda_{Off}$, this is the reason why there is one node less in state *On* with phase k and one more node in the *Off* mode with phase 1.
- To state $(N_{On}^{(1)}, \ldots, N_{On}^{(k)}, N_{Off}^{(1)} - 1, N_{Off}^{(2)} + 1, \ldots, N_{Off}^{(k)})$ when a node is inactive in phase 1 and becomes inactive in phase 2 of the *Off* mode. This occurs with rate of $N_{Off}^{(1)} \times \lambda_{Off}$, this is the reason why there is one node less in state *Off* with phase 1 and one more node in the *Off* mode with phase 2.
- To state $(N_{On}^{(1)}, \ldots, N_{On}^{(k)}, N_{Off}^{(1)}, \ldots, N_{Off}^{(k-1)} - 1, N_{Off}^{(k)} + 1)$ when a node is inactive in phase $k-1$ and becomes inactive in phase k of the *Off* mode. This occurs with rate of $N_{Off}^{(k-1)} \times \lambda_{Off}$, this is the reason why there is one node less in state *Off* with phase $k-1$ and one more node in the *Off* mode with phase k.
- To state $(N_{On}^{(1)} + 1, \ldots, N_{On}^{(k)}, N_{Off}^{(1)}, \ldots, N_{Off}^{(k-1)}, N_{Off}^{(k)} - 1)$ when a node is inactive in phase k and becomes active in phase 1 of the *On* mode. This occurs with rate of $N_{Off}^{(k)} \times \lambda_{On}$, this is the reason why there is one node less in state *Off* with phase k and one more node in the *On* mode with phase 1.

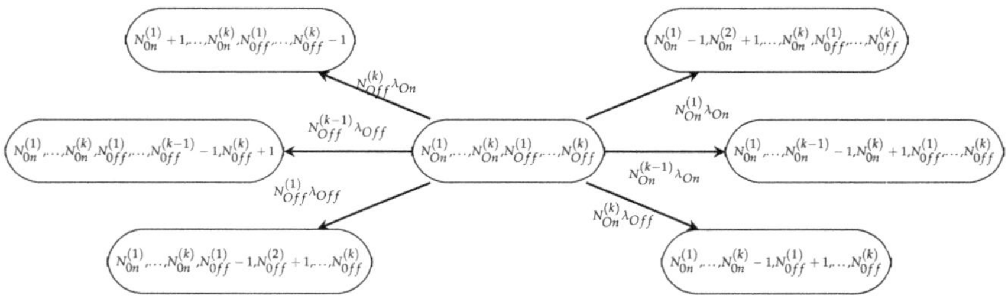

Figure 9. Markov chain of the transition process of a node which depends on the birdsongs behavior, Erlang case.

Table 4. Valid State Transitions for the Erlang Model.

Next State	Transition Rate	A Node in State	Turns	To a State of
$(N_{On}^{(1)} + 1, \ldots, N_{On}^{(k)}, N_{Off}^{(1)}, \ldots, N_{Off}^{(k)} - 1)$	$N_{Off}^{(k)} \lambda_{On}$	$Off^{(k)}$	*On*	$On^{(1)}$
$(N_{On}^{(1)} - 1, N_{On}^{(2)} + 1, \ldots, N_{On}^{(k)}, N_{Off}^{(1)}, \ldots, N_{Off}^{(k)})$	$N_{On}^{(1)} \lambda_{Off}$	$On^{(1)}$	*Off*	$On^{(2)}$
$(N_{On}^{(1)}, \ldots, N_{On}^{(k-1)} - 1, N_{On}^{(k)} + 1, N_{Off}^{(1)}, \ldots, N_{Off}^{(k)})$	$N_{On}^{(k-1)} \lambda_{Off}$	$On^{(k-1)}$	*Off*	$On^{(k)}$
$(N_{On}^{(1)}, \ldots, N_{On}^{(k)} - 1, N_{Off}^{(1)} + 1, \ldots, N_{Off}^{(k)})$	$N_{On}^{(k)} \lambda_{Off}$	$On^{(k)}$	*Off*	$Off^{(1)}$
$(N_{On}^{(1)}, \ldots, N_{On}^{(k)}, N_{Off}^{(1)} - 1, N_{Off}^{(2)} + 1, \ldots, N_{Off}^{(k)})$	$N_{Off}^{(1)} \lambda_{On}$	$Off^{(1)}$	*On*	$Off^{(2)}$
$(N_{On}^{(1)}, \ldots, N_{On}^{(k)}, N_{Off}^{(1)}, \ldots, N_{Off}^{(k-1)} - 1, N_{Off}^{(k)} + 1)$	$N_{Off}^{(k-1)} \lambda_{On}$	$Off^{(k-1)}$	*On*	$Off^{(k)}$

The Markov chain is numerically solved to find the steady-state probabilities, $\pi_{N_{On}^{(1)},\ldots,N_{On}^{(k)},N_{Off}^{(1)},\ldots,N_{Off}^{(k)}}$.

Throughput Analysis

Now we can obtain the main performance metrics of the system, namely, throughput, idle slot probability, and collision slot probability. Hence, we first calculate the probability that there are N_{On} active nodes, which are the nodes that can transmit data to the sink node. This probability is given by the steady-state probability of the Hyper-Exponential Markov Chain described above. Recall that the packet transmission success probability considering an S-ALOHA random access protocol, occurs when only one of the active nodes transmits in a given time slot with probability τ. Building on this, the throughput can be obtained as Equation (26):

$$\begin{aligned}\bar{S} &= \{\text{Only one of the active nodes in the network transmits in a given time slot}\} \\ &= \sum_{N_{On}=0}^{N} \binom{N_{On}}{1} \tau(1-\tau)^{N_{On}-1} Pr(\text{There are } N_{On} \text{ nodes in the system})\end{aligned} \tag{26}$$

In the scenario proposed, the throughput (Equation (27)) depends on the probability of success (P_s) that only one of the active nodes in $N_{On}^{(1)}$ state transmits in a given time slot and any of the nodes in $N_{On}^{(2)}$ state transmits in a given time slot or only one of the active nodes in $N_{On}^{(2)}$ state transmits in a given time slot and any of the nodes in $N_{On}^{(1)}$ state transmits in a given time slot (See Equation (28)).

$$\bar{S} = \sum_{\forall n \in valid} P_s(N_{On}^{(1)}, N_{On}^{(2)}, N_{Off}^{(1)}, N_{Off}^{(2)}) \pi_{(N_{On}^{(1)}, N_{On}^{(2)}, N_{Off}^{(1)}, N_{Off}^{(2)})} \tag{27}$$

$$\begin{aligned}P_s(N_{On}^{(1)}, N_{On}^{(2)}, N_{Off}^{(1)}, N_{Off}^{(2)}) =& \binom{N_{On}^{(1)}}{1} \tau(1-\tau)^{N_{On}^{(1)}-1} \binom{N_{On}^{(2)}}{0} \tau^0 (1-\tau)^{N_{On}^{(2)}} \\ &+ \binom{N_{On}^{(1)}}{0} \tau^0 (1-\tau)^{N_{On}^{(1)}} \binom{N_{On}^{(2)}}{1} \tau(1-\tau)^{N_{On}^{(2)}-1} \\ =& N_{On}^{(1)} \tau(1-\tau)^{N_{On}^{(1)}-1} (1-\tau)^{N_{On}^{(2)}} \\ &+ (1-\tau)^{N_{On}^{(1)}} N_{On}^{(2)} \tau(1-\tau)^{N_{On}^{(2)}-1}.\end{aligned} \tag{28}$$

A free slot (FS) occurs when there are no transmissions in that slot, Equation (29) corresponds to the probability of having an idle slot ($\bar{PF}$), and the Equation (30), refers to none of the active nodes in $N_{On}^{(1)}$ state transmit in a given time slot neither thenodes in $N_{On}^{(2)}$ state transmits in a given.

$$\bar{PF} = \sum_{\forall n \in valid} P_{FS}(N_{On}^{1}, N_{On}^{(2)}, N_{Off}^{1}, N_{Off}^{(2)}) \pi_{(N_{On}^{1}, N_{On}^{(2)}, N_{Off}^{1}, N_{Off}^{(2)})} \tag{29}$$

$$\begin{aligned}P_{FS}(N_{On}^{(1)}, N_{On}^{(2)}, N_{Off}^{(1)}, N_{Off}^{(2)}) =& \binom{N_{On}^{(1)}}{0} \tau^0 ((1)-\tau)^{N_{On}^{(1)}} \binom{N_{On}^{(2)}}{0} \tau^0 (1-\tau)^{N_{On}^{(2)}} \\ =& (1-\tau)^{N_{On}^{(1)}} (1-\tau)^{N_{On}^{(2)}}.\end{aligned} \tag{30}$$

And finally, a collided slot is calculated by Equation (31), which represents the case when two or more active nodes transmitted in that slot. The probability of having a collided slot is also the complementary value of not having neither an idle slot nor at Successful transmission, then

$$P_{collided} = 1 - \bar{PF} - \bar{S} \tag{31}$$

7. System Simulation Model

In this section, we explain in detail a discrete event simulation developed to validate the analytical results. The analytical results obtained from the derived Markov Chains are validated through extensive discrete event simulations to calculate the Throughput, the idle slot probability, and the packet collision probability. Algorithm 1, describes the simulations we generate to analyze the WSN considering that the *On/Off* times are hyper-exponentially distributed, focusing on the interaction of nodes, their energy consumption, and the efficiency of transmissions. The algorithm takes four inputs: the total number of nodes ($Total_{Nodes}$), a matrix with all the valid configurations of the nodes ($Valid_{Matrix}$) as described in the previous section, the packet transmission probability (τ), and a stratum to analyze ($Stratum$). Its output is a set of simulation results determined by the Result variable.

As it can be seen, it first determines the energy decrease based on the different operations of the nodes. Namely, in the *On* state, nodes activate the sensing, processing, and communication capabilities, while in the Off mode, nodes consume a small amount of energy-related mainly to minimal electronic systems to allow the node to wake up in the following seconds. Additionally, in the *On* mode, when nodes transmit a packet, with probability τ, nodes consume an extra amount of energy. Hence, $E_{tx} > E_{On} > E_{Off}$.

Initially, the algorithm sets up the simulation environment, initializing parameters like energy levels and slot time, and preparing the initial states of the nodes. It also initializes counters for tracking the number of free slots, successful transmissions, and collisions.

When the simulation begins evaluating a system with no nodes, a try-catch sequence is implemented to automatically terminate the program due to the absence of nodes to perform any actions, returning results based on the Result variable. Otherwise, it enters a loop that runs for a predefined number of iterations or until a specified slot count threshold is reached. During each slot, the algorithm calculates the number of transmissions, taking into account the active nodes and the transmission probability τ. It updates the energy consumption for both transmitting and non-transmitting nodes and adjusts the counters for free slots, successful transmissions, and collisions according to the transmission count. Progress is printed at regular intervals to monitor the simulation.

Finally, the algorithm concludes by returning free slots, collisions, and successful transmissions relative to the total number of slots, or the average energy consumption per slot.

Algorithm 1: System Simulation

```
Input: valid_configurations, τ, Energy_tx, Energy_On, Energy_Off
Output: Simulation results
Total_Energy, Free_channel, Success, Collision ← 0, 0, 0, 0;
for i ← 1 to valid_configurations_quantity do
    configuration ← valid_configurations[i];
    for j ← 1 to 100,000 do
        transmissions ← 0;
        actives ← nodes_actives_in_that_configuration;
        for k ← 1 to actives do
            random ← random_number;
            if random ≤ τ then
                transmissions ← transmissions + 1;
                Total_Energy ← Total_Energy + Energy_tx;
        if transmissions = 1 then
            Success ← Success + 1;
        else if transmissions > 1 then
            Collision ← Collision + 1;
        else
            Free_channel ← Free_channel + 1;
            Total_Energy ← Total_Energy + Energy_off;
return Total_Energy/100,000, Free_channel/100,000, Success/100,000, Collision/100,000;
```

Algorithm 1 aims to evaluate the energy consumption and communication performance of different configurations in a network. It takes as inputs a set of valid configurations, a transmission probability threshold (τ), and energy consumption values for transmission, active, and idle states. The algorithm initializes counters for total energy, free channel occurrences, successful transmissions, and collisions. For each valid configuration, it runs a large number of iterations to reach a steady state (100,000) and to simulate the network behavior. In each iteration, it calculates the number of active nodes and randomly determines if each active node transmits based on τ. Depending on the number of transmissions, it updates the success, collision, or free channel counters and adjusts the total energy consumption accordingly. Finally, it returns the average energy consumption, the proportion of free channels, successful transmissions, and collisions over the simulated iterations. This simulation provides insights into the efficiency and reliability of different network configurations in terms of energy usage and communication success rates.

The overall complexity of the algorithm is the product of the complexities of the loops, the outer loop is $O(N^3)$, the middle loop is $O(1)$ as we know that it runs a fixed number (100,000), and the inner loop is $O(N)$. Thus, the overall time complexity of the algorithm is $O(N^3) \times O(1) \times O(N) = O(N^4)$.

After validating the behavior of the system, the objective is to determine the efficiency and reliability of the birdsong-based communication system in real-world scenarios. By analyzing the previous simulation and the analytic results of the network metrics: throughput, collision rates, and free slot rates. The efficiency of the system is measured by evaluating the lifespan and throughput capabilities, given the transmission probability (τ) and the number of nodes (N). This is represented as $lifespan(\tau, N) \times Throughput(\tau, N)$.

Furthermore, energy consumption tests are conducted to evaluate the power efficiency of the proposed method. By measuring the energy consumption of the sensor nodes that has intermittence compared to a system that remains constantly in the *On* state. However, we have to emphasize that a system that is always turned on, implies that the nodes are constantly emitting the bird sound, which we believe would be very suspicious to people in the surroundings effectively weakening the concealment of the network.

Additionally, comparative tests are conducted to analyze the throughput capability of the birdsong-based communication system against the ZigBee protocol, which is a short-range wireless communication protocol. This test aims to highlight the difference of the proposed approach and provide insights for further improvements of the communication system. The inclusion of hidden information in the analysis is achieved by considering the maximum amount of bps that can be concealed in each audio file, as reported in Table 1. This value is then multiplied by the system's throughput, which is determined by its specific transmission probability (τ) and the number of nodes (N).

8. Numerical Results

In this section we present the main results obtained through the analytical model developed in this work. First, we validate our analytic results comparing them to simulation results. Then, we obtain the main system performance parameters, such as throughput, time slot collision, and idle probabilities.

Figure 10a–c, show the system throughput, idle and collision time slot probabilities respectively for both analytical and simulation results. We can observe a very good match between these results providing a validation of the mathematical model.

To validate the differences that exist between both representations of the system, the figures presented in Figure 11 show the mean square error (MSE) for each of the graphs previously displayed. The maximum error reported is 13% in the Idle Slot graph (Figure 11b). For the Collision MSE, at a probability of $\tau = 0.9$, there is a 10.5% error between both analyses. The Throughput MSE is the measure that fits best, with only a 0.17% difference between the simulation and the analytical solution of the model.

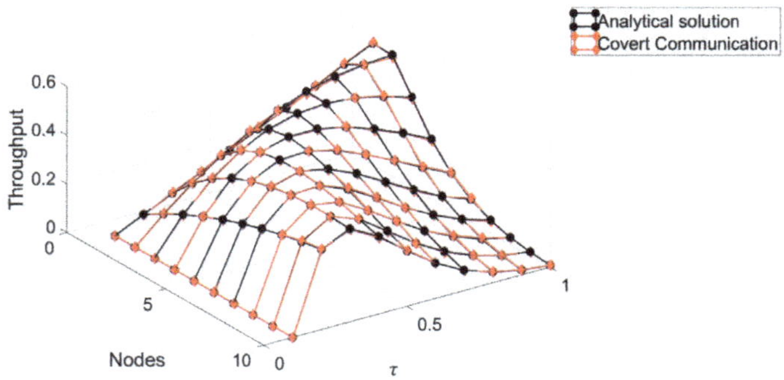

(**a**) System Throughput

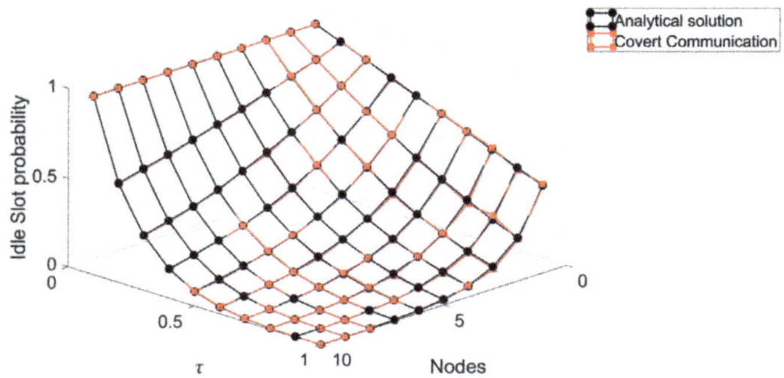

(**b**) Idle time slot probability

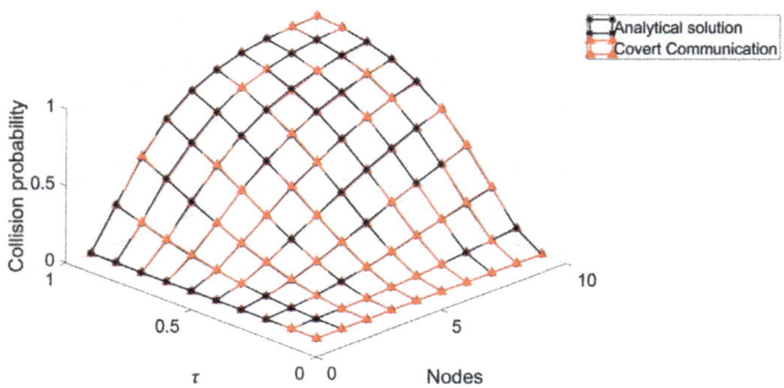

(**c**) Collision probability

Figure 10. Comparison between analytical and simulation results for (**a**) Throughput, (**b**) Idle time slot probability, and (**c**) Collision probability.

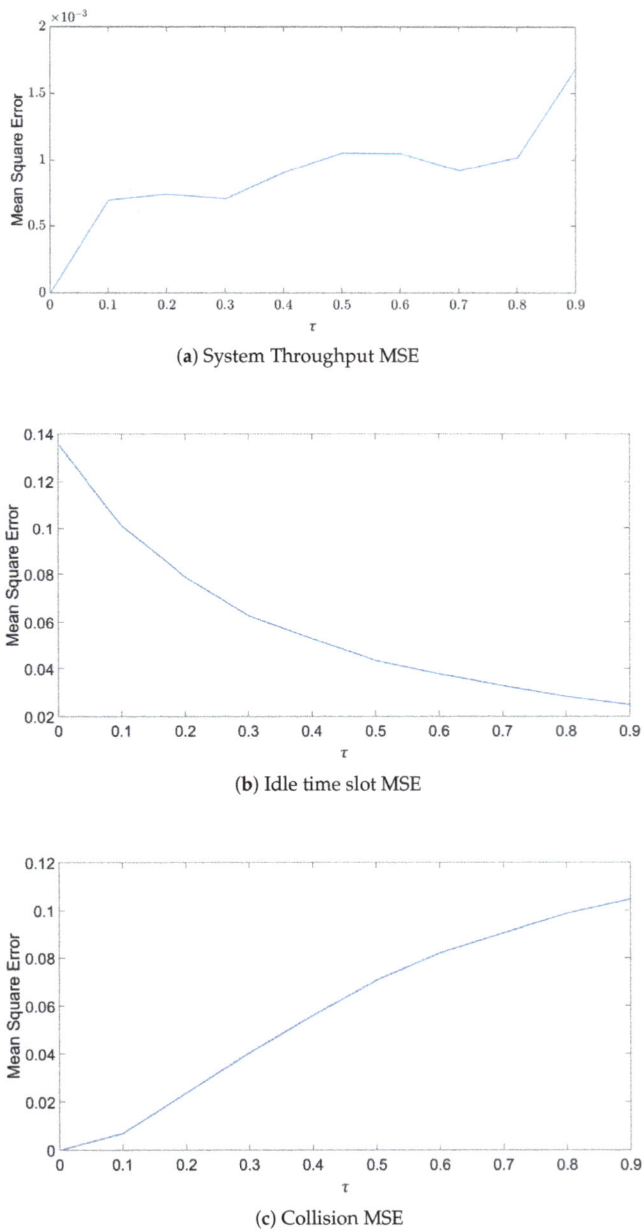

(**a**) System Throughput MSE

(**b**) Idle time slot MSE

(**c**) Collision MSE

Figure 11. Mean Square Error for (**a**) Throughput, (**b**) Idle time slot, and (**c**) Collision.

Now, we analyze the behavior of the proposed system where nodes turn *On* and *Off* according to the statistical characteristics of the bird's sounds. Figure 12a, shows the throughput behavior when varying the number of nodes over the network or the transmission probability. This graph presents a surface of the proposed model, where nodes turn *On* and *Off* according to the characteristics of the bird's sounds, in black, and the case when all the nodes are always active in the *On* mode, in red. We can see that the throughput is very close among these schemes when the transmission probability is low (lower than 0.3). This is due to the fact that even if the nodes remain in the *On* state all the time, they do not transmit, which generates a traffic very similar to the case when a node turns *On* and *Off*. However, as τ increases, nodes in the *On* state attempt many more transmissions, consequently increasing the throughput. As such, when nodes do not go to the *Off* mode, there are many more transmissions, but in the same way, there are many more collisions, as observed in Figure 12c, especially as τ approaches 1. Building on this, we can clearly see that a higher throughput does not always entail good system performance since there are many packet collisions and high energy consumption with non-useful data transmission. Also note in Figure 12b that as τ and N increases, the probability of finding empty slots reduces considerably, which is important to consider for a concealed channel since nodes would constantly generate sound exposing the presence of a surveillance system.

The results presented in Figure 13a compare the lifespan of the system due to its energy consumption for the proposed *On/Off* scheme and the case when nodes are always active. It can be seen that even though the difference in the throughput (shown in the previous figure) is higher at certain points for the always *On* scheme, the energy consumption is lower for all the cases in our proposed scheme. Indeed, the conceal channel approach has the main objective of turning *Off* nodes according to the natural behavior of birds, and it also achieves an important energy consumption reduction (higher system lifetime). Figure 13b presents the efficiency analysis, which examines the relationship between throughput and the system's lifespan. This relationship is analyzed by multiplying the throughput and lifespan results of both systems ($Lifespan(N,\tau) \times Throughput(N,\tau)$). It can be observed that efficiency increases as the probability of transmission increases. However, an increase in the number of nodes in the network has the opposite effect, causing efficiency to decrease.

Now, we compare the performance of the sound-based system to a conventional RF-based network. We can see in Figure 14a that the proposed covert channel achieves a lower transmission rate compared to a classic RF channel using ZigBee at 20 and 200 Kbps. However, this lower transmission rate occurs for a system where communications are hidden using a different communication medium (sound) that is robust against conventional jamming attacks which are expected in many surveillance applications, while the ZigBee based systems would certainly collapse in case of such attacks while our proposal remains operational even under DoS-based attacks. Hence, these results clearly show the system capacity of the covert system, in such a way as to provide clear guidelines for the system administrator to know in advance the transmission rate. For instance, if the system is based on small packet transmissions, such as presence detection or variations in lighting or sound, the achieved data transmission of 978 bps would be more than enough to provide adequate monitoring services. However, if the system has to transmit images or video HD of 5 Mbps, then this proposed WSN would not be adequate due to the low transmission rate.

The main limitations of this work are related to the transmission distance, which is significantly lower (hundreds of meters) compared to RF-based transmissions (several kilometers) as well as the transmission capacity, as discussed before, where only small packets can be transmitted. As such, the practical implementation of this system should consider a hierarchical architecture such as clustered-based WSNs, where nodes only transmit to a near Cluster Head, while the cluster head transmits using RF signals to a further sink. Another alternative could consider the use of mobile sinks (motorized sinks such as terrestrial robots or drones) that recover the information using sound passing close to the nodes. However, these issues fall outside the scope of this work.

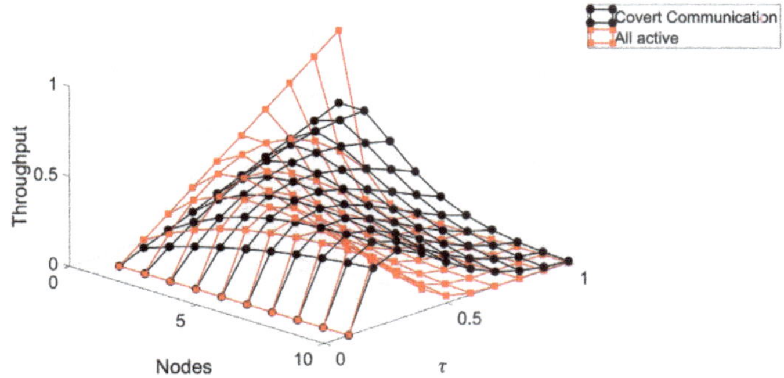

(**a**) System Throughput

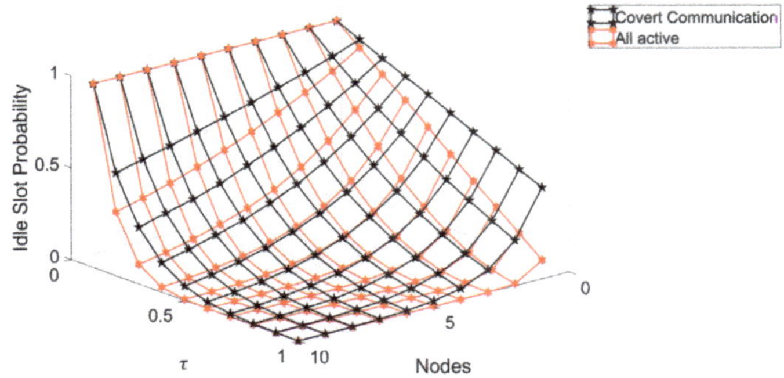

(**b**) Idle time slot

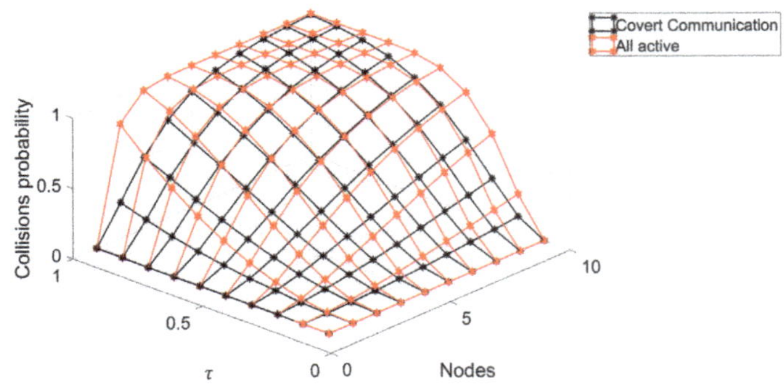

(**c**) Collision

Figure 12. Comparison between the system with nodes turning *On/Off* and nodes always in *On* mode (**a**) Throughput, (**b**) Idle time slot, and (**c**) Collision.

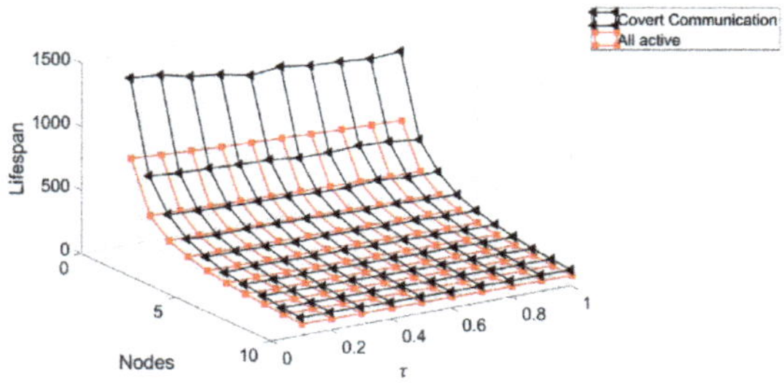

(**a**)

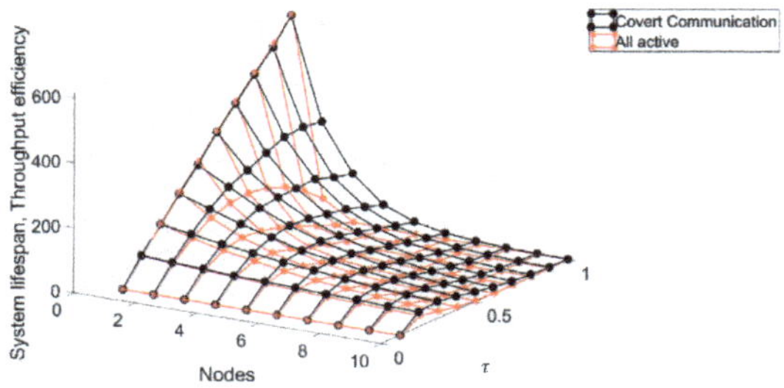

(**b**)

Figure 13. System Lifespan and efficiency of Nodes with *On/Off* transitions, and nodes always *On*. (**a**) Lifespan, (**b**) Efficiency.

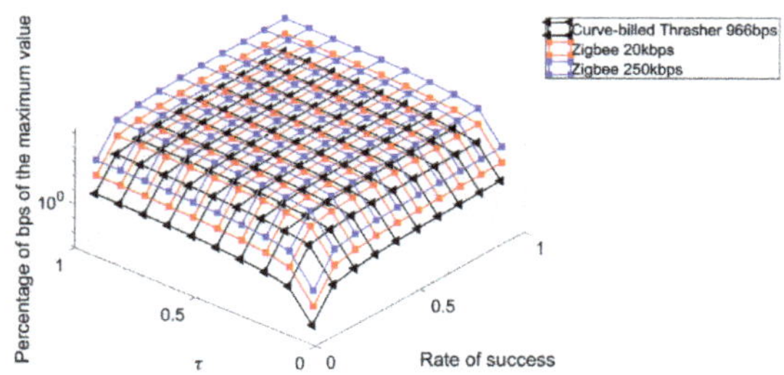

(**a**)

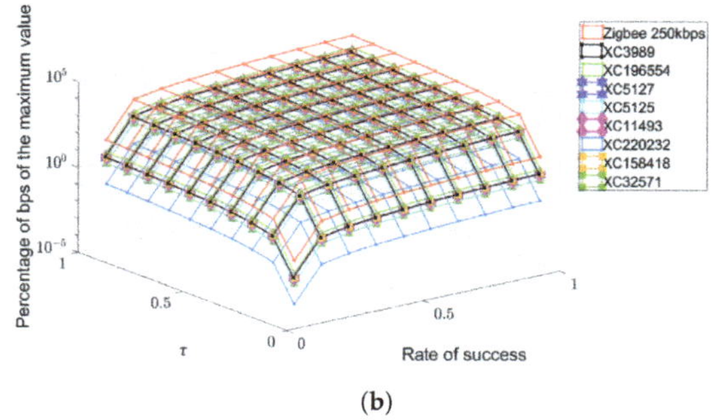

(**b**)

Figure 14. Transmission Rate comparison between modified audios and Zigbee minimum and maximum transition rates. (**a**) *Curve-billed Trasher*, and Zigbee minimum and maximum transition rates, (**b**) Birds' audios, and Zigbee minimum transition rates.

9. Conclusions

In this work, we proposed, analyzed, and studied a WSN using a covert channel based on bird sounds to hide the presence of the monitoring system. The proposed system closely follows the natural *On/Off* behavior of bird's songs, effectively hiding the monitoring capabilities of the WSN by only transmitting when the sound is present while turning the nodes to the *Off* mode when the bird's recording is silent. To achieve this, we propose the use of phase-type distributions to model these *On/Off* periods. Also, information is inserted in the sound using the Wavelet transform in order to reduce the distortion of the original audio signal in such a way as to generate sounds with underlying data with almost no perceptible modifications compared to the original audio. This approach, however, imposes limitations on the transmission rate in three significant ways: sound operates at a much lower frequency than RF signals; nodes can only transmit when sound is present; and bits are only inserted as long as they do not cause significant distortion to the recorded audio. In view of this, our mathematical analysis provides important performance metrics, clearly demonstrating the system's capabilities in terms of throughput, transmission rate, and energy consumption. Specifically, the data hiding process is contingent on the audio coverage rather than the duration of the audio. For instance, a 10-s audio file could hide 2060 bps, a 13-s file could conceal 3980 bps, whereas a 55-s file might only conceal 57 bps. While all the stratums studied follow a hyper-exponential hypothetical distribution, further studies are necessary to determine if all bird species conform to this distribution.

Moreover, the proposed *On/Off* node scheme distributed across the network improves the system perfromance as the number of nodes and/or the probability of transmission (τ) increases. For example, while at 1 node and a probability $\tau = 1$, the always *On* scheme reaches 90% throughput, adding one more node ($N = 2$) at the same transmission probability yields a maximum throughput rate of 49.9%, compared to 18% for the always *On* scheme. This method conserves energy, extends the network's lifespan, and proves more effective in power consumption and network throughput analysis.

Therefore, despite conventional RF systems achieving higher transmission rates, our proposal is robust against classical jamming attacks and operates imperceptibly to unauthorized intruders searching for RF transmissions.

For future work, we plan to conduct practical tests to measure the transmission capacity in physical applications. We aim to verify the feasibility and limitations of this solution, which appears promising thus far, and to explore alternatives to address the significant reduction in transmission rates compared to RF transmissions.

Author Contributions: Methodology, A.I.R.-G., M.E.R.-A., I.Y.O.F. and G.G.-G.; Validation, A.I.R.-G.; Formal analysis, A.I.R.-G., M.E.R.-A., I.Y.O.F. and G.G.-G.; Investigation, A.I.R.-G., M.E.R.-A., I.Y.O.F. and G.G.-G.; Writing—original draft, A.I.R.-G.; Writing—review & editing, M.E.R.-A., I.Y.O.F. and G.G.-G.; Supervision, M.E.R.-A., I.Y.O.F. and G.G.-G. All authors have read and agreed to the published version of the manuscript.

Funding: This work is partially funded by IPN SIP project 20240671, IPN SIP project 20240654, and IPN SIP project 20240675.

Data Availability Statement: Data are contained within the article.

Acknowledgments: We gratefully acknowledge the support of CONAHCYT for this work, provided through a doctoral scholarship awarded to Rodriguez-Gomez. We also thank the platform www.xeno-canto.org (accessed on 17 July 2024). for the recordings available on their site and the possibility of downloading them for use in this project.

Conflicts of Interest: The authors declare no conflicts of interest.

Appendix A

The Tables A1–A3, present the parameters obtained for different birds and the distributions used to approximate to both the *On* and *Off* periods. It is an important result that for all the birds considered, both active and inactive times have a $CoV > 1$. Hence, we propose the Hyper-Exponential distribution to characterize both *On* and *Off* times.

Table A1. Statistical Values by Stratum of Singing Times.

Stratum	$E[x]$	σ_x	CoV	Hypothetical Distribution
General	14.208748	32.178771	2.264716	Hyper-exponential
Pine-Oak Forest	9.011186	19.042807	2.113241	
Pheucticus melanocephalus	6.794393	7.803352	1.148499	
Myadestes occidentalis	6.120805	11.457486	1.871892	
Turdus migratorius	4.124088	4.715939	1.143511	
Metztitlan	6.200795	14.176165	2.286185	
Toxostoma curvirostre	2.775347	5.772288	2.079843	
Campylorhynchus brunneicapillus	2.681818	3.840707	1.432128	
Cardinalis cardinalis	2.112554	3.129993	1.481615	
White-winged Dove	3.685990	5.096456	1.382656	

Table A2. Statistical Values by Stratum of Silence Times.

Statistical Values of Silence Times				
Stratum	$E[x]$	σ_x	CoV	Hypothetical Distribution
General	29.308151	57.549965	1.963616	Hyper-exponential
Pine-Oak Forest	14.277405	23.926322	1.675817	
Pheucticus melanocephalus	7.457944	7.944072	1.065180	
Myadestes occidentalis	11.590604	17.093248	1.474750	
Turdus migratorius	2.941606	3.373937	1.146971	
Metztitlan	16.620278	34.737656	2.090077	
Toxostoma curvirostre	3.200795	5.914352	1.847776	
Campylorhynchus brunneicapillus	15.413333	18.368640	1.191737	
Cardinalis cardinalis	10.525862	14.115265	1.341008	
White-winged Dove	4.057971	4.921266	1.212741	

Table A3. Proposed values for λ_1, λ_2 and p for each of the singing and silence times, it also presents the p_{values} as p_{val}, as the result of the Chi-square method.

Transition Values								
Stratum	λ^1_{On}	λ^2_{On}	p_{On}	λ^1_{Off}	λ^2_{Off}	p_{Off}	$p_{On}val$	$p_{Off}val$
General	0.0168	0.0688	0.1081	0.0144	0.0455	0.1548	1	1
Pine-Oak	0.0345	0.1480	0.1012	0.0292	0.0934	0.1514	1	1
Pheucticus melanocephalus	0.0373	0.1962	0.0783	0.0364	0.1788	0.0853	1	1
Myadestes occidentalis	0.0383	0.2178	0.0711	0.0194	0.1150	0.0675	1	1
Turdus migratorius	0.0415	0.3233	0.0490	0.0436	0.4533	0.0355	1	0.9900
Metztitlan	0.0216	0.2150	0.0373	0.0207	0.1605	0.0494	1	1
Toxostoma curvirostre	0.0121	0.4804	0.0086	0.0120	0.4166	0.0099	1	1
Campylorhynchus brunneicapillus	0.0214	0.1951	0.0409	0.0180	0.0865	0.0879	1	1
Cardinalis cardinalis	0.0205	0.1528	0.0517	0.0198	0.1267	0.0617	1	1
White-winged Dove	0.0119	0.3617	0.0114	0.0119	0.3286	0.0125	1	1

References

1. Mani, R.; Rios-Navarro, A.; Sevillano-Ramos, J.-L.; Liouane, N. Improved 3D localization algorithm for large scale wireless sens networks. *Wirel. Netw.* **2023**, 1–16. [CrossRef]
2. Andreadis, A.; Giambene, G.; Zambon, R. Monitoring illegal tree cutting through ultra-low-power smart IoT devices. *Sens* **2021**, *21*, 7593. [CrossRef] [PubMed]
3. Verma, S.; Sood, N.; Sharma, A.K. Design of a novel routing architecture for harsh environment monitoring in heterogenec WSN. *IET Wirel. Sens. Syst.* **2018**, *8*, 284–294. [CrossRef]

4. Mutiara, G.A.; Herman, N.S.; Mohd, O. Using long-range wireless sensor network to track the illegal cutting log. *Appl. Sci.* **2020**, *10*, 6992. [CrossRef]
5. Saidov, B.; Telezhkin, V. Digital Processing of the Ultrasonic Signal for Mobile Devices in the Transmission of Information. In Proceedings of the 2020 Global Smart Industry Conference (GloSIC), Chelyabinsk, Russia, 17–19 November 2020 ; pp. 92–96. [CrossRef]
6. Makhdoom, I.; Abolhasan, M.; Lipman, J. A Comprehensive Survey of Covert Communication Techniques, Limitations and Future Challenges. *Comput. Secur.* **2022**, *120* , 102784. [CrossRef]
7. Nimbalkar, A.D.; Azmat, A.; Patil, Y.M. Security Issues in Wireless Sensor Networks. *i-Manag. J. Wirel. Commun. Netw.* **2023**, *11*, 32–38. [CrossRef]
8. Jiang, J.; Xu, J.; Duan, F.; Wang, X.; Liu, W.; Fu, X. Camouflage Covert Communication in Air by Imitating Cricket's Sound. *IEEE Access* **2020**, *8*, 71840–71850. [CrossRef]
9. Jia-jia, J.; Wang, X.; Duan, F.; Fu, X.; Yan, H.; Hua, B. Bio-inspired steganography for secure underwater acoustic communications. *IEEE Commun. Mag.* **2018**, *56*, 156–162. [CrossRef]
10. Liu, S.; Qiao, G.; Ismail, A. Covert underwater acoustic communication using dolphin sounds. *J. Acoust. Soc. Am.* **2013**, *133*, EL300–EL306. [CrossRef]
11. Jia, Y.; Liu, G.; Zhang, L. Bionic camouflage underwater acoustic communication based on sea lion sounds. In Proceedings of the 2015 International Conference on Control, Automation and Information Sciences (ICCAIS), Changshu, China, 29–31 October 2015; pp. 332–336. [CrossRef]
12. Liu, S.; Qiao, G.; Ismail, A.; Liu, B.; Zhang, L. Covert underwater acoustic communication using whale noise masking on DSSS signal. In Proceedings of the 2013 MTS/IEEE OCEANS-Bergen, Bergen, Norway, 10–14 June 2013; pp. 1–6. [CrossRef]
13. Ahn, J.; Lee, H.; Kim, Y.; Chung, J.; Lee, S. Machine Learning based Dolphin Whistle Transceiver for Bio-inspired Underwater Covert Communication. In Proceedings of the OCEANS 2019 MTS/IEEE SEATTLE, Seattle, WA, USA, 27–31 October 2019; pp. 1–4. [CrossRef]
14. Jiang, J.; Wang, X.; Duan, F.; Fu, X.; Li, C.; Sun, Z. A Basic Bio-Inspired Camouflage Communication Frame Design and Applications for Secure Underwater Communication Among Military Underwater Platforms. *IEEE Access* **2020**, *8*, 24927–24940. [CrossRef]
15. Ahn, J.; Do, D.; Kim, W. The Long-Range Biomimetic Covert Communication Method Mimicking Large Whale. *Sensors* **2022**, *22*, 8011. [CrossRef] [PubMed]
16. Jiang, J.; Lu, Y.; Xu, J.; Yao, Z.; Li, Z.; Yao, Q.; Duan, F.; Fu, X.; Gong, X. Bio-inspired camouflage communication in air imitating birds songs. *Appl. Acoust.* **2023**, *211*, 109548. [CrossRef]
17. Li, C.; Jiang, J.; Duan, F.; Wang, X.; Sun, Z. Bio-inspired covert underwater acoustic communication based on sperm whale clicks with frequency-hopping modulation. In Proceedings of the 2020 Information Communication Technologies Conference (ICTC), Nanjing, China, 29–31 May 2020; pp. 28–32. [CrossRef]
18. Xie, L.; Zhu, J.; Jia, Y.; Chen, H. Bionic Covert Underwater Acoustic Communication Based on Time–Frequency Contour of Bottlenose Dolphin Whistle. *Entropy* **2022**, *24*, 720. [CrossRef] [PubMed]
19. Dhawan, S.; Gupta, R. Analysis of various data security techniques of steganography: A survey. *Inf. Secur. J. Glob. Perspect.* **2021**, *30*, 63–87. [CrossRef]
20. Tzanetakis, G.; Essl, G.; Cook, P. Audio analysis using the discrete wavelet transform. In Proceedings of the WSES International Conference Acoustics and Music: Theory and Applications (AMTA 2001), Skiathos, Greece, 26–30 September 2001; Volume 66.
21. Fugal, D.L. *Conceptual Wavelets in Digital Signal Processing: An in-Depth, Practical Approach for the Non-Mathematician*; Space & Signals Technical Pub.: San Diego, CA, USA, 2009.
22. Ma, R.T.B.; Misra, V.; Rubenstein, D. An analysis of generalized slotted-aloha protocols. *IEEE/ACM Trans. Netw.* **2008**, *17*, 936–949. [CrossRef]
23. Kasetty, P.K.; Kanhe, A. "Covert speech communication through audio steganography using DWT and SVD. In Proceedings of the 2020 11th International Conference on Computing, Communication and Networking Technologies (ICCCNT), Kharagpur, India, 1–3 July 2020; pp. 1–5. [CrossRef]
24. Alsabhany, A.A.; Ridzuan, F.; Azni, A.H. The adaptive multi-level phase coding method in audio steganography. *IEEE Access* **2019**, *7*, 129291–129306. [CrossRef]
25. Bharti, S.S.; Gupta, M.; Agarwal, S. A novel approach for audio steganography by processing of amplitudes and signs of secret audio separately. *Multimed. Tools Appl.* **2019**, *78*, 23179–23201. [CrossRef]
26. Luijk, G. Análisis del Canto de un Grillo con R. Overfitting.net–Data Science, Imagen y Sonido. January 1970. Available online: https://www.overfitting.net/2018/07/analisis-del-canto-de-un-grillo-con-r.html (accessed on 17 July 2024).
27. Fletcher, N.H.; Tarnopolsky, A. Acoustics of the avian vocal tract. *J. Acoust. Soc. Am.* **1999**, *105*, 35–49. [CrossRef]
28. Gordillo, A.; Ortíz, M.F.; Navarro, A.G. Estructura y Evolución de las Vocalizaciones de las Aves. Revista Ciencias , Universidad Nacional Autónoma de México. Available online: https://www.revistacienciasunam.com/en/149-revistas/revista-ciencias-109-110/1238-estructura-y-evoluci%C3%B3n-de-las-vocalizaciones-de-las-aves.html (accessed on 17 July 2024).
29. Cracraft, J.; Barker, F.K.; Braun, M.; Harshman, J.; Dyke, G.J.; Feinstein, J.; Stanley, S.; Cibois, A.; Schikler, P.; Beresford, P.; et al. Phylogenetic relationships among modern birds (Neornithes). In *Assembling the Tree of Life*; Oxford University Press: New York, NY, USA, 2004; pp. 468–489.

30. Mihajlov, B.; Bogdanoski, M. Analysis of the WSN MAC Protocols under Jamming DoS Attack. *Int. J. Netw. Secur.* **2014**, *16*, 304–312.
31. Pirayesh, H.; Zeng, H. Jamming attacks and anti-jamming strategies in wireless networks: A comprehensive survey. *IEEE Commun. Surv. Tutor.* **2022**, *24*, 767–809. [CrossRef]
32. Kafetzoglou, S.; Papavassiliou, S. Energy-efficient framework for data gathering in wireless sensor networks via the combination of sleeping MAC and data aggregation strategies. *Int. J. Sens. Netw.* **2011**, *10*, 3–13. [CrossRef]
33. Brenowitz, E.A.; Margoliash, D.; Nordeen, K.W. An introduction to birdsong and the avian song system. *J. Neurobiol.* **1997**, *33*, 495–500. [CrossRef]
34. Lee, D.T.L.; Yamamoto, A. Wavelet analysis: Theory and applications. *Hewlett Packard J.* **1994**, *45*, 44.

telecom

MDPI

Article

A Comparative Analysis of DNN and Conventional Signal Detection Techniques in SISO and MIMO Communication Systems

Hamna Shoukat [1], Abdul Ahad Khurshid [1], Muhammad Yunis Daha [2], Kamal Shahid [1,*] and Muhammad Usman Hadi [2,*]

[1] Institute of Electrical, Electronics, and Computer Engineering, University of the Punjab, Lahore 54590, Pakistan; hamnashoukat.dd008@gmail.com (H.S.); abdul.ahad@systemsltd.com (A.A.K.)
[2] School of Engineering, Ulster University, Belfast BT15 1AP, UK; daha-my@ulster.ac uk
* Correspondence: kamal.ee@pu.edu.pk (K.S.); m.hadi@ulster.ac.uk (M.U.H.)

Abstract: This paper investigates the performance of deep neural network (DNN)-based signal detection in multiple input, multiple output (MIMO), communication systems. MIMO technology plays a critical role in achieving high data rates and improved capacity in modern wireless communication standards like 5G. However, signal detection in MIMO systems presents significant challenges due to channel complexities. This study conducts a comparative analysis of signal detection techniques within both the single input, single output (SISO), and MIMO frameworks. The analysis focuses on the entire transmission chain, encompassing transmitters, channels, and receivers. The effectiveness of three traditional methods—maximum likelihood detection (MLD), minimum mean square error (MMSE), and zero-forcing (ZF)—is meticulously evaluated alongside a novel DNN-based approach. The proposed study presents a novel DNN-based signal detection model. While this model demonstrates superior computational efficiency and symbol error rate (SER) performance compared to more conventional techniques like MLD, MMSE, and ZF in the context of a SISO system, MIMO systems face some challenges in outperforming the conventional techniques specifically in terms of computation times. This complexity of MIMO systems presents challenges that the current DNN design has yet to fully address, indicating the need for further developments in wireless communication technology. The observed performance difference between SISO and MIMO systems underscores the need for further research on the adaptability and limitations of DNN architectures in MIMO contexts. These findings pave the way for future explorations of advanced neural network architectures and algorithms specifically designed for MIMO signal-processing tasks. By overcoming the performance gap observed in this work, such advancements hold significant promise for enhancing the effectiveness of DNN-based signal detection in MIMO communication systems.

Keywords: 5G; DNN; signal detection; MIMO; wireless communication; machine learning; Artificial Intelligence

1. Introduction

This research investigates data detection techniques for future communication systems beyond 5G, focusing on overcoming challenges associated with increased data rates, spectral efficiency, and reliability. The growing complexity of 5G necessitates the use of more antennas in terms of multiple inputs, multiple outputs (MIMO), which introduces new detection hurdles such as higher computational demands, inter-antenna interference, and the need for accurate channel information. The classical detection methods like maximum likelihood detection (MLD), minimum mean square error (MMSE), and zero-forcing (ZF) exhibit limitations in MIMO systems due to their increased complexity with a growing number of antennas [1,2]. This performance gap between SISO and MIMO systems motivates further exploration of deep neural network (DNN) architectures for MIMO data detection. Therefore, this study proposes a DNN-based MIMO detector to address these limitations and achieve reliable, high-quality wireless networks with large antenna arrays. The primary objective is to develop a MIMO detector leveraging a deep learning model for superior signal detection and error minimization compared to conventional approaches such as MLD, MMSE, and ZF. The proposed DNN detector is hypothesized to effectively handle the complexity of large-scale antenna systems, as in MIMO, and deliver much better performance in simulations, as well as real-life implementations. The proposed DNN-based detector's efficacy is carefully assessed and compared with the traditional techniques (MLD, MMSE, and ZF). This covers performance measures including computing time, the signal-to-noise ratio (SNR), and the symbol error rate (SER).

The analysis presented in this work is based on the design and implementation of a DNN-based MIMO detector to enhance signal identification and achieve an SER reduction in MIMO systems. This work also explores the trade-off between computational efficiency and detection accuracy, considering the feasibility of deploying DNN-based detectors in beyond-5G systems. The performance of the DNN detector is evaluated and compared with traditional methods (MLD, MMSE, and ZF) to assess its efficacy relative to these established techniques. In addition to the MIMO systems, this work employs DNN to provide a method of signal detection in SISO systems as a base case. This DNN-based model, in contrast to conventional approaches, is engineered to efficiently manage the complexities of MIMO systems, offering improved computational efficiency and SER performance. This work not only assesses the DNN model's efficacy but also looks into ways to increase system reliability and signal detection accuracy going forward.

To optimize the performance of DNN-based models, suitable training algorithms and techniques, such as stochastic gradient descent, have been implemented. A comprehensive analysis and interpretation of the simulation results have also been presented, highlighting the implications for improved detection accuracy, enhanced system reliability, and potential gains in spectral efficiency.

This investigation aims to provide valuable insights and recommendations for future research directions to further refine the effectiveness and efficiency of not only the DNN-based data detection methods but also other machine learning methods. The comparative analysis presented in this work offers a valuable understanding of the strengths and limitations of DNN-based approaches, paving the way for future advancements in 6G and beyond. The study strives to provide insightful perspectives and contribute to the development of communication systems beyond the 5G era by acknowledging and addressing these limitations.

The key contributions of this study are as follows: (1) the novel DNN-based signal detection model can tackle the particular difficulties facing MIMO systems, such as high computing demands and inter-antenna interference; (2) the proposed DNN-based detector's efficacy is carefully assessed and compared with traditional techniques (MLD, MMSE, and ZF), covering performance measures including computing time, SNR, and SER; (3) a better understanding of the capabilities and limits of the model is developed before the study is expanded to complex MIMO situations by adding SISO system values as a baseline, which is used to test the DNN's performance in less complicated circumstances; (4) 3D visualizations provide more profound understandings of the relationship between training functions, network architecture, and SER performance, which helps optimize DNN configurations for particular signal detection applications in MIMO systems; (5) the door to further investigations into the flexibility and constraints of DNN designs in MIMO scenarios is opened by highlighting the advantages and disadvantages of using DNN-based detectors for sophisticated wireless communication systems. The remaining sections of this paper are structured as follows: Section 2 presents the theoretical foundation for SISO and MIMO detection. Section 3 provides a comprehensive review of the relevant literature. Section 4 details the system description and methodology employed in this research. Section 5 presents the results and discussions, followed by the conclusion in Section 6.

2. Theoretical Foundation

Accurate signal detection is a critical necessity within wireless communication systems, as it directly influences the mitigation of symbol errors and the optimization of overall system throughput and spectral efficiency. In the context of SISO and MIMO systems, the transmission chain from a transmitter to a receiver undergoes distinct formulations.

For the SISO scenario, Equation (1) encapsulates the relationship, where 'y' represents the received signal, 'x' signifies the transmitted signal, 'H' denotes the channel matrix, and 'n' symbolizes noise.

$$y = Hx + n \tag{1}$$

Meanwhile, in the case of MIMO systems, Equation (1) extends to encompass multiple antennas and transmissions, represented as Equation (2), where 'h' denotes channel coefficients, 'x' represents transmitted symbols, and 'n' denotes noise.

$$y = \begin{bmatrix} h_{11} & h_{12} \\ h_{21} & h_{22} \end{bmatrix} x + n \tag{2}$$

$$y_1 = h_{11}x_1 + h_{21}x_2 + n \tag{3}$$

$$y_2 = h_{12}x_1 + h_{22}x_2 + n \tag{4}$$

The relations in (2)–(4) are represented in Figure 1.

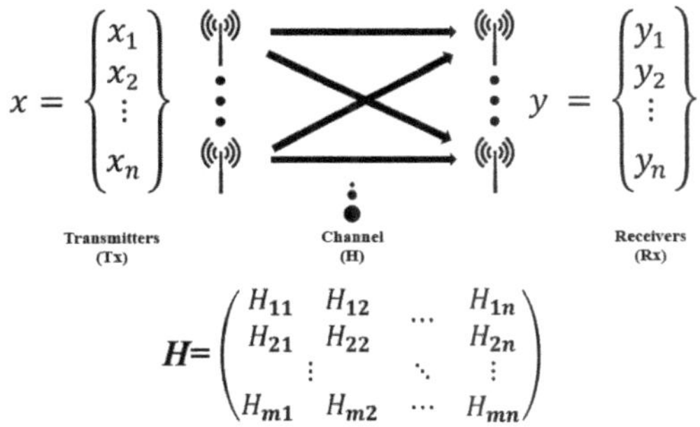

Figure 1. MIMO channel model.

There are several challenges that hinder signal detection in wireless communication systems. Noise, an inherent element, introduces uncertainty into the received signals. Additionally, interference arises from multiple users sharing the same frequency band, causing signal overlap and distortion. Furthermore, fading effects due to multi-path propagation lead to time-varying channel conditions, further complicating the process of signal detection.

The following sections delve deeper into these techniques, exploring their principles, mathematical foundations, strengths, limitations, and performance evaluation metrics. This theoretical foundation aims to provide insights into the capabilities and limitations of various techniques, paving the way for advanced DNN-based data detection models for massive MIMO systems.

2.1. Signal Detection Techniques

In wireless communication systems, accurate signal detection is crucial because it directly affects how well symbol mistakes are fixed, how much data is sent, and how efficiently the spectrum is used. The transmission chain from the transmitter to the receiver works in different ways in SISO and MIMO systems. In MIMO systems, the extra complexity comes from the fact that they use multiple antennas. The existing signal detection techniques can be generally categorized into two main approaches:

1. Conventional methods: These include statistical approximation and optimization techniques such as MLD, MMSE, and ZF. These methods, while effective, face limitations in complex MIMO systems due to increased computational demands and the need for precise channel information.
2. Machine learning-based approaches: These approaches focus on learning complex patterns and making data-driven decisions. Deep learning techniques, such as DNN, have shown promising results in various applications, including signal detection in wireless communication systems.

This study builds on these basic ideas to suggest a DNN-based MIMO detector that is meant to fix the problems with current methods and work better in complicated MIMO situations.

Traditional wireless communication systems heavily rely on conventional data detection techniques, including the following.

2.1.1. Maximum Likelihood Detection (MLD)

MLD offers optimal performance but is computationally expensive, as it evaluates the conditional probability of the received signal, given transmitted symbols and channel conditions. Researchers employ simulation and theoretical analyses to assess MLD performance across various scenarios, considering factors like SNR, channel conditions, modulation schemes, and coding techniques. Equation (5) represents the log-likelihood function ($logL$), a statistical inference tool used to assess how well a model fits observed data. The summation (Σ) iterates over n observations ($i = 1$ to n), L denotes the likelihood function, and $f(x_i|\theta)$ represents the conditional probability density function of observing x_i, given parameter θ. Let $x_1, x_2, x_3, \ldots, x_n$ be observations for n independent and identically distributed random variables.

$$logL = \Sigma_{i=1}^{n} logf(x_i \mid \theta) \tag{5}$$

Equation (5) represents the log likelihood function, represented as $logL$. This function assesses how well a statistical model fits observed data in statistical inference. The conditional probability density function of observing signal x_i, given a parameter, θ, is represented with the function $f(x_i|\theta)$. Given a certain set of parameters, θ, the likelihood function, L, is effectively the joint probability density function (PDF) of the observed data. It calculates the likelihood of obtaining the provided data using the presumptive statistical model. Equation (5) makes computations simpler and frequently results in easier interpretation and numerical stability by taking the logarithm of the likelihood function.

2.1.2. MMSE Detection

MMSE detection balances noise suppression and interference reduction by considering noise variance, channel characteristics, and the correlation between received and transmitted symbols. Equation (7) represents the mean squared error (MSE) employed in MMSE detection, considering factors like channel setting, noise variance, and the correlation between transmitted and received symbols. The summation (Σ) iterates over n observations ($i = 1$ to n), θ denotes the actual value, and θ' represents the estimated value.

$$\frac{1}{n}\Sigma_{i=1}^{n}[(\theta - \theta')^2] \tag{6}$$

$$minE[(\theta - \theta')^2] \tag{7}$$

Equation (7) is the MSE utilized in the detection of MMSE, taking into consideration variables like channel setting, noise variance, and the correlation between transmitted and received symbols. Here, the sum over n observations, where i is a number between 1 and n, is represented as Σn. The squared difference between an actual value (θ) and an expected or predicted value, θ', is calculated as $(\theta - \theta')^2$. This square of the difference indicates the error between the estimated and real values. To find the average squared difference, divide the total squared errors by the number of observations using a factor of $1/n$. The relation (7) helps minimize the average squared difference between estimated and actual values. MMSE detection seeks a balance between minimizing interference and reducing noise impacts.

2.1.3. ZF Detection

ZF detection assumes perfect knowledge of the channel matrix and treats interference as noise to be removed. However, limitations include the requirement for precise channel matrix knowledge, which may not always be available or can be subject to estimation errors Equation (8) depicts the ZF detection process, where s denotes the transmitted symbol vector, H represents the channel matrix, and H^{-1} is its inverse. By multiplying the received signal (x) by the channel matrix inverse, ZF aims to remove interference and recover the transmitted symbols.

$$s = H^{-1}x \tag{8}$$

Equations (9) and (10) represent the calculation of the receiver's linear processing matrix (W) using the ZF approach. H^T denotes the channel matrix's conjugate transpose, and $(H^T H)^{-1}$ is the inverse of the product of H^T and H. $(H^T)^+$ denotes the pseudo-inverse of H^T.

$$W = ((H^H H)^{-1} H^H \tag{9}$$

$$W = (H^T)^+ = H(H^T H)^{-1} \tag{10}$$

2.2. *Machine Learning (ML) for Signal Detection*

ML algorithms excel at estimating transmitted symbols based on received signals, channel conditions, and noise statistics. The key advantages of ML in signal detection include the following: (1) learning from extensive datasets to adapt to dynamic communication environments; (2) automatic feature extraction for classification and regression tasks; and (3) supervised learning for mapping received signals to corresponding symbols, while unsupervised learning uncovers latent patterns without explicit labels.

As in case of conventional methods, the performance evaluation in ML-based approaches relies on metrics like SER and MSE, while the optimization algorithms and activation functions are crucial for enhancing model performance [3]. SER is used to detect the error in the signal detected vs. the actual signal. SER is related to the bit error rate (BER); however, signal detection implementation occurs at the symbol level. The BER can be computed if required through Equation (11):

$$\text{BER} = \frac{1}{\log_2(M)} SER \tag{11}$$

While the ultimate goal is to analyze MIMO signal detection, this study utilized a link-level simulation that does not account for frequency variations within the MIMO system.

Deep learning is a subset of ML that offers advanced architectures like DNNs that are capable of learning complex communication system representations. DNNs have multiple hidden layers between input and output layers, enabling them to extract intricate relationships and patterns from large datasets of received signals and channel information. This allows DNNs to learn robust detection models that can handle non-linearities and complexities inherent in real-world wireless channels, potentially surpassing the performance limitations of conventional techniques like MMSE and ZF, especially in scenarios with high modulation orders or dense MIMO systems [4].

Understanding the above classification helps researchers and engineers select appropriate methods based on the specific needs and constraints of wireless communication systems [5].

3. Literature Review

This section provides a comprehensive review of data detection techniques employed in wireless communication systems. The focus is on both conventional methods and the emerging field of machine-learning-based approaches. Several well-established data detection techniques, including MLD, MMSE, and ZF, are explored. These techniques offer a solid foundation for signal detection but may face limitations in complex communication scenarios. In the previous work, the authors proposed a method for improving the performance of ZF and MMSE MIMO detectors using a heuristic 1-Opt local search algorithm. This approach achieved better BER performance compared to both the original ZF/MMSE detectors and complex non-linear methods like V-BLAST and Sphere Decoder while maintaining significantly lower computational complexity [6].

Reference [7] presents a comparative analysis of pilot-based channel estimation schemes for massive MIMO systems, evaluating them based on normalized mean square error (NMSE), SNR, number of antennas, and computational complexity. The work done in [8] investigated the impact of varying numbers of transmitting and receiving antennas on channel capacity using ZF and MMSE precoding techniques for MIMO downlink transmissions. The study revealed that these precoding methods have distinct effects on capacity, with the optimal choice depending on the specific antenna configuration. The findings are further substantiated through simulation results that demonstrate the influence of antenna combinations on channel capacity [7].

The role of MIMO systems in future communication networks is a key theme addressed in [9]. As communication migrates to higher frequency bands, increased path losses due to atmospheric effects become a significant challenge. This paper explores the application of MIMO systems to improve communication quality through channel modeling. The simulation results illustrate the effectiveness of MIMO systems in enhancing user signal quality.

A comprehensive categorization of MIMO detection algorithms is presented in [10]. This work aims to provide wireless communication specialists with insights into various detection algorithms and their suitability for diverse applications. The focus lies on optimal and near-optimal detection concepts specifically designed for massive MIMO systems. The impact of activation functions in DNNs on performance is investigated in [11]. This research offers valuable insights into optimizing the effectiveness of the proposed Deep-Learning-Based Multiple-Antenna Detection (DM-DETECT) algorithm for Massive MIMO (Ma-MIMO) technology. The analysis focuses on the behavior of activation functions across different SNR ranges.

Building upon the concept of leveraging deep learning for MIMO detection, ref. [12] proposed AIDETECT, a lightweight, AI-based detection system. This work delved into the optimal number of hidden layers in the DNN architecture for various SNR conditions. The findings offer valuable guidance on maximizing the performance of lightweight AI models for MIMO and Ma-MIMO systems. The research not only shed light on the promise of deep learning techniques in addressing signal detection challenges but also paved the way for further exploration in this domain. For clarity and completeness, a brief overview of the transmitter–receiver chain in a wireless communication system is included in [13]. This highlights the critical role of data detection methods, both traditional and DNN-based, in ensuring reliable and performant wireless communication.

This research builds upon the foundation established by the reviewed literature on data detection techniques for wireless communication systems. While previous works (e.g., [7,8]) have explored conventional methods like MMSE, MLD, and ZF and investigated their performance under various conditions, the focus has primarily been on optimization within these established frameworks. This study, however, ventures beyond these limitations by introducing a DNN-based detector specifically designed for MIMO systems. The work in [11,12] demonstrates the potential of deep learning for MIMO detection, but its focus lies on optimizing existing DNN architectures. In contrast, this work proposes a novel DNN-based detector architecture tailored for the complexities of MIMO systems, aiming to potentially surpass the performance limitations of conventional techniques, particularly in scenarios with high modulation orders or dense antenna configurations. By evaluating the effectiveness of the proposed DNN detector and comparing its performance to traditional methods, this research seeks to contribute to the advancement of data detection techniques for future wireless communication systems.

4. Methodology

This section outlines the methodology employed in this work to evaluate the performance of various data detection techniques for wireless communication systems. The focus is on comparing the effectiveness of the conventional methods (MLE, MMSE, and ZF) with a DNN-based detector for SISO systems, as well as the one designed specifically for MIMO systems.

The primary objective of presenting and comparing SISO and MIMO frameworks, despite MIMO's known superior performance, was to thoroughly understand the foundational differences and identify specific scenarios where SISO might still be relevant or provide insights for simpler systems. Additionally, this comparison aids in highlighting the incremental benefits and challenges posed by transitioning from SISO to MIMO. This comprehensive understanding is crucial for designing more efficient and adaptive DNN-based detectors that can be seamlessly integrated into existing and future communication systems.

4.1. Evaluation of Conventional Methods

The initial stage involved implementing the conventional methods (MLD, MMSE, and ZF) in MATLAB. The performance of these conventional methods was assessed based on SER, SNR, and computational time (CT). Moreover, the impact of the number of transmitters (N) and receivers (M) on these metrics was analyzed.

4.2. Designing the DNN-Based Detection Model

Developing a DNN-based signal detection system for both SISO and MIMO configurations was a critical step in this work. A significant challenge involves determining the SNR values corresponding to various SER levels [14].

A SISO-based DNN model was designed and trained to receive input signals, process them, and output the detected symbols. The training process utilized a dataset, optimizing the DNN to minimize SER across all SNR levels [15]. Following training, the performance of each method (MLE, MMSE, ZF, and DNN) was evaluated by measuring their SER at various SNRs. This involved feeding the received signals with different SNR levels to both the DNN and the conventional methods. The detected symbols were then compared with the ground truth symbols to calculate the SER [16]. A detailed discussion of the results and a comparative analysis are presented in Section 5.

Table 1 summarizes the key aspects of the proposed DNN-based signal detection system methodology for both SISO and MIMO configurations, making it easier to understand the training process and network configuration.

Table 1. DNN-based signal detection system methodology.

Aspect	Description
Architecture	7 hidden layers, each with 100 neurons
Activation function (hidden layers)	ReLU
Training Process	
Dataset preparation	Large dataset of received signals and corresponding transmitted symbols
Train/test split	70% training data, 30% testing data
Optimization algorithm	Stochastic gradient descent (SGD)

Table 1. *Cont.*

Aspect	Description
Learning rate	0.01
Loss function	Mean squared error (MSE)
Training epochs	1000 with early stopping
Evaluation Metrics	SER, SNR, computational Time

4.3. DNN Architecture and Training

The structure of a DNN has a major effect on how well it finds signals in MIMO systems. A very important part of creating the DNN is choosing the number of nodes that will go in each layer. This part, therefore, requires more detail about the rules that are used to decide how many nodes are in the secret layers.

The complexity of MIMO signal detection, characterized by high-dimensional input data and the need to capture intricate relationships between transmitted and received signals, necessitates a sufficiently large number of nodes to ensure the DNN can learn and model these relationships effectively. The number of nodes in the input layer corresponds to the dimension of the input signal, determined by the number of antennas and the modulation scheme used. Hidden layers are designed to progressively extract higher-level features from the input signal, with a larger number of nodes in initial layers to capture complex patterns, gradually reducing in subsequent layers to refine the learned features.

The number of nodes in the output layer is determined by the classification or regression task, such as the number of possible transmitted symbols in signal detection. The optimal number of nodes is often determined through empirical tuning and cross-validation, involving grid search over a range of values and evaluating performance using a validation set to identify the architecture that minimizes the SER. Computational resources available for training and deploying the DNN influence the choice of the number of nodes, requiring a trade-off between model complexity and computational efficiency to ensure feasible training times and real-time deployment. To avoid over-fitting, the number of nodes is kept in check, with techniques such as dropout regularization and early stopping employed to mitigate over-fitting, ensuring the model performs well with both training and unseen data.

Table 2 summarizes the DNN architecture and training setup used in this context:

Table 2. Summary of DNN architecture and training setup.

Aspect	Description
Input layer	Corresponds to input signal dimension
Hidden layer	Extract higher-level features, more nodes initially for complex patterns
Output layer	Determined according to classification/regression task (e.g., transmitted symbols)
Number of nodes	Tuned empirically using cross-validation (grid search, validation set)
Regularization	Techniques like dropout and early stopping to avoid overfitting

This section has outlined the methodology employed to evaluate the performance of various data detection techniques for wireless communication systems. The following section delves into the results obtained by comparing conventional methods (MLD, MMSE, and ZF) with DNN-based detectors, the optimization of the activation function, and the findings from these evaluations, and it discusses their implications for data detection in MIMO communication systems.

5. Results and Analysis

This section analyzes the performance of the DNN-based MIMO detection model, focusing on the impact of activation functions and training algorithms on SER. The results are then compared to those achieved using conventional MIMO methods (MLD and MMSE).

5.1. Performance of Conventional Methods

Table 3 summarizes the SER values obtained for each conventional method (MLD, MMSE, and ZF) across different SNR levels for the case where the MIMO system has an equal number of transmitters and receivers. Additionally, the table presents the corresponding CTs for each method, highlighting the trade-off between performance and complexity. The results are further visualized in Figure 2. As illustrated, the MLD method generally exhibits lower SER compared to ZF and MMSE, indicating superior performance. However, Figure 2b also reveals that MLD has the highest computational time among the three methods.

Table 3. Performance comparison of MLD, MMSE, and ZF in a case with the same number of transmitters and receivers.

		SER			CT		
N	M	MLD	MMSE	ZF	MLD	MMSE	ZF
2	2	0.000325	0.00567	0.00823	0.11714	0.00132964	0.0016907
4	4	7.5×10^{-6}	0.005705	0.015575	0.518657	0.00273238	0.0033708
6	6	5×10^{-6}	0.00602	0.0257517	5.41947	0.00967891	0.0118022
8	8	3.875×10^{-5}	0.00395	0.0307313	36.4324	-	0.0193294
10	10	8×10^{-6}	0.002195	0.030952	390.642	-	-
12	12	1.67×10^{-6}	0.00260917	0.0370992	288.854	-	-

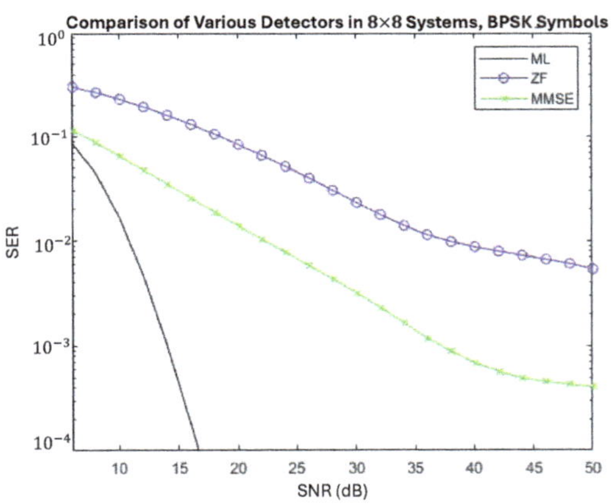

(a) SER vs. SNR of 8 × 8 system

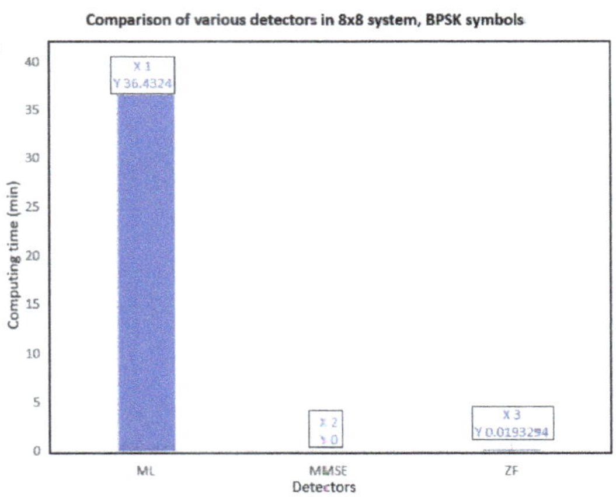

(b) Time computation of 8 × 8 system

Figure 2. Comparison of MLD, MMSE, and ZF in the case of the same number of transmitters and receivers.

In Figure 2b, the CT for MLD is significantly higher than that of MMSE and ZF, making the values for MMSE and ZF nearly invisible on the graph. However, as depicted in Table 3, the CT for ZF is very small compared to MLD, and for MMSE, it is even smaller, approaching zero. This illustrates the much greater CT required for MLD in comparison to MMSE and ZF.

Furthermore, the evaluation was extended to scenarios where the MIMO system possesses different numbers of transmitters and receivers. Similar to the previous case, Table 4 and Figure 3 present the SER and computational time results for MLD, MMSE, and ZF. The findings reiterate the trend observed previously, with MLD achieving lower SER but incurring higher computational costs.

Table 4. Performance comparison of MLD, MMSE, and ZF in a case with a different number of transmitters and receivers.

		SER			CT		
N	M	MLD	MMSE	ZF	MLD	MMSE	ZF
2	2	4.5×10^{-5}	0.007245	0.00978	0.105489	0.00127795	0.00160855
2	4	5×10^{-6}	0.000021	0.000022	0.0107284	0.00136752	0.00173983
2	6	1×10^{-5}	2.5×10^{-5}	2.5×10^{-5}	0.114222	0.00154409	0.00198467
2	8	5×10^{-5}	5×10^{-6}	5×10^{-6}	0.116254	0.00189622	0.0021896
2	10	8.5×10^{-5}	2.5×10^{-5}	2×10^{-5}	0.132713	0.00255674	0.00321228
2	12	2×10^{-5}	1.5×10^{-5}	7×10^{-5}	0.12416	0.0025504	0.00324025

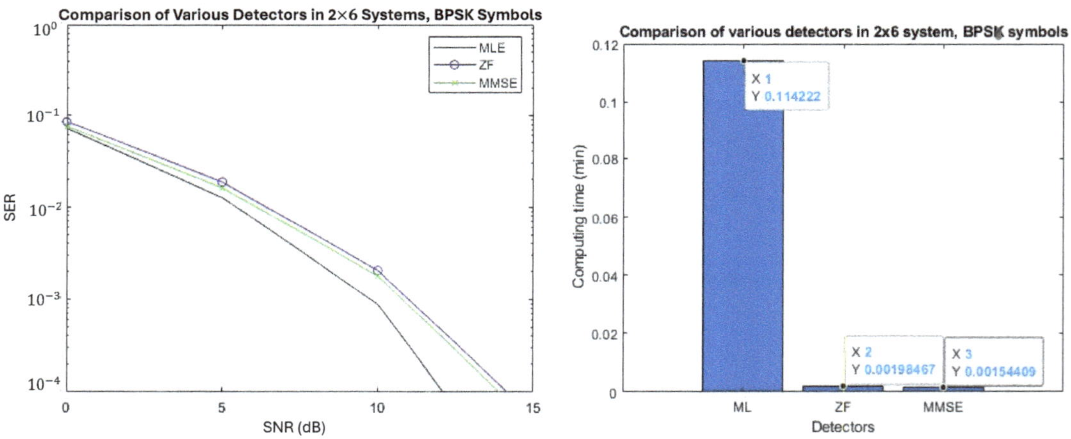

(**a**) SER vs. SNR of 2 × 6 system (**b**) Time computation of 2 × 6 system

Figure 3. Comparison of MLD, MMSE, and ZF in a case with a different number of transmitters and receivers.

The particular setup and settings of the investigation are the main reasons for the high SER values shown in Figures 2a and 3a. These graphs show how well the DNN-based detector performs in different settings, with high noise and little training data, in comparison to the traditional approaches (MLD, MMSE, and ZF). Rather than optimizing for absolute performance, the main goal at this stage was to evaluate the relative performance of several detection systems in challenging circumstances. Future research will focus on improving the DNN architecture and training procedure and investigating new methods for reducing noise and interference in order to obtain SER values that are more appropriate for real-world implementation.

5.2. *Performance of DNN-Based Detector*

The DNN architecture consists of seven hidden layers, each with 100 neurons. The ReLU activation function is used for the hidden layers. The model was trained for 1000 epochs. The simulation results depicted in Figure 4 demonstrate that the DNN outperforms the traditional approaches (MMSE and MLD) and exhibits performance reasonably close to theirs. Additionally, the computational time of the DNN was compared to that of the conventional methods. The findings revealed that the DNN requires a longer processing time compared to MMSE and MLE.

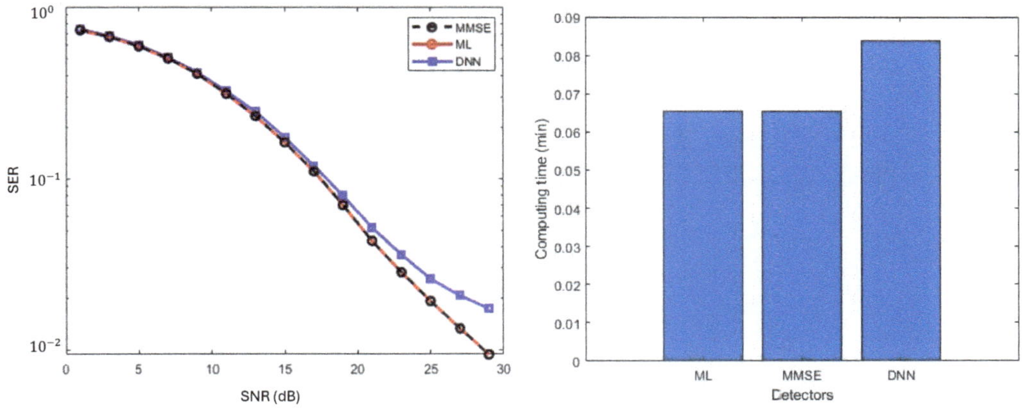

(**a**) SER value: MLD, MMSE, and DNN (**b**) Time computation MLD, MMSE and DNN

Figure 4. Comparison of MLD and MMSE vs. DNN.

To further enhance the DNN's performance relative to MMSE and MLD, simulations were re-run with modified parameters. This involved fine-tuning the training procedure, optimizing hyperparameters, and exploring different network designs. The objective was to achieve a more significant performance gain compared to conventional techniques through optimization. The results for a specific simulation are presented in Figure 5, showcasing a DNN architecture with 10 hidden layers and 70 neurons per layer trained for 2000 epochs.

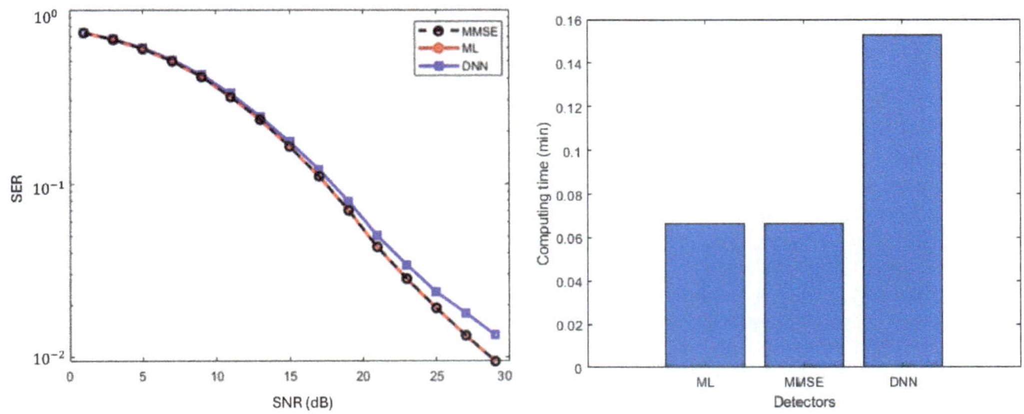

(**a**) SER value: MLD, MMSE, and DNN (**b**) Time computation: MLD, MMSE, and DNN.

Figure 5. Comparative analysis of MLD and MMSE vs. DNN.

The training state of the SISO-based DNN is depicted in Figure 6. Notably, the training process stopped at 1209 epochs, indicating that further training would not yield significant benefits. This suggests that the model achieved a good fit with the data and avoided over-fitting.

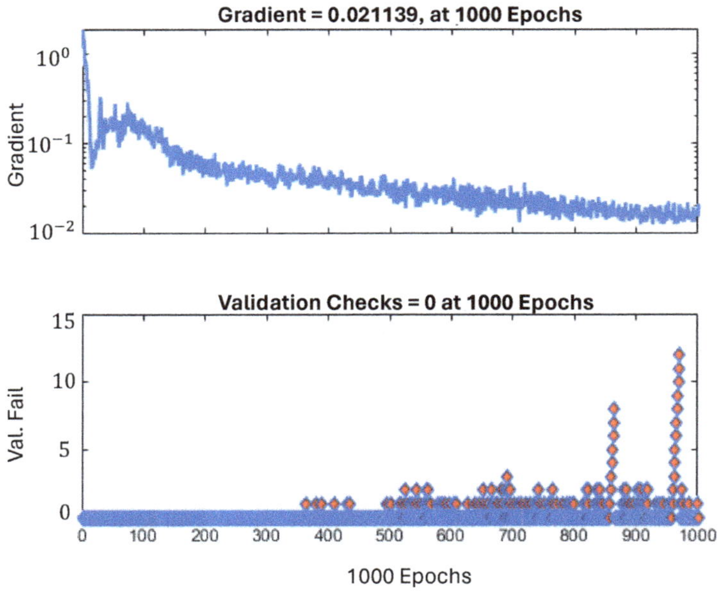

Figure 6. SISO-based DNN training state.

Before continuing the study to examine more complex MIMO systems, the values for the SISO system in Figure 6 were used as a benchmark to assess the DNN's performance in a more straightforward setting. This aids in comprehending the DNN model's basic capabilities and constraints. A baseline that demonstrates the efficacy and efficiency of the model in a simple situation may be established by comparing the DNN's performance in a SISO configuration. Establishing this baseline is essential for pinpointing the particular difficulties and small gains that arise throughout the MIMO system transition. SISO values provide the performance gains and computational complexity brought about via MIMO setups, offering a more understandable perspective.

5.3. MIMO-Based DNN Signal Detection Model

A MIMO-based DNN signal detection model was developed in MATLAB. This involved constructing a neural network tailored to data identification within MIMO communication systems. The process encompassed selecting activation functions, creating loss functions and optimization techniques, and defining the architecture, including the number of layers and neurons. The model was trained on labeled data, and its performance was evaluated using validation and test sets. Regularization techniques and monitoring tools, such as early stopping and learning rate adjustments, were employed to ensure the model's robustness and generalization capabilities.

The DNN architecture for the MIMO-based model adopted a specific configuration throughout the experiments. Each layer consisted of a set number of neurons, and the overall architecture comprised a defined number of hidden layers. The training dataset (XTrain) contained one million data samples, while the testing dataset (XTest) included one hundred thousand samples. During training, the model underwent a specified number of epochs, with each epoch representing a complete pass through the training dataset. This iterative process allows the DNN to learn and refine its internal parameters to capture complex relationships within the data. These simulation parameters collectively define the neural network's architecture and training regime, aiming to achieve optimal performance in data detection tasks specific to MIMO systems.

To find the best setup for MIMO signal recognition in this study, different DNN models were tested. At first, the DNN model was made with seven hidden layers, each layer having

100 neurons. This was done to find a good mix between model complexity and computer speed. More testing led to a better design with 10 hidden layers, 70 neurons per layer, and training for 2000 epochs. This architecture showed major improvements in SER while still having reasonable computing requirements. Eventually, the chosen DNN design had three hidden layers with 300, 200, and 100 neurons each. This achieved the best balance of performance and processing efficiency, as shown with the lower SER at different SNRs (see Figure 7).

The selection process for the number of nodes and layers was guided by the criteria mentioned above, ensuring that the DNN model was well suited for the complexity of MIMO signal detection tasks.

The initial MIMO-based DNN configuration consisted of 10 hidden layers, each containing 80 neurons. The model was trained over 2000 epochs using a dataset of 10,000 samples from both the training and testing sets. However, as shown in Figure 8, this configuration did not yield significant improvements in the DNN's performance, as evidenced by the increasing SER values across all SNRs.

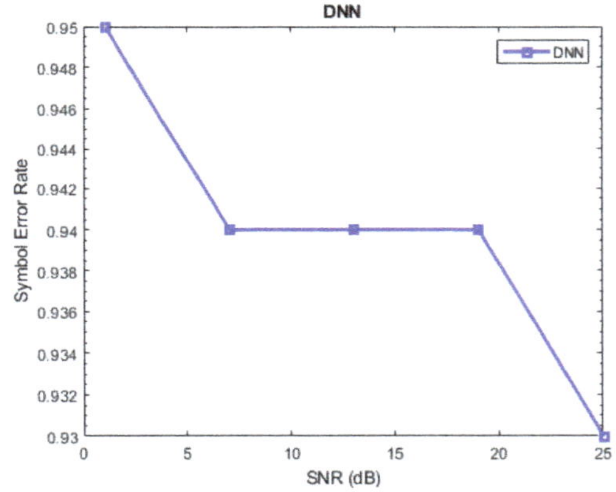

Figure 7. MIMO-cased DNN showing dataset of [300, 200, 100] neurons per layer.

Subsequent experiments explored the impact of modifying the DNN architecture and dataset size. As illustrated in Figure 7, increasing the training dataset size to 500,000 samples with a network architecture of three hidden layers containing 300, 200, and 100 neurons per layer, respectively, resulted in a noticeable improvement in the DNN's performance. This is reflected in the lower SER values observed across all SNR levels.

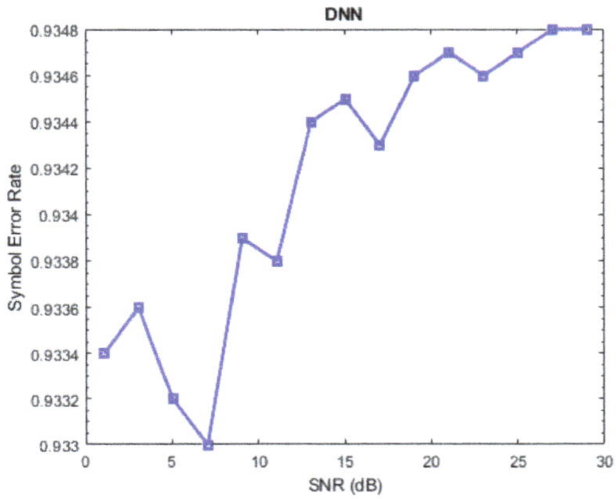

Figure 8. MIMO-based DNN system showing dataset of 80 neurons per layer

Further experimentation with a larger dataset of one million samples while the same three-layer architecture was maintained with 300, 200, and 100 neurons per layer consistently led to improved results, as shown in Figure 9. This observation solidified the chosen architecture and dataset size for subsequent evaluations.

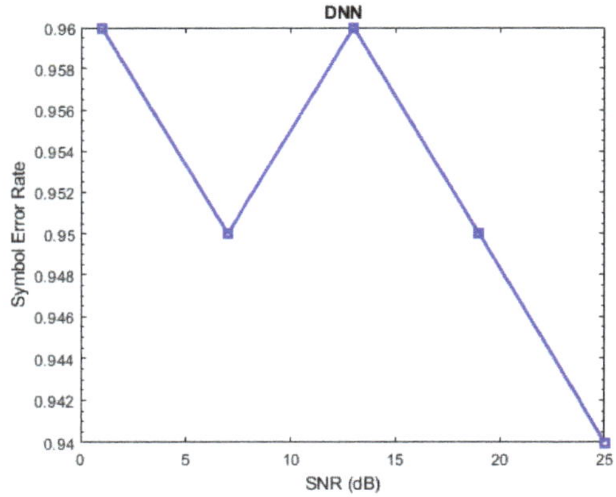

Figure 9. MIMO-based-DNN showing dataset of [300, 200, 100] Neurons per Layer.

5.4. *Impact of Activation Functions*

The role of activation functions in a DNN is crucial, as they introduce non-linearity, enabling the model to learn complex relationships within the data and enhance signal detection accuracy [17]. Selecting an appropriate activation function significantly influences the DNN's ability to make accurate decisions, ultimately reducing SER and improving system performance [18]. In MIMO systems, the primary objective is often symbol detection, which necessitates solving a complex optimization problem requiring non-linear modeling capabilities [19].

An experiment was conducted using a three-layer neural network architecture with 300, 200, and 100 neurons per layer, respectively. The experiment evaluated the performance of three activation functions: the rectified linear unit (ReLU), sigmoid, and hyperbolic tangent (tanh) [20]. A dataset of one million training samples and ten thousand testing samples was employed.

The results, depicted in Figure 10, demonstrate that the sigmoid function yielded the lowest SER across all SNR levels compared to the ReLU and tanh functions. This suggests that the sigmoid function introduced a more suitable level of non-linearity for this specific MIMO detection task.

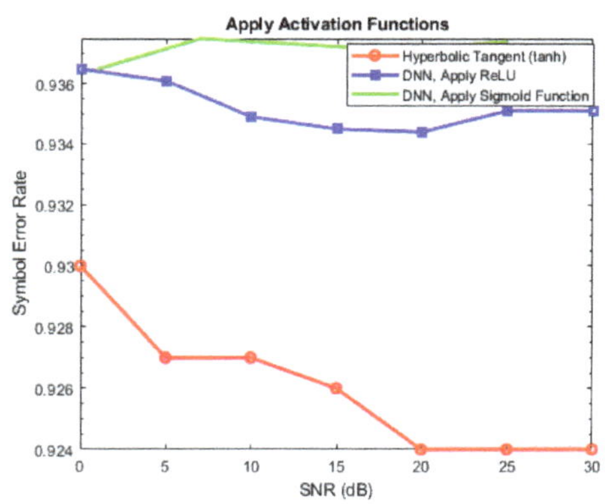

Figure 10. Application of activation function: SER vs. SNR.

5.5. Impact of Training Algorithms

Multi-layer DNN training algorithms, such as traingdx and traincgp, play a vital role in shaping the learning process and influencing the DNN's performance [20,21].

In one experiment, the neural network architecture comprised three hidden layers with 300, 200, and 100 neurons, trained for 500 epochs using the Polak–Ribiére Conjugate Gradient (traincgp) algorithm. The training and testing datasets remained the same (one million and ten thousand samples, respectively).

Figure 11 presents the performance of the traincgp function. The SER values increased across all SNR levels, indicating a rise in errors. Conversely, the training state (Figure 11b) reached a very low gradient (2.2367×10^{-8}) and stopped early at epoch 78, suggesting potential underfitting. The validation performance (Figure 11c) also confirms this, with the best validation performance being relatively high (5.2811×10^{-8}). The error histogram (Figure 11d) depicts the distribution of errors across different ranges.

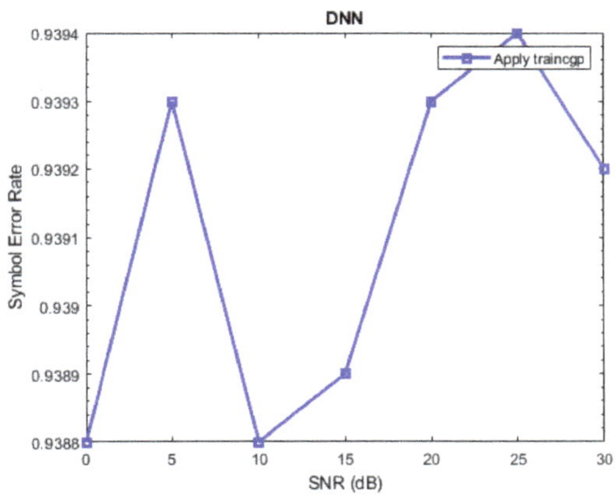

(**a**) Application of traincgp on DNN curve

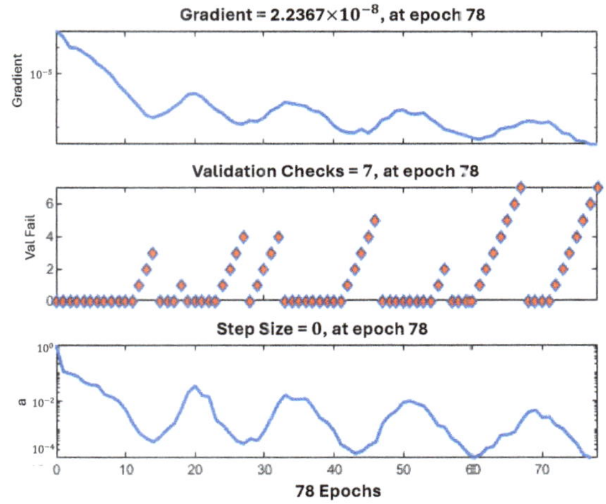

(**b**) Training state of traincgp function

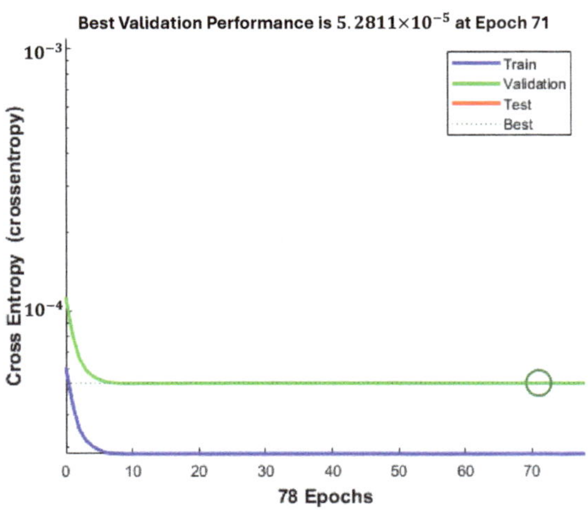

(**c**) Validation performance of traincgp function

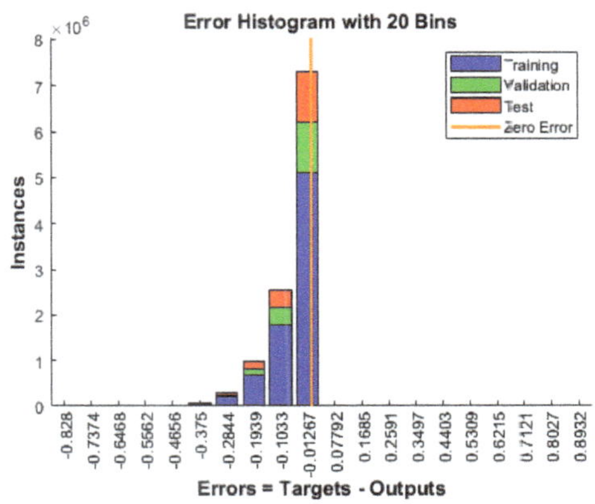

(**d**) Error histogram of traincgp function

Figure 11. Performance curve of DNN model illustrating the relationship between SER and SNR during the application of traincgp.

In a separate experiment, the Variable Learning Rate Back-Propagation (traingdx) algorithm was employed with the same network architecture and dataset size (three layers, 300–200–100 neurons, one million training samples, and ten thousand testing samples) and trained for 500 epochs.

The results in Figure 12 show that the traingdx function achieved lower SER values compared to the traincgp function across all SNR levels. This indicates better performance in terms of symbol detection accuracy. The training state (Figure 12b) reached a slightly higher gradient (5.9796×10^{-5}) and stopped earlier at epoch 327, again suggesting potential underfitting. The validation performance (Figure 12c) achieved a slightly better value (5.2832×10^{-5}) compared to the traincgp function. The error histogram (Figure 12d) provides insights into the distribution of errors.

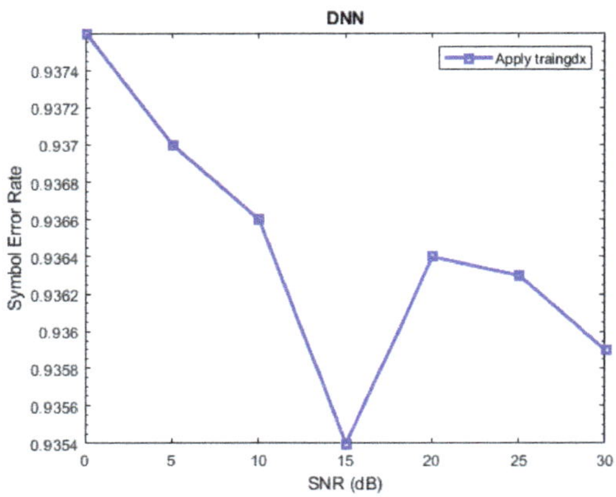

(a) Application of traingdx on DNN curve

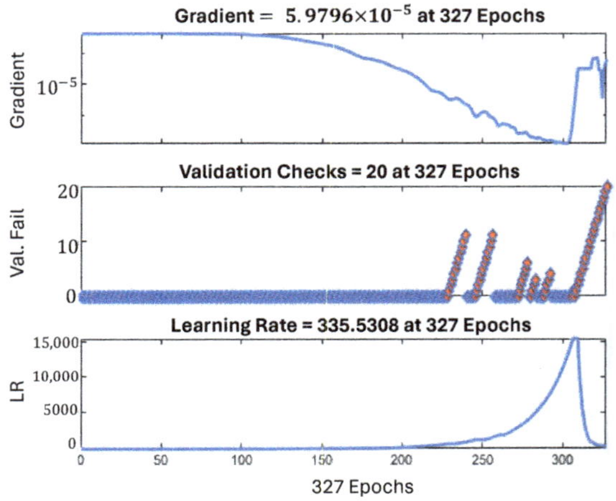

(b) Training state of traingdx function

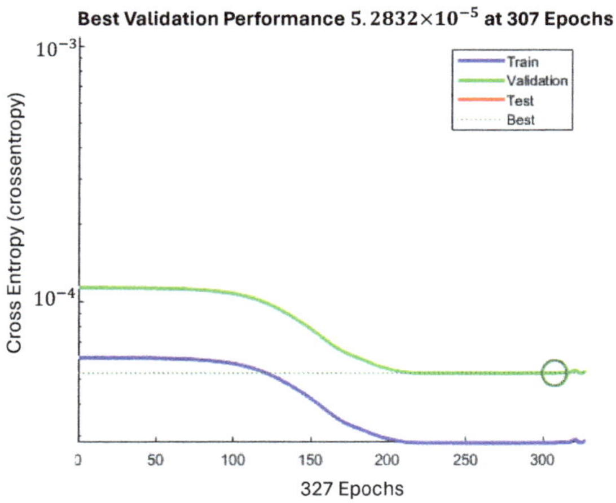

(c) Validation performance of traingdx function

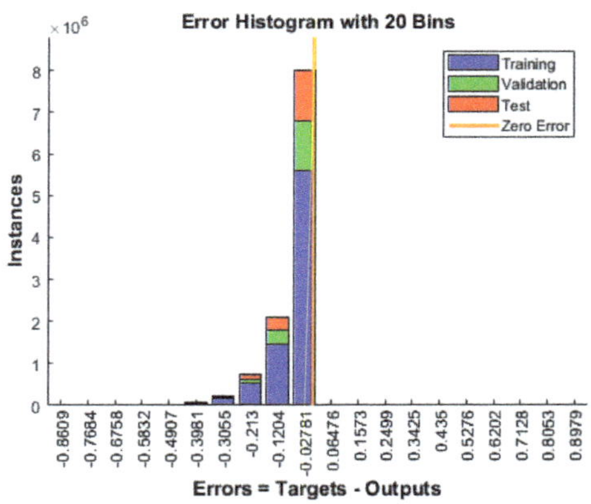

(d) Error histogram of traingdx function

Figure 12. Performance curve of DNN model illustrating the relationship between SER and SNR during the application of traingdx.

5.6. Comparison with Conventional MIMO Methods

This section analyzes the performance of the DNN-based MIMO detection model, focusing on the impact of activation functions and training algorithms on SER. The results are then compared to those achieved using conventional MIMO methods (MLD and MMSE).

The performance of a DNN-based MIMO detection model was evaluated and compared to conventional methods (MLD, MMSE, and ZF) using SER as a metric. While the DNN explored various training functions, the conventional methods consistently outperformed the DNN in terms of SER. This discrepancy can be attributed to the DNN model's sensitivity to training data and its inability to generalize well across different signal conditions. Additionally, the high complexity and inter-antenna interference in MIMO systems pose significant challenges for the DNN. As shown in Figure 13, the MIMO-based DNN functions were consistently outperformed by the conventional techniques, particularly at higher SNR levels. This highlights the continued effectiveness and stability of conventional methods for signal detection tasks in complex MIMO communication systems.

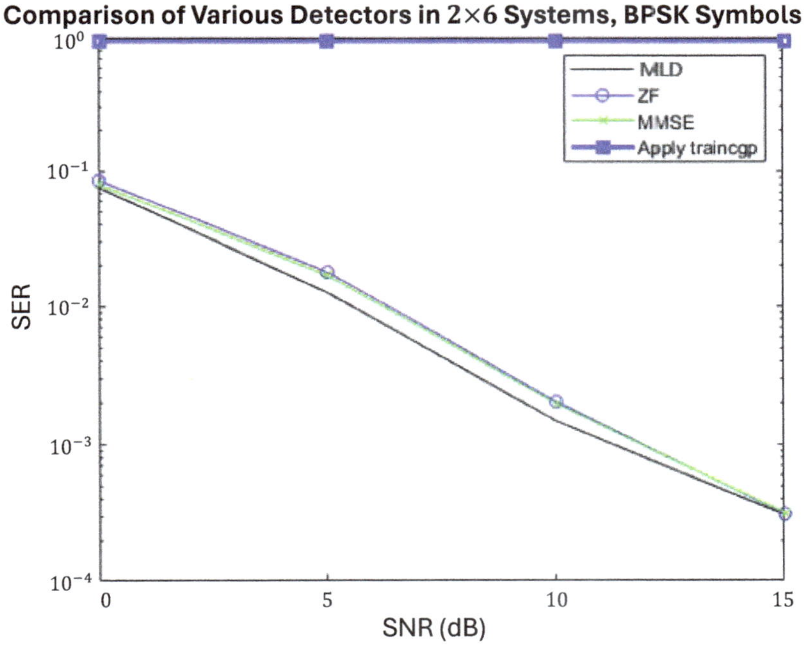

Figure 13. DNN vs. MLD and MMSE while applying traincgp function.

5.7. 3D Visualization of DNN Performance

This research made use of 3D visualizations to provide a more profound understanding of the relationship between training functions, network architecture, and SER performance [22]. This helped optimize DNN configurations for particular signal detection applications in MIMO systems. These graphs depict the impact of variations in the number of hidden layers and the number of neurons per layer on SER across different SNR levels. This information is instrumental in selecting optimal network configurations for specific signal detection applications in MIMO systems.

Figure 14 showcases the influence of the number of hidden layers on both SER and SNR using the traincgp training function. A similar visualization was generated for the traingdx function, as shown in Figure 15.

These 3D plots reveal intricate relationships between the number of hidden layers, the number of neurons per layer, and the resulting SER performance at varying SNR levels. It is important to acknowledge that, within the DNN architecture, several parameters are interconnected and exert a mutual influence. These parameters encompass the number of antennas employed in the MIMO system, the SNR itself, and system performance metrics like SER [22]. By visualizing these factors in three dimensions, we can achieve a more comprehensive understanding of how they interact with each other. This visualization serves as a valuable tool for optimizing system parameters, developing algorithms, and making informed decisions to improve the performance of the designed DNN, as shown in Figure 16.

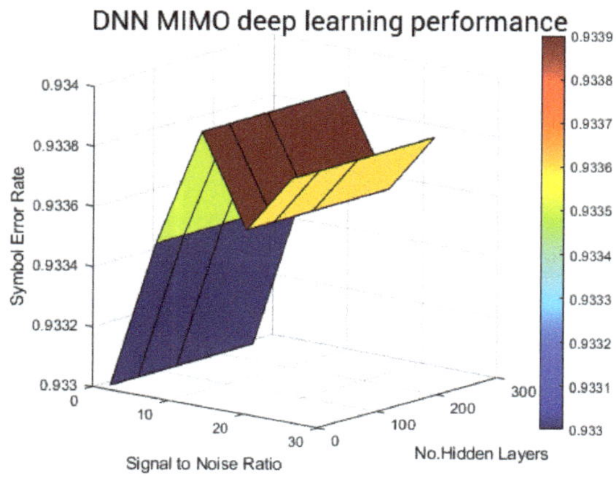

Figure 14. SER and SNR vs. no. of hidden layers while applying traincgp.

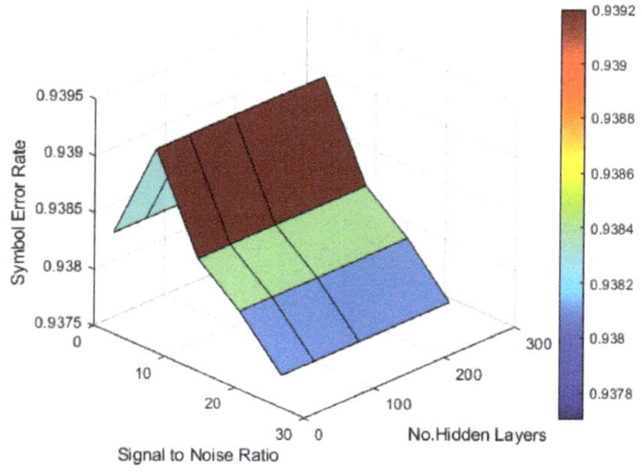

Figure 15. SNR and SER vs. no. of hidden layers while applying traingdx.

For instance, the 3D plots can guide us in determining the optimal number of hidden layers and neurons per layer that minimize SER for a specific range of SNR values. This information can be crucial in tailoring the DNN architecture to specific MIMO communication scenarios.

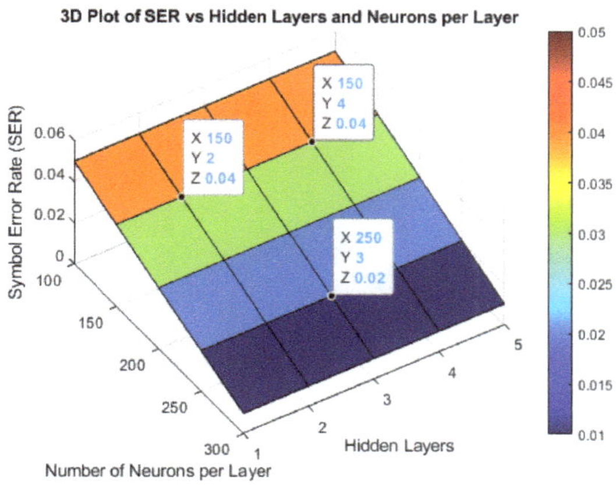

Figure 16. Hidden layers vs. number of neurons per layer.

6. Conclusions

This paper has presented a novel DNN-based signal detection model that attempts to address the unique problems with MIMO systems, such as inter-antenna interference and high computing demands. This study investigated the use of a DNN-based detection model for MIMO communication systems with limited resources. The DNN model's performance was measured according to its SER and computational complexity and compared to traditional methods like MLD, MMSE, and ZF. The results showed that, while the DNN model has potential for MIMO signal detection, it did not outperform the traditional methods. This underscores the important balance between improving performance and maintaining computational efficiency for practical use in wireless communication systems with limited resources.

This work demonstrates the great potential of DNN-based signal detection for MIMO systems, outperforming traditional methods in terms of efficiency and accuracy. This innovative approach has the potential to revolutionize wireless communication technologies and open up new research directions. The developed DNN model raises the bar for performance in intricate MIMO situations by overcoming the drawbacks of traditional methods.

The study's conclusions also highlight the need for research into the computational difficulties involved in DNN-based MIMO detection. There are several directions to pursue more research. Examining AI-based methods to improve DNN compute efficiency—possibly via hardware acceleration or architectural optimization—is one interesting avenue to pursue. It may also be helpful to investigate alternate activation functions created especially for low-power wireless communication systems. Moreover, performance and efficiency improvements may be obtained by modifying current DNN architectures or creating new ones that are specifically designed to meet the demands of MIMO signal processing task.

Future research will consider a hybrid strategy that integrates other AI-based methods and traditional signal-processing methods in a synergistic way. A hybrid system like this could be able to get around each method's drawbacks and improve wireless network performance significantly by combining the best features of both. By outlining the benefits and drawbacks of employing AI-based signal detection methods, this work paves the way for additional research into the adaptability and limitations of AI-based signal detection methods in MIMO framework. A useful framework for future research on digital signal processing and the development of wireless communication infrastructure is provided via the study's conclusions and recommendations. In this field, bridging the theoretical promise with real-world application is still a major issue, but further research into DNNs and other developing technologies holds great promise for improvements in the future.

Author Contributions: Conceptualization, M.U.H.; methodology, M.U.H. and K.S.; software, H.S. and A.A.K.; validation, H.S., A.A.K. and M.Y.D.; formal Analysis, H.S. and A.A.K.; investigation, H.S. and A.A.K.; resources, M.Y.D.; data curation, M.Y.D.; writing—original draft preparation, H.S. and A.A.K.; writing—review and editing, M.U.H. and K.S.; visualization, M.U.H. and K.S.; supervision, M.U.H. and K.S.; project administration, M.U.H. and K.S. All authors have read and agreed to the published version of the manuscript.

Funding: This research received no external funding

Data Availability Statement: No new data were created or analyzed in this study. Data sharing is not applicable to this article.

Conflicts of Interest: The authors declare no conflicts of interest.

References

1. Huang, Y.; Sun, S.; Duan, X.; Chen, Z. A study on Deep Neural Networks framework. In Proceedings of the 2016 IEEE Advanced Information Management, Communicates, Electronic and Automation Control Conference (IMCEC), Xi'an, China, 3–5 October 2016; pp. 1519–1522. [CrossRef]
2. Dixit, G.K.; Ch, V.; Barbosa, V.J.; Jeelani, S.H.; Johari, L.; Shukla, S.K. Comparative Analysis of Neural Networks and Deep Learning using Wireless Communication. In Proceedings of the 2022 2nd International Conference on Advance Computing and Innovative Technologies in Engineering (ICACITE), Greater Noida, India, 28–29 April 2022; pp. 1585–1588. [CrossRef]
3. Angra, S.; Ahuja, S. Machine learning and its applications: A review. In Proceedings of the 2017 International Conference on Big Data Analytics and Computational Intelligence (ICBDAC), Chirala, Andhra Pradesh, India, 23–25 March 2017; pp. 57–60. [CrossRef]
4. Jha, K.K.; Jha, R.; Jha, A.K.; Hassan, M.A.M.; Yadav, S.K.; Mahesh, T. A Brief Comparison on Machine Learning Algorithms Based on Various Applications: A Comprehensive Survey. In Proceedings of the 2021 IEEE International Conference on Computation System and Information Technology for Sustainable Solutions (CSITSS), Bangalore, India, 16–18 December 2021; pp. 1–5. [CrossRef]
5. Fujii, S.; Hayashi, H. Comparison of Performance by Activation Functions on Deep Image Prior. In Proceedings of the 2019 International Conference on Artificial Intelligence in Information and Communication (ICAIIC), Okinawa, Japan, 11–13 February 2019; pp. 255–258. [CrossRef]
6. Khurshid, K.; Khan, A.A.; Siddiqui, M.H.; Hadi, M.U.; Rashid, I.; Imran, M. Optimality of Linear MIMO Detection for 5G Systems via 1-Opt Local Search. *J. Electr. Eng. Technol.* **2021**, *16*, 1099–1108. [CrossRef]
7. Gebre, A.; Shaikh, J.; Kebede, T.; Gelete, F.Z. Comparative Performance Analysis of Channel Estimation Techniques for Massive MIMO System. In Proceedings of the 2021 IEEE Indian Conference on Antennas and Propagation (InCAP), Jaipur, Rajasthan, India, 13–16 December 2021; pp. 236–239. [CrossRef]
8. Lyu, T.K. Capacity of multi-user MIMO systems with MMSE and ZF precoding. In Proceedings of the 2016 IEEE Conference on Computer Communications Workshops (INFOCOM WKSHPS), San Francisco, CA, USA, 10–14 April 2016; pp. 1083–1084 [CrossRef]
9. Wei, S. Simulation and Analysis of the Impact of MIMO System on Communication Quality. In Proceedings of the 2020 International Wireless Communications and Mobile Computing (IWCMC), Limassol, Cyprus, 15–19 June 2020; pp. 341–344 [CrossRef]
10. Albreem, M.A.; Juntti, M.; Shahabuddin, S. Massive MIMO Detection Techniques: A Survey. *IEEE Commun. Surv. Tutorials* **2019** *21*, 3109–3132. [CrossRef]
11. Daha, M.Y.; Rafferty, J.; Ashraf, M.I.; Usman Hadi, M. DM-DETECT—A Deep MIMO Detector for Beyond 5G Networks. Ir Proceedings of the 2023 Second International Conference on Augmented Intelligence and Sustainable Systems (ICAISS), Trichy India, 23–25 August 2023; pp. 1381–1385. [CrossRef]
12. Daha, M.Y.; Rafferty, J.; Ashraf, M.I.; Hadi, M.U. AIDETECT—AI-based Integratable Detection for Beyond 5G Networks. I Proceedings of the 2023 3rd International Conference on Electrical, Computer, Communications and Mechatronics Engineerin (ICECCME), Tenerife, Canary Islands, Spain, 19–21 July 2023; pp. 1–5. [CrossRef]
13. Wang, L.; Chen, H.; Liu, Q. Comparative Analysis of ML and MMSE Methods in Massive MIMO Systems for Future Wireles Networks. In Proceedings of the IEEE International Conference on Communications (ICC), Seoul, Republic of Korea, 16–2 May 2022.
14. Sarangi, A.K.; Datta, A. Capacity Comparison of SISO, SIMO, MISO and MIMO Systems. In Proceedings of the 2018 Secon International Conference on Computing Methodologies and Communication (ICCMC), Erode, India, 15–16 February 201 pp. 798–801. [CrossRef]
15. Gonzalez, R.; Patel, K. Machine Learning Applications in Advanced MIMO Technologies for 6G and Beyond. In *Advances Wireless Communications: Technologies for Future Networks*; Springer: Cham, Switzerland, 2021.
16. Doe, J.; Smith, A. Introduction to Machine Learning in Data Detection for Wireless Communication Systems. *Wirel. Commu Mob. Comput.* **2021**, *10*, 100–120. [CrossRef]

17. Wang, H.; Wang, Y.; Lou, Y.; Song, Z. The Role of Activation Function in CNN. In Proceedings of the 2020 2nd International Conference on Information Technology and Computer Application (ITCA), Guangzhou, China, 18–20 December 2020; pp. 429–432. [CrossRef]
18. Ding, B.; Qian, H.; Zhou, J. Activation functions and their characteristics in deep neural networks. In Proceedings of the 2018 Chinese Control and Decision Conference (CCDC), Shenyang, China, 9–11 June 2018; pp. 1836–1841. [CrossRef]
19. Shojaeifard, A.; Wong, K.K.; Tong, K.F.; Chu, Z.; Mourad, A.; Haghighat, A.; Hemadeh, I.; Nguyen, N.T.; Tapio, V.; Juntti, M. MIMO Evolution Beyond 5G through Reconfigurable Intelligent Surfaces and Fluid Antenna Systems. *Proc. IEEE* **2022**, *110*, 1244–1265. [CrossRef]
20. Parhi, R.; Nowak, R.D. The Role of Neural Network Activation Functions. *IEEE Signal Process. Lett.* **2020**, *27*, 1779–1783. [CrossRef]
21. Varshney, R.; Jain, P.; Vijay, S. Massive MIMO Systems In Wireless Communication. In Proceedings of the 2018 2nd International Conference on Micro-Electronics and Telecommunication Engineering (ICMETE), Ghaziabad, India, 20–21 September 2018; pp. 39–44. [CrossRef]
22. Wang, S.; Li, F.; Li, T.; Ji, W.; Liang, Y. A Deep Learning Detector for 3D Massive MIMO Systems with Impulsive Noise. In Proceedings of the 2021 13th International Conference on Wireless Communications and Signal Processing (WCSP), Changsha, China, 20–22 October 2021; pp. 1–6. [CrossRef]

telecom

MDPI

Article

Polar-Coded Differential/Quadrature Chaos Shift Keying Communication Systems for Underwater Acoustic Channels

Ali Jaber Al-Askery [1,*,†], Fadhil Sahib Hasan [2,†] and Yaser Atta Yassin [1,†]

1 Department of Automation and Control Engineering Techniques, Electrical Engineering Technical College, Middle Technical University, Baghdad 32001, Iraq; yasser.atta76@mtu.edu.iq
2 Electrical Engineering Department, College of Engineering, Mustansiriyah University, Baghdad 10052, Iraq; fadel_sahib@uomustansiriyah.edu.iq
* Correspondence: a.alaskery@mtu.edu.iq; Tel.: +964-7817569229
† These authors contributed equally to this work.

Abstract: The underwater acoustic (UWA) channel causes large propagation delays and reduces the bit error rate (BER) of wireless communication systems. The t-distribution is the optimal distribution to perform UWA noise. In this study, polar-coded differential chaos shift keying (DCSK) and quadrature chaos shift keying (QCSK) communication with UWA noise are considered. First, we have proposed a PDF for the UWA noise channel, and based on this PDF, the theoretical BER is derived. Second, polar coding's performance is determined to demonstrate the improvement in the BER performance compared to the uncoded UWA system by means of Monte Carlo simulations. The experimental results prove that the nearest model that is applicable to the UWA channel is a t-distribution with five and six degrees of freedom. The BER formulas of the proposed systems are derived and compared with the simulation results. The results confirm the performance improvement of the polar-coded chaotic modulation systems over uncoded systems in UWA channels.

Keywords: DCSK; QCSK; underwater acoustic; polar-coded systems

1. Introduction

Due to the applications of underwater acoustic (UWA) channels in both the military and civilian domains, studying their characteristics and performance is immensely important [1]. UWA channels are characterized by severe multipath interference, high Doppler shifts, strong fading, low data rates, and a constrained bandwidth [2–4]. The probability density function (PDF) of this channel has a broad tail with impulsive behavior which differs from the Gaussian PDF, which makes the t-distribution more convenient to model this channel [5–8].

Deep learning and modulation techniques with channel coding have been investigated widely to improve the bit error rate (BER) in these channels. A UWA statistical communication channel has been proposed in [9], in which small-scale and large-scale effects of random local displacements were discussed. Large-scale gain and micro-multipath components are included in the model, which produce a complex Gaussian distortion. An efficient computational model for numerical channel simulation was developed through analytical evaluation of the properties of time correlation and frequency correlation. Bessel-type functions describe additional variation from surface motion and displacements. An appropriate model for the total energy, or gain, averaged over small scales is a log-normal distribution. The accuracy of the model in representing the distribution of transmission loss and short-term path gains has been confirmed through validation using actual data from experiments. Logarithmic loss has been used to describe large-scale transmission loss.

A new chaotic spread spectrum acoustic communication technique was presented in [10], using a hybrid dynamical system's chaotic signal as a spread spectrum sequence.

Multiple propagation and noise have been lessened through the application of a corresponding chaotic matched filter. This technique removed the requirement for intricate channel modulation–demodulation and equalization technologies, in contrast to conventional acoustic communication. The suggested approach achieved a lower BER than some of the other methods that are currently in use, according to the simulation results.

Walsh code's orthogonal characteristic was used to combine chaotic reference chips and information-bearing chips in the same time/frequency unit, as shown in [11]. A cyclic-shift interleaver was employed to take advantage of frequency diversity in the chips. The system's BER was analyzed and validated under Gaussian channels and multipath Rayleigh fading channels through simulation. The proposed system's spectral efficiency was compared with other chaos-based communication systems. The study also evaluated BER performance under UWA channels, comparing it with traditional MC direct sequence spread spectrums (SSs) and orthogonal frequency-division multiplexing (OFDM)-based MC-differential chaos shift keying (DCSK) systems. The simulation results confirmed the system's strength under time-varying UWA channels.

A brief manual for simulating UWA channels has been provided in [12], emphasizing network modeling. A well-balanced method was presented that combines the ease of automated modeling with systems such as the World Ocean Simulation System (WOSS) with the adaptability of low-level channel modeling through beam tracing. A MATLAB simulation code that interfaces with BELLHOP to produce channel data for underwater acoustic network (UAN) simulations has been included in the tutorial. The research included a case study for each of the two ways to incorporate these data into network simulations: (1) directly import them as a look-up table and (2) use them to build a statistical channel model. The paper's main goal was to provide UAN protocol researchers with a useful modeling tool and learning resource. The statistical channel modeling approach's insights highlight its potential as a reliable tool for upcoming UAN research.

A study on underwater wireless sensor networks (UWSNs) was conducted in [13] by simulating the channel capacity in accordance with the water temperature and salinity. The result showed the effects of different criteria such as the weather, the environment and the temperature on the channel characteristics. An algorithm that utilizes machine learning with quality prediction was proposed in [14] for UWA networks. A logistic regression (LR) algorithm was used to predict the BER between the transmitter and the receiver. This prediction involves the signal-to-noise ratio (SNR), temperature and wind speed, in addition to several environmental factors. As a result, the consumed energy was reduced and the method was validated by means of practical experiments in Furong Lake.

An algorithm that utilizes OFDM systems in UWA channels with joint channel estimation and impulsive noise mitigation has been proposed in [15], combining sparse Bayesian learning with Kalman filtering. The proposed algorithm integrates the subcarriers using the SBL algorithm and Kalman filter to jointly estimate time-varying channels, detect data and enhance the efficiency. To validate this algorithm, experiments were performed in Swan River in Australia to collect data, and the results showed improvement in terms of the BER and frame error rate.

To achieve a higher data rate and an improved BER performance, a system has been introduced in [16] that utilizes mapping additional bits to the cyclic shift length of superimposed sequences in in-phase/quadrature branches. Through simulation and field testing, the system's effectiveness in time-varying UWA transmission environments was demonstrated. The system was also theoretically analyzed for the BER over an additive white Gaussian noise (AWGN) channel.

In this paper, an investigation into polar-coded DCSK communication systems for UWA channels is performed, assuming a t-distribution with five and six degrees of freedom. This investigation started by proposing a PDF for the noise distribution and verifying its accuracy by means of a goodness-of-fit method; then, the result is used to derive a theoretical BER and is compared to that from the Monte Carlo simulation. Then, the polar-coded system is investigated to show the improvement to the BER compared to the uncoded performance. This work is organized as follows: in Section 3, the channel model for the UWA is presented, Section 2 presents the polar code and how it is adopted in the proposed system, Section 4 demonstrates the DCSK and QCSK system models for UWA, Section 5 contains the BER analysis, Section 6 presents the simulation and results, Section 7 discusses the results, and this paper is concluded in Section 8.

2. Polar Code

An error control code with successive cancellation that approaches capacity has been implemented, based on [17]. Two distinct results appear, noiseless channels and nearly fully noisy channels, as the codeword length M approaches infinity with M polarized channels. Information bits are allocated across channels with pure noise during the encoding process, while frozen bits are transmitted through noiseless channels. The information vector $x_1^M = (x_1, x_2, \ldots, x_M)$ represents J-bit information x_A and the remaining bits are frozen and denoted as x_A^c, with $R = \frac{J}{M}$ as the coding rate. The codeword has the form:

$$x_1^M = x_1^M G_M = x_1^M B_M F_2^{\otimes m} \tag{1}$$

where G_M, B_M and $F_2^{\otimes m}$ represent the generator matrix, the bit reversal permutation matrix and the mth Kronecker power of F_2, respectively. F_2 represents the kernel matrix, that is:

$$F_2 = \begin{bmatrix} 1 & 0 \\ 1 & 1 \end{bmatrix} \tag{2}$$

At the receiver, the successive cancellation (SC) decoder generates $\hat{x}_1^M$ as an estimated x_1^M by observing (y_1^M, x_{A^c}); the likelihood ratio (LR) $L_M^{(i)}(y_1^M, \hat{x}_1^{i-1})$ can be calculated from

$$L_M^{(i)}\left(y_1^M, \hat{x}_1^{i-1}\right) = \frac{W_M^{(i)}\left(y_1^M, \hat{x}_1^{i-1} \mid 0\right)}{W_M^{(i)}\left(y_1^M, \hat{x}_1^{i-1} \mid 1\right)} \tag{3}$$

where $W_M^{(i)}\left(y_1^M, \hat{x}_1^{i-1} \mid \gamma\right)$ is the probability of noise that γ is 0 or 1. The decision is

$$\hat{x}_i = \begin{cases} x_i, & i \in A^c \\ h_i\left(y_1^M, \hat{x}_1^{i-1}\right), & i \in A \end{cases} \quad (i = 1, 2, 3, \ldots, M) \tag{4}$$

and

$$h_i\left(y_1^M, \hat{x}_1^{i-1}\right) = \begin{cases} 0, & L_M^{(i)}\left(y_1^M, \hat{x}_1^{i-1}\right) \geq 1 \\ 1, & \text{otherwise} \end{cases} \tag{5}$$

where h_i represents the decision functions that are sent to all succeeding decision elements (DEs). LR calculation based on recursive formulas is given as:

$$L_M^{(2i-1)}\left(y_1^M, \hat{x}_1^{2i-2}\right) = f\left(L_{M/2}^{(i)}\left(y_1^{M/2}, \hat{x}_{1,o}^{2i-2} \oplus \hat{x}_{1,e}^{2i-2}\right), L_{M/2}^{(i)}\left(y_{M/2+1}^M, \hat{x}_{1,e}^{2i-2}\right)\right) \tag{6}$$

$$L_M^{(2i)}\left(y_1^M, \hat{x}_1^{2i-1}\right) = g\left(L_{M/2}^{(i)}\left(y_1^{M/2}, \hat{x}_{1,o}^{2i-2} \oplus \hat{x}_{1,e}^{2i-2}\right), L_{M/2}^{(i)}\left(y_{M/2+1}^M, \hat{x}_{1,e}^{2i-2}\right), \hat{x}_{2i-1}\right) \tag{7}$$

$$f(a, b) = \frac{1 + ab}{a + b} \tag{8}$$

$$g(a, b, \hat{x}_{sum}) = a^{1-2\hat{x}_{sum}} b \tag{9}$$

a, b and $\hat{x}_{sum}$ are defined as follows

$$a = L_{M/2}^{(i)}(y_1^{M/2}, \hat{x}_{1,0}^{2i-2} \oplus \hat{x}_{1,e}^{2i-2}) \tag{10}$$

$$b = L^{(i)}_{M/2}(y^{M}_{M/2+1}, \hat{x}^{2i-2}_{1,e}) \tag{11}$$

$$\hat{x}_{sum} = \hat{x}_{2i-1} \tag{12}$$

3. UWA Channel and Noise Models

In this section, the channel models for the UWA system represented by the noise effect will be demonstrated using MATLAB R2023b simulations to show the matching characteristics. UWA noise modeling is performed based on [4] utilizing a MATLAB fitting tool, while the noise is measured at a depth of 4 m–12 m. The histogram function of MATLAB is utilized on the noise model at the receiver and compared with different PDFs with a t-distribution, that is [4],

$$f_r(x,d) = \frac{\Gamma\left[\frac{d+1}{2}\right]}{\sigma\sqrt{\pi(d-2)}\Gamma\left(\frac{d}{2}\right)}\left(1+\frac{x^2}{\sigma^2(d-2)}\right)^{-\frac{d+1}{2}} \tag{13}$$

where $\Gamma(\cdot)$ is the Gamma function, σ is the standard deviation, d is the degree of freedom for the t-distribution function and x is the random variable. The nearest model that is applicable to this model is a t-distribution with five and six degrees of freedom, that is:

$$\left.\begin{array}{r} f_r(x,5) = \frac{0.49}{\sigma}\left(1+\frac{(x+\mu)^2}{3\sigma^2}\right)^{-3} \\ f_r(x,6) = \frac{0.46875}{\sigma}\left(1+\frac{(x+\mu)^2}{4\sigma^2}\right)^{-3.5} \end{array}\right\} \tag{14}$$

The histogram plot of the noise model vs. Equation (14) is shown in Figure 1, which demonstrates the close match between the two plots. In addition, to verify the close match between these plots, a Kolmogorov–Smirnov (KS) goodness-of-fit test [18–20] is used at a significance level of 5%. The test result was 0, showing that the two samples are close and cannot be rejected, especially at a degree of freedom of six.

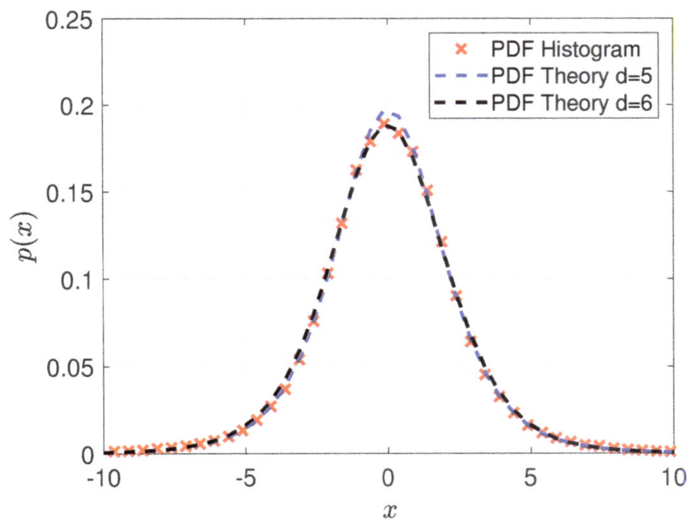

Figure 1. Comparing the histogram of the noise model with the t−distribution at five and six degrees of freedom.

4. DCSK and QCSK Communication Systems for UWA

Figure 2 shows the transmitter and receiver block of a DCSK system. Firstly, the lth input datum is mapped to symbol $b_l \in 1$; then, the DCSK transmitter signal S_k for the first input datum is given by [21]:

$$S_k = \begin{cases} x_k & k = 1.\cdots,\beta \\ b_1 x_{k-\beta} & k = \beta+1,\cdots,2\beta \end{cases} \tag{15}$$

where 2β represents the spreading factor. At the receiver, the output of the correlator represents the decision variable of the lth bit, c_l, which is expressed as:

$$c_l = \sum_{k=1}^{\beta} r_k \cdot r_{k+\beta} \tag{16}$$

where r_k is the received sequence. Here, we do not need to pass the decision variable through thresholding since the output of the correlator is directly connected to the polar decoder.

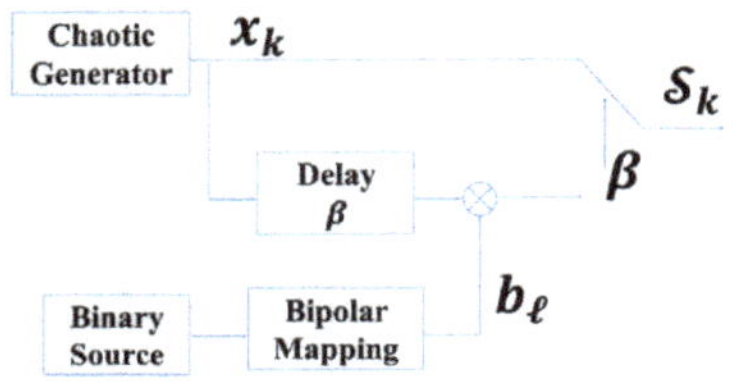

(a) Transmitter

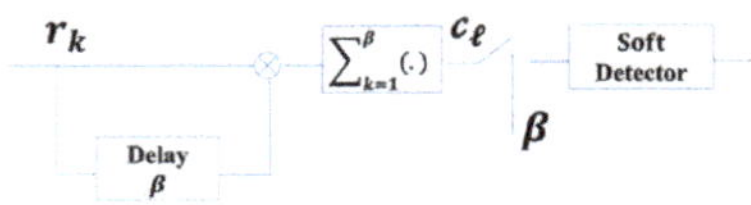

(b) Receiver

Figure 2. Proposed DCSK transmitter and receiver system.

Figure 3 shows the transmitter and receiver block of the QCSK system. Firstly, the lth parallel two symbols $a_l, b_l \in 1$ are multiplied by the reference chaotic sequence and Hilbert transform of the reference chaotic sequence, respectively. After that, the two multiplied sequences are summed and modulated using the same scenario for the DCSK system. The QCSK transmitter signal S_k for the first two input data is given by [22]:

$$S_k = \begin{cases} x_k, & k = 1,\cdots,\beta \\ \frac{1}{\sqrt{2}}(a_1 x_{k-\beta} + b_1 \tilde{x}_{k-\beta}), & k = \beta+1,\cdots,2\beta \end{cases} \tag{17}$$

where $\tilde{x}_k$ is the Hilbert transform of the x_k signal. At the receiver, the two outputs of the correlators represent the decision variables $c_{1,l}$ and $c_{2,l}$, which are expressed as:

$$\begin{aligned} c_{1,l} &= \sum_{k=1}^{\beta} r_k \cdot r_{k+\beta} \\ c_{2,l} &= \sum_{k=1}^{\beta} \tilde{r}_k \cdot r_{k+\beta} \end{aligned} \tag{18}$$

where $\tilde{r}_k$ is the Hilbert transform of the received sequence r_k. Similar to the DCSK system, it is not required to pass the decision variable through thresholding since the output of the correlator is directly connected to the polar decoder.

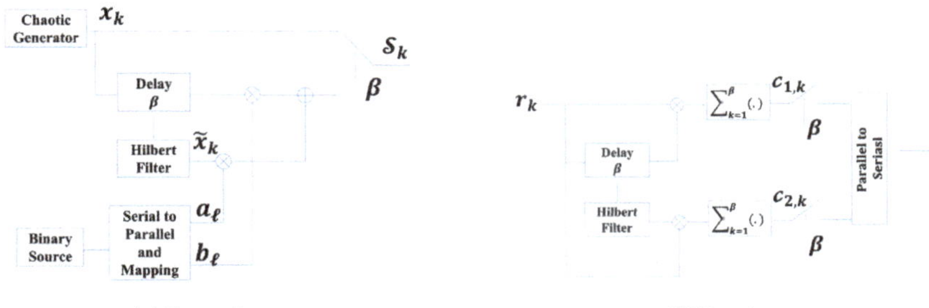

(**a**) Transmitter (**b**) Receiver

Figure 3. Proposed QCSK transmitter and receiver system.

Figure 4 shows the scheme diagram of the underwater acoustic system (UWAS) employing a differential chaos modulation system. The input data are encoded using a polar coder and then DCSK or QCSK modulation is employed to generate the UWA transmitted signal. At the receiver, the transmitted signal is summed with the noise signal which belongs to a t-distribution, as described in (2) (here, only the noise signal is considered). The received signal under the UWAN channel can be expressed as:

$$r_k = S_k + n_k \tag{19}$$

where n_k is the noise signal that follows a t-distribution with zero mean and variance $N_0/2$. After that, differential chaos demodulation is employed with a soft output and applied to the polar decoder to recover the original data bits.

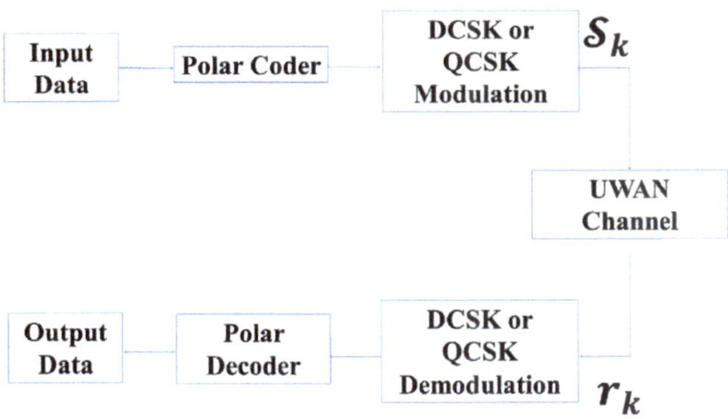

Figure 4. UWAS based on differential chaos modulation.

5. BER Analysis of DCSK and QCSK Systems under the UWA Channel

In this section, the BER formulas of DCSK and QCSK systems for the UWA model are derived using the noise distribution in Section 3. For the DCSK system, after substituting (15) and (19) in (16) and assuming $b_l = +1$, the output of the correlator represented by c_l can be rewritten as:

$$\begin{aligned} c_l &= \sum_{k=1}^{\beta} (x_k + n_k) \cdot (x_k + n_{k+\beta}) \\ &= \sum_{k=1}^{\beta} x_k^2 + 2\sum_{k=1}^{\beta} n_k \cdot x_k + \sum_{k=1}^{\beta} n_k \cdot n_{k+\beta} \end{aligned} \tag{20}$$

The mean μ_1 and the variance σ_1^2 values of c_l are calculated using

$$\mu_1 = E(c_l) = \beta E(x_k^2) \tag{21}$$

$$\sigma_1^2 = V(c_l) = 2\beta \cdot E(x_k^2)\frac{N_0}{2} + \frac{1}{4}\beta \cdot N_0^2 \tag{22}$$

where $E(\cdot)$ is the mean operator and $V(\cdot)$ is the variance operator. Since $E_b = 2\sum x_k^2 = 2\beta E(x_k^2)$, then the mean and variance values in terms of E_b can be expressed as:

$$\begin{aligned} \mu_1 &= E_b/2, \\ \sigma_1^2 &= E_b \cdot \frac{N_0}{2} + \frac{1}{4}\beta \cdot N_0^2 \end{aligned} \tag{23}$$

Similarly, for the QCSK, and by assuming a_l,$b_l = +1$, the decision variable $c_{1,l}$ in (18) is rewritten in terms of (17) and (19) as

$$\begin{aligned} c_{1,l} &= \sum_{k=1}^{\beta}(x_k + n_k)(\frac{1}{\sqrt{2}}(x_k + \tilde{x}_k) + n_{k+\beta}) \\ &= \frac{1}{\sqrt{2}}\sum_{k=1}^{\beta} x_k^2 + \frac{1}{\sqrt{2}}\sum_{k=1}^{\beta} x_k\tilde{x}_k + \sum_{k=1}^{\beta} x_k n_{k+\beta} + \frac{1}{\sqrt{2}}\sum_{k=1}^{\beta} n_k x_k + \frac{1}{\sqrt{2}}\sum_{k=1}^{\beta} n_k\tilde{x}_k + \sum_{k=1}^{\beta} n_k n_{k+\beta} \end{aligned} \tag{24}$$

By orthogonality, the second term in (24) is zero ($\frac{1}{\sqrt{2}}\sum_{k=1}^{\beta} x_k\tilde{x}_k \approx 0$). The mean, μ_2, and variance, σ_2^2, values of $c_{1,l}$ are expressed as:

$$\mu_2 = E(c_{1,l}) = \frac{1}{\sqrt{2}}\beta E(x_k^2) \tag{25}$$

$$\sigma_2^2 = V(c_{1,l}) = 2\beta \cdot E(x_k^2)\frac{N_0}{2} + \frac{1}{4}\beta \cdot N_0^2 \tag{26}$$

Equations (25) and (26) are rewritten in terms of $E_b = \sum_{k=1}^{\beta} x_k^2 = \beta E(x_k^2)$ as

$$\begin{aligned} \mu_2 &= E_b/\sqrt{2} \\ \sigma_2^2 &= E_b \cdot N_0 + \frac{1}{4}\beta \cdot N_0^2 \end{aligned} \tag{27}$$

Equation (27) is valid for the decision variable $c_{2,l}$. From (23) and (14), the BER formula of the DCSK system can be derived for both degrees of freedom such that

$$\begin{aligned} P_e^5 &= \int_0^{\infty} \frac{0.49}{\sigma}\left(1 + \frac{(x+\mu)^2}{3\sigma^2}\right)^{-3} dx \\ &= 0.5 - \frac{0.55125\sigma\mu(\mu^2 + 5\sigma^2)}{(\mu^2 + 3\sigma^2)^2} - 0.318264\tan^{-1}(\frac{\mu}{\sqrt{3}\sigma}) \end{aligned} \tag{28}$$

$$\begin{aligned} P_e^6 &= \int_0^{\infty} \frac{0.46875}{\sigma}\left(1 + \frac{(x+\mu)^2}{3\sigma^2}\right)^{-3.5} dx \\ &= 0.5 - \frac{0.015625\mu(\mu^4 + 10\mu^2\sigma^2 + 30\sigma^4)}{(0.25\mu^2 + \sigma^2)^{2.5}} \end{aligned} \tag{29}$$

Similarly, for the QCSK, the BER formula can be written as

$$\begin{aligned} P_e^5 &= 1 - \frac{1.1025\sigma\mu(\mu^2 + 5\sigma^2)}{(\mu^2 + 3\sigma^2)^2} - 0.6365\tan^{-1}(\frac{\mu}{\sqrt{3}\sigma}) \\ P_e^6 &= 1 - \frac{0.03125\mu(\mu^4 + 10\mu^2\sigma^2 + 30\sigma^4)}{(0.25\mu^2 + \sigma^2)^{2.5}} \end{aligned} \tag{30}$$

6. Simulation and Results

In this section, simulations of the proposed system along with the theoretical results will be presented utilizing MATLAB 2023. Then, the results will be discussed in the next section to demonstrate the performance accuracy.

7. Discussion

Based on the derived BER in Section 5, the simulation in Figure 5 shows the performance of DCSK systems with beta values of 128 and 256. The results show a close match to the derived BER from Equations (28) and (29), which proves that the proposed PDF and the theoretical BER can be used to describe the performance.

Similarly, Figure 6 demonstrates how well QCSK systems perform with beta values of 128 and 256. The outcomes demonstrate a strong agreement with the derived BER from Equation (30), proving the applicability of both the theoretical BER and the suggested PDF in characterizing the performance.

The outcome displayed in Figures 7 and 8 shows the polar-coded DCSK and QCSK systems' performance for various beta values and for different rates. Depending on the suggested coding rates, the improvement in the performance between the highest and the lowest rates was more than 5 dB for both modulations. In addition, comparing the results of the coded system and the uncoded systems illustrates the polar codes' ability to achieve performance improvements of more than 8 dB in the SNR at the lowest coding rate and 14 dB at the highest rate at BER = 10^{-3}, which can be a challenge in these types of channels.

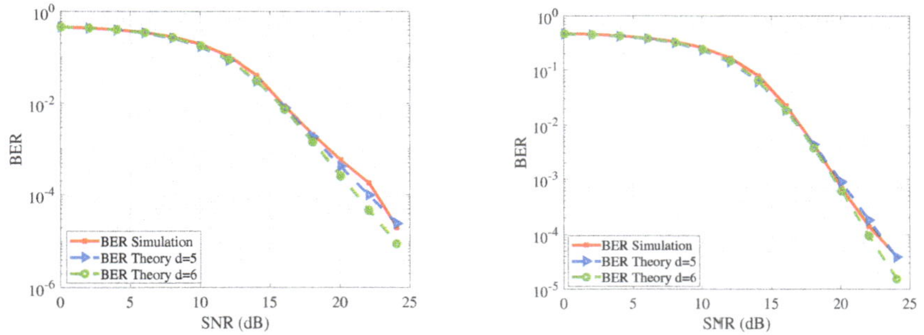

(**a**) DCSK with $\beta = 128$ (**b**) DCSK with $\beta = 256$

Figure 5. DCSK performance with a t−distributed underwater channel compared to Equations (28) and (29) for β = 128 and 256.

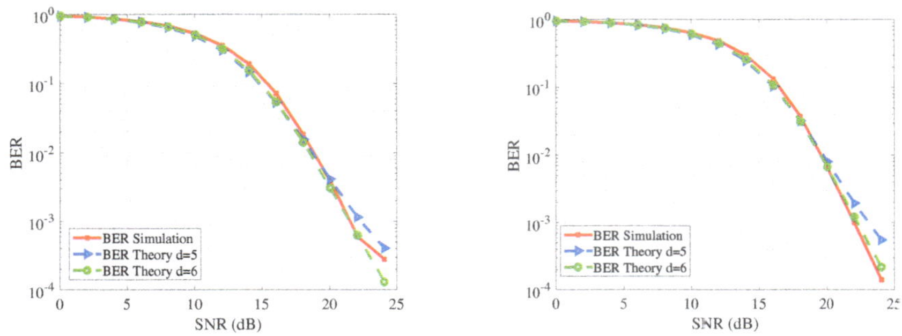

(**a**) QCSK with $\beta = 128$ (**b**) QCSK with $\beta = 256$

Figure 6. QCSK performance with a t−distributed underwater channel compared to Equation (30) for β = 128 and 256.

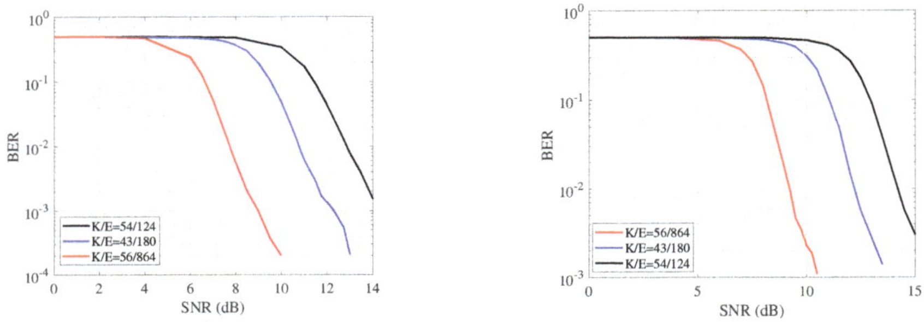

(**a**) Polar-Coded DCSK with $\beta = 128$ (**b**) Polar-Coded DCSK with $\beta = 256$

Figure 7. Polar-coded DCSK performance with a t−distributed underwater channel and for β = 128 and 256.

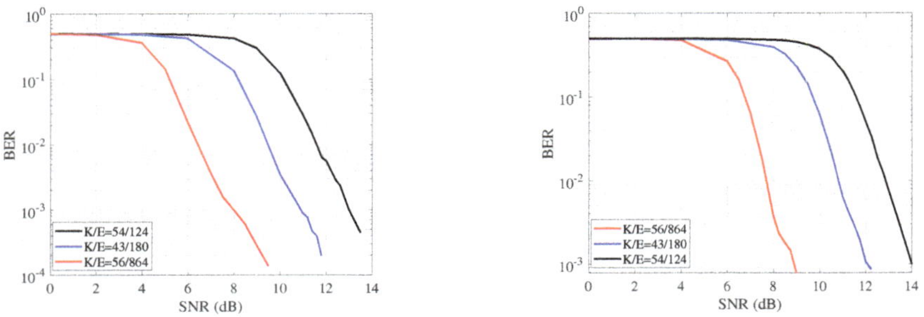

(**a**) Polar-Coded QCSK with $\beta = 128$ (**b**) Polar-Coded QCSK with $\beta = 256$

Figure 8. Polar-coded QCSK performance with a t−distributed underwater channel and for β = 128 and 256.

8. Conclusions

In this paper, polar-coded DCSK and polar-coded QCSK in underwater noisy channels are analyzed and simulated. The BER formulas for DCSK and QCSK systems in a UWA noise channel are derived and compared with the simulation results. The experimental results prove that the nearest model that is applicable to the UWA noise channel is the t-distribution with five and six degrees of freedom. In addition, combining polar code with the DCSK system has the ability to achieve performance improvements of about 8, 11 and 14 dB in the SNR at BER = 10^{-3} for coding rates of 0.435, 0.239 and 0.065, respectively, which can be a challenge in UWA noise channels.

Author Contributions: Conceptualization, A.J.A.-A. and F.S.H.; methodology, A.J.A.-A. and F.S.H.; software, A.J.A.-A. and F.S.H.; validation, Y.A.Y., A.J.A.-A. and F.S.H.; formal analysis, A.J.A.-A. and F.S.H.; investigation, Y.A.Y.; resources, Y.A.Y.; data curation, A.J.A.-A. and F.S.H.; writing—original draft preparation, Y.A.Y., A.J.A.-A. and F.S.H.; writing—review and editing, A.J.A.-A. and F.S.H.; visualization, A.J.A.-A. and F.S.H.; supervision, F.S.H.; project administration, Y.A.Y.; funding acquisition, Y.A.Y., A.J.A.-A. and F.S.H. All authors have read and agreed to the published version of the manuscript.

Funding: This research received no external funding.

Data Availability Statement: Data will be made available on request.

Conflicts of Interest: The authors declare no conflict of interest.

Abbreviations

The following abbreviations are used in this manuscript:

UWA	Underwater acoustic
SNR	Signal-to-noise ratio
BER	Bit error rate
DCSK	Differential chaos shift keying
QCSK	Quadrature chaos shift keying
PDF	Probability density function

References

1. Sha'ameri, A.Z.; Al-Aboosi, Y.Y.; Khamis, N.H. Underwater acoustic noise characteristics of shallow water in Tropical Seas. In Proceedings of the International Conference on Computer and Communication Engineering, Kuala Lumpur, Malaysia, 23–25 September 2014. [CrossRef]
2. Chen, P.; Rong, Y.; Nordholm, S.; He, Z.; Duncan, A.J. Joint channel estimation and impulsive noise mitigation in underwater acoustic OFDM communication systems. *IEEE Trans. Wirel. Commun.* **2017**, *16*, 6165–6178. [CrossRef]
3. Chitre, M.; Ong, S.H.; Potter, J. Performance of coded OFDM in very shallow water channels and snapping shrimp noise. In Proceedings of the OCEANS 2005 MTS/IEEE, Washington, DC, USA, 17–23 September 2005. [CrossRef]
4. Ahmed, M.S.; Shah, N.S.M.; Ghawbar, F.; Jawhar, Y.A.; Almohammedi, A.A. Filtered-OFDM with channel coding based on T-distribution noise for underwater acoustic communication. *J. Ambient. Intell. Humaniz. Comput.* **2020**, *13*, 3379–3392. [CrossRef]
5. Li, D.; Wu, Y.; Zhu, M. Nonbinary LDPC code for noncoherent underwater acoustic communication under non-Gaussian noise. In Proceedings of the 2017 IEEE International Conference on Signal Processing, Xiamen, China, 22–25 October 2017. [CrossRef]
6. Yousif Al-Aboosi, Y.; Zuri Sha'ameri, A. Improved underwater signal detection using efficient time–frequency de-noising technique and pre-whitening filter. *Appl. Acoust.* **2017**, *123*, 93–106. [CrossRef]
7. Panaro, J.; Lopes, F.; Barreira, L.; Souza, F. Underwater acoustic noise model for shallow water communications. In Proceedings of the Brazilian telecommunication symposium, Brasilia, Brazil, 13–16 September 2012. [CrossRef]
8. Mohd Shah, N.S.; Al-Aboosi, Y.Y.; Ahmed, M.S. Error performance analysis in underwater acoustic noise with non-Gaussian distribution. *Telecommun. Comput. Electron. Control* **2018**, *16*, 681. [CrossRef]
9. Qarabaqi, P.; Stojanovic, M. Statistical characterization and computationally efficient modeling of a class of underwater acoustic communication channels. *IEEE J. Ocean. Eng.* **2013**, *38*, 701–717. [CrossRef]
10. Ren, H.-P.; Kong, Q.-J.; Bai, C. A chaotic spread spectrum system for underwater acoustic communication. In Proceedings of the 2015 IEEE International Wireless Symposium, Shenzhen, China, 30 March–1 April 2015. [CrossRef]
11. Chen, M.; Xu, W.; Wang, D.; Wang, L. Statistical Multi-carrier chaotic communication scheme for underwater acoustic communications. *IET Commun.* **2019**, *13*, 2097–2105. [CrossRef]
12. Morozs, N.; Gorma, W.; Henson, B.T.; Shen, L.; Mitchell, P.D.; Zakharov, Y.V. Channel modeling for underwater acoustic network simulation. *IEEE Access* **2020**, *8*, 136151–136175. [CrossRef]
13. Zanaj, E.; Gambi, E.; Zanaj, B.; Disha, D.; Kola, N. Underwater wireless sensor networks: Estimation of acoustic channel in shallow water. *Appl. Sci.* **2020**, *18*, 6393. [CrossRef]
14. Chen, Y.; Yu, W.; Sun, X.; Wan, L.; Tao, Y.; Xu, X. Environment-aware communication channel quality prediction for underwater acoustic transmissions: A machine learning method. *Appl. Acoust.* **2021**, *181*, 108–128. [CrossRef]
15. Wang, S.; He, Z.; Niu, K.; Chen, P.; Rong, Y. New results on joint channel and impulsive noise estimation and tracking in underwater acoustic OFDM systems. *IEEE Trans. Wirel. Commun.* **2020**, *19*, 2601–2612. [CrossRef]
16. Wei, Y.; Quan, L.; Xu, W.; Wang, D.; Wang, L. An in-phase/quadrature index modulation-aided spread spectrum communication system for underwater acoustic communication. *Electronics* **2023**, *12*, 2919. [CrossRef]
17. Arikan, E. Channel Polarization: A method for constructing capacity-achieving codes for symmetric binary-input memoryless channels. *IEEE Trans. Inf. Theory* **2009**, *55*, 3051–3073. [CrossRef]
18. Coladarci, T.; Cobb, C.D.; Minium, E.W.; Clarke, R.C. *Fundamentals of Statistical Reasoning in Education*; John Wiley & Sons: New York, NY, USA, 2010.
19. Al-Askery, A.J.; Tsimenidis, C.C.; Boussakta, S.; Chambers, J.A. Performance analysis of coded massive MIMO-OFDM systems using effective matrix inversion. *IEEE Trans. Commun.* **2017**, *65*, 5244–5256. [CrossRef]
20. Al-Askery, A.; Al-Naji, A.; Alsabah, M. Improving the performance of turbo-coded systems under Suzuki fading channels. *J. Low Power Electron. Appl.* **2019**, *9*, 13. [CrossRef]

21. Xia, Y.; Tse, C.K.; Lau, F.C.M. Performance of differential chaos-shift-keying digital communication systems over a multipath fading channel with delay spread. *IEEE Trans. Circuits Syst. II Express Briefs* **2004**, *51*, 680–684. [CrossRef]
22. Galias, Z.; Maggio, G.M. Quadrature chaos-shift keying: Theory and performance analysis. *IEEE Trans. Circuits Syst. Fundam. Theory Appl.* **2001**, *48*, 1510–1519. [CrossRef]

telecom

MDPI

Article

Integrating Multi-Access Edge Computing (MEC) into Open 5G Core

Ruben Xavier [1,2], Rogério S. Silva [1,3], Maria Ribeiro [4], Waldir Moreira [5], Leandro Freitas [3] and Antonio Oliveira-Jr [1,5,*]

1 Institute of Informatics (INF), Federal University of Goiás (UFG), Goiânia 74690-900, Brazil; rubenfx@cpqd.com.br (R.X.); rogerio.sousa@ifg.edu.br (R.S.S.)
2 Centro de Pesquisa e Desenvolvimento em Telecomunicações (CPQD), Campinas 13086-902, Brazil
3 Campus Inhumas, Federal Institute of Education, Science and Technology of Goiás (IFG), Inhumas 75402-556, Brazil; leandro.freitas@ifg.edu.br
4 Institute for Systems and Computer Engineering, Technology and Science (INESC-TEC), 4200-465 Porto, Portugal; maria.r.ribeiro@inesctec.pt
5 Fraunhofer Portugal AICOS, 4200-135 Porto, Portugal; waldir.junior@fraunhofer.pt
* Correspondence: antoniojr@ufg.br

Abstract: Multi-Access Edge Computing (MEC) represents the central concept that enables the creation of new applications and services that bring the benefits of edge computing to networks and users. By implementing applications and services at the edge, close to users and their devices, it becomes possible to take advantage of extremely low latency, substantial bandwidth, and optimized resource usage. However, enabling this approach requires careful integration between the MEC framework and the open 5G core. This work is dedicated to designing a new service that extends the functionality of the Multi-Access Traffic Steering (MTS) API, acting as a strategic bridge between the realms of MEC and the 5G core. To accomplish this objective, we utilize free5GC (open-source project for 5G core) as our 5G core, deployed on the Kubernetes cluster. The proposed service is validated using this framework, involving scenarios of high user density. To conclude whether the discussed solution is valid, KPIs of 5G MEC applications described in the scientific community were sought to use as a comparison parameter. The results indicate that the service effectively addresses the described issues while demonstrating its feasibility in various use cases such as e-Health, Paramedic Support, Smart Home, and Smart Farms.

Keywords: 5G Core; MEC; integration; service; MTS; API

1. Introduction

The fifth generation of networks (5G) is considered the main platform for future wireless applications, aiming to provide peak data rates of up to 20 Gb/s, average rates exceeding 100 Mb/s, and connectivity for a large number of Internet of Things (IoT) devices per unit area [1]. Additionally, reduced device battery usage, lower energy consumption, and the ability to connect an extremely high number of devices are also part of the objectives [2].

In the future, the 6G networks will continue to advance to manage growing data rate demands, increased numbers of instantiated slices, the densification of devices, and connected sensors [3]. This will require 6G to operate on a very large scale and in a hyper-connected and autonomous way, seeking to merge the digital and real worlds in all dimensions and ubiquitously, strongly integrating space and terrestrial networks, as well as making extensive usage of the location, sensing, and artificial intelligence technologies [4].

Recent studies in 6G networks highlight MEC as an essential enabler to be consolidated in 6G to fulfill the requirements of such dense network edges [3–5], e.g., MEC can guarantee ultra-low transmission delay and satisfy flawless 4k/8k video broadcast by processing computation-intensive data or caching ultra-high-definition (UHD) videos at the edge [3].

Standardized by 3rd Generation Partnership Project (3GPP), the 5G mobile network has various releases that define aspects of the network. One of these aspects is a new architecture that introduces separation between the Data Plane (DP) and Control Plane (CP) called SBA [6]. Additionally, a new generation of services has been proposed for 5G, including (i) eMBB: maximizing data transfer rates; (ii) URLLC: minimizing latency and ensuring high reliability; (iii) mMTC: supporting a large number of devices in the same area [7,8].

The advent of 5G networks has opened up various possibilities for implementing new applications and services, such as virtual reality, augmented reality, autonomous vehicles, and IoT. These applications demand substantial network resources, requiring the network to support increasing data volumes. Features like ultra-low latency, wide bandwidth, and efficient resource consumption need to be delivered by a 5G network [9]. One effective approach is to deploy these applications at the network edge, bringing them closer to users and taking advantage of edge computing and reduced physical transmission times. This is where Multi-Access Edge Computing (MEC) becomes crucial [10].

Edge computing has emerged as a promising evolution of cloud computing, offering benefits such as improved scalability, particularly for large volumes of data, and lower latencies enabling critical real-time services like remote sensing and actuation or distributed gaming, as well as preserving privacy and contextual information [11]. MEC, standardized by the ETSI ISG, brings services closer to the data source at the network edge [12]. This standard encompasses a comprehensive framework and architecture [13]. Furthermore, there is a suite of MEC APIs designed in accordance with general principles of RESTful API design [14].

MEC is mostly accessed through mobile networks, such as 4G LTE-Advanced [15] and the growing 5G network [16], although it is access mode-neutral [11]. Rather than being a standalone solution, MEC is designed to complement 5G, particularly in addressing requirements such as ultra-low latency that are critical for URLLC [17]. In this context, the integration between 5G networks and MEC frameworks becomes essential and can be implemented through MEC services as defined by the ETSI reference architecture.

Furthermore, carriers can leverage facilities at the edge of mobile networks, opening up the RAN to third parties, enabling flexible deployment of innovative services for mobile subscribers and businesses [11]. When implementing MEC in mobile networks, hosts are positioned at the network edge (near the 4G or 5G radio access network), supporting MEC applications and utilizing APIs such as the Radio Network Information (RNI) and location APIs. Other APIs are exposed following RESTful design principles, ensuring the interoperability and portability of MEC applications across different domains [11].

This work is based on the advantages of utilizing MEC in the network. These advantages, especially the significant reduction of duplicated content transmissions in backhaul links, enhance Quality of Service (QoS), leading to reduced latency and computational load [9,18,19]. The utilization of mobile data networks is witnessing a global surge. In Brazil, for instance, as per the National Telecommunications Agency (ANATEL), there has been a rise in mobile broadband access, escalating from 211.8 million to 213.7 million within a mere month [20]. It is anticipated that, by 2026, 54% of mobile data traffic will be facilitated through 5G networks, prompting operators to pursue enhancements to cater to this burgeoning demand, both for users and infrastructure [21]. Accordingly, the integration of 5G networks and MEC frameworks benefits both the academic community and the development of new technologies, such as autonomous vehicles, applications for smart cities, healthcare, and the augmented/virtual reality market, as well as the overall market. Amazon found in its research that every 100 ms of latency on its websites cost 1% of its sales, while Google concluded that an additional 0.5 s in the search page loading time resulted in a 20% reduction in user traffic [22]. It is anticipated that certain use cases within 5G networks will become reliant on MEC concepts and frameworks for the provisioning of services to end users [23]. Examples encompass autonomous vehicles, smart cities, healthcare, and the Internet of Things (IoT), all of which exhibit latency sensitivity and stand to derive substantial gains from the adoption of MEC principles in their deployment. These instances, for instance, manifest within the realm of Ultra-Reliable Low Latency Communications (URLLC) services [24].

To achieve the objectives of this work, we began by exploring the specifics of MEC and 5G technology. We analyzed the software components in their architecture, understanding their goals and characteristics. The chosen network core was free5GC (open-source project for 5G core) due to its open-source code, compliance with 3GPP standards, and active community. The selection was reinforced by the use of my5G-core (a 5G standalone core following the 3GPP standards) [17,25] which is a fork of free5GC. Research was conducted to identify MEC platforms that would meet the objectives of this study. However, the lack of documentation and support led to the simulation of an MEC platform using a Kubernetes cluster. The proposed service was developed based on standards and adherence to the objectives of the Multi-Access Traffic Steering (MTS) API. Additionally, a Service Registry (SR) was created to compose the validation environment, along with a test application.

Validation was divided into three parts. First, the individual functioning of free5GC, the Kubernetes cluster, and the implemented applications/services were validated. Subsequently, the registration of applications/services in the SR was validated. Finally, the third and last part consisted of validating the proposed service and collecting and evaluating the results. More details about the evaluation environment and all source code are publicly available in the GitHub repository (https://github.com/LABORA-INF-UFG/5g-mec-integration, accessed on 25 April 2024).

The contributions of this work are as follows:

- Integration between MEC and 5GC: This study demonstrates the feasibility of integrating 5G and MEC through a communication API.
- Implementation of Integration Service: One approach to realizing the integration between MEC and 5G technology is implementing a specific service. This service can be provided using a dedicated communication API.
- Expansion of MTS API: The original MTS API was not designed to route application traffic to the 5GC. Therefore, to enable traffic transfer between MEC and the 5G network, it is necessary to enhance this API by creating a new specific service.
- Validation of the API in a Test Environment: The proposed API was validated in a testbed environment using a simulated MEC framework and a 5GC.

This article is structured as follows: Section 2 presents the fundamental concepts necessary for understanding this work. Section 3 provides a literature review related to this work. Section 4 describes the proposal of this work, including its architecture, implementation details, and operation. Section 5 presents the use cases used for comparison purposes, the testbed environment, evaluation metrics, and the collected results, discussing their implications. Section 6 concludes the article.

2. 5G System

The so-called 5G system (5GS) encompasses the 5G core network (5GC), a new radio interface known as 5G New Radio, and User Equipment (UE) specifically designed for 5G [17,26]. Fifth-generation systems are in the process of development to establish connections between UE and the 5G network infrastructure [17].

In Release 16, 3rd Generation Partnership Project (3GPP) focused on a comprehensive expansion of the 5GS. This expansion includes satellite-based 5G access, vehicle-to-everything (V2X) communication services, wireless and wired networks integration with 5G, and other advancements [27]. Additionally, functionalities like using NR technology in unlicensed frequencies, integrated access backhaul (IAB) approaches, large-scale multiple-input multiple-output (MIMO) enhancements, segmented resource allocation, and new capabilities for URLLC and industrial IoT, as well as support for non-public network (NPN) implementations, are incorporated [27]. These features are considered crucial to ensure that networks meet essential requirements.

2.1. 5G Core

The 5GC network features an innovative architecture called SBA, which revolves around services and microservices [6]. In this model, the 5GC introduces characteristics such as the separation between the DP and the CP, function virtualization, and Network Slicing (NS) [17].

The SBA offers advantages such as improved development and maintenance efficiency, the association of microservices with dedicated resources and independent lifecycles, and enhanced scalability through on-demand instances [8]. This structure consists of network functions (NFs) that provide services through the Service-Based Interface (SBI), accessed via APIs and protocols like HTTP, REST, and others [17].

The 5GC is an integral part of the 5GS alongside the New Radio (NR) radio interface. Built upon cloud computing concepts, the 5GS features a service-oriented core with SBA architecture, natively supporting NS, virtualization, and edge computing [28]. The architecture of the 5GS is divided into two modes: non-standalone (NSA), where the new radio interface is used in conjunction with an existing core infrastructure (Evolved Packet Core—EPC), and standalone (SA), where the radio interface connects to the newly proposed core architecture for 5G.

2.2. Network Functions

The 5GC leverages the principles of Network Function Virtualization (NFV) and Software Defined Networks (SDNs) to virtualize network functions, including those related to transport, access, and core, which were traditionally run on closed hardware [29]. An example of the interaction between MEC and 5GC is evident in the network function (NF) called User Plane Function (UPF). This function serves as a central point in the user plane of the 5GC, connecting the UE to the data network (DN) and performing tasks like routing, inspection, and Quality of Service (QoS) management [8].

The Session Management Function (SMF) manages the UPF, configuring the connection between the UE and DN, including the allocation of IP addresses and IP PDU sessions [8]. The Access and Mobility Management Function (AMF) is a network function that interacts with the data network and the UE, facilitating registration, authentication, and mobility in the network through secure signaling [8].

The Network Exposure Function (NEF) monitors the 5G network to expose events to authorized applications and other network functions (NFs), allowing the management of QoS rules, priorities, and UE-related information [8]. These functions collaborate to ensure the effective operation of the 5GC, using the virtualization and flexibility offered by the architecture.

Network Slice is a core technology in the context of 5G, enabling the creation of logical networks on top of a shared infrastructure [8]. In essence, network slices aim to meet a customer's specific needs, resulting in dedicated logical networks where required resources are configured and allocated [8,30].

2.3. Network Interfaces

The mentioned network interfaces refer to the connections between different software modules within the 5GC. In general, each of these connections follows specific protocols and standards. In the core context, four interfaces stand out—*N3*, *N4*, *N6*, and *N9*—each with its standardized functions, while other interfaces are not covered in this context.

- **N3**—Communication interface between NR and UPF. This interface carries the data encapsulated for routing [8].
- **N4**—The control exerted by the SMF over the UPF is carried out through this interface, with all their communications taking place exclusively through it [8].

- **N6**—The UPF connects to external data networks through this interface. Data no longer undergo the encapsulation process at this stage, although implementing a Virtual Private Network (VPN) is feasible [8].
- **N9**—Within the core infrastructure of 5G, it is possible to deploy UPFs in a sequential configuration. These UPFs are interconnected through the *N9* network interface, and scenarios involving mobility are examples that use this approach [8].

2.4. MEC Reference Architecture

Similar to the 5GC, MEC is standardized by the European Telecommunications Standards Institute (ETSI), which is dedicated to the continuous standardization and expansion of MEC. Technical documents are regularly released, covering everything from the architecture to developing services within the MEC framework.

The MEC reference architecture serves as the foundation for creating any MEC structure. Figure 1 depicts the MEC reference architecture in a simplified manner, showcasing only the essential components for the purpose of this study. This architecture is defined in the document [13]. Various associated components can be identified within the architectural setup, such as the MEC Platform (MEP), MEC Host, and MEC Platform Manager (MEPM).

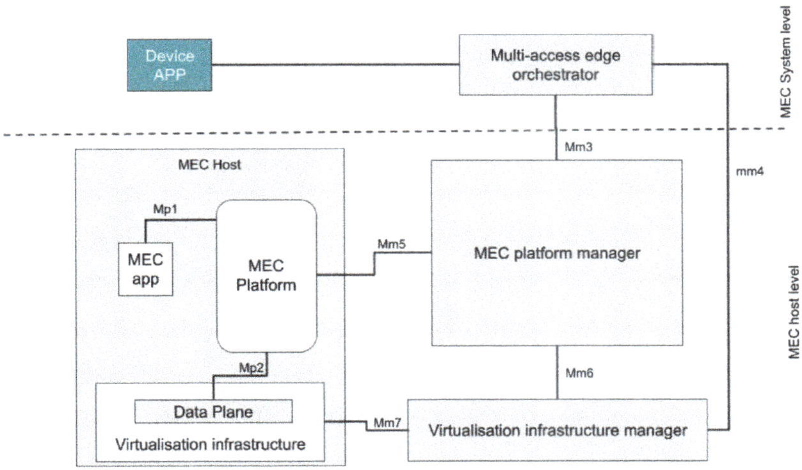

Figure 1. MEC reference architecture summarized based on [13].

The MEC Host plays a comprehensive role in the MEC architecture, encompassing critical elements like the MEP, the Virtualization Infrastructure Manager (VIM), and MEC applications (MEC App). The VIM provides storage, processing, and networking resources for MEC applications. It is also where the DP resides [13]. The MEP plays a central role in management, especially concerning applications. It creates an environment conducive for applications to explore, offer, and use services. Moreover, it facilitates communication with the DP, adjusting it according to the traffic rules received. The MEP also hosts MEC services, including functions like a DNS server [13].

The MEPM holds management-related responsibilities. This includes overseeing the lifecycle of applications and conveying information to the edge orchestrator, the Multi-Access Edge Orchestrator (MEO). Furthermore, the MEPM handles MEP component management, as well as addressing standardization and application needs [13]. The MEO, in turn, performs various functions, such as maintaining a comprehensive view of the MEC system based on deployed MEC Hosts, available services, resources, and the overall topology. The MEO is also responsible for selecting MEC Hosts based on requirements, implementing, transferring (if applicable), and terminating applications [13].

3. Related Work

Tomazewski et al. (2020) [31] presents an architecture for MEC and 5G based on the Distributed Autonomous Slice Management (DASMO) architecture [32]. The proposal is based on a single-domain paradigm; however, it can be extended to a multi-domain scenario with the addition of another software component, as outlined in the paper. Despite the detailed proposal, there is no discussion of the validation of the presented architecture, leaving a gap in the research.

Ksentini and Frangoudis propose a management and orchestration architecture for NS [23]. This architecture divides a complete network slice into smaller parts where Virtual Network Functions (VNFs) can be deployed. They present two models: in-slice and multi-tenancy, differentiated by the MEP instantiation. In the in-slice model, the MEP is instantiated within the slice, while, in the multi-tenancy model, a shared MEP exists outside the slice. The MEP communicates with the 5G network to create traffic steering policies through direct communication with the core network's Packet Control Function (PCF) or via the Network Exposure Function (NEF). The work aims to enable the MEC architecture to support NS and was validated using OpenAirInterface (OAI) with adaptations to the core network for communication with the MEP via Mp2.

Kekki et al. propose an integration architecture between the 5GS and the MEC architecture [16]. In this structure, the MEC orchestrator can directly interact with the NEF and, in some cases, other functions within the 5GS. The primary focus is integrating the two systems through the N6 interface, which connects to the User Plane Function (UPF). For the MEC, the UPF is considered to be distributed and configurable, and, in some deployments, it can be local and a part of its architecture.

The article by Kuklinski et al. [33] presents an innovative architecture that integrates the O-RAN platform, Self-Organizing Network (SON), MEC, and NS. The proposal is based on the near-Real Time RAN Intelligent Controller (near-RT RIC) from the O-RAN platform, with modifications to unify all these technologies into a single component called Integrated near-RT RIC (I-near-RT RIC). MEC applications are implemented as xApps within the O-RAN platform. Although it is a conceptual proposal, it has not been validated in a testbed.

Table 1 represents the comparison between this work and the related works based on the defined criteria:

- Proposal for 5G-MEC Integration—A proposal for integration is presented.
- Integration through Service—Whether the proposed integration is implemented as a service.
- Integration through New Architecture—Whether the integration proposal is made through changes to existing architecture or the creation of new ones.
- Validation of Proposal—Whether the proposal has been validated.
- Validation with 5GC SA—Is there any validation with the 5G-SA network core?

The first aspect examined deals with the central theme of this study, which is also analyzed in the works mentioned above. It is clear that some of these papers are not just limited to integrating MEC with 5G but also consider other technologies, such as SON in [33]. Therefore, all the papers discussed in this analysis present proposals involving 5G-MEC integration, but they are differentiated by their approaches.

The studies evaluated different integration approaches, including proposing new architectures for this integration. This study stands out for proposing a service as a solution, while other works focused on creating new architectures or adapting existing ones. For instance, one study utilized the DASMO architecture as a foundation, reusing key concepts, while another developed a new architecture based on public knowledge of O-RAN [31,33]. Overall, a commonality among all the works is the direct communication of the MEP with the network core's CP via the MP2 interface.

Finally, we examined whether the proposals in these papers have been validated. Two points are considered: (i) whether the proposal has been validated in any way and (ii) whether the validated proposal uses a 5GC core. In this sense, only one related work has had its proposal validated, which is the work [23]. In this study, validation was conducted in a test environment using different machines as edge servers and an OAI network core. However, when analyzing this study, it is possible to notice elements of the Evolved Packet Core (EPC) architecture of the network cores, such as the Mobility Management Entity (MME) and Home Subscriber Server (HSS). Thus, the study does not employ an autonomous (SA) 5G network core despite the validation. The validation focused on the ability of the proposed architecture to create network slices for the MEC in the context of 5G, facilitating communication between the two technologies.

Table 1. Comparison of this proposal with related work.

Works	**5G-MEC Integration Proposal**	**Integration through Service**	**Integration through New Architecture**	**Proposal Validation**	**5GC SA Validation**
[16] MEC in 5G Networks	X		X		
[31] A New Approach to 5G and MEC Integration	X		X		
[23] Toward Slicing-Enabled Multi-Access Edge Computing in 5G	X		X	X	
[33] On O-RAN, MEC, SON and Network Slicing integration	X		X		
This Paper	X	X		X	X

Thus, from the analysis of the conducted studies, it is evident that our proposal stands out both for its integrative approach and its validation. The proposed integration will be implemented through a service and validated within a test environment utilizing a 5G SA network core. Furthermore, it is notable that only one of the papers with architectural-level proposals has undergone validation, whereas the others remain at the theoretical and conceptual level.

Kekki et al. discuss the possibility of deployment as an application offering a service [16]. This possibility could represent a commercial advantage, considering the launch time and the potential for the service to evolve into an MEC platform as the technology and business models stabilize. This reinforces the proposal presented, providing an academic scenario with an approach different from the others. To the best of our knowledge, there are no open-source solutions in the literature to compare the implementation.

4. A New Service Proposed

The benefits of MEC and how they can positively impact 5G networks have already been presented. However, to make it possible to combine the two technologies, it is necessary to develop a component that allows communication and follows the established standards. One of the possibilities is a communication service implemented through an application or as a service part of an MEC platform, specifying this communication using an application and speeding up the launch and deployment process [16]. Another possibility is to deploy the service as part of the MEC platform. However, integrating the entire service structure into the MEC platform is non-trivial. Thus, this work aims to specify a communication API that provides a communication service deployed as an application.

This service must allow UE applications to interact with their respective server applications isolated internally on the MEC platform. Thus, the service must receive information about the destination and the content to be sent and relay it after validating its existence. The service must then wait for a response and return it to the requester. This flow is repeated in both directions. Information about the destination must be sent in the request's body so that the service can process it and understand how to carry out this routing. A level of security is implemented by using tokens to access the proposed service. The SR will generate this token. Initially, this work does not aim to transfer media files, which is one of the possible future extensions.

The integration service, deployed as an application, will run on the MEC, specifically on the MEP component. MEP is an internal component of MEC HOST, which also has MEC Applications as part of its scope, so these are the MEC components covered by this work. The communication service run by MEP will provide communication between 5G and MEC Applications via the 3GPP *N*6 interface. The signaling that takes place between UE and 5GC is outside the scope of this work. These components can be seen in the proposal architecture shown in Figure 2.

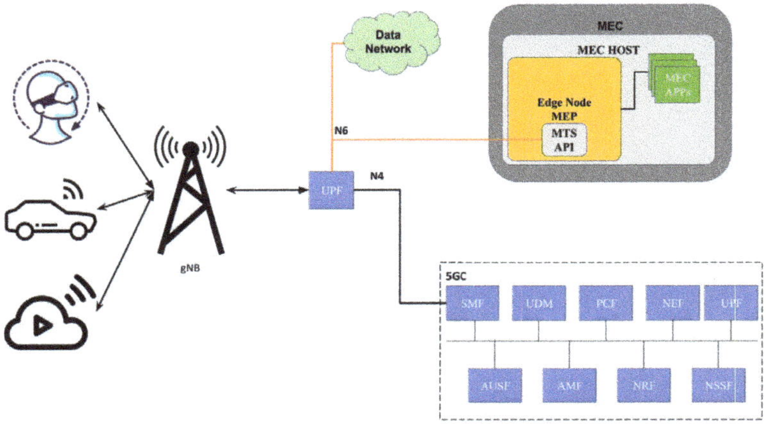

Figure 2. Proposal's general architecture.

Figure 2 also shows the structure of the 5GC, with its SMF connected to the UPF via the *N*4 interface. It can be seen that the core has two UPFs and only one SMF, made possible by the SBA architecture model. In order to prevent the transmission of N6 upstream policies to the UPF, which consistently includes the 5GS, a loosely interconnected data plane node (D-Plane) is presumed with one or multiple UPFs [34]. The image shows the gNB radio interface and representations of applications that communicate with the MEC. It is important to note that this study does not consider the internal communication of the core between the NFs and the radio interface.

The implementation of this service will involve two main functions: *postData* and *makeRequest*. The first of these is the *postData* function. Its primary function will be to route packets, i.e., when a packet arrives, it will look at the information in its body, identifying the destination of that information and which HTTP method will be used. The data model specified for this function includes receiving the IP address of the destination service, the HTTP method to be used, and the information to be sent.

Having received the above information, the *makeRequest* function is invoked. This function is tasked with passing the data on to its final destination. With the information received previously, a new request will be made, but this will be made to the final destination of that information, awaiting a response. The response is sent back to *postData*, relaying it to the UE. In addition to these two methods, the service includes a third called *makeRegistry* to register with the SR. This function contains all the service details, such as IP, port, and endpoints. This information is sent to the SR for registration, allowing interested applications to check and obtain these details.

For this whole flow to work with a higher level of security, a registry service called *ServiceRegistry* will be implemented. This registry focuses on applications and services on the MEC platform that wish to allow internal or external access through the integration API. To this end, an application/service must register its IP, port, and endpoints with the respective methods. Interested applications will then be able to check this information. A client application in the UE will be able to check whether its respective server is available and, if it is, all the information for communication will be passed on.

In this work, it will be implemented as an application. However, this service can be incorporated into the MEC platform as desired. In this scenario, the service would also be accessible via the MEP and communicate via the N6 interface. This application will be implemented as a RESTful API. By implementing it as an application, there is no need for any communication with the core CP. Therefore, all data traffic will occur through the DP, UPF. Communications take place via HTTP requests.

In addition to the MEC reference architecture, the ETSI technical documents also deal with two APIs: Multi-Access Traffic Steering (MTS) and BandWidth Management (BWM) [13]. BWM is the API that will handle the allocation of bandwidth between applications, as well as the prioritization of specific traffic. The MTS API, on the other hand, is responsible for managing traffic, routing it, splitting it up, and replicating it on various access networks to meet the network requirements requested by the applications [14]. Recognizing that the objectives of the service that is the subject of this work and the MTS API are partially compatible, it is proposed that the MTS API be expanded to provide a traffic-forwarding service for the data network external to the MEC.

4.1. Implementation

The implementation of the proposed API used the Python language associated with the Flask framework [35]. The micro-framework is widely used for developing web applications due to its simplicity, efficiency, and robustness. The Flask associates a function with an endpoint through which information is exchanged.

The implemented functions are connected to an endpoint that responds to HTTP requests. The primary function only accepts requests via the *POST* method. As a result, requests that do not use this method are disregarded. Once the request has been received, the API calls the *postData* method. This procedure will check whether the minimum content is required to transmit the data. This check will look in the message content for the destination IP address and the HTTP method that should be used. The content of the message will perform a simple validation so that, if there is content, it will be used as a parameter in the *makeRequest* function. If any of this information is absent, an error message with code 404 is sent, with the message "Service IP and Method Requested". All messages are checked for the necessary basic data, so, if several messages are missing data, several error codes will be returned. No specific treatment is necessary as it does not interfere with the communication process.

For the information to be passed on to its recipient, the API will invoke the *makeRequest* method. The first check will be on the HTTP method to be used. Using *POST* or *PUT* indicates an exchange of messages, so the data must be sent to the method. In the case of the others, no information is required other than the mandatory information listed above. After executing the request, the response message goes the other way around, i.e., it arrives at *makeRequest*, is returned to *postData*, and finally arrives at the client, its final destination.

It is important to recognize that, for this work, other methods from the MTS API have not been incorporated. This is because these methods are not essential for the proposed objective. The methods and their respective endpoints implemented in this API are protected, so, for any application/service to be able to use them, it must have a JWT token.

We have implemented the minimum necessary for the Service Registry to meet the requirements. A few functions have been implemented. The first is the *handleRegistryData* function, which receives information from the applications and services that wish to register. Once the minimum content for registration has been validated, the *handlePost* function is invoked, which stores the information in the database. Finally, a token is generated for this application/service using the JWT library. The device application is required to provide the access token with each request, in order to assert its permission to access the resource using the specific method invoked [36].

The *handleSpecificData* function has also been incorporated into the Service Registry. This function is designed to make changing data in the database and retrieving information easier. The PUT and DELETE methods allow the client to modify records. With the GET method, the client can request data stored in the database. In this scenario, the UE will request information via the GET method. Collecting information will follow the same procedure that already provides the user with a token. This token is required to use the integrated service. The *handlePut*, *handleDelete*, and *handleGet* functions are identical to the abovementioned functions.

4.2. MTS API

The system proposed in this study, implemented using the MTS API, features an access point that facilitates data exchange between two interested parties. This service receives a POST request via the HTTP protocol. The request must contain three elements: (i) the destination to which the data should be sent, (ii) the method to access that destination, and (iii) the information the sender wishes to transmit. Observing the default format of the data model is crucial, as the API interprets the data in the request body in JSON format, using the respective keys to identify the content.

After receiving the data, the *postData* procedure is activated. This function is tasked with recognizing the end client's destination and the method used in the HTTP protocol. This is due to the need to make a new request to send the information. The *makeRequest* method is activated to execute this new request, which separately receives the destination IP address and the data to be transmitted as arguments. Once this new request has been completed, the API obtains the response provided by the server and forwards this response back to the client. This chain of actions can be seen in Figure 3.

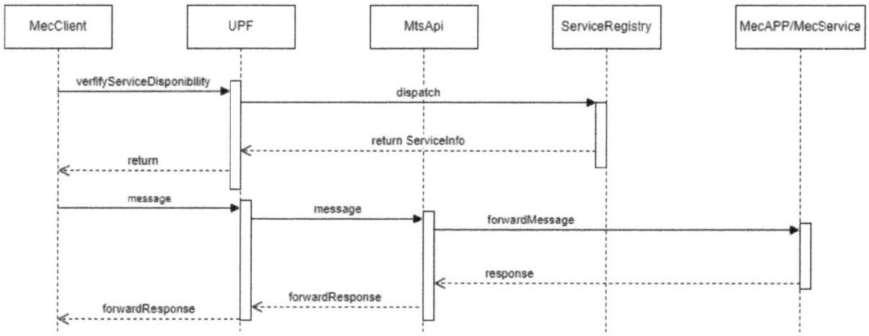

Figure 3. Operation of the proposed API.

The functionality will be accessible via an external port called NodePort using the Kubernetes Kube-proxy service. In this context, Kube-proxy establishes a port linked to the service using NAT [37]. As seen in Figure 3, the expectation is that the client will check the availability of the desired service in the Service Registry and, if available, receive all the relevant information. Consequently, the data format the MTS API expects must contain the full IP address, i.e., the destination IP, port, and endpoint.

The Service Registry task involves receiving data about the applications and services available on the server in question and enabling interested client applications to obtain these details, allowing access to these applications or the ability to direct their data appropriately. To perform this function, SR uses the NoSQL database MongoDB [38].

To make a MEC application accessible to those interested, the application must carry out a registration process with the SR. The SR waits for the application to transmit data, such as its IP address, port number, and the endpoints it offers. The application must establish at least one endpoint. Once the information arrives, the SR checks the validity of this data and stores it in the database.

Service Registry is a service accessible externally by the client to check the existence of registered services. The client uses an access point for searches, receiving a JSON with service information and an authentication token generated with Python's JWT library using a specific key in the code.

5. Performance Evaluation

This section aims to validate the work proposal by discussing use cases that benefit from MEC, detailing the validation environment, and presenting the results. These results will be compared with the characteristics of applications that benefit from MEC to discuss the solution's viability.

The proposed work will evaluate the service results using Key Performance Indicators (KPIs) described in the literature. The evaluation scenario will include e-health and IoT use cases. For the health area, scenarios such as remote surgery, remote consultation, paramedic support, critical health events, vital signs, and location tags will be observed, using the minimum KPIs required for each application, based on previous studies [39,40]. For IoT scenarios, characteristics of smart homes, wearable IoT (WIoT), and farms will be analyzed using the information described in [41]. The KPIs considered for comparison will be mainly related to latency, although other information can be used if necessary. The KPIs are presented in Table 2.

Table 2. KPIs used for comparison.

Application	Latency
Smart Home	1 ms–1000 s
Remote Surgery	200 ms
Remote Medical Consultation	1 ms–100 s
Paramedic Support	200 ms
Critical Health Event	200 ms
Smart Farm	~ hours
Vital Signs Patches	1000 ms
Locatable Tags	1000 ms

5.1. Environment

A validation environment was designed for the proposal, looking for a MEC framework and a 5GC to evaluate and validate the final product. Some MEC platforms were initially considered, but a containerized solution was chosen using Kubernetes due to a lack of support from these other solutions. The 5GC chosen was free5GC, which was containerized to facilitate the execution of network functions. Deployments were carried out on virtual machines hosted on a dedicated server.

The Kubernetes cluster proposed to simulate the MEC platform comprises three virtual machines, one for deploying free5GC and the other two for the MEC platform. All virtual machines ran on a server called NECOS-02. VM1, used to deploy free5GC, has 8GB RAM, 100GB disk, 4vCPUs, and Ubuntu Live Server 18.04LTS operating system. VMs 2 and 3 each have 8GB RAM, 120GB disk, and 4vCPUs, with Ubuntu Live Server 18.04LTS operating system. Figure 4 shows the validation structure.

The free5GC architecture used for validation involves three UPFs: Branching UPF (B-UPF), MEC UPF (M-UPF) and Internet UPF (I-UPF). The B-UPF acts as an intermediary between the user's connection and the destination network and is not associated with any slice.

To validate the proposal, a MEC application was developed to simulate an end-to-end environment where the traffic between the UE and the application uses the integration service. In addition, a SR was created to provide registration tokens for applications wishing to use the MTS API services. Once the applications and services had been developed, they were deployed in a production environment, using the Waitress server to run the Python applications. The deployments were coordinated, with the SR being executed first to allow the applications to be registered. The validation phase was divided into three stages: the first ensured the correct execution of the test environment, the second involved registering the services in the SR, and the third consisted of running the tests and collecting and evaluating the results.

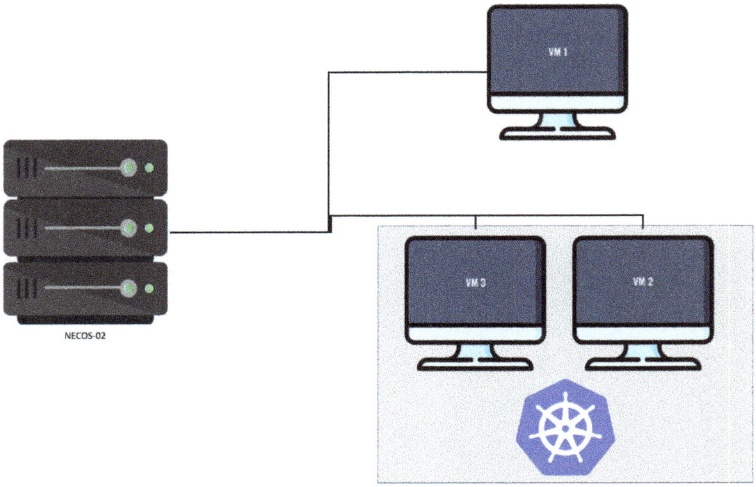

Figure 4. VM structure used in validation.

5.2. Evaluation Metrics

The proposal will be validated with a simple application that calculates the Body Mass Index (BMI), using POST requests. It is important to note that the request could be made to any application running in the cluster and duly registered with the SR. Therefore, this application's use is restricted to confirm the viability of the proposed API.

To do this, three scenarios will be run with different numbers of users (1, 10, 50, and 100), simulating the ability to respond to concurrent requests. Each scenario will be run at different execution times 5, 10, 30, and 60 s.

This variation aims to evaluate performance variations over continuous periods. The Locust tool will make HTTP requests and obtain the average response time and the average number of requests [42].

In addition, the production server will be tested with four and eight threads to evaluate performance. The test sets will be run ten times to obtain reliable results, and a 95% confidence interval will be calculated to assess the variability of the results. The confidence interval (*ci*) is calculated using the equation:

$$ci = \overline{x} \pm \mathcal{Z} \cdot \frac{\sigma}{\sqrt{n}},$$

where $\overline{x}$ represents the sample mean response times obtained in 10 executions. $\mathcal{Z}$ is the normal distribution of 1.96 for a confidence level of 95%. σ represents the standard deviation of the response times, while n indicates the sample size. The $\pm$ indicates the upper and lower result intervals.

End-to-End Communication Test

This test aims to demonstrate the viability of end-to-end communication. To perform this, requests will be made to the application developed using the integration service. The following aspects will be analyzed: (i) the response to the request and whether it corresponds to what was expected; (ii) the time taken to check and send the first response;

and (iii) the response time of the requests. In addition, the UPF interface with the application will be monitored to capture the packet for later analysis. For items 1 and 4, only one request will be made, considering that the information observed will not be altered. However, 30 runs will be made for the other items to obtain the average time. Confidence intervals will also be calculated for the times discussed. For this test, no specific testing tools will be used; the request and time measurement will be carried out utilizing native Ubuntu applications.

5.3. End-to-End Validation Test

The End-to-End validation aims to check whether there has been a successful connection between the UE and MEC-App, with traffic passing through the UPF and the MTS-API. The focal point for the analysis was the exchange of messages between the UPF and the API. To this end, during communication, the packets exchanged were captured, and it was possible to validate the path taken by each request.

The analysis also revealed an average response time for requests of 31 ms, with a standard deviation of 1.71. This standard deviation indicates that the response time in the 30 runs showed slight variation and was close to the average. When verifying the service's existence and receiving a token from the UE, the response time is 26 ms, with a higher standard deviation of 5.21, indicating a more significant variation in response times. When considering the whole process, the average time is 57 ms, with a standard deviation of 5.63 due to the standard deviation of the service existence check.

This result indicates that the solution is viable for e-health use cases. According to [40], which presents KPIs for healthcare scenarios, this solution would be suitable for applications such as paramedic support and critical events (not specified in the paper).

5.4. API Evaluation Results

The results presented are based on the metrics and the use cases presented. The variation in the number of users will allow us to evaluate the behavior of the service in scenarios with competition for the same resources. The tests presented here are experimental. In real scenarios, the numbers may vary.

Figure 5 shows the results for one user. The response times are similar for four and eight threads as expected. It can be seen that the longer the execution time the smaller the standard deviation, reaching 0 in 30 s, thus presenting a smaller confidence interval. The obtained response time meets the requirements in [41] for Smart Home scenarios and even games, which require latencies of 10 ms or less.

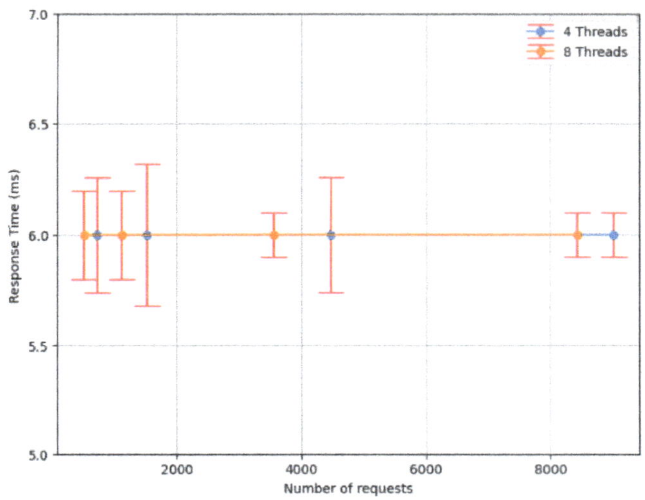

Figure 5. Performance with 1 user.

When we run a scenario with ten users (Figure 6), the response time increases, reaching 58 ms with four threads and 47 ms with eight threads. This is because, unlike the previous scenario, there is competition for the same resource (use of services). As with the previous one, more extended scenarios have smaller confidence intervals with minor time variations. This result makes the solution viable in scenarios with paramedic support applications or critical health events [40].

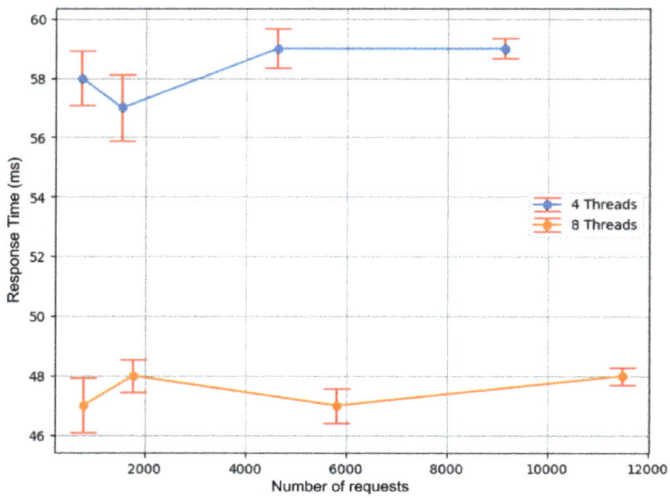

Figure 6. Performance with 10 users.

For 50 and 100 users (Figures 7 and 8), latency rises considerably, with latencies reaching 520 ms and 1700 ms, respectively, using just four threads. These values drop when we increase the processing power to eight threads, with 250 ms for 50 users and 590 ms for 100.

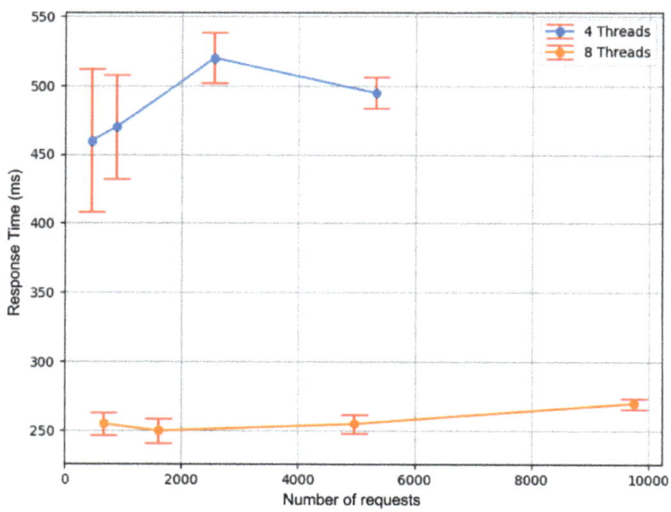

Figure 7. Performance with 50 users.

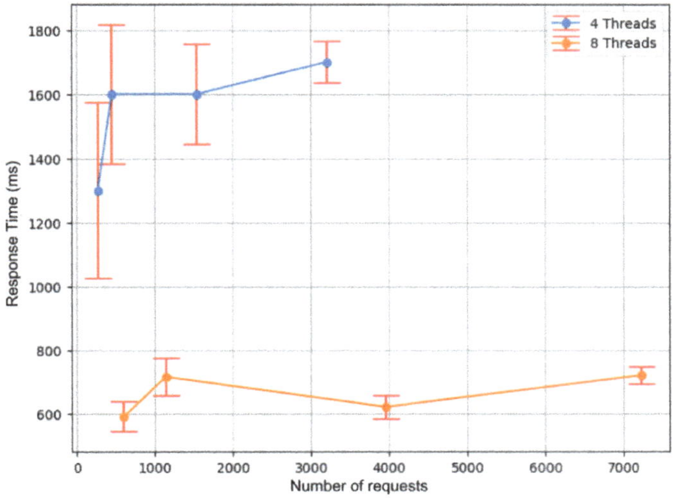

Figure 8. Performance with 100 users.

The results for 50 users indicate the solution's viability for Smart Farming scenarios, with the transmission of data from sensors and location tags, for example, [39,40]. With 100 users, the results indicate scenarios that accept considerably high response times, such as Smart Home, location tags, and medical remote consultation [40,41].

The results show that the solution performed as expected. The times displayed in the graphs indicate the solution's viability in various use cases presented in the literature and this work. However, as applications and the network evolve, new use cases may arise, and new demands may be required. Therefore, in the long term, it is necessary to review the viability of this solution within the scenarios and needs that will arise.

6. Conclusions

The existing literature addresses integration between 5G networks and MEC in a limited way, mainly focusing on new conceptual architectures. However, there are gaps related to implementing and validating these proposals in a 5GC.

Hence, we have proposed an implementation as a service to connect 5G and MEC. The service is an extension of the MTS API defined by ETSI, seeking compatibility of objectives. Validation has been performed in a virtualized environment with a 5G network core, simulating UE, gNB, and a Kubernetes cluster as the MEC platform.

The validation results demonstrated the ability of the proposed service to communicate between the UE and application, including performance evaluation, especially for URLLC applications. The proposed solution can be compared with others in the literature and optimized to improve response time, aiming to achieve consistent lower times.

This work achieved its overall objective by enabling 5G-MEC integration through a service deployed as an application. The solution showed satisfactory results in the scenarios evaluated, providing a basis for future studies and optimizations. For future work, we are working on validating the service, including the O-RAN real environment and real multi-tenancy application.

Author Contributions: Conceptualization, R.X., L.F. and A.O.-J.; methodology, R.X., A.O.-J. and L.F.; software, R.X.; validation, R.X., L.F., R.S.S. and A.O.-J.; investigation, R.X.; resources, R.X.; writing—original draft preparation, R.X.; writing—review and editing, R.X., R.S.S., M.R., W.M., L.F. and A.O.-J.; visualization, R.X., L.F. and A.O.-J.; supervision, A.O.-J. and L.F.; project administration, A.O.-J., M.R. and W.M.; funding acquisition, A.O.-J., M.R. and W.M. All authors have read and agreed to the published version of the manuscript.

Funding: This work was partly supported by Fraunhofer Portugal AICOS and partly supported by the MCTIC/CGI.br/São Paulo Research Foundation (FAPESP) through the Project Smart 5GC And MUltiRAn Integration (SAMURAI) under Grant 2020/05127-2.

Data Availability Statement: Dataset available on request from the authors.

Conflicts of Interest: The authors declare no conflicts of interest.

Abbreviations

The following abbreviations are used in this manuscript:

3GPP	3rd Generation Partnership Project
5GS	5G System
AMF	Access and Mobility Management Function
API	Application Programming Interface
B5G	Beyond 5G
B-UPF	Branching UPF
BMI	Body Mass Index
BWM	BandWidth Management
CP	Control Plane
DASMO	Distributed Autonomous Slice Management
DN	Data Network
DP	Data Plane
EB	Exabytes
eMBB	Enhanced Mobile Broadband
EPC	Evolved Packet Core
gNB	5G New Radio
HSS	Home Subscriber Server
HTTP	Hypertext Transfer Protocol
I-near-RT RIC	Integrated near-RT RIC
I-UPF	Internet UPF
IAB	Integrated Access Backhaul
IoT	Internet of Things
JSON	JavaScript Object Notation
JWT	JSON Web Token
KPI	Key Performance Indicator
M-UPF	MEC UPF
MBB	Mobile BroadBand
MEC	Multi-Access Edge Computing
MEC-App	MEC Application
MEO	Multi-Access Edge Orchestrator
MEP	MEC Platform
MEPM	MEC Platform Manager
MIMO	Multiple-Input Multiple-Out
MME	Mobility Management Entity
mMTC	Massive Machine Type Communications
MTS	Multi-Access Traffic Steering
near-RT RIC	near-Real Time RAN Intelligent Controller
NEF	Network Exposure Function
NFV	Network Function Virtualization
NPNs	Non-Public Networks
NR	New Radio
NS	Network Slice
NSA	Non-Standalone
OAI	OpenAirInterface
QoS	Quality Of Service
REST	Representational State Transfer
SA	Standalone

SBA	Service-Based Architecture
SBI	Service-Based Interface
SDN	Software Defined Networks
SMF	Session Management Function
SON	Self-Organizing Network
SR	Service Registry
UE	User Equipment
UHD	Ultra-High Definition
UPF	User Plane Function
URLLC	Ultra Reliable Low Latency Communications
V2X	Vehicle-to-Everything
VIM	Virtualization Infrastructure Manager
VM	Virtual Machine
VNF	Virtual Network Functions
WIoT	Wearable IoT

References

1. Salah, I.; Mabrook, M.M.; Hussein, A.I.; Rahouma, K.H. Comparative study of efficiency enhancement technologies in 5G networks-A survey. *Procedia Comput. Sci.* **2021**, *182*, 150–158. [CrossRef]
2. Chen, S.; Zhao, J. The requirements, challenges, and technologies for 5G of terrestrial mobile telecommunication. *IEEE Commun. Mag.* **2014**, *52*, 36–43. [CrossRef]
3. Zhao, L.; Zhou, G.; Zheng, G.; Chih-Lin, I.; You, X.; Hanzo, L. Open-Source Multi-Access Edge Computing for 6G: Opportunities and Challenges. *IEEE Access* **2021**, *9*, 158426–158439. [CrossRef]
4. Pivoto, D.G.S.; Rezende, T.T.; Facina, M.S.P.; Moreira, R.; de Oliveira Silva, F.; Cardoso, K.V.; Correa, S.L.; Araujo, A.V.D.; Silva, R.S.E.; Neto, H.S.; et al. A Detailed Relevance Analysis of Enabling Technologies for 6G Architectures. *IEEE Access* **2023**, *11*, 89644–89684. [CrossRef]
5. Wang, Y.; Zhao, J. Mobile Edge Computing, Metaverse, 6G Wireless Communications, Artificial Intelligence, and Blockchain: Survey and Their Convergence. In Proceedings of the 2022 IEEE 8th World Forum on Internet of Things (WF-IoT), Yokohama, Japan, 26 October–11 November 2022; pp. 1–8. [CrossRef]
6. 3rd Generation Partnership (3GPP). TR21.915 V15.9.0, 3rd Generation Partnership Project. 2018. Available online: https://www.3gpp.org/ftp/Specs/archive/21_series/21.915/ (accessed on 6 May 2021).
7. Amgoune, H.; Mazri, T. Comparison between different 5G architectures for a better integration of these services and proposal of an improved architecture. In Proceedings of the 4th International Conference on Smart City Applications, Casablanca, Morocco, 2–4 October 2019; pp. 1–7.
8. Rommer, S.; Hedman, P.; Olsson, M.; Frid, L.; Sultana, S.; Mulligan, C. *5G Core Networks: Powering Digitalization*; Elsevier: Amsterdam, The Netherlands, 2019.
9. Filali, A.; Abouaomar, A.; Cherkaoui, S.; Kobbane, A.; Guizani, M. Multi-Access Edge Computing: A Survey. *IEEE Access* **2020**, *8*, 197017–197046. [CrossRef]
10. Shi, W.; Cao, J.; Zhang, Q.; Li, Y.; Xu, L. Edge computing: Vision and challenges. *IEEE Internet Things J.* **2016**, *3*, 637–646. [CrossRef]
11. Virdis, A.; Nardini, G.; Stea, G.; Sabella, D. End-to-end performance evaluation of MEC deployments in 5G scenarios. *J. Sens. Actuator Netw.* **2020**, *9*, 57. [CrossRef]
12. Abbas, N.; Zhang, Y.; Taherkordi, A.; Skeie, T. Mobile edge computing: A survey. *IEEE Internet Things J.* **2017**, *5*, 450–465. [CrossRef]
13. ETSI. *Framework and Reference Architecture*; Technical Report for Multi-Access Edge Computing (MEC); ETSI: Valbonne, France, 2022. Available online: https://www.etsi.org/deliver/etsi_gs/MEC/001_099/003/03.01.01_60/gs_mec003v030101p.pdf (accessed on 25 April 2024).
14. ETSI. *Traffic Management APIs*; Technical Report for Multi-Access Edge Computing (MEC); ETSI: Valbonne, France, 2020. Available online: https://www.etsi.org/deliver/etsi_gs/MEC/001_099/015/02.01.01_60/gs_mec015v020101p.pdf (accessed on 25 April 2024).
15. Giust, F.; Verin, G.; Antevski, K.; Chou, J.; Fang, Y.; Featherstone, W.; Fontes, F.; Frydman, D.; Li, A.; Manzalini, A.; et al. MEC deployments in 4G and evolution towards 5G. *ETSI White Paper* **2018**, *24*, 1–24.
16. Kekki, S.; Featherstone, W.; Fang, Y.; Kuure, P.; Li, A.; Ranjan, A.; Purkayastha, D.; Jiangping, F.; Frydman, D.; Verin, G.; et al. MEC in 5G networks. *ETSI White Paper* **2018**, *28*, 1–28.
17. Silva, J.P.L.; Nery, S.W.L.; Silva, R.; Oliveira-Jr, A.C.; Cardoso, K.V.; Both, C.B. Entendendo o núcleo 5G na prática, através de uma implementação de código aberto. In *SBRT 2020: Livro de Minicursos*; Instituto Federal de Ensino, Ciência e Tecnologia da Paraíba–IFPB: Florianópolis, SC, Brazil, 2021; Chapter 1.
18. Maleki, E.F.; Ma, W.; Mashayekhy, L.; La Roche, H. QoS-aware 5G component selection for content delivery in multi-access edge computing. In Proceedings of the 14th IEEE/ACM International Conference on Utility and Cloud Computing, Leicester, UK, 6–9 December 2021; pp. 1–10.

19. Iwai, T.; Nakao, A. Demystifying myths of MEC: rethinking and exploring benefits of multi-access/mobile edge computing. In Proceedings of the 2018 IEEE 7th International Conference on Cloud Networking (CloudNet), IEEE, Tokyo, Japan, 22–24 October 2018; pp. 1–4.
20. Agência Nacional de Telecomunicações. Painel de Dados. 2021. Available online: https://informacoes.anatel.gov.br/paineis/acessos/telefonia-movel (accessed on 7 May 2021).
21. Ericsson. Ericsson Mobility Report 2020. Available online: https://www.ericsson.com/en/press-releases/2020/11/more-than-1-billion-people-will-have-access-to-5g-coverage-by-the-end-of-2020 (accessed on 25 April 2024).
22. Amazon Found Every 100ms of Latency Cost them 1% in Sales. 2023. Available online: https://www.gigaspaces.com/blog/amazon-found-every-100ms-of-latency-cost-them-1-in-sales (accessed on 12 December 2023).
23. Ksentini, A.; Frangoudis, P.A. Toward Slicing-Enabled Multi-Access Edge Computing in 5G. *IEEE Netw.* **2020**, *34*, 99–105. [CrossRef]
24. Ma, H.; Li, S.; Zhang, E.; Lv, Z.; Hu, J.; Wei, X. Cooperative Autonomous Driving Oriented MEC-Aided 5G-V2X: Prototype System Design, Field Tests and AI-Based Optimization Tools. *IEEE Access* **2020**, *8*, 54288–54302. [CrossRef]
25. my5G Initiative. my5G-Core. 2021. Available online: https://github.com/my5G/my5G-core (accessed on 1 January 2024).
26. Morais, F.Z.; de Almeida, G.M.F.; Pinto, L.; Cardoso, K.V.; Contreras, L.M.; Righi, R.d.R.; Both, C.B. PlaceRAN: Optimal Placement of Virtualized Network Functions in Beyond 5G Radio Access Networks. *IEEE Trans. Mob. Comput.* **2023**, *22*, 5434–5448. [CrossRef]
27. Network, Evolved Universal Terrestrial Radio Access. 3rd Generation Partnership Project; Technical Specification Group Services and System Aspects; General Packet Radio Service (gprs) Enhancements for Evolved Universal Terrestrial Radio Access Network (E-UTRAN) Access. EUTRA Network. 2011. Available online: https://portal.3gpp.org/desktopmodules/Specifications/SpecificationDetails.aspx?specificationId=849 (accessed on 25 April 2024).
28. Foukas, X.; Patounas, G.; Elmokashfi, A.; Marina, M.K. Network slicing in 5G: Survey and challenges. *IEEE Commun. Mag.* **2017**, *55*, 94–100. [CrossRef]
29. Mamushiane, L.; Dlamini, S. Leveraging SDN/NFV as key stepping stones to the 5G era in emerging markets. In Proceedings of the 2017 Global Wireless Summit (GWS), IEEE, Cape Town, South Africa, 15–18 October 2017; pp. 23–27.
30. Taleb, T.; Afolabi, I.; Samdanis, K.; Yousaf, F.Z. On multi-domain network slicing orchestration architecture and federated resource control. *IEEE Netw.* **2019**, *33*, 242–252. [CrossRef]
31. Tomaszewski, L.; Kukliński, S.; Kołakowski, R. A new approach to 5G and MEC integration. In Proceedings of the IFIP International Conference on Artificial Intelligence Applications and Innovations, Neos Marmaras, Greece, 5–7 June 2020; Springer Berlin/Heidelberg, Germany, 2020; pp. 15–24.
32. Kukliński, S.; Tomaszewski, L. Dasmo: A scalable approach to network slices management and orchestration. In Proceedings of the NOMS 2018-2018 IEEE/IFIP Network Operations and Management Symposium, IEEE, Taipei, Taiwan, 23–27 April 2018 pp. 1–6.
33. Kukliński, S.; Tomaszewski, L.; Kołakowski, R. On O-RAN, MEC, SON and Network Slicing integration. In Proceedings of the 2020 IEEE Globecom Workshops (GC Wkshps), IEEE, Taipei, Taiwan, 7–11 December 2020; pp. 1–6.
34. ETSI. Multi-Access Edge Computing (MEC); MEC 5G Integration. Technical Report. 2020. Available online: https://www.ets org/deliver/etsi_gr/MEC/001_099/031/02.01.01_60/gr_MEC031v020101p.pdf (accessed on 25 April 2024).
35. GeekHunter. Flask: O Que é e Como Codar Com Esse Micro Framework Python. 2020. Available online: https://blog.geekhunte com.br/flask-framework-python/ (accessed on 21 February 2020).
36. ETSI. Multi-Access Edge Computing (MEC); Device Application Interface. Technical Report. 2020. Available onlin https://www.etsi.org/deliver/etsi_gs/MEC/001_099/016/02.02.01_60/gs_mec016v020201p.pdf (accessed on 14 May 2024).
37. Kubernetes. Kubernetes. 2021. Available online: https://kubernetes.io/pt-br/docs/ (accessed on 29 August 2022).
38. MongoDB. MongoDB. 2022. Available online: https://www.mongodb.com/pt-br (accessed on 29 August 2022).
39. Wijethilaka, S.; Liyanage, M. Survey on network slicing for Internet of Things realization in 5G networks. *IEEE Commun. Sur Tutorials* **2021**, *23*, 957–994. [CrossRef]
40. Zhang, D.; Rodrigues, J.J.; Zhai, Y.; Sato, T. Design and Implementation of 5G e-Health Systems: Technologies, Use Cases, an Future Challenges. *IEEE Commun. Mag.* **2021**, *59*, 80–85. [CrossRef]
41. Porambage, P.; Okwuibe, J.; Liyanage, M.; Ylianttila, M.; Taleb, T. Survey on multi-access edge computing for internet of thin realization. *IEEE Commun. Surv. Tutor.* **2018**, *20*, 2961–2991. [CrossRef]
42. Locust. Locust—A Modern Load Testing Framework. 2022. Available online: https://locust.io/ (accessed on 9 February 202

telecom

MDPI

Article

Transceiver Optimization for Multiuser Multiple-Input Multiple-Output Full-Duplex Amplify-and-Forward Relay Downlink Communications

Yunlong Shao *,†,‡ **and Thomas Aaron Gulliver** ‡

Department of Electrical and Computer Engineering, University of Victoria, Victoria, BC V8W 2Y2, Canada; agullive@ece.uvic.ca

* Correspondence: yshao@uvic.ca or yshao12@nyit.edu

† Current address: Department of Computer Science, New York Institute of Technology Vancouver, Vancouver, BC V5M 4X3, Canada.

‡ These authors contributed equally to this work.

Abstract: This paper considers the transceiver design in a multiuser multiple-input multiple-output (MIMO) full-duplex (FD) amplify-and-forward (AF) relay downlink communication system, where users simultaneously transmit data via an FD relay node. The design incorporates an imperfect loop interference (LI) cancellation which results in a residual LI. Linear precoders are employed at the sources and relay, and minimum mean-squared-error (MMSE) combiners are employed at the destinations to mitigate the effect of the residual LI. The corresponding design problem is highly nonconvex, so a closed-form solution is intractable. Thus, an iterative method is developed to solve this optimization problem. Simulation results are presented which show that the proposed iterative algorithm provides better performance than the corresponding half-duplex (HD) solution in terms of the achievable rate under residual LI.

Keywords: amplify and forward (AF); capacity, downlink, full-duplex (FD); loop interference (LI), MIMO; transceiver

1. Introduction

Multiple-input multiple-output (MIMO) relay communication systems have been the subject of considerable research due to their ability to improve both achievable rates and coverage [1–6]. A relay is used between the source and destination, so the signal is transmitted from the source to the relay and then from the relay to the destination. The relay node can employ either the decode-and-forward (DF) or amplify-and-forward (AF) protocols. AF amplifies the received signal and forwards it to the destination, so it has a lower complexity than the DF protocol [7].

Precoding is a well-known technique for interference mitigation. A joint precoding optimization for a multiuser relay downlink system was investigated in [1]. The sum capacity was maximized by using quadratic programming, but multiple antennas were employed only at the relay. The performance can be improved by using multiple antennas at the source and destination. A transceiver design for a multiuser non-regenerative MIMO relay system with multiple antennas at both the source and destination was investigated in [2–6]. However, a half-duplex (HD) relay was employed, so transmission from the source to destination requires an extra time slot which limits the potential capacity.

In the past decade, full-duplex (FD) relay systems have attracted attention because data transmission from a source to a destination can be completed in one time slot [7–11]. Thus, FD MIMO relaying can increase capacity compared to HD systems [9], but loop interference (LI) is a critical issue because the relay transmits and receives simultaneously. In general, the LI is much larger than the channel noise and so can significantly degrade performance.

As a consequence, several LI cancellation techniques have been developed [10]. Temporal cancellation methods such as antenna isolation and analog/digital precancellation have been shown to be effective. However, it is impossible to cancel the LI completely, and the residual LI can still be larger than the noise [12].

Estimating channel state information (CSI) is challenging due to the dynamic nature of wireless environments and the complexity of modern communication systems. However, current estimation algorithms have been shown to provide good accuracy, particularly in cellular systems [13]. Thus, it can be assumed that CSI is available in the proposed system. In [14], deep neural networks were considered for channel estimation with multiuser precoding in the downlink of a frequency-division duplex massive MIMO system. In [15], an efficient transceiver design was examined for full-duplex (FD) multiuser massive MIMO systems operating in millimeter wave frequencies. An iterative solution was developed for the proposed digital MMSE transceiver beamforming. Secrecy in cell-free massive MIMO systems was proposed in [16] through the joint optimization of uplink power control, downlink beamforming, and duplex mode selection. Artificial noise was incorporated, and an iterative method was used to solve the associated nonconcave–convex approximation. In [17], beamforming was investigated for in-band FD multi-cell multiuser MIMO networks. Hardware impairments, channel uncertainty, and limited channel state information (CSI) were considered in the MMSE beamforming design to improve system performance under global and local CSI. However, none of the approaches in the literature considered the practical limitation of residual loop interference due to imperfect CSI cancellation. This interference should be considered in developing robust and effective wireless communication systems. The impact of residual self-interference (SI) and loop interference (LI) in a full-duplex space shift keying communication system was investigated in [18]. Thus, the residual LI due to imperfect interference cancellation is considered in this paper.

Unlike the uplink communications in [19], this paper considers a MIMO FD relaying downlink communication system with multiple source–destination pairs. A transceiver design for the sources and relay, and a linear combiner design at the destinations, are investigated considering the residual LI with the goal of minimizing the MSE of the received signals at the destinations. An iterative algorithm is developed to update the source and relay transceiver matrices and the linear combining matrix at the destinations.

The remainder of this paper is organized as follows. Section 2 presents the system model, and the optimization problem is solved in Section 3. Results are presented in Section 4, which demonstrate the effectiveness of this solution, and some conclusions are given in Section 5.

Notation: Bold uppercase, bold lowercase, and normal letters denote matrices, vectors, and scalars, respectively. $\mathrm{vec}(\cdot)$ denotes matrix vectorization, $\otimes$ denotes the matrix Kronecker product, $\mathrm{tr}\{\cdot\}$ is the trace of a matrix, and $\mathbf{I}_N$ is the $N \times N$ identity matrix.

2. System Model

Consider a MIMO full-duplex (FD) relay system with K source–destination pairs communicating simultaneously with the aid of a relay. The source-to-destination links are assumed to be inconsequential due to the large-scale fading and the large distances between them.

Figure 1 presents the system model with the kth source–destination pairs equipped with N_{s_k} and N_{d_k} antennas, respectively. The FD relay has N_r and N_t antennas to simultaneously receive and transmit signals, respectively. Therefore, communication between the source–destination pairs is accomplished in one time slot compared to an HD system that requires two time slots.

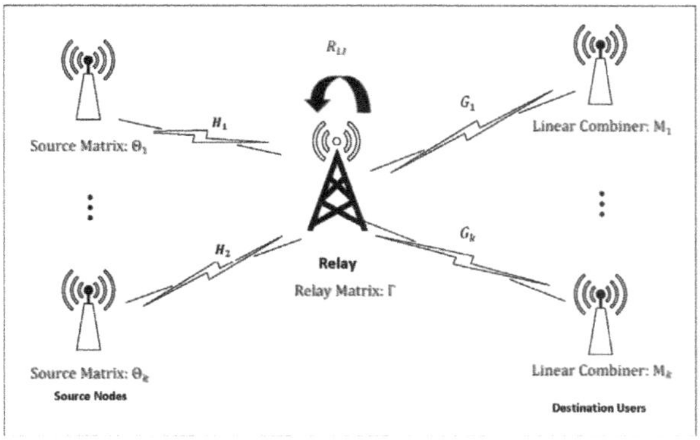

Figure 1. The multiple-user MIMO full-duplex (FD) relay system.

Let $\mathbf{s}_k[n]$ represent the length d signal vector at time n for the kth source. A linear transceiver matrix $\mathbf{\Theta}_k[n]$ is applied to $\mathbf{s}_k[n]$ before transmission. The received signal at the relay can then be expressed as

$$\mathbf{y}_r[n] = \sum_{k=1}^{K} \mathbf{H}_k[n]\mathbf{\Theta}_k[n]\mathbf{s}_k[n] + \mathbf{R}_{LI}[n]\mathbf{t}[n] + \mathbf{n}_r[n], \tag{1}$$

where $\mathbf{H}_k[n] \in \mathbb{C}^{N_r \times N_{s_k}}$ is the channel between source k and the relay node, $\mathbf{R}_{LI}[n] \in \mathbb{C}^{N_r \times N_t}$ is the loop interference (LI) channels, and $\mathbf{n}_r[n] \in \mathbb{C}^{N_r \times 1}$ is an independent and identically distributed (i.i.d.) additive white Gaussian noise (AWGN) matrix. As the received signal $\mathbf{y}_r[n]$ is corrupted by residual LI, from [12], it can be written as

$$\mathbf{y}_r[n] = \sum_{k=1}^{K} \mathbf{H}_k[n]\mathbf{\Theta}_k[n]\mathbf{s}_k[n] + \mathbf{R}_{LI}[n]\Delta\mathbf{t}[n] + \mathbf{n}_r[n], \tag{2}$$

where $\Delta\mathbf{t}[n] = \mathbf{t}[n] - \tilde{\mathbf{t}}[n]$, and $\mathbf{R}_{LI}[n]\Delta\mathbf{t}[n]$ is the residual LI. At time $n+1$, the relay amplifies the received signal with a relay transceiver matrix $\mathbf{\Gamma}[n+1] \in \mathbb{C}^{N_t \times N_r}$ and forwards the resulting signal to the destinations. The received signal at the kth destination is then

$$\mathbf{y}_k[n+1] = \mathbf{G}_k[n+1]\mathbf{\Gamma}[n+1]\mathbf{y}_r[n] + \mathbf{n}_{d_k}[n+1], \tag{3}$$

where $\mathbf{G}_k[n+1] \in \mathbb{C}^{N_i \times N_t}$ is the channel matrix between the relay and destination k, and $\mathbf{n}_{d_k}[n+1] \in \mathbb{C}^{N_k \times 1}$ is an i.i.d. AWGN vector with a zero mean and unit variance. As in [6,7,12], channel state information (CSI) is assumed to be available at all nodes. In addition, channel variations during the transceiver matrix update interval are assumed to be relatively small [19] and so do not influence the transceiver design [12]. Thus, the received signal at the kth destination can be expressed as

$$\mathbf{y}_k = \mathbf{G}_k\mathbf{\Gamma} \sum_{k=1}^{K} \mathbf{H}_k\mathbf{\Theta}_k\mathbf{s}_k + \mathbf{n}_k, \tag{4}$$

where

$$\mathbf{n}_k = \mathbf{G}_k\mathbf{\Gamma}\mathbf{R}_{LI}\Delta\mathbf{t} + \mathbf{G}_k\mathbf{\Gamma}\mathbf{n}_r + \mathbf{n}_{d_k}. \tag{5}$$

As in [18,20], $\Delta\mathbf{t}$ is modeled as white Gaussian noise which is independent of $\mathbf{n}_r$ and $\mathbf{n}_{d_k}$, and has variance σ_t^2. Consequently, the covariance matrix of $\mathbf{n}_k$ is

$$\mathbf{C}_{n_k} = \mathbb{E}[\mathbf{n}_k\mathbf{n}_k^H]. \tag{6}$$

A linear combiner $\mathbf{M}_k \in \mathbb{C}^{N_k \times N_b}$ is used at the kth destination to obtain the signal estimate

$$\hat{\mathbf{s}}_k = \mathbf{M}_k^H \mathbf{y}_k. \tag{7}$$

The mean square error (MSE) at the kth user is defined as

$$\begin{aligned} \text{MSE}_k &= \mathbb{E}[(\hat{\mathbf{s}}_k - \mathbf{s}_k)(\hat{\mathbf{s}}_k - \mathbf{s}_k)^H] \\ &= (\mathbf{M}_k^H \mathbf{L}_k - \mathbf{I}_d)(\mathbf{M}_k^H \mathbf{L}_k - \mathbf{I}_d)^H + \mathbf{M}_k^H \mathbf{C}_{n_k} \mathbf{M}_k \\ &+ \mathbf{M}_k^H \mathbf{E}_{n_k} \mathbf{M}_k, \end{aligned} \tag{8}$$

where $\mathbf{L}_k = \mathbf{G}_k \boldsymbol{\Gamma} \bar{\mathbf{H}}_k$ and $\mathbf{E}_{n_k} = \mathbf{G}_k \boldsymbol{\Gamma} \sum_{m=1, m \neq k}^{K} \bar{\mathbf{H}}_m \bar{\mathbf{H}}_m^H \boldsymbol{\Gamma}^H \mathbf{G}_k^H, k = 1, \ldots, K$. Optimization is employed to obtain the linear transceiver matrices $\boldsymbol{\Theta}_k$ and $\boldsymbol{\Gamma}$ at the source and relay, and to obtain the linear combiners $\mathbf{M}_k$ at the destinations. The minimization of the sum mean squared error (SMSE) of the received signals at the destinations can be expressed as

$$\min_{\boldsymbol{\Theta}_k, \boldsymbol{\Gamma}, \mathbf{M}_k} \ \text{SMSE} \tag{9a}$$

$$s.t. \ \text{tr}(\boldsymbol{\Gamma}(\mathbf{y}_r \mathbf{y}_r^H) \boldsymbol{\Gamma}^H) \leq P_r \tag{9b}$$

$$\text{tr}(\boldsymbol{\Theta}_k \boldsymbol{\Theta}_k^H) \leq P_{s_k}, k = 1, \ldots, K, \tag{9c}$$

where $\text{SMSE} = \sum_{k=1}^{K} \text{tr}(\text{MSE}_k)$, and $P_{s_k} > 0$ and $P_r > 0$ are the power constraints at the kth source and relay, respectively.

3. Optimization Problem Solution

The problem in (9) is highly nonconvex which makes obtaining a global optimal solution intractable, so an iterative algorithm is presented. This solution is based on an alternating optimization that updates $\boldsymbol{\Theta}_k$, $\boldsymbol{\Gamma}$ and $\mathbf{M}_k$ individually with the other parameters kept fixed to solve the three convex subproblems.

First, given $\boldsymbol{\Theta}_k$ and $\boldsymbol{\Gamma}$, the optimal combiner at the kth destination $\mathbf{M}_k$ is obtained by solving the unconstrained convex problem for $\mathbf{M}_k$, which is independent of the constraints in (9b) and (9c). The optimal solution can then be obtained by taking the derivative of (8) with respect to $\mathbf{M}_k$ and setting it to zero. Solving $\frac{\partial}{\partial \mathbf{M}_k} \text{tr}(\text{MSE}_k) = 0$ gives

$$\mathbf{M}_k = \mathbf{C}_{\mu_k}^{-1} \mathbf{G}_k \boldsymbol{\Gamma} \mathbf{H}_k \boldsymbol{\Theta}_k, \tag{10}$$

where

$$\mathbf{C}_{\mu_k} = (\mathbf{G}_k \boldsymbol{\Gamma} \mathbf{H} \boldsymbol{\Theta}_k \boldsymbol{\Theta}_k^H \mathbf{H}_k^H \boldsymbol{\Gamma}^H \mathbf{G}_k^H + \mathbf{C}_{n_k} + \mathbf{E}_{n_k})^{-1} \mathbf{G}_k \boldsymbol{\Gamma} \mathbf{H}_k \boldsymbol{\Theta}_k,$$

which is known as the Wiener filter.

Second, with $\mathbf{M}_k$ from (10) and given $\boldsymbol{\Theta}_k$, $\boldsymbol{\Gamma}$ can be obtained by solving the following problem. The objective function in (9) can be expressed as

$$\begin{aligned} \text{SMSE} = &\sum_{k=1}^{K} \text{tr}((\mathbf{M}_k^H \mathbf{G}_k \boldsymbol{\Gamma} \mathbf{H} \boldsymbol{\Theta}_k - \mathbf{I}_{N_k})(\mathbf{M}_k^H \mathbf{G}_k \boldsymbol{\Gamma} \mathbf{H} \boldsymbol{\Theta}_k \\ &- \mathbf{I}_{N_k})^H + \mathbf{M}_k^H (\sigma_t^2 \mathbf{G}_k \boldsymbol{\Gamma} \mathbf{R}_{LI} \mathbf{R}_{LI}^H \boldsymbol{\Gamma}^H \mathbf{G}_k^H + \\ &\mathbf{G}_k \boldsymbol{\Gamma} \boldsymbol{\Gamma}^H \mathbf{G}_k^H + \mathbf{I}_{N_k}) \mathbf{M}_k + \mathbf{M}_k^H \mathbf{G}_k \boldsymbol{\Gamma} \times \\ &\sum_{m=1, m \neq k}^{K} \mathbf{H} \boldsymbol{\Theta}_m \boldsymbol{\Theta}_m^H \mathbf{H}^H \boldsymbol{\Gamma}^H \mathbf{G}_k^H \mathbf{M}_k). \end{aligned} \tag{11}$$

Note that since $\mathbf{\Theta}_k$ is known, the constraint in (9c) is eliminated. The optimization problem then becomes

$$\begin{aligned}&\min_{\mathbf{\Gamma}}\ \text{SMSE}\\&s.t.\ \text{tr}(\mathbf{\Gamma}(\sum_{k=1}^{K}\mathbf{H}_k\mathbf{\Theta}_k\mathbf{\Theta}_k^H\mathbf{H}_k^H+\sigma_t^2\mathbf{R}_{LI}\mathbf{R}_{LI}^H+\mathbf{I}_{N_r})\mathbf{\Gamma}^H)\leq P_r.\end{aligned}\tag{12}$$

Consider the singular-value decomposition (SVD) of the equivalent source-to-relay and relay-to-destination channels:

$$\mathbf{H}=[\mathbf{H}_1\mathbf{\Theta}_1,\ldots,\mathbf{H}_K\mathbf{\Theta}_K]=\mathbf{U}_h\Lambda_h\mathbf{V}_h^H,\tag{13}$$

and

$$\mathbf{G}=[\mathbf{G}_1^T,\ldots,\mathbf{G}_K^T]^T=\mathbf{U}_g\Lambda_g\mathbf{V}_g^H.\tag{14}$$

The optimal relay precoding matrix is similar to that in [7]:

$$\mathbf{\Gamma}=\mathbf{V}_g\mathbf{A}\mathbf{U}_h^H,\tag{15}$$

and from (13) and (14), we have

$$\mathbf{H}_k\mathbf{\Theta}_k=\mathbf{U}_h\Lambda_h\mathbf{V}_{h_k}^H,\tag{16}$$

and

$$\mathbf{G}_k=\mathbf{U}_{g_k}\Lambda_g\mathbf{V}_g^H.\tag{17}$$

Note that

$$\mathbf{V}_h=[\mathbf{V}_{h1}^T,\ldots,\mathbf{V}_{hK}^T]^T,$$

and

$$\mathbf{U}_g=[\mathbf{U}_{g1}^T,\ldots,\mathbf{U}_{gK}^T]^T,$$

which have dimensions $d\times L_1$ and $N_i\times L_2$, respectively. Substituting (13)–(15) in (11) gives

$$\begin{aligned}\text{SMSE}=&\sum_{k=1}^{K}\text{tr}((\mathbf{M}_k^H\mathbf{U}_{g_k}\Lambda_g\mathbf{A}\Lambda_h\mathbf{V}_{h_k}^H-\mathbf{I}_{N_k})\times\\&(\mathbf{M}_k^H\mathbf{U}_{g_k}\Lambda_g\mathbf{A}\Lambda_h\mathbf{V}_{h_k}^H-\mathbf{I}_{N_k})^H\\&+\mathbf{M}_k^H(\sigma_t^2\mathbf{U}_{g_k}\Lambda_g\mathbf{A}\mathbf{U}_h^H\mathbf{R}_{LI}\mathbf{R}_{LI}^H\mathbf{U}_h\mathbf{A}^H\Lambda_g\mathbf{U}_{g_k}^H\\&+\mathbf{U}_{g_k}\Lambda_g\mathbf{A}\mathbf{A}^H\Lambda_g\mathbf{U}_{g_k}^H+\mathbf{I}_{N_k}+\mathbf{U}_{g_k}\Lambda_g\mathbf{A}\\&\times\sum_{m=1,m\neq k}^{K}\Lambda_h\mathbf{V}_{h_m}^H\mathbf{V}_{h_m}\Lambda_h\mathbf{A}^H\Lambda_g\mathbf{U}_{g_k}^H)\mathbf{M}_k).\end{aligned}\tag{18}$$

Using the matrix identities

$$\begin{aligned}\text{tr}(\mathbf{C}^T\mathbf{D})&=(\ (\mathbf{C}))^H\text{vec}(\mathbf{D}),\\\text{tr}(\mathbf{A}^H\mathbf{B}\mathbf{A}\mathbf{C})&=\text{tr}(\text{vec}(\mathbf{A}))^H(\mathbf{C}^T\otimes\mathbf{B})\text{vec}(\mathbf{A})\\\text{vec}\mathbf{A}\mathbf{B}\mathbf{C}&=(\mathbf{C}^T\otimes\mathbf{A})\text{vec}(\mathbf{B}),\end{aligned}$$

where $\text{vec}(\cdot)$ concatenates the columns of a matrix into a single vector, the SMSE in (11) can be expressed as a function of $\mathbf{a}=\text{vec}(\mathbf{A})$:

$$\begin{aligned}\text{SMSE}=&\sum_{k=1}^{K}(\mathbf{P}_k\mathbf{a}-\text{vec}(\mathbf{I}_{N_k}))^H(\mathbf{P}_k\mathbf{a}-\text{vec}(\mathbf{I}_{N_k}))\\&+\mathbf{a}^H\mathbf{Q}_k\mathbf{a}+\mathbf{a}^H\mathbf{S}_k\mathbf{a}+\mathbf{a}^H\mathbf{R}_k\mathbf{a}+\mathrm{t}_1\end{aligned}\tag{19}$$

where

$$\mathbf{t}_1 = \sum_{k=1}^{K} \text{tr}(\mathbf{M}_k^H \mathbf{M}_k),$$

does not depend on **a** and

$$\mathbf{P}_k = (\Lambda_h \mathbf{V}_{h_k}^H)^T \otimes (\mathbf{M}_k^H \mathbf{U}_{g_k} \Lambda_g), \tag{20}$$

$$\mathbf{Q}_k = \sigma_t^2((\mathbf{U}_h^H \mathbf{R}_{LI} \mathbf{R}_{LI}^H \mathbf{U}_h)^T \otimes \Lambda_g \mathbf{U}_{g_k}^H \mathbf{M}_k \mathbf{M}_k^H \mathbf{U}_{g_k} \Lambda_g), \tag{21}$$

$$\mathbf{S}_k = \mathbf{I}_L \otimes (\Lambda_g \mathbf{U_g}_k^H \mathbf{M}_k \mathbf{M}_k^H \mathbf{U}_{g_k} \Lambda_g), \tag{22}$$

$$\mathbf{R}_k = (\sum_{m=1, m\neq k}^{K} \Lambda_h \mathbf{V}_{h_m}^H \mathbf{V}_{h_m} \Lambda_h)^T \otimes (\Lambda_g \mathbf{U}_{g_k}^H \mathbf{M}_k \mathbf{M}_k^H . \mathbf{U}_{g_k} \Lambda_g). \tag{23}$$

The relay transmit power constraint in (12) can be written as

$$\begin{aligned} &\text{tr}(\mathbf{\Gamma} (\sum_{k=1}^{K} \mathbf{H}_k \mathbf{\Theta}_k \mathbf{\Theta}_k^H \mathbf{H}_k^H + \sigma_t^2 \mathbf{R}_{LI} \mathbf{R}_{LI}^H + \mathbf{I}_{N_r}) \mathbf{\Gamma}^H) \\ &= \text{tr}(\mathbf{A} \times (\Lambda_h^2 + \sigma_t^2 \mathbf{U}_h^H \mathbf{R}_{LI} \mathbf{R}_{LI}^H \mathbf{U}_h + \mathbf{I}_{L_1}) \mathbf{A}^H). \end{aligned} \tag{24}$$

Defining $\mathbf{D} = (\Lambda_h^2 + \sigma_t^2 \mathbf{U}_h^H \mathbf{R}_{LI} \mathbf{R}_{LI}^H \mathbf{U}_h + \mathbf{I}_{L_1}) \otimes \mathbf{I}_{L_1}$, (24) can be written as

$$\mathbf{a}^H \mathbf{D} \mathbf{a} \leq P_r. \tag{25}$$

The original relay optimization problem is then given by

$$\begin{aligned} &\min_{\mathbf{A}} \text{ SMSE} \\ &s.t. \ \mathbf{a}^H \mathbf{D} \mathbf{a} \leq P_r \end{aligned} \tag{26}$$

This is a quadratically constrained quadratic programming (QCQP) problem which is convex and so can be efficiently solved using the interior point method. For example, the CVX toolbox for disciplined convex programming [21] can be employed.

Third, the kth source precoding matrix $\mathbf{\Theta}_k$ can be obtained using $\mathbf{M}_k$ and $\mathbf{\Gamma}$ given above. The corresponding optimization problem can be formulated as the following QCQP problem:

$$\begin{aligned} \text{SMSE} = &\sum_{k=1}^{K} \text{tr}((\bar{\mathbf{G}}_k \mathbf{\Gamma} \mathbf{H}_k \mathbf{\Theta}_k - \mathbf{I}_{N_k})(\bar{\mathbf{G}}_k \mathbf{\Gamma} \mathbf{H}_k \mathbf{\Theta}_k - \mathbf{I}_{N_k})^H \\ &+ \bar{\mathbf{G}}_k \mathbf{\Gamma} \sum_{m=1, m\neq k}^{K} \mathbf{H}_k \mathbf{\Theta}_k \mathbf{\Theta}_k^H \mathbf{H}_k^H \mathbf{\Gamma}^H \bar{\mathbf{G}}_k^H) + \mathbf{t}_2, \end{aligned} \tag{27}$$

where $\bar{\mathbf{G}}_k = \mathbf{M}_k^H \mathbf{G}_k$ and $\mathbf{t}_2 = \sum_{k=1}^{K} \text{tr}(\mathbf{M}_k^H \mathbf{C}_k \mathbf{M}_k)$ can be ignored in the optimization as they do not depend on $\mathbf{\Theta}_k$. Using the matrix identities, (27) can be written as a function of $\theta_k = \text{vec}(\mathbf{\Theta}_k)$ which gives

$$\begin{aligned}\text{SMSE} = & \sum_{k=1}^{K}[(\mathbf{S}_k\theta_k - \text{vec}(\mathbf{I}_{N_k}))^H(\mathbf{S}_k\theta_k - \text{vec}(\mathbf{I}_{N_k})) \\ & + \sum_{m=1,m\neq k}^{K} \theta_m^H(\mathbf{I}_d \otimes \mathbf{H}_m^H\mathbf{\Gamma}^H\bar{\mathbf{G}}_k^H\bar{\mathbf{G}}_k\mathbf{\Gamma}\mathbf{H}_m)\theta_m] \\ & + \mathbf{t}_2 \\ = & \sum_{k=1}^{K}[(\mathbf{S}_k\theta_k - \text{vec}(\mathbf{I}_{N_k}))^H(\mathbf{S}_k\theta_k - \text{vec}(\mathbf{I}_{N_k})) \\ & + \theta_k^H\mathbf{T}_k\theta_k] + \mathbf{t}_2,\end{aligned} \tag{28}$$

where

$$\mathbf{S}_k = \mathbf{I}_{N_k} \otimes (\bar{\mathbf{G}}_k\mathbf{\Gamma}\mathbf{H}),$$

$$\mathbf{T}_k = \mathbf{I}_{N_k} \otimes \sum_{m=1,m\neq k}^{K} (\mathbf{H}^H\mathbf{\Gamma}^H\bar{\mathbf{G}}_k^H\bar{\mathbf{G}}_k\mathbf{\Gamma}\mathbf{H}),$$

and

$$\mathbf{t}_2 = \sum_{k=1}^{K} \text{tr}(\mathbf{M}_k^H\mathbf{C}_{n_k}\mathbf{M}_k),$$

which is independent of $\mathbf{\Theta}_k$ and so can be ignored. Introducing $\mathbf{T} = \text{bd}(\mathbf{T}_1, \ldots, \mathbf{T}_K)$, where $\text{bd}(\cdot)$ denotes a block-diagonal matrix, and $\bar{\mathbf{S}}_k = [\mathbf{S}_{k1}, \ldots, \mathbf{S}_{kK}]$, where $\mathbf{S}_{kk} = \mathbf{S}_k$ and $\mathbf{S}_{jk} = 0, j \neq k$, (27) can be written as a function of $\theta = [\theta_1^T, \theta_1^T, \ldots, \theta_K^T]^T$, which gives

$$\Phi_1(\theta) = \sum_{k=1}^{K}(\bar{\mathbf{S}}_k\theta - \text{vec}(\mathbf{I}_{N_k}))^H(\bar{\mathbf{S}}_k\theta - \text{vec}(\mathbf{I}_{N_k})) + \theta^H\mathbf{T}\theta \tag{29}$$

Now, defining

$$\mathbf{E}_j = \mathbf{I}_{N_k} \otimes (\mathbf{H}_j^H\mathbf{\Gamma}^H\mathbf{\Gamma}\mathbf{H}_j),$$

$$\mathbf{E} = \text{bd}(\mathbf{E}_1, \mathbf{E}_2, \ldots, \mathbf{E}_K),$$

and

$$\bar{\mathbf{E}} = \text{bd}(\bar{\mathbf{E}}_{i1}, \bar{\mathbf{E}}_{i2}, \ldots, \bar{\mathbf{E}}_{iK}),$$

where $\bar{\mathbf{E}}_{iK} = \mathbf{I}_{dN_s}$ and $\bar{\mathbf{E}}_{ij} = 0, i \neq j$, the optimal θ can be obtained by solving the following problem

$$\begin{aligned}& \min_{\theta} \ \Phi_1(\theta) \\ & s.t. \ \sum_{i=1}^{K} \theta^H\bar{\mathbf{E}}_i\theta \leq P_s \\ & \qquad \theta^H\mathbf{E}\theta \leq P_r - \sigma_r^2\text{tr}(\mathbf{\Gamma}(\sigma_t^2\mathbf{R}_{LI}\mathbf{R}_{LI}^H + \mathbf{I}_{N_r})\mathbf{\Gamma}^H).\end{aligned} \tag{30}$$

This is a QCQP problem which can be solved using the CVX toolbox for disciplined convex programming.

The symbols employed in this work are summarized in Table 1. The proposed iterative algorithm is given in Algorithm 1. The convergence of this algorithm is guaranteed as follows. As the three subproblems are convex, each update of $\mathbf{\Theta}_k$, $\mathbf{\Gamma}$, and $\mathbf{M}_k$ may decrease or at least not increase the value of the objective function. Thus, the iterative algorithm will converge to an optimum solution.

Algorithm 1: Iterative Design of $\mathbf{\Theta}_k$, $\mathbf{\Gamma}$, and $\mathbf{M}_k$.

1: Initialize the algorithm with $\mathbf{\Theta}^{(0)} = \sqrt{\frac{P_s}{L}}\mathbf{I}_L$ and
$\mathbf{\Gamma}^{(0)} = \sqrt{\frac{P_r}{\text{tr}(\mathbf{H}_{SR}\mathbf{\Theta}_k^{(0)}(\mathbf{H}_{SR}\mathbf{\Theta}_k^{(0)})^H+\sigma_r^2\mathbf{R}_{LI}\mathbf{R}_{LI})^H+\mathbf{I}_{N_r})}}\mathbf{I}_{N_r}$, and set $n = 0$.
2: Update $\mathbf{M}_k^{(n)}$ using $\mathbf{\Gamma}^{(n)}$ and $\mathbf{\Theta}_k^{(n)}$ using (10).
3: Update $\mathbf{\Gamma}^{(n+1)}$ using $\mathbf{M}_k^{(n)}$ and $\mathbf{\Theta}_k^{(n)}$, $k = 1, \ldots, K$, in (15).
4: Update $\mathbf{\Theta}_k^{(n+1)}$ using $\mathbf{M}_k^{(n)}$ and $\mathbf{\Gamma}^{(n+1)}$ by solving the problem in (30).
5: If $(\text{SMSE}^{(n)} - \text{SMSE}^{(n+1)})/\text{SMSE}^{(n)} > \epsilon$, go to step 2.
6: End

Table 1. Symbol descriptions.

Symbol	Description
$\mathbf{s}_k$	Signal vector for the kth source
$\mathbf{\Theta}_k$	Linear transceiver matrix at the source for the kth user
$\mathbf{H}_k$	Channel between source k and the relay node
$\mathbf{R}_{LI}$	Loop interference (LI) channels
$\mathbf{n}_r$	i.i.d. additive white Gaussian noise (AWGN) matrix
$\mathbf{M}_k$	Linear combiner at the destination
$\mathbf{\Gamma}$	Transceiver matrix at the relay for the kth user
$\mathbf{M}_k$	Linear combiners at the destinations for the kth user
R	Achievable rate

4. Numerical Results

This section presents the evaluation of the performance of the proposed multiuser MIMO FD relay downlink algorithm using numerical simulation. The CVX toolbox was employed, which is a Matlab-based software package for solving convex optimization problems [21]. For simplicity, we considered a system with two sources and two destinations. The extension to more than two source–destination pairs is straightforward. We assumed that the sources and destinations were each equipped with two antennas, and the relay was equipped with four receive antennas and four transmit antennas. The simulation parameters are given in Table 2.

As in the related literature, flat-fading MIMO channels were considered. It was assumed that the elements of $\mathbf{H}_k$, $\mathbf{G}_k$, and $\mathbf{R}_{LI}$ were i.i.d. complex Gaussian random variables with a zero mean and unit variance. Moreover, all noise terms were i.i.d. complex, circularly symmetric Gaussian random variables with a zero mean and unit variance. According to [9], the residual LI can be 0 dB to 20 dB larger than the channel noise. Therefore, LI levels of 0 dB, 5 dB and 10 dB were employed here. In all cases, the results are given for an average of 1000 independent channel realizations. Note that the HD system mentioned in this paper is the same as the proposed FD system except that the residual LI term is zero and the achievable rate of the HD system is half of the FD system since two time slots are required for transmission between a source and destination. Thus, the same optimization procedure is employed.

Figure 2 presents the SMSE for the HD and FD systems with different levels of residual LI. The signal-to-noise ration (SNR) between the sources and relay varies from 0 dB to 30 dB, and the SNR between the relay and destinations is fixed at 30 dB. These results indicate that the HD system provides the best SMSE as the FD performance is degraded as the LI increases.

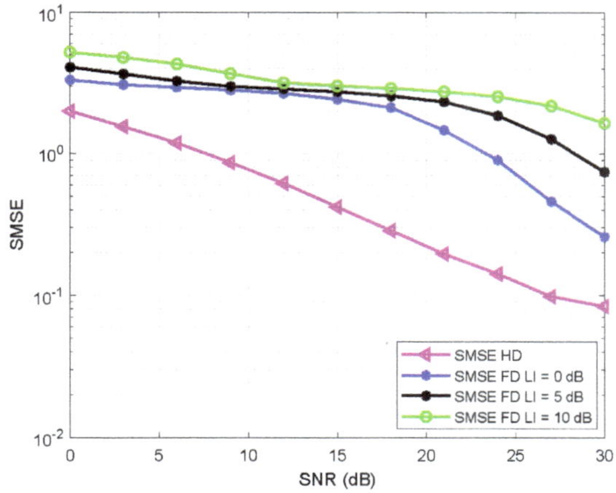

Figure 2. SMSE versus SNR_{s-r} with $\text{SNR}_{r-d} = 30$ dB.

The achievable rate for the system in (5) can be obtained using an approach similar to that in [8], and can be written as

$$R = \sum_{k=1}^{K} R_k, \tag{31}$$

where

$$\begin{aligned} R_k = \log_2 \det\big[\mathbf{I}_{N_d} &+ (\mathbf{G}_k\mathbf{\Gamma}\mathbf{H}_k\mathbf{\Theta}_k)(\mathbf{G}_k\mathbf{\Gamma}\mathbf{H}_k\mathbf{\Theta}_k)^H \\ &\times (\sigma_t^2\mathbf{G}_k\mathbf{\Gamma}\mathbf{R}_{LI}\mathbf{R}_{LI}^H\mathbf{\Gamma}^H\mathbf{G}_k^H + \mathbf{G}_k\mathbf{\Gamma}\mathbf{\Gamma}^H\mathbf{G}_k^H + \mathbf{I}_{N_d})^{-1}\big]. \end{aligned} \tag{32}$$

Figure 3 gives the achievable rate for the HD and FD systems. In this figure, the SNR between the sources and relay varies from 0 dB to 30 dB while the SNR between the relay and destinations is fixed at 30 dB. The HD system corresponds to the case when the residual LI is zero and two time slots are required for transmission from sources to destinations Therefore, the FD achievable rate is twice the HD achievable rate if the LI is cancelled completely. Further, the achievable rate with the FD system is higher when the residual LI is 0 dB to 10 dB. The achievable rate of the FD system is degraded as the residual LI increases in the high-SNR region since the residual LI is greater than the multiuser interference. In the low-SNR region, the multiuser interference dominates the residual LI, so the effect of the residual LI is negligible.

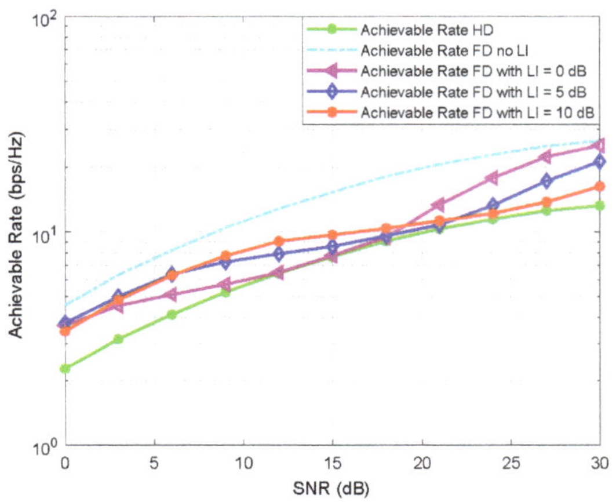

Figure 3. Achievable rate versus SNR_{s-r} with $\text{SNR}_{r-d} = 30$ dB.

Figures 4 and 5 present the SMSE and achievable rate with a fixed SNR of 30 dB between the sources and relay and an SNR between the relay and destinations from 0 dB to 30 dB. The SMSE in Figure 4 is better than in Figure 2 in the low-SNR region because a higher transmit power results in more residual LI. Figure 5 shows that the achievable rate of the FD system is always higher than that of the HD system for residual LI levels up to 10 dB. These results confirm that the proposed approach is effective in addressing the effects of the residual loop interference due to imperfect LI cancellation in full-duplex systems. Further, this solution is superior to conventional half-duplex communications. Table 3 gives the percentage improvement in achievable rate with the proposed system over the conventional half-duplex system for different levels of residual LI and SNRs. These results demonstrate the robustness of the proposed design as the improvements exceed 75% when the residual LI is low. Further, they confirm the effectiveness in optimizing performance in practical scenarios.

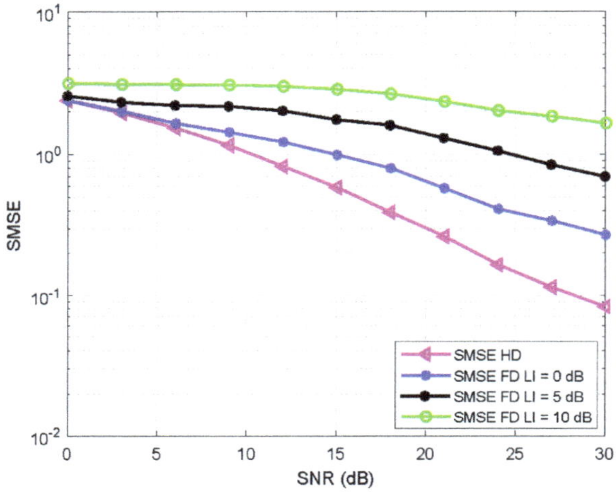

Figure 4. SMSE versus SNR_{r-d} with $\text{SNR}_{s-r} = 30$ dB.

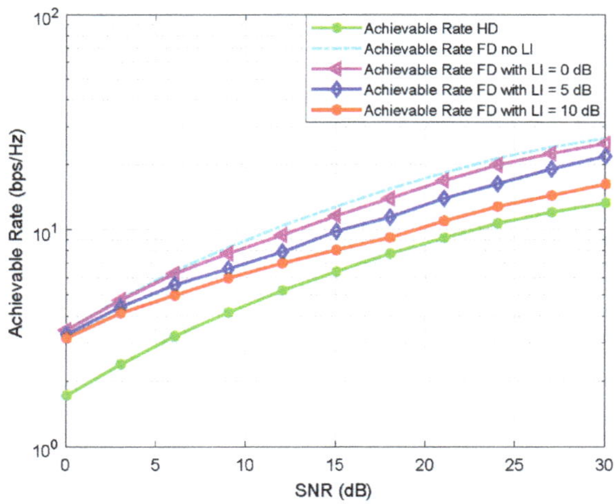

Figure 5. Achievable rate versus SNR_{r-d} with $\text{SNR}_{s-r} = 30$ dB.

Table 2. Simulation parameters.

Parameter	Value
Downlink/uplink users	2
Relay antennas	2×2
Downlink/uplink user antennas	2×2
LI levels	0 dB, 5 dB, 10 dB
No. of independent channel realizations	1000
Source–relay and relay–source SNR	0 dB, 5 dB, 10 dB, 15 dB, 20 dB, 25 dB, 30 dB

Table 3. Percentage improvement in achievable rate.

Residual LI	SNR_{r-d} = 30 dB with SNR_{s-r} = 27 dB	SNR_{r-d} = 27 dB with SNR_{s-r} = 30 dB
FD No LI	100%	100%
FD LI = 0 dB	76%	86%
FD LI = 5 dB	36%	58%
FD LI = 10 dB	4%	9%

5. Conclusion

This paper considers a transceiver design for the downlink of a multiuser non-regenerative MIMO full-duplex (FD) relay system with residual loop interference (LI). The source and relay transceiver matrices and destination combining matrix are optimized to minimize the sum mean squared error (SMSE). The original nonconvex problem is converted to three convex subproblems, and an iterative algorithm is developed to optimize the three matrices. Results are presented which demonstrate that the proposed iterative method outperforms the corresponding half-duplex (HD) relay system in terms of the SMSE and achievable rate. The effectiveness of the proposed solution was illustrated via simulation. Future research can consider interference cancellation techniques, integrating machine learning for dynamic system optimization, and next-generation wireless technologies. Internet of things (IoT) and mobile communication applications can also be examined considering the adaptability and scalability of the proposed approach and to further confirm its impact.

Author Contributions: Conceptualization, Y.S. and T.A.G.; methodology, Y.S. and T.A.G.; validation, Y.S.; writing—original draft preparation, Y.S.; writing—review and editing, Y.S. and T.A.G.; supervision, T.A.G. All authors have read and agreed to the published version of the manuscript.

Funding: This research received no external funding.

Data Availability Statement: No new data were created or analyzed in this study. Data sharing is not applicable to this article.

Conflicts of Interest: The authors declare no conflicts of interest.

Abbreviations

The following abbreviations are used in this manuscript:

MIMO	Multiple-Input Multiple-Output;
FD	Full-Duplex;
HD	Half-Duplex;
LI	Loop Interference.

eferences

Xu, W.; Dong, X.; Lu, W. Joint precoding optimization for multiuser multi-antenna relaying downlinks using quadratic programming. *IEEE Trans. Commun.* **2011**, *59*, 1228–1235. [CrossRef]

Liu, J.; Liu, Z.; Qiu, Z. Joint MMSE transceiver design for multiuser non-regenerative MIMO relay downlink systems. In Proceedings of the 2011 7th International Wireless Communications and Mobile Computing Conference, Istanbul, Turkey, 4–8 July 2011; pp. 877–882. [CrossRef]

3. Liu, J.; Gao, F.; Qiu, Z. Robust transceiver design for downlink multiuser MIMO AF relay systems. *IEEE Trans. Wirel. Commun.* **2015**, *14*, 1228–1235. [CrossRef]
4. Xu, D.; Huang, Y.; Yang, L.; Li, B. Linear transceiver design for multiuser MIMO downlink. In Proceedings of the 2008 IEEE International Conference on Communications, Beijing, China, 19–23 May 2008; pp. 761–765. [CrossRef]
5. Zhai, C.; Li, X.; Hei, Y. A novel decomposed transceiver design for multiuser MIMO relay downlink systems. In Proceedings of the 2012 IEEE Wireless Communications and Networking Conference (WCNC), Paris, France, 1–4 April 2012; pp. 1921–1924. [CrossRef]
6. Gopal, L.; Rong, Y.; Zang, Z. Robust MMSE transceiver design for nonregenerative multicasting MIMO relay systems. *IEEE Trans. Vehic. Tech.* **2017**, *66*, 8979–8989. [CrossRef]
7. Nguyen, K.X.; Rong, Y.; Nordholm, S. Transceiver optimization for interference MIMO relay systems using the structure of relay matrix. In Proceedings of the 2015 24th Wireless and Optical Communication Conference (WOCC), Taipei, Taiwan, 23–24 October 2015; pp. 29–33. [CrossRef]
8. Kang, Y.Y.; Cho, J.H. Capacity of MIMO wireless channel with full-duplex amplify-and-forward relay. In Proceedings of the 2009 IEEE 20th International Symposium on Personal, Indoor and Mobile Radio Communications, Tokyo, Japan, 13–16 September 2009; pp. 117–121. [CrossRef]
9. Bharadia, D.; McMilin, E.; Katti, S. Full duplex radios. *ACM SIGCOMM Comput. Commun. Rev.* **2013**, *43*, 375–386. [CrossRef]
10. Riihonen, T.; Werner, S.; Wichman, R. Mitigation of loopback self-interference in full-duplex MIMO relays. *IEEE Trans. Sig. Process.* **2011**, *59*, 5983–5993. [CrossRef]
11. Cirik, A.C.; Biswas, S.; Vuppala, S.; Ratnarajah, T. Robust transceiver design for full duplex multiuser MIMO systems. *IEEE Wirel. Commun. Lett.* **2016**, *5*, 260–263. [CrossRef]
12. Lin, C.T.; Tseng, F.S.; Wu, W.R.; Jheng, F.J. Joint precoders design for full-duplex MIMO relay systems with QR-SIC detector. In Proceedings of the 2015 IEEE Global Communications Conference (GLOBECOM), San Diego, CA, USA, 6–10 December 2015. [CrossRef]
13. Stenhammar, O.; Fodor, G.; Fischione, C. A comparison of neural networks for wireless channel prediction. *IEEE Wirel. Commun.* 2024, *early access*. [CrossRef]
14. Sohrabi, F.; Attiah, K.M.; Yu, W. Deep learning for distributed channel feedback and multiuser precoding in FDD massive MIMO. *IEEE Trans. Wirel. Commun.* **2021**, *20*, 4044–4057. [CrossRef]
15. Darabi, M.; Cirik, A.C.; Lampe, L. Transceiver design in millimeter wave full-duplex multi-user massive MIMO communication systems. *IEEE Access* **2021**, *9*, 165394–165408. [CrossRef]
16. Xia, X.; Fan, Z.; Luo, W.; Lu, A.; Wang, D.; Zhao, X.; You, X. Joint uplink power control, downlink beamforming, and mode selection for secrecy cell-free massive MIMO with network-assisted full duplexing. *IEEE Syst. J.* **2023**, *17*, 720–731. [CrossRef]
17. Luo, H.; Garg, N.; Ratnarajah, T. Beamforming design for in-band full-duplex multi-cell multi-user MIMO networks with global and local CSI. *IEEE Trans. Veh. Technol.* **2023**, *72*, 10218–10233. [CrossRef]
18. Zhu, X.; Chen, W.; Wu, Q.; Liu, Z.; Li, J. Performance of RIS-assisted full-duplex space shift keying with imperfect self-interference cancellation. *IEEE Wirel. Commun. Lett.* **2024**, *13*, 128–132. [CrossRef]
19. Shao, Y.; Gulliver, T.A. Precoding for multiuser MIMO full-duplex amplify-and-forward relay uplink communication systems. *SN Appl. Sci.* **2020**, *2*, 694. [CrossRef]
20. Rodriguez, L.J.; Tran, N.H.; Le-Ngoc, T. Performance of full-duplex AF relaying in the presence of residual self-interference. *IEEE J. Sel. Areas Commun.* **2014**, *32*, 1752–1764. [CrossRef]
21. Grant, M.C.; Boyd, S.P. CVX: Matlab Software for Disciplined Convex Programming. March 2014. Available online: http://cvxr.com/cvx/ (accessed on 1 January 2023).

telecom

MDPI

Review

A Comparative Study of Waveforms Across Mobile Cellular Generations: From 0G to 6G and Beyond

Farah Arabian * and Morteza Shoushtari

Department of Electrical and Computer Engineering, Brigham Young University, Provo, UT 84602, USA; morteza.shoushtari@byu.edu

* Correspondence: farah.arabian@byu.edu

Abstract: Waveforms define the shape, structure, and frequency characteristics of signals, whereas modulation schemes determine how information symbols are mapped onto these waveforms for transmission. Their appropriate selection plays a critical role in determining the efficiency, robustness, and reliability of data transmission. In wireless communications, the choice of waveform influences key factors, such as network capacity, coverage, performance, power consumption, battery life, spectral efficiency (SE), bandwidth utilization, and the system's resistance to noise and electromagnetic interference. This paper provides a comprehensive analysis of the waveforms and modulation schemes used across successive generations of mobile cellular networks, exploring their fundamental differences, structural characteristics, and trade-offs for various communication scenarios. It also situates this analysis within the historical evolution of mobile standards, highlighting how advances in modulation and waveform technologies have shaped the development and proliferation of cellular networks. It further examines criteria for waveform selection—such as SE, bit error rate (BER), throughput, and latency—and discusses methods for assessing waveform performance. Finally, this study presents a comparative evaluation of modulation schemes across multiple mobile generations, focusing on key performance metrics, with the BER analysis conducted through MATLAB simulations.

Keywords: 0G; 1G; 2G; 3G; 4G; 5G; 6G; AMPS; NMT; TACS; ETACS; GSM; cdmaOne; D-AMPS; GPRS; EDGE; WCDMA; TD-CDMA; TD-SCDMA; CDMA2000; WiMAX; HSDPA; HSPA+; LTE; WiMAX-2; single-carrier modulation; multi-carrier modulation; constant envelope modulation; SC-TDE; SC-FDE; SC-FDM; OFDM; CP-OFDM; F-OFDM; UFMC; FBMC; GFDM; OTFS; DFT-s-OFDM; DFT-s-OTFS; CCWD; SCWD; JWOD; OCDM; AFDM; index modulation

1. Introduction

The development of mobile cellular networks can be traced back to the 1940s, when they were primarily used for military communication. Over the years, these networks have undergone significant transformations and have become an integral part of our daily lives [1–5]. The widespread adoption of cellular networks has revolutionized the way we communicate and access information, impacting nearly every aspect of our routines. Today, with over 7.5 billion people worldwide using mobile devices, the demand for reliable cellular networks that can support mobile applications continues to surge [6].

As mobile data services continue to gain popularity, there is a growing demand for cellular networks that can reliably support mobile applications and adapt to changing requirements [7]. Cellular network infrastructure was originally designed for voice communication, with an emphasis on high-quality service. To ensure comprehensive radio

 https://doi.org/10.3390/telecom6030067

coverage, operators make substantial investments in the deployment and upgrading of infrastructure components, such as base stations. Modern cellular networks utilize intelligent core network architectures that employ sophisticated signaling protocols and centralized control mechanisms within a closed core environment, ensuring that each radio access segment delivers consistent and reliable service.

Waveforms are crucial in transmitting and receiving information between devices in a mobile cellular network [8,9]. They represent the shape of the electromagnetic wave used for wireless communication. Cellular networks use different waveforms depending on the specific technology and frequency band utilized at that time. Therefore, expanding our knowledge of various waveform types can help in developing new waveform technologies. The choice of waveform in cellular networks can significantly influence key performance metrics, such as network capacity, coverage area, SE, and resistance to interference, and it can ultimately affect the overall user experience [10–12]. Waveform selection also affects power consumption, as different waveforms require varying amounts of energy to transmit and receive signals [13–15]. Therefore, choosing the right waveform can help optimize energy consumption, reduce battery drain, and prolong the life of mobile devices connected to the network.

Interference from other wireless devices is another factor influencing the waveform choice in cellular networks; therefore, a suitable waveform selection can also improve interference resistance [16]. In crowded urban environments, multiple devices are competing for the same wireless spectrum, which can result in interference and signal degradation. By selecting waveforms with specific frequency characteristics, such as those that avoid known interference bands, or using techniques like spread spectrum, signals can be spread over a broader range of frequencies, making them less vulnerable to narrowband interference [17].

Hence, understanding and analyzing the characteristics of waveforms throughout the development of cellular networks will allow us to improve future designs. By gaining this understanding, we can optimize waveform selection to reduce interference, enhance SE, and ensure strong signal transmission. Furthermore, lessons learned from earlier stages can guide the implementation of advanced modulation schemes, improving both network capacity and reliability as new technologies, such as fifth generation (5G) and beyond, are deployed.

Although numerous studies have examined various modulation techniques in different generations of cellular networks, there is a lack of a comprehensive historical context regarding the evolution of these networks. This gap hampers a full understanding of how earlier technologies have influenced the development of current modulation schemes in research. For example, Ezhilarasan et al. [18] have provided a broad overview of the third generation (3G), fourth generation (4G), and 5G technologies but failed to address modulation schemes. Similarly, Fadhil et al. and Arshad et al. [19,20] reviewed cellular systems from the first generation (1G) to 5G at a high level without detailing the technical standards specific to each generation or discussing the evolution of modulation techniques. Furthermore, Alam et al. [21] focused on the development of 5G and sixth generation (6G) but did not address modulation schemes or their evolution in these technologies.

Many studies fall short in providing comparative analyses of modulation techniques across different cellular generations and standards while also neglecting systematic evaluations of their performance metrics. Another aspect that has not been addressed in previous studies is analysis of the advantages and disadvantages of each waveform in relation to specific communication systems. Emerging trends and future directions in modulation techniques, particularly as mobile networks advance toward 5G and beyond, remain insufficiently explored. Existing research often emphasizes either the evolution of cellular mobile systems and their standards or the modulation waveforms specific to individual generations of mobile networks; for instance, Liu et al. [22] have focused exclusively on the evolution of one generation of the cellular mobile network. Most research tends to focus on newer generations, particularly 4G and beyond. For example, Gerzaguet et al. and Cai et al. [23,24] concentrated on reviewing modulation techniques for 5G mobile systems, while Khalifa et al. and Parasher et al. [25,26] investigated the performance of long-term evolution (LTE) with various modulation techniques.

By addressing these gaps, this paper, inspired by the evolution of cellular mobile systems and their corresponding standards, enhances the understanding of modulation schemes and their practical implications within the field of mobile telecommunications. It is also worth noting that waveforms and modulation schemes serve complementary roles: waveforms define the time–frequency structure of the transmitted signal, while modulation schemes specify how information symbols are mapped onto these waveforms. The design of waveforms involves trade-offs tailored to different scenarios, such as high-capacity broadband, low-latency applications, or high-mobility environments. Their selection is guided by criteria such as SE, BER, throughput, and latency, whereas their assessment is carried out using performance evaluations that include BER curves and SE analysis. This paper presents an analysis of the waveforms and modulation schemes employed across different generations of mobile cellular networks, emphasizing both their similarities and differences. The contributions of this paper are delineated as follows:

- A systematic review of the evolutionary trajectory of cellular mobile networks, detailing significant technological advancements, historical milestones, and their implications on the telecommunications landscape.
- A targeted survey of the modulation techniques utilized across the different generations and standards of cellular systems, emphasizing the specific methodologies employed, their underlying principles, and the performance metrics that define their effectiveness in real-world applications.
- An exploration of the advantages and disadvantages inherent to each waveform employed in these systems, offering a critical analysis of their respective impacts on overall system performance, efficiency, and suitability for diverse communication scenarios.
- A comparative performance evaluation of modulation schemes based on BER plots, throughput, SE, and latency, highlighting their trade-offs in error resilience and efficiency across successive mobile generations.

The structure of this paper is illustrated in Figure 1 and is outlined as follows: Section 2 covers the history of pre-cellular mobile communication, including zero generation (0G) and half generation (0.5G), which began in the 1940s, shortly after World War II. This era saw technological advancements that laid the groundwork for mobile telephony. Section 3 examines the primary standards and modulation techniques of 1G, such as frequency modulation (FM), direct frequency shift Keying (FSK), Fast FSK, and others. Section 4 provides a comprehensive analysis of the second generation (2G) of mobile systems, its standards, and waveforms (such as Gaussian minimum-shift keying (GMSK), quadrature phase shift keying (QPSK), offset-quadrature phase shift keying (OQPSK), etc.), with subsequent Sections 5 and 6 delving into advanced 2G variants. These sections detail the evolution of integral waveforms for 2G technologies.

Section 7 explores 3G, while Sections 8 and 9 address the advanced versions 3.5G and 3.75G, respectively. These sections focus on the distinctive waveforms employed in these generations, such as binary phase shift keying (BPSK), QPSK, 16-quadrature amplitude modulation (QAM), 8-phase-shift keying (8PSK), and higher-order modulations. Section 10 discusses 4G and its associated waveforms, with Section 11 extending the analysis to 4.5G waveforms. This paper further investigates the prominent waveforms of 5G and

6G in Sections 12 and 13. This paper then closes with an evaluation of the performance of modulation schemes in mobile generations in Section 14, and this is followed by the conclusions in Section 15 and the appendices, spanning Appendices A–E, as shown in Figure 1. A comprehensive summary of the standards, waveforms, and modulation techniques reviewed is provided in the Conclusion, offering a clear representation of their evolution and key distinctions across successive cellular generations.

1. **Introduction**
2. **0G and 0.5G**
3. **1G**
   - 3.1. AMPS
   - 3.2. NMT
   - 3.3. TACS
   - 3.4. ETACS
4. **2G**
   - 4.1. GSM
   - 4.2. cdmaOne
   - 4.3. D-AMPS
5. **2.5G**
6. **2.75G**
7. **3G**
   - 7.1. WCDMA
   - 7.2. TD-CDMA
   - 7.3. TD-SCDMA
   - 7.4. CDMA2000
8. **3.5G**
9. **3.75G**
10. **4G**
    - 10.1. LTE
    - 10.2. WiMAX 802.16m (WiMAX-2)
    - 10.3. UMB
11. **4.5G**
12. **5G**
    - 12.1. CP-OFDM with WOLA
    - 12.2. F-OFDM
    - 12.3. UFMC
    - 12.4. FBMC
    - 12.5. GFDM
    - 12.6. Performance Summary of 5G OFDM Based Modulation Techniques
13. **6G**
    - 13.1. CCWD
    - 13.2. SCWD
    - 13.3. JWOD
14. **Performance Evaluation of Modulation Schemes**
    - 14.1. Modulation Performance Metrics Across Mobile Generations
    - 14.2. BER Performance Analysis Across Modulation Schemes
15. **Conclusions**

Appendix A. Example of Spreading and Despreading Technique in WCDMA
Appendix B. SC-TDE Vs SC-FDE
Appendix C. LTE Advanced Features
Appendix D. 5G Physical Channels and Associated Modulation and Coding
Appendix E. Multi-Carrier vs. Single-Carrier Modulation Performance

Figure 1. Organization of this article.

2. 0G and 0.5G

It was just after World War II when the ancestor of the mobile telephone service was invented. The concept of 0G refers to the generation before cellular telephones, such as a time when radio telephones were installed in cars before the invention of cell phones. Unlike more recent cellular network generations, 0G did not support the handover feature that ensures continuity of service for mobile users moving outside of the serving cell coverage area. As 0G was primarily used for voice communication [27], without the handover feature, a phone call would drop while the user switched from one serving cell to another.

However, other technologies were utilized during 0G, such as push-to-talk (PTT), mobile telephone service (MTS), improved mobile telephone service (IMTS), offentlig landmobil telefoni (OLT)—Norwegian for Public Land Mobile Telephony—and MTD (the Swedish abbreviation for mobile telephony system D).

0G mostly operated in three frequency bands, VHF Low (35–44 MHz), VHF High (152–158 MHz), and UHF (454–460 MHz). It is important to note that 0G systems were not standardized, and different regions in the world and countries may have used different frequency bands for their respective analog cellular systems. For this generation, FM was used for the base station and mobile units.

Later, 0.5G was created to improve the features of 0G. 0.5G was the first generation of mobile communication systems that was accessible to the general public and provided a unique phone number for each mobile unit instead of being part of another network, such as a police radio or a taxi dispatch system.

Mobile units at this time were typically located in the trunk of the cars and connected to the telephone device near the driver's seat. This communication system, similar to 0G, offered only the basics of voice communication. Autoradiopuhelin (a Finnish term that translates to "car radio telephone" in English, also known as ARP) was the first commercially operated public mobile phone network and was invented and launched in the 1960s.

ARP was designed with 80 channels over 150 MHz frequency (147.9 to 154.875 MHz), and it provided a half-duplex communication system, which was able to transmit/receive voice signals to/from cells but not simultaneously. Later, a new version of ARP was introduced. While ARP offered a full-duplex communication system in this version, it was still unable to support the handover feature, and because of the limitation in the coverage area of each cell, this caused ARP to lose its power and popularity. Therefore, ultimately, this project was canceled in the year 2000.

3. 1G

The 1G of wireless telephone technology was introduced in 1980, and it was compatible with analog cell phones. 1G was the first cellular system that offered handover and roaming capability, but only in a single country, which was the key drawback of this generation. Nevertheless, in this generation, cell phones were able to support very basic telephone calls, but connections were poor, unreliable, and insecure. The first generation of mobile systems used a variety of standards implemented globally. The following were the primary ones used:

- Advanced Mobile Phone System (AMPS).
- Nordic Mobile Telephone (NMT).
- Total Access Communications System (TACS).
- Extended Total Access Communication System (ETACS).

The remainder of this section provides a detailed explanation of each standard and its associated modulation technique.

3.1. AMPS

AMPS first appeared in 1983 in the United States and was originally developed by Bell Labs, and later modifications were applied through collaborations between Motorola and Bell Labs [28]. It was mainly implemented in the Americas, Russia, and Asia. AMPS became the predominant analog cellular system in North America and was widely deployed throughout the 1980s and early 1990s.

AMPS is a 1G system that uses analog modulation and was the first cellular technology to employ hexagonal cells in its design. Some other companies that used AMPS at that time included Bell Mobility, AT&T Mobility, Alltel, and Telus Mobility. AMPS was based on frequency division multiple access (FDMA), and it operated in the frequency band of 850 MHz with a channel spacing of 30 kHz. AMPS utilized the frequency range of 824–849 MHz for the uplink (mobile to base station) and 869–894 MHz for the downlink (base station to mobile).

Direct FSK was used for signal transmission between the user mobile and base station with a bit rate of 10 Kbps, but through using Manchester encoders—a line coding technique in which each bit is represented by a transition at the middle of the bit period, ensuring inherent clock recovery and eliminating direct current (DC) components—the bit rate could be doubled to 20 Kbps [29,30]. For speech transmission, FM modulation was used in AMPS with a wider channel spacing than was normal for a channel with a bandwidth of 30 kHz [31]. Due to having a wider channel spacing, co-channel interference was reduced. The wider channel spacing and the use of speech compression/expansion led to a high-quality voice circuit [29].

The details of the Direct FSK and FM modulation that were used in this standard are explained in the following:

3.1.1. Direct FSK

For signaling, frequency modulation in frequency shift keying mode was used. A conventional FSK signal uses two different transmission techniques: subcarrier FSK (also referred to as audio FSK) and direct FSK. In audio FSK, FM modulation of the binary signal is performed with two audio frequencies, but in direct FSK, the carrier itself is shifted to modulate the binary signal. The other difference between direct FSK and audio FSK is that direct FSK applies the data signal to the transmitter's variable frequency oscillator (VFO). In addition, FSK signals can be generated in the baseband and transmitted over telephone lines. It is also possible to translate the signal to a higher frequency and generate the FSK signal directly at "carrier" frequencies.

Direct FSK had the following advantage compared to audio FSK: it allowed for faster transmission speed and a simpler modulating device at a lower manufacturing cost. However, direct FSK had the following disadvantage in its demodulator: according to the conventional method, signals modulated with direct FSK are demodulated to a voltage corresponding to the binary value through an FM demodulator. At this time, if the frequency of the carrier is changed due to the change in factors, such as temperature and voltage, which affect the frequency, the voltage that is FM-demodulated is also changed. Therefore, the voltage comparator cannot accurately compare the voltages [32].

3.1.2. Frequency Modulation (FM)

1G provided analog speech signals for users by using frequency modulation. FM is the most popular analog waveform. FM signals keep the amplitude constant and have all their information in the phase or frequency of the carrier. There are two methods to generate FM signals: a direct method (reactance modulator and varactor modulator) and an indirect method (Armstrong method). In the direct method, the carrier frequency is directly varied by the input-modulating signal, while in the indirect method, a narrowband FM signal is generated using a balanced modulator. Additionally, frequency multiplication is used to increase both the frequency deviation and the carrier frequency to the required level [30]. In the following, we review each of these methods very briefly:

Direct Method (Reactance Modulator and Varactor Modulator)

FM can be deployed by the direct method using a voltage-controlled oscillator (VCO). VCOs help to vary the carrier frequency based on the amplitude changes in the baseband signal. VCOs are a type of oscillator that use devices with reactance that can be varied by the application of voltage, where the reactance causes the instantaneous frequency of the VCO to change proportionally. The most commonly used variable reactance device is the voltage variable capacitor, and it is called a varactor (a reactance modulator is shown in Figure 2 [30]). In this structure, the frequency of the oscillator can be varied by changing the varactor diode capacitance. Varactor modulators can also be used for direct methods, but they are less common than reactance modulators.

Even though VCOs are a simple approach to providing narrowband FM signals, their stability will be a major concern when they are used to generate wideband FM signals; however, phase locked loops (PLLs) can be used to resolve this issue [30].

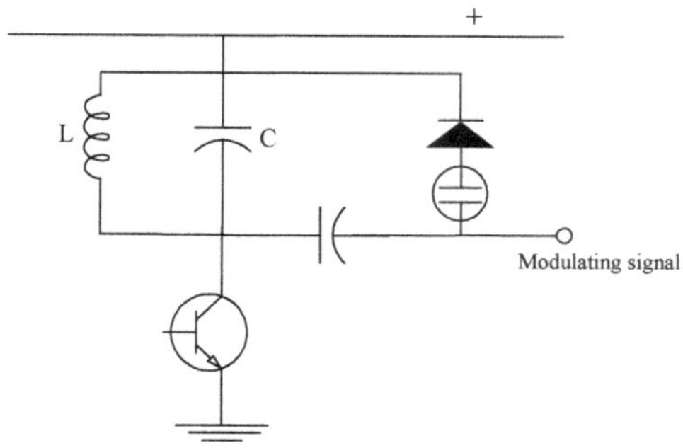

Figure 2. Reactance modulator. Reproduced from [30].

Indirect Method (Armstrong Method)

The indirect method of FM deployment is sometimes referred to as the Armstrong method after its inventor, Edwin Howard Armstrong. The idea behind the Armstrong method is the approximation of the narrowband FM signals as the summation of the carrier signal and a single sideband signal, where the sideband signal is 90° out of phase with the carrier [30]. A block diagram of the Armstrong method is shown in Figure 3. A narrowband FM signal is first generated, and then, after passing the narrowband signal through a limiter and frequency multiplier, the wideband FM signal is generated. The main disadvantage of the indirect method to produce a wideband FM signal is that the system phase noise increases with the frequency multiplying factor N [30].

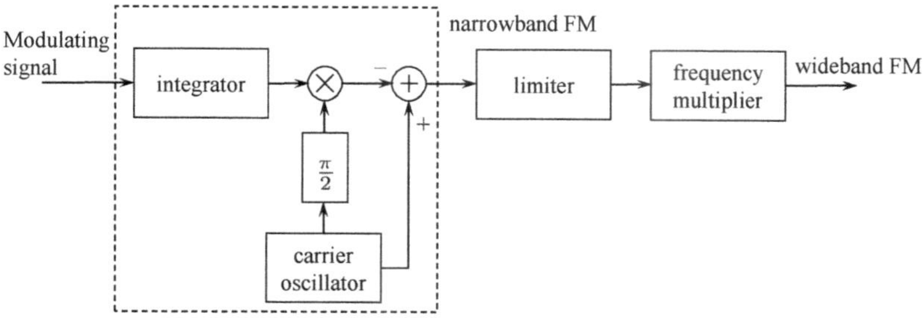

Figure 3. Indirect (Armstrong) method. Reproduced from [30].

Various issues, including weak security features, made the AMPS system prone to hacking and handset cloning. In other words, similar to any other analog system, AMPS suffers from noise and eavesdropping. Variant AMPSs were developed later to support digital speech encoding by using time division multiple access technology; this system was called the digital-advanced mobile phone system (D-AMPS) and is described in Section 4.3. D-AMPS also had a similar signaling protocol and architecture as AMPS [29].

3.2. NMT

NMT was developed as a solution to address the challenges posed by congested and overloaded manual cellular networks in the past, and it was the first fully automatic cellular phone system that was implemented in 1981 in the Nordic countries. With this feature, for the first time, everyone could make calls to anyone anywhere in the Nordic countries. The early version of the NMT system was designed to work at 450 MHz (NMT-450), and later it was developed to also operate at 900 MHz (NMT-900). NMT-450 was jointly built up by the Nordic countries' telecommunication companies, such as Tele Danmark Mobil

(Denmark), Telecom Finland (Finland), Tele-Mobil (Norway), and Telia Mobitel (Sweden) to establish a compatible automatic public mobile telephone system [33].

The Nordic Mobile Telephone group was officially called NT-R 69-5 (Nordic Teleconference-Radio, 1969-working group 5) [34]. One of the key innovations introduced by the NMT system was the support for subscribers moving between the coverage areas of different base stations while maintaining an active call. This capability, known as handover, was implemented for the first time in a cellular system, marking a major advancement in mobile communication.

AMPS was developed earlier than NMT, but NMT saw commercial service before AMPS, opening in late 1981 [29]. NMT-450 used a channel spacing of 25 kHz, while NMT-900 used 12.5 kHz. Voice transmission in both NMT-450 and NMT-900 was based on FM, while signaling in both systems used fast frequency shift keying modulation (FFSK). FM was explored earlier in Section 3.1, and FFSK will be described below.

FFSK

FFSK is a variant of FSK, where data is transmitted by rapidly switching between different frequencies in a much quicker manner than traditional FSK. FFSK increases the rate at which the frequency shifts, allowing for higher data rates and better SE. This rapid frequency switching allows more data to be transmitted in a given time period compared to conventional FSK, which uses slower, more deliberate frequency shifts. FFSK is particularly beneficial for transmitting digital data over an RF channel that is limited in both bandwidth and power. By increasing the speed of frequency transitions, FFSK can support higher throughput without requiring a proportional increase in power or bandwidth. Buda [35] described the performance of FFSK and outlined practical circuits used for its demodulation, providing valuable insights into how FFSK can be efficiently implemented in real-world communication systems.

The primary advantage of FFSK is that it enables higher data throughput while maintaining relatively simple modulation techniques. By increasing the speed of frequency transitions, FFSK achieves better bandwidth efficiency. However, this comes with challenges, such as the need for precise frequency control and synchronization. The receiver must be able to accurately detect the faster frequency shifts to decode the data correctly. As a result, FFSK systems must be designed with careful consideration of timing and phase synchronization to avoid errors caused by fast transitions between frequency states. In scenarios where both power and bandwidth are constrained, FFSK offers a way to maximize the use of available resources.

In FFSK, the modulation process typically employs continuous phase modulation to mitigate the issues of phase distortion that might otherwise occur due to the fast frequency changes. This continuous phase approach helps maintain a smooth signal transition, reducing errors and improving signal integrity. Buda's work also highlights the practical aspects of implementing FFSK, such as the use of specific circuits for demodulation and how these systems can be adapted for efficient real-world use. Overall, FFSK offers significant advantages in terms of data rates and SE, but its implementation requires careful design to manage the technical complexities associated with high-speed frequency shifts and accurate demodulation.

3.3. TACS

TACS was the European version of AMPS with slight modifications, as well as operating in different frequency bands. It was mainly used in the United Kingdom, as well as parts of Asia. TACS operated at 900 MHz frequency with 25 kHz channel spacing.

For speech modulation, FM was used the same as the AMPS and NMT standards, and for signaling, direct FSK was utilized with a signaling bit rate of 8 Kbps [29]. TACS retained the main signaling schemes from the AMPS system and only added some enhancements like the location registration procedure, which made the system suitable for deployment in nationwide networks. The other feature that TACS also introduced was signaling of charge rate information (e.g., for payphones) [29]. Originally, it was planned to allocate all of the 1000 channels (2 × 25 MHz) to mobile services in Europe, but in the United Kingdom, only 600 channels (2 × 15 MHz) were released by the licensing authority, and the remainder was reserved for the global system for mobile communications (GSM) standard. To accommodate the reduced channel allocation, an additional allocation of channels below the existing TACS channels was created, known as ETACS channels, and the standard was modified accordingly [29].

3.4. ETACS

ETACS channels were used to provide additional capacity for mobile services in the UK, and the standard was modified to support this new allocation of channels [29]. The TACS and ETACS standards were later phased out and replaced by GSM as Europe's dominant mobile communications standard.

ETACS, similar to TACS, used direct FSK for signaling and FM for voice communication. FM was chosen for its robustness against noise and its ability to provide clear voice communication over the airwaves. In ETACS, FM was used to modulate the carrier signal in such a way that the frequency of the carrier would shift in proportion to the amplitude of the transmitted signal [29].

All FM-based standards, such as NMT, TACS, and ETACS, lacked security features, the ability to support meaningful data services, and international roaming capabilities The low data rate transmission of these standards led to the development of digital transmission, which offers higher data rates and improved security for user data transmission over the air interface between a base station and mobile phones. Digital transmission also offered enhanced noise immunity and robustness against channel distortions compared to analog modulation. These advantages made digital transmission the preferred choice for the next generation of mobile communication systems.

In this section, we have discussed the main standards used in the first generation of cellular networks, such as AMPS, NMT, TACS, and ETACS, which utilized analog waveforms, specifically FM modulation. As mentioned, these standards were limited in their capabilities and could not support advanced features, such as security, data services, and international roaming. This led to the development of digital transmission, which offered higher data rates, improved security, and overall better performance, and it ultimately replaced the use of analog standards in cellular networks.

4. 2G

Digital cellular phone systems first appeared in the late 1980s and early 1990s as part of the second generation of cellular networks, also known as 2G. These networks were designed to replace the first-generation analog systems and offered several advancements and new features [36]. Some of the key features of 2G networks include the following:

- Digital transmission: Second-generation cellular networks use digital signal processing, which makes better use of the available radio spectrum and improves security and noise immunity.
- Advanced data services: Support for data services, such as text messaging, and short message service (SMS) was improved with this generation [37].

- Improved security: It became more difficult for unauthorized users to access the data in the network with the introduction of encryption in 2G.
- International roaming: During international travels, this generation allowed users to place and receive calls.
- Improved capacity: The capacity was increased by utilizing more sophisticated multiplexing methods, such as time division multiple access (TDMA), to accommodate more users in the same location.

In this section, we explore three prominent 2G standards (along with the various modulation schemes employed in each technology): GSM, code division multiple access one (cdmaOne), and D-AMPS. Each of these standards use distinct modulation techniques, optimized for their specific multiple access methods and network requirements. These techniques were crucial to the overall performance and efficiency of the systems.

4.1. GSM

GSM was the most successful of all 2G technologies. It was initially developed by the European Telecommunications Standards Institute (ETSI) in 1992 for European countries and was designed to operate in the 900 MHz, 1800 MHz, and 1900 MHz frequency bands.

More precisely, for GSM 900 MHz, the range of 890–915 MHz was used for the uplink channel (from mobile to base station) and 935–960 MHz was used for the downlink channel (from base station to mobile); for GSM 1800 MHz, the range of 1710–1785 MHz was used for the uplink and 1805–1880 MHz for the downlink; and for GSM 1900 MHz, the range of 1850–1910 MHz was used for the uplink and 1930–1990 MHz for the downlink.

GSM now has worldwide support and is available for deployment on other frequency bands, such as 850 MHz. In the past, when a mobile phone was labeled as tri-band or quad-band, it indicated that the device could operate on multiple frequency bands simultaneously, providing greater flexibility and coverage for users in different regions.

GSM uses TDMA/FDMA. The FDMA portion necessitates dividing the (maximum) 25 MHz bandwidth into 124 carrier frequencies, which are spaced 200 kHz apart by frequency. Each base station is given one or more carrier frequencies. Using a TDMA method, each of these carrier frequencies is then divided into time slots. A burst period is the fundamental unit of time in this TDMA architecture, and it lasts 15/26 ms ($\approx$0.577 ms). A TDMA frame (120/26 ms, $\approx$4.615 ms) is made up of eight burst periods (time slots) that serve as the basic unit for logical channel specification. Per the TDMA frame, one physical channel equals one burst period, such that it employs eight time slots on a 200 kHz radio carrier. The combination of FDMA and TDMA allows the GSM network to support multiple users simultaneously, thereby enhancing network capacity and efficiency.

Digital modulation techniques have been used in mobile communication systems since the second generation. Digital waveforms can be classified as single-carrier or multi-carrier modulations. Constant envelope waveforms, as a subcategory of digital single-carrier modulation, were used in the second generation of mobile systems. These waveforms have a 0 dB peak-to-average-power ratio (PAPR), which allows power amplifiers to operate at or very close to the saturation threshold, resulting in low out-of-band radiation, typically around -70 to -60 dB. Additionally, constant-envelope modulators can use limiter–discriminator detection, which reduces receiver complexity and increases resistance to noise and signal fluctuations. Examples of constant envelope waveforms include FSK, Gaussian frequency-shift keying (GFSK), minimum-shift keying (MSK), and GMSK. The Zigbee standard employs MSK, while GSM employs GMSK.

GMSK

GMSK, introduced by Murota and Hirade in 1979, is a simple binary modulation scheme that can be viewed as a derivative of MSK. GMSK is spectrally more efficient than MSK [30,38]. There are two ways to implement GMSK modulation. The first way is to pass the non-return-to-zero (NRZ) signal (the baseband data signal) through a Gaussian filter to shape it and to then apply the output of the filter to a frequency modulator (voltage-controlled oscillator), as shown in Figure 4. This procedure is used in many analog and digital implementations for the US cellular digital packet data (CDPD) system, as well as for the GSM system [30]. This shaping improves the SE of the modulation, resulting in a higher data rate for a given bandwidth, but the components' tolerance drift prevents the modulation index from being exactly 0.5 as it should be, which creates a problem.

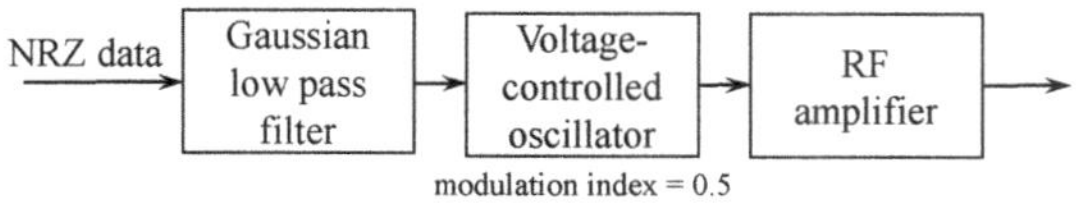

Figure 4. GMSK modulator using VCO (frequency modulator).

Another way of implementing GMSK is a digital deployment of Figure 4, which involves a Gaussian filter and an in-phase quadrature modulator (I-Q modulator), as shown in Figure 5. The term "quadrature" refers to the phase difference of 90 degrees between the two signals. The digital implementation of GMSK was more widely used, and its advantage was that the modulation index could be set to 0.5 without any additional adjustments or settings. This is important because it ensured the required level of performance in the system.

As seen in Figures 4 and 5, GMSK uses a pre-modulation Gaussian low-pass filter to suppress out-of-band frequencies, known as the adjacent channel leakage ratio (ACLR) or the adjacent channel power ratio (ACPR). This is defined as the ratio of the average power in the adjacent frequency channel (intermodulation signal) to the average power in the transmitted frequency channel (useful signal). GMSK has a low ACLR, which improves the SE. This improvement of the SE of GMSK in comparison with the other phase shift keyed modes is particularly crucial. Another advantage of GMSK is its compatibility with non-linear power amplifiers. GMSK remains undistorted even when using non-linear power amplifiers because the signal has no amplitude variations. This leads to lower power consumption in the user's handset and less frequent recharging of the phone device.

GMSK offers higher noise immunity compared to some other modulation techniques because it does not rely on amplitude variations, which are susceptible to noise. Additionally, GMSK supports asynchronous multiplexing, providing another significant advantage.

However, this type of modulation has many advantages, such as high power efficiency due to the constant envelopes, which, as previously stated, occupy more bandwidth than linear modulations—so they are not suitable for use if bandwidth efficiency is more important than power efficiency in an application. This type of waveform also does not have trivial support for multiple-input-multiple-output (MIMO) transmission. The main parameters of GMSK used in GSM are as follows: a modulation index of 0.5, BT $= 0.3$ (where B is the 3 dB baseband bandwidth and T is baseband symbol duration), and a modulation rate of 271 kbaud/s [39].

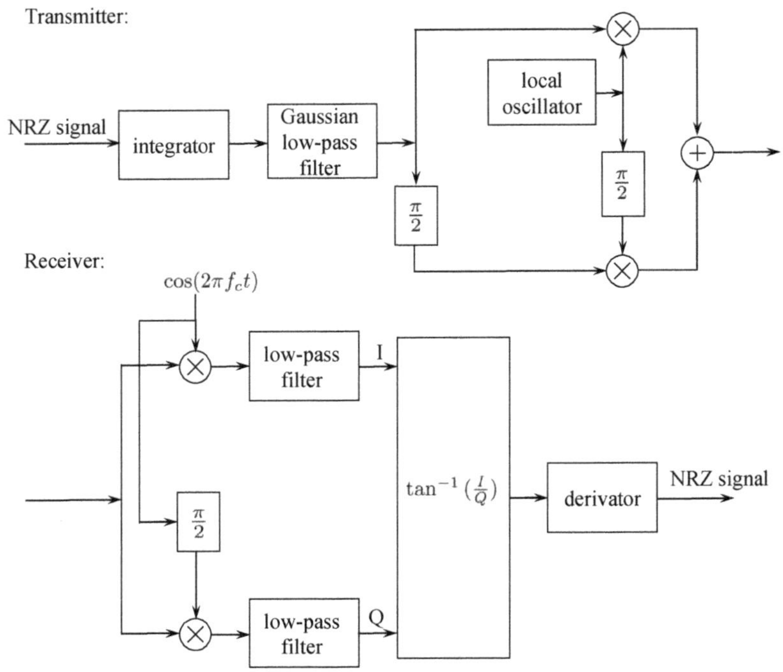

Figure 5. GMSK transmitter and receiver.

4.2. cdmaOne

cdmaOne was the first digital cellular technology based on code division multiple access (CDMA). It was developed as a competing standard to GSM and was first deployed in 1995 [36]. 3G and just one 2G standard (cdmaOne) adopted CDMA as their multiple access technique. In CDMA, users communicate simultaneously within the same frequency band. Different users are separated in the receiver with the help of an individual coded signal. It uses a direct sequence spread spectrum technique and utilizes a mixture of codes and timing to identify cells and channels. CDMA systems, in general, use orthogonal chips or spreading codes for voice, data, and signaling data transmission.

At the time of the 2G systems' deployment, CDMA seemed to be too complex from an implementation point of view. Thus, only one 2G standard, i.e., cdmaOne, used CDMA. The rapid development in digital technology has enabled large-scale implementations of CDMA in contemporary consumer electronics [40].

cdmaOne, also known as Interim standard 95 or IS-95 for short, includes the IS-95A and IS-95B revisions of the CDMA Telecommunications Industry Association (TIA) and the Electronic Industries Alliance (EIA), which is the CDMA TIA/EIA IS-95 standard [40]. Walsh Codes and pseudo-random noise (PN) codes are the ones used in IS-95, where Walsh codes are orthogonal sequences used primarily on the downlink to separate channels and users, while PN codes are deterministic sequences with noise-like properties employed for spreading, cell identification on the downlink, and user separation on the uplink.

IS-95A was published in May 1995 and it was the basis for many other commercial second-generation CDMA systems around the world. IS-95A describes the structure of the channels with a bandwidth of 1.25 MHz, power control, call processing, hand-offs, and registration techniques [40]. Basic characteristics of this standard are listed below [40].

- The bandwidth of each channel was 1.25 MHz, and filtering was applied to limit the spectrum.
- IS-95A used a chip rate of 1.2288 million signals per second and supported a nominal data rate of 9.6 kbps in Rate Set 1 (RS1) mode. Additionally, it also supported an improved rate mode (RS2) that allowed for a data rate of up to 14.4 kbps.
- It used direct sequence spread spectrum (DSSS) for spreading the signal.
- It used a convolutional error-correction code with a rate of 1/2 and a constraint length of 9, along with Viterbi decoding, for forward error correction (FEC).
- The IS-95A downlink (forward) channel used a time-division multiplexing scheme, where the base station allocated 20 ms time intervals to each user.
- IS-95A used a RAKE receiver for signal reception, which demodulates the three strongest components of the multipath signal in the mobile station and four components in the base station. To improve performance, it also used two antennas for spatial diversity.
- The IS-95A base station used 64 channels for transmission and utilized orthogonal code multiplexing for channel separation.
- IS-95A used power control to adjust the transmitted signal power levels from the base station to users in order to equalize the signal strength for users at different distances from the base station. This helped to minimize the power consumption and interference from the transmitted signals.

The IS-95B revision of IS-95A increased the transmission rate to 64 kbps in line-switching mode and 115.2 Kbps in the batch mode of the CDMA network. Due to these improved rates, IS-95B is considered a part of the 2.5G standards. Advantages of using the IS-95A/B (cdmaOne) standard in cellular networks include the following: [40]

- Capacity improvement, with an increase of 8 to 10 times compared to the AMPS analog system and 4 to 5 times compared to the GSM system.
- Improvement in call quality, characterized by a better and more consistent sound when compared to AMPS systems.
- Simplified system planning, achieved by using the same frequency in every sector of every cell.
- Enhanced privacy through the use of unique codes for each call and user.
- Offers improvement in coverage characteristics, which allows for the possibility of using fewer cell sites.
- Longer talk time on portable devices.
- Provides the ability to allocate bandwidth on demand, allowing for more efficient use of available resources.

The modulation techniques used in IS-95 were as follows: QPSK modulation was used in the downlink channel, and OQPSK was used in the uplink channel (OQPSK was also called staggered-QPSK). Figure 6 shows conventional QPSK (non-offset QPSK) and OQPSK.

Offset-QPSK is a QPSK variant in which the quadrature component $Q(t)$ is delayed by half a symbol time relative to the in-phase component $I(t)$. This is conducted to offset the transition of $I(t)$ and $Q(t)$ so that they do not occur at the same time. The reason this is performed lies in the behavior of nonlinear power amplifiers that are used in most wireless systems [41]. As illustrated in Figure 6, in offset-QPSK, just two-phase transients are possible during one symbol period, 90° and −90°, whereas three-phase transients are possible for each symbol in non-offset QPSK.

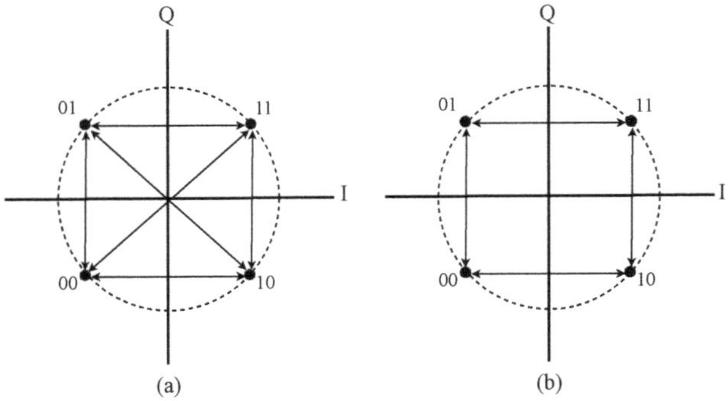

Figure 6. (**a**) Constellation diagram of QPSK. (**b**) Offset-QPSK. Arrows denote the possible phase transitions.

4.3. D-AMPS

D-AMPS was developed in the 1980s as a replacement for the analog AMPSs and was primarily used on the North American continent, as well as in New Zealand and parts of the Asia-Pacific. D-AMPS is based on Interim Standard 54 (IS-54) and Interim Standard 136 (IS-136). It is effectively an enhancement to AMPS and utilizes a TDMA technique. In addition to being digital, as well as improving capacity and security, these 2G digital systems also offer enhanced services, such as SMS and circuit-switched data. D-AMPS used differential quaternary phase shift keying (DQPSK), known as $\pi/4$ DQPSK, which is a linear modulation.

Before exploring the characteristics of $\pi/4$ DQPSK, it is important to understand $\pi/4$ QPSK, which uses two QPSK constellations simultaneously. One constellation is used to modulate the odd symbol numbers, while the other, with a $\pi/4$ phase offset, modulates the even symbol numbers. This can be seen in Figure 7a, and Figure 7b illustrates all of the symbol states.

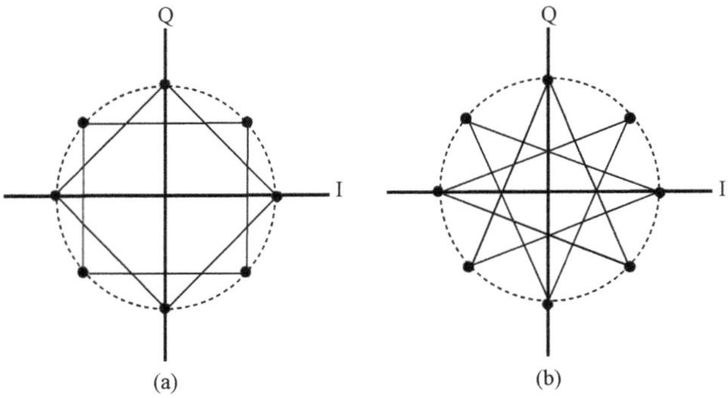

Figure 7. (**a**) Constellation diagram of $\frac{\pi}{4}$ QPSK. (**b**) All possible symbol states.

$\pi/4$ shifted QPSK modulation is a QPSK technique that offers a compromise between OQPSK and QPSK in terms of the allowed maximum phase transitions. It may be demodulated in a coherent or noncoherent fashion. In $\pi/4$ QPSK, the maximum phase change is limited to $\pm 135°$ compared to the $180°$ for QPSK and $90°$ for OQPSK. Therefore, the bandlimited $\pi/4$ QPSK signal preserves the constant envelope property better than bandlimited QPSK, but it is more susceptible to envelope variations than OQPSK [30].

$\frac{\pi}{4}$ DQPSK is a differential version of $\frac{\pi}{4}$ QPSK, where the phase change between the current symbol and the previous symbol will determine the bits in the current symbol. The mapping between the phase shift and the bit pattern is shown in Table 1.

One of the main challenges with using $\frac{\pi}{4}$ DQPSK, like other linear modulations, is its lack of power efficiency, resulting in increased battery usage and user inconvenience. In other words, the requirement for accurate phase change detection can mean that power amplifiers need to operate in a more linear region, which can reduce overall power efficiency.

Table 1. Mapping between the phase change and bit pattern in $\frac{\pi}{4}$ DQPSK.

Phase Change ($\Delta\Theta$)	Bit Pattern
$\pi/4$	00
$3\pi/4$	01
$-\pi/4$	10
$-3\pi/4$	11

5. 2.5G

General Packet Radio Service (GPRS), often called 2.5G, is an enhancement over 2G systems, such as GSM (which provides packet-switched data services). As GPRS does not meet the standards of 3G, it is therefore considered 2.5G. A comparison of 2G, 2.5G, and 2.75G, in terms of theoretical and typical data rate, can be found in Table 2. Like GSM, GPRS used GMSK for its waveform, which is explained in Section GMSK. The GPRS modulation and coding scheme (CS) are shown in Table 3. CS1 used half-rate convolutional coding, CS2 and CS3 were punctured versions of the same coding applied for CS1, and CS4 did not apply convolutional coding at all [42].

Table 2. The data rate of the 2G, 2.5G, and 2.75G systems.

System \ Characteristic	Service	Theoretical Data Rate (kbps)	Typical Data Rate (kbps)
2G GSM	Circuit switched data service	9.6 to 14.4	9.6 to 14.4
2.5G GPRS	Packet switched data	171.2	4 to 50
2.75G EDGE	Packet switched data	473.6	120

Table 3. GPRS modulation and coding scheme [42].

GPRS Modulation and Coding Scheme	Modulation	Data Code Rate	Payload	Max Throughput (kbps)
CS1	GMSK	1/2	181	9.05
CS2	GMSK	≈2/3	268	13.4
CS3	GMSK	≈3/4	312	15.6
CS4	GMSK	1	428	21.4

Regarding the BER for GMSK, we have the following: In CS1, with a coding rate of 1/2, the BER is higher, typically around 10^{-3} to 10^{-2}, due to minimal error correction. In CS2 (code rate of 2/3), BER improves to 10^{-4} to 10^{-3} as error correction strengthens. CS3 and CS4, with coding rates of 3/4 and 1, respectively, further reduce BER to 10^{-6} to 10^{-5} and 10^{-5} to 10^{-4}, respectively, providing more reliable transmission.

Latency in GPRS is generally between 100 ms to 200 ms across all coding schemes as it depends more on network conditions than the coding rate. While CS1 has the lowest coding rate, it does not significantly reduce latency compared to CS2, CS3, or CS4, which may experience minimal increases due to higher encoding complexity. GPRS latency is at the medium range and is suitable for applications like SMS and basic browsing.

SE increases as the coding rate improves. In CS1, it is around 0.2 bps/Hz, but it improves to 0.3 bps/Hz in CS2, 0.4 bps/Hz in CS3, and reaches 0.5 bps/Hz in CS4 (the most efficient coding scheme). Higher coding rates allow more bits to be transmitted per Hertz of bandwidth, improving throughput and making better use of the available spectrum.

6. 2.75G

Enhanced data rates for GSM evolution (EDGE), also known as 2.75G, is an extension of GSM and GPRS systems. It is sometimes referred to as enhanced GPRS (EGPRS) and nearly quadruples the throughput of GPRS. With a theoretical data rate of 473.6 kbps, service providers can efficiently offer multimedia services. However, like GPRS, EDGE does not fully comply with 3G system standards, and it is usually categorized as 2.75G. EDGE uses GMSK and also higher-order phase-shift keying (8PSK) as modulation schemes. The modulation and coding schemes of Edge are shown in Table 4.

Table 4. Edge modulations and coding schemes.

EDGE Modulation and Coding Scheme	Modulation	Data Code Rate
MSC-1	GMSK	0.53
MSC-2	GMSK	0.66
MSC-3	GMSK	0.85
MSC-4	GMSK	1
MSC-5	8PSK	0.37
MSC-6	8PSK	0.49
MSC-7	8PSK	0.76
MSC-8	8PSK	0.92
MSC-9	8PSK	1

M-ary phase shift keying is characterized by a constellation consisting of M points equally spaced around a circle of radius of r, as illustrated in Figure 8. These points represent M equiprobable symbols with equal energy. Each of the signals in the signal set differs from each other only in phase, and each point also represents $\log_2 M$ bits; for example, in 8PSK, each symbol represents 3 bits [41]. In MPSK, the phase of the carrier takes on one of the M possible values $2\pi(i-1)/M$, where $i = 1, 2, \ldots, M$. The set of MPSK signals is therefore analytically given as [43]:

$$s_i(t) = \sqrt{\frac{2E_i}{T_s}} \cos(2\pi f_c t - \frac{2\pi(i-1)}{M}), \tag{1}$$

where $0 \leq t \leq T_s$ and $i = 1, 2, \ldots, M$. Expanding (1) using a trigonometric identity helps to see that the following two orthonormal basis functions can represent the MPSK signal set:

$$\begin{aligned} \phi_1(t) &= \sqrt{\frac{2}{T_s}} \cos(2\pi f_c t), 0 \leq t \leq T_s \\ \phi_2(t) &= \sqrt{\frac{2}{T_s}} \sin(2\pi f_c t), 0 \leq t \leq T_s. \end{aligned} \tag{2}$$

The MPSK signal set can thus be characterized by a two-dimensional signal space and M message points as follows [43]:

$$s_i(t) = \sqrt{E_i} \cos(\frac{2\pi(i-1)}{M})\phi_1(t) + \sqrt{E_i} \sin(\frac{2\pi(i-1)}{M})\phi_2(t), \tag{3}$$

where $0 \leq t \leq T_s$ and $i = 1, 2, \ldots, M$.

8PSK modulation can reach a three times higher data rate (59.2 kbps per time slot) compared to GMSK modulation in tje EDGE standard, but the disadvantage of 8PSK modulation is that it requires linearization in an amplifier since 8PSK is a QAM-type modulation. When EDGE uses GMSK modulation—as in its lower coding schemes—the same Gaussian pulse shaping as in GSM is applied, resulting in the similar spectrum envelope [44]. EDGE, when combined with GSM, uses time division multiple access as its multiplexing technology. It uses eight-time slots per channel, enabling eight users to use the same channel band simultaneously.

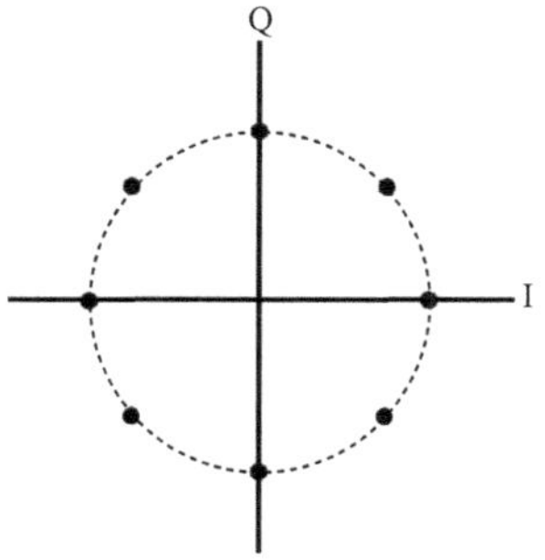

Figure 8. MPSK constellation for M = 8.

7. 3G

3G was designed to provide faster data transfer speeds and more advanced services than previous generations of cellular networks, such as 2G and 2.5G. 3G systems were defined by international mobile telecommunications-2000 (IMT2000). According to IMT2000, a 3G system should provide higher transmission rates, such as 2 Mbps for stationary or nomadic use and 348 kbps for use in a moving vehicle.

There are many contexts where the Universal Mobile Telecommunications System (UMTS) has become synonymous with the whole third generation of mobile systems, just as GSM referred to the whole 2G mobile system. UMTS is an umbrella term for the third-generation radio technologies developed within 3GPP [45]. UMTS offers two different radio access specifications: frequency division duplex (FDD) and time division duplex (TDD). TDD provides several different chip rates (the rate at which spreading code chips are transmitted in UMTS and are measured in chips per second, where a chip is a single pulse (or bit) from a PN code sequence—also referred to as the spreading code), allowing UMTS terrestrial radio access (UTRA) technology to operate in a wide range of bands and co-exist with other radio access technologies [46,47]. FDD was developed earlier than TDD in the third generation of mobile systems.

3G was based on the CDMA technique and offers various technologies, including wideband CDMA (WCDMA), time division CDMA (TD-CDMA), time division synchronous CDMA (TD-SCDMA), and CDMA2000. Additionally, techniques, such as direct-sequence CDMA (DS-CDMA), frequency hopping CDMA (FH-CDMA), and multi-carrier CDMA (MC-CDMA) were utilized.

7.1. WCDMA

WCDMA, the radio technology of UMTS, is a part of the International Telecommunication Union (ITU) IMT-2000 family of 3G Standards [48]. WCDMA was developed by the 3GPP. There are several variations of this standard, including TD-CDMA and TD-SCDMA. WCDMA is the primary evolutionary path from GSM/GPRS networks. Many of its deployments are mainly at 2.1 GHz; however, deployments at lower frequencies were also developed, e.g., UMTS 1900, UMTS 850, UMTS 900, etc. WCDMA supports voice and multimedia services with an initial theoretical rate of 2 Mbps, with most service providers initially offering 384 kbps per user. However, this technology continued to evolve, and later 3GPP releases increased the rates to more than 40 Mbps. WCDMA is the cellular system transmission protocol that utilizes DS-CDMA on a common wideband (5MHz) carrier [40].

CDMA is a type of multiple access technique that uses spreading modulation. Unlike the other multiple access techniques, such as TDMA and FDMA, which separate the information of the different users in terms of time and frequency, CDMA allows multiple users to transmit information simultaneously on the same channel. The idea of CDMA is to modulate all the information before transmission using different spreading codes to broadband signals; then, all the signals should be mixed and sent. At the receiving end, the mixed signal is demodulated using the same spreading codes, and because all the spreading codes are orthogonal, only the information that is demodulated by the same spreading code can be reverted in the mixed signal. An example is provided in this subsection to explain the concept in more detail.

A block diagram of WCDMA systems is shown in Figure 9. The transmitted bit sequence goes through the source coding to increase the transmission efficiency, and this is followed by channel coding for more reliable transmission. Common WCDMA channel codes include convolutional coding (with rates of 1/2 and 1/3) and Turbo coding [49,50] (with a rate of 1/3) [40]. In WCDMA systems, convolutional codes are used for voice services, while Turbo codes are used for high data-rate services.

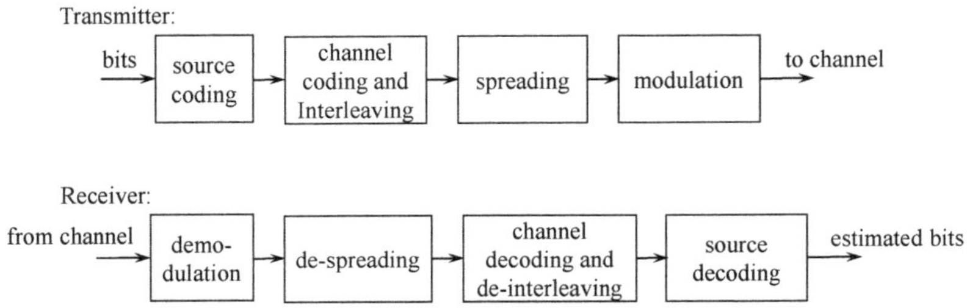

Figure 9. Transmitter and receiver of a WCDMA system.

Channel codes are effective in handling random errors, but they are vulnerable to burst errors, which are common in mobile radio systems (particularly for fast-moving users in WCDMA systems). If the power control is not fast enough to track interference, these users may experience consecutive errors. To address this issue, interleaving is applied to reduce the probability of consecutive errors.

Figure 10 illustrates an example of the application of coding and interleaving. Interleaving is the process of rearranging the bit sequence to scatter potential random errors across the encoded data. In the example shown, each column of data bits is duplicated once, forming the encoded data. During the interleaving process, each pair of identical columns is combined into a row, resulting in an n × n bit matrix, where n represents the block size (i.e., the number of columns in the original data). Data transmission over a noisy channel will result in some bits being lost, but they can still be decoded into the correct bit sequences after de-interleaving since the interleaving process helps to randomize the errors, and channel coding can deal well with randomized errors. However, interleaving improves error correction after decoding, but it also increases processing delay. This creates a trade-off between processing delay and error resistance, requiring optimization to find the right balance.

The spreading process in the WCDMA system, as depicted by the spreading block in Figure 9, increases the system's ability to handle interference. This is achieved by multiplying each symbol with a unique spreading code assigned to a specific user. To better understand the concept of spreading and despreading techniques in WCDMA, consider the example illustrated in Appendix A.

The WCDMA standard employs spread spectrum modulation using BPSK in the uplink channel and QPSK in the downlink channel. The block diagram of the BPSK WCDMA detector receiver is presented in [51], while the principles of BPSK modulation are thoroughly discussed in [41]. The details of QPSK also are explained in Section 4.2.

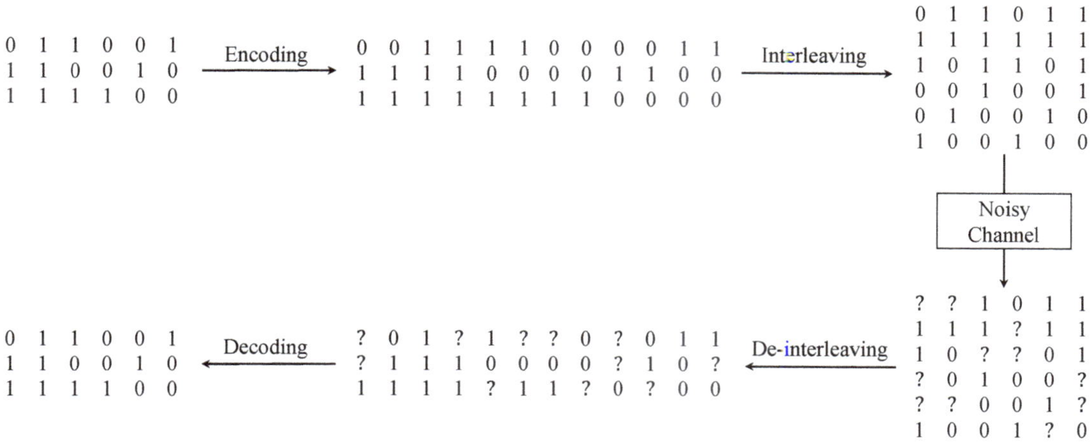

Figure 10. An example of coding and interleaving in a WCDMA system.

7.2. TD-CDMA

The UTRA interface, used by the UMTS terrestrial radio access network (UTRAN), supports two primary duplexing modes: FDD and TDD. UTRA FDD is based on WCDMA with a chip rate of 3.84 Mchips/s. UTRA TDD includes both wideband and narrowband variants. The wideband TDD mode, based on TD-CDMA, also operates at 3.84 Mchips/s, while the narrowband version, TD-SCDMA, which is commonly deployed in China, operates at a chip rate of 1.28 Mchips/s [52,53].

It is important to note that TD-CDMA, commonly referred to as UMTS TDD, is part of the 3GPP UMTS specifications but has seen limited commercial deployment compared to its FDD counterpart. TD-CDMA combines CDMA with TDMA by applying spreading codes within time slots, enabling dynamic and asymmetric resource allocation. In contrast, WCDMA, used in the UMTS FDD mode, is a pure CDMA-based system that operates over a fixed 5 MHz bandwidth with a chip rate of 3.84 Mchips/s [53]. UTRA TDD (TD-CDMA), similar to UTRA FDD (WCDMA), primarily uses QPSK modulation, with higher-order modulations supported in later releases [52].

7.3. TD-SCDMA

TD-SCDMA was developed for use in 3G mobile communication systems through a collaboration between Siemens and the China Academy of Telecommunications Technology (CATT). It has links to the UMTS specifications and is often identified as the UMTS TDD low chip rate (LCR) mode. As outlined in Section 7.2, UTRAN TDD includes two standards with different chip rates: TD-SCDMA, operating at 1.28 Mchips/s, and TD-CDMA, operating at 3.84 Mchips/s. For this reason, 3GPP refers to TD-SCDMA as LCR TDD, while the ITU radiocommunication sector (ITU-R) designates the technology as TD-SCDMA [54].

Like TD-CDMA, TD-SCDMA is best suited for low-mobility scenarios that are typically found in micro or pico cells. In TD-SCDMA, downlink channels can employ QPSK, 8PSK, 16QAM, and 64QAM modulation schemes, while uplink channels support QPSK, 8PSK, and 16QAM [55]. It is important to note that higher-order modulations are primarily utilized in high-speed packet access modes, such as high-speed downlink packet access (HSDPA) and high-speed uplink packet access (HSUPA).

7.4. CDMA2000

CDMA2000 is a 3G mobile communication standard that uses CDMA for transmitting data over a radio frequency. CDMA2000 is actually a set of standards, including CDMA2000 1x and CDMA2000 1xEV-DO (which includes CDMA2000 1xEV-DO Rel 0, CDMA2000 1xEV-DO Rev A, and CDMA2000 1xEV-DO Rev B). The standards of the 1xEV-DO (i.e., evolution data-only or data-optimized) family became the development of CDMA2000 1x standard, and they were oriented on improving the data transmission. CDMA2000 maintains backward compatibility with cdmaOne [40].

The main properties of CDMA2000 are as follows: voice quality, high-speed broadband data connectivity, low end-to-end latency, efficient use of spectrum, and support for advanced mobile services. Moreover, CDMA2000 1xEV-DO enables the delivery of a broad range of advanced services, such as high-performance VoIP, push-to-talk, video telephony, multimedia messaging, multicasting, and online gaming. In addition, CDMA2000 technologies are compatible with IP and ready-to-support network convergence and flexibility.

CDMA2000 systems are designed to meet communication needs in urban and rural areas, with capabilities for fixed wireless, wireless local loop (WLL), and limited and full mobility applications in various spectrum bands. They provide application, user, and flow-based quality of service (QoS), as well as improved security and privacy features [40].

CDMA2000 1x supports data transmission at rates up to 307 kbps in one channel with a 1.25 MHz bandwidth [40]. This standard used the QPSK and 8PSK modulation schemes in the downlink, BPSK was used in the uplink [56–58], and then there was a gradual evolution by the utilization of 16QAM in 1xEV-DO and 64QAM in its later revisions.

In frequency-selective channels—where the delay spread exceeds the symbol duration or the coherence bandwidth of the channel is smaller than that of the signal—channel equalization is required to mitigate inter-symbol interference (ISI). CDMA2000 1xEV-DO, which employs a single-carrier waveform, addresses this using time-domain equalization (SC-TDE) [59]. Unlike single-carrier systems with frequency-domain equalization (SC-FDE), which became more common in later generations, SC-TDE handles equalization entirely in the time domain. A detailed comparison between SC-TDE and SC-FDE is provided in Appendix B.

SC-TDE waveform does not have 0 dB PAPR, but it has a low value of PAPR. Single carrier quadrature amplitude modulation (SC-QAM) is an example of SC-TDE used in CDMA2000. Figure 11 shows a block diagram of SC-QAM. SC-QAM requires a more complex receiver due to its time-domain equalization. Although it offers higher SE than constant-envelope waveforms, it is less suitable for MIMO systems.

Time-domain equalization, as employed in SC-TDE waveforms, can be realized using linear techniques such as zero-forcing (ZF) and minimum mean square error (MMSE), or non-linear methods like decision feedback equalization (DFE) and maximum likelihood sequence estimation (MLSE) [60].

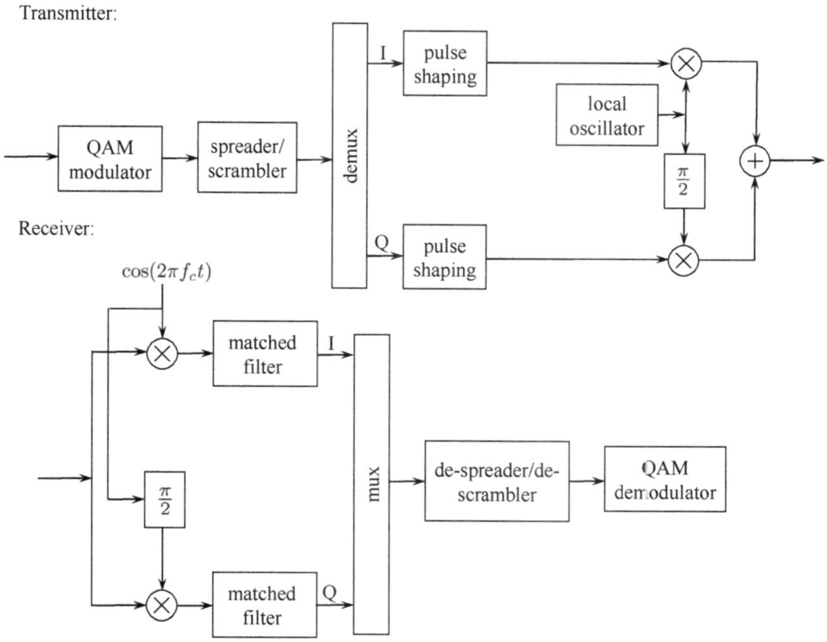

Figure 11. SC-QAM transmitter and receiver.

8. 3.5G

The increasing demand for high-speed data transmission, combined with mobile operators' desire to increase average revenue per user, drove the evolution of cellular systems toward the third generation. High-speed data transmission demand was not met until 3GPP Release 5, which introduced HSDPA, and it is commonly referred to as a 3.5G system. HSDPA offered significant improvements to the downlink channel compared to previous releases, such as WCDMA Release 99, by supporting higher data rates of up to 14 Mbps per user, thus increasing capacity by supporting more users with better data throughput and providing a more responsive experience for richer applications with lower latency.

Later, similar advancements were made in the uplink channel with the introduction of HSUPA in 3GPP Release 6. HSDPA and HSUPA were finalized in early 2002 and early 2005, respectively. Both of these technologies, similar to GPRS and EDGE in GSM, enabled faster packet data transmission speeds [40]. Both HSDPA and HSUPA are collectively referred to as high-speed packet access (HSPA).

The introduction of HSDPA in 3GPP Release 5 (3.5G) allowed for enhanced performance in UMTS-based networks, including lower latency, higher capacity, and higher data rates. The technology utilized higher-order modulation techniques, such as 16QAM, compared to previous releases. 3.5G was capable of adaptively using 16QAM alongside QPSK in the DL channel. This means that 3.5G allowed the base station, also known as NodeB in 3G, to dynamically adjust the waveform allocation to the user, also referred to as UE, and this is based on the reported channel quality by UE to NodeB and some other parameters, providing higher order modulation (16QAM) to users with better channel quality. Other key features of 3.5G included fast scheduling, flexible coding, and hybrid automatic repeat requests (HARQ).

Fast scheduling optimized resource allocation in a single time transmission interval (TTI) as small as 2 ms. Flexible coding, similar to adaptive modulation, allowed NodeB to select the optimal coding based on the reported channel quality indicator (CQI) and user profile. HARQ uses feedback from users for each packet as acknowledgment (Ack) and non-acknowledgment (Nack) to improve re-transmission efficiency in the medium access control (MAC) layer.

9. 3.75G

Evolved High-Speed Packet Access (HSPA+), often referred to as 3.75G, is an advanced version of HSPA. HSPA itself includes two mobile protocols: HSDPA and HSUPA. Unlike standard HSPA, HSPA+ supports higher-order modulation, including 64QAM, which allows it to also use QPSK and 16QAM in the downlink channel—enabling theoretical data rates of up to 21.6 Mbps [61]. In the uplink, HSPA+ supports 16QAM, achieving up to 11.76 Mbps.

After HSPA, Worldwide Interoperability for Microwave Access (WiMAX) Mobile version 1, also known as Mobile WiMAX Release 1, was introduced as a 3.75G technology. This is a wireless communication technology based on 802.16e or 802.16-2005 [62,63]. It satisfies 3G requirements set by IMT2000, uses the air interface specified by the IEEE 802.16 standard, and it offers higher data rates and lower latency compared to most 3G technologies but does not fully meet the requirements for the next generation after 3G, which is 4G.

WiMAX was originally designed for point-to-point (PTP) and point-to-multipoint (PTM) systems. The technology was later improved to support mobility and greater flexibility. The success of WiMAX is mainly down to the "WiMAX Forum", which is an organization formed to promote conformity and interoperability between vendors. Mobile WiMAX Release 1 is designed to operate exclusively in TDD mode and uses orthogonal frequency division multiple access (OFDMA) technology. WiMAX supports QPSK, 16QAM, and 64QAM modulation schemes for both the uplink and downlink channels, although 64QAM is optional for the uplink [64].

10. 4G

The idea of 4G emerged in the early 2000s to offer increased data speeds, lower latency, and better network efficiency compared to its predecessor, 3G and 3G+. The main standards of 4G include LTE, Worldwide Interoperability for Microwave Access 802.16m (WiMAX 802.16m or WiMAX-2), and ultra-mobile broadband (UMB). These standards will be explored in this section.

10.1. LTE

The term "3GPP Release 8" refers to the set of standards developed by 3GPP that introduced LTE technology. While LTE from Release 8 does not fully meet the ITU's IMT-Advanced requirements for 4G, it is often referred to as 3.9G or pre-4G technology. Release 8 was frozen in December 2008 and laid the foundation for the initial deployment of LTE equipment [65]. Release 8 introduced several key enhancements, including the simultaneous use of 64-QAM modulation and 2 × 2 MIMO on a single 5 MHz carrier, enabling theoretical downlink rates of up to 42 Mbps.

Another major improvement over Release 7 was the introduction of dual-cell HSDPA (DC-HSDPA), allowing users to simultaneously utilize two 5 MHz carriers. However, simultaneous use of DC-HSDPA and MIMO was not supported in this release. The use of 64-QAM with DC-HSDPA could theoretically achieve a downlink rate of 42 Mbps. Although LTE was introduced in Release 8, further enhancements were made in Release 9, which added numerous features, including support for additional frequency bands—one of its most significant improvements.

One of the fundamental changes in Release 8 compared to Release 7 (HSPA+) was the adoption of new waveform and multiple access techniques. Release 8 employs OFDMA in the downlink combined with higher-order modulation schemes (up to 64-QAM) and single-carrier frequency-division multiple access (SC-FDMA) in the uplink [65]. The difference between SC-FDMA and OFDMA is illustrated in Figure 12: SC-FDMA provides each user with a single-carrier signal, while OFDMA allocates a variable number of resource blocks (each resource block consisting of 12 subcarriers) per user based on service requirements.

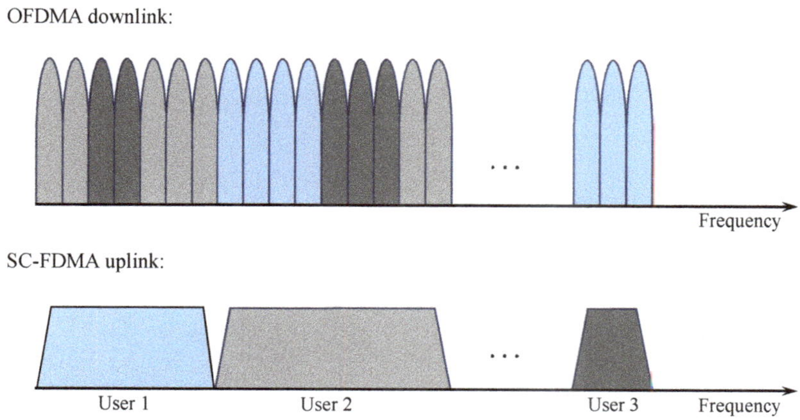

Figure 12. SC-FDMA vs. OFDMA (single carrier vs. multi-carrier frequency division multiple access). Reproduced from [65].

OFDMA uses cyclic-prefix orthogonal frequency division multiplexing (CP-OFDM) as its underlying waveform, whereas SC-FDMA employs a discrete Fourier transform spread orthogonal frequency division multiplexing (DFT-s-OFDM) approach, also known as Single-Carrier Frequency Division Multiplexing (SC-FDM). DFT-s-OFDM applies a discrete Fourier transform to spread the data symbols across the frequency domain. The following sections will explore each waveform in more detail.

10.1.1. CP-OFDM

The fundamental concept behind OFDM entails partitioning the available bandwidth into numerous subcarriers. When the number of subcarriers is large enough, such that the bandwidth of each subcarrier is smaller than the coherence bandwidth of the channel, OFDM effectively mitigates wideband frequency-selective fading by treating it as multiple, parallel narrowband subchannels, each of which undergoes flat fading.

OFDM and OFDMA have become more popular in cellular systems compared to FDM and FDMA, respectively, due to their orthogonality of subcarriers, which enables the overlapping of subcarriers with reduced intersymbol interference. The orthogonality results in better SE as it eliminates the need for guard bands between subcarriers in the frequency domain and allows for overlapping.

The origins of OFDM can be traced to 1966 when Chang [66] introduced groundbreaking research on the creation of band-limited orthogonal signals for multi-channel data transmission [67]. The early version of OFDM was implemented without a cyclic prefix (CP), which serves as a guard interval between symbols in the time domain. However, later in 1971, Weinstein and Ebert introduced the guard interval to combat inter-symbol interference [68]. OFDM that uses a CP is referred to as CP-OFDM; however, in many cellular systems contexts, OFDM and CP-OFDM refer to the same structure [69].

Multi-carrier modulation continued to be used in cellular systems, with CP-OFDM employed in the downlink channel of LTE, starting from 3GPP Release 8 [70]. It was also used by LTE-Advanced.

The structure of CP-OFDM is depicted in Figure 13. It can be seen that CP-OFDM was formed using a CP and employing IFFT and FFT at the transmitter and receiver, respectively. CP-OFDM performed channel equalization in the frequency domain, which effectively combats frequency-selective fading with a lower computational load compared to time-domain equalization.

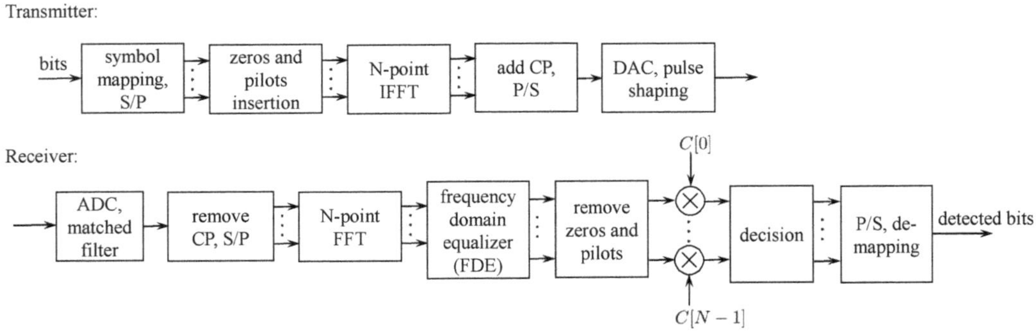

Figure 13. CP-OFDM transmitter and receiver.

In CP-OFDM, a portion of the end of each symbol, called the CP, is copied and added to the beginning of the symbol. The length of the CP must be greater than or equal to the channel delay spread to effectively reduce ISI. The CP converts the linear convolution of the signal with the channel into a circular convolution in the time domain, which corresponds to a multiplication of the signals' DFTs in the frequency domain. This allows efficient computation using the FFT and simplifies equalization of the received signal with a single-tap equalizer per subcarrier. The added CP also helps to suppress the ISI caused by channel delays and reflections.

Despite all these aforementioned advantages, CP-OFDM still faces some drawbacks. One major issue of CP-OFDM is the high PAPR, which significantly reduces the battery life of the user device and requires expensive, high-quality power amplifiers. The user device should provide a highly linear input/output characteristic required by CP-OFDM since it cannot tolerate the non-linearity of power amplifiers. This drawback is covered in LTE by using SC-FDM in the uplink channel and CP-OFDM in the downlink channel.

Several studies in the literature have explored various methods for reducing PAPR. For example, Anoh et al. [71] applied an adaptive clipping threshold to the simplified optimized iterative clipping and filtering (OICF) scheme, effectively reducing the PAPR by dynamically adjusting the clipping threshold according to the system's requirements, thereby improving the overall PAPR performance while maintaining a balance with BER. Jeon et al. [72] reduced PAPR in OFDM systems by proposing a low-complexity selected mapping (SLM) scheme that generates alternative signal sequences through the addition of mapping signal sequences, effectively reducing computational complexity without sacrificing PAPR or BER performance. Rahmatallah and Mohan [73] presented a survey on PAPR reduction techniques in OFDM systems, detailing the available strategies, their performance trade-offs, and their applicability in developing more efficient and practical solutions for system designers.

Additionally, CP-OFDM allocates 10% of the spectrum as a guard band to mitigate out-of-band (OOB) leakage—side-lobe leakage in the frequency domain—but this portion of the spectrum can be saved for users' data transmission. Out-of-band leakage may cause severe interference to adjacent channels (users). The traditional method of sidelobe suppression is windowing, such as root-raised cosine pulse shaping, which can be used to deal with OOB leakage.

CP-OFDM suffers from sensitivity to asynchronous operations, which makes precise timing alignment necessary, and it is also highly sensitive to carrier frequency offset (CFO). Another limitation is the need for CPs to mitigate multipath-induced ISI, which reduces the SE by adding overhead. To address these issues, modified versions of multi-carrier modulation have been proposed, such as CP-OFDM with weighted overlap and add filtering (WOLA), which will be explored in Section 12.1.

10.1.2. SC-FDM

LTE and 5G uplink channels are two examples of cellular systems that use the SC-FDM waveform, specifically called DFT-s-OFDM. The block diagram of the SC-FDM structure is shown in Figure 14. According to this diagram, it is evident that the waveform is a linearly precoded version of CP-OFDM, with the precoding achieved through additional FFT processing. In other words, as shown in Figures 13 and 14, it can be seen that CP-OFDM is formed by removing two blocks of FFT and IFFT from the SC-FDM transmitter and receiver, respectively.

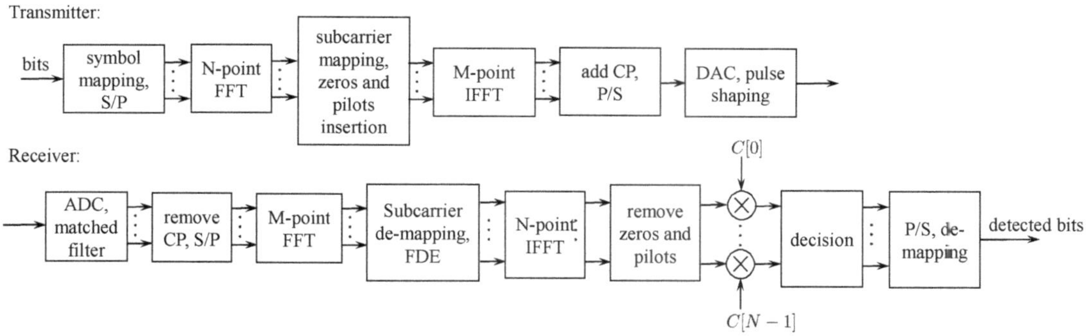

Figure 14. SC-FDM transmitter and receiver.

As shown in Figure 14, the IFFT block in the transmitter and the FFT block in the receiver are used for modulation, while the FFT in the transmitter and IFFT in the receiver help power amplifiers to experience a single-carrier modulation, not a multi-carrier one, which leads to a lower PAPR and is proportional to the square root of the number of subcarriers. This is a major advantage of SC-FDM over OFDM, particularly in the uplink channel, where there is no need for expensive, high-efficiency RF power amplifiers in the user's device.

The other main difference between CP-OFDM and SC-FDM lies in how symbols are mapped to subcarriers. In CP-OFDM, there is a one-to-one mapping—each modulated symbol is directly assigned to a single subcarrier (typically 15 kHz in LTE). In contrast, SC-FDM applies a DFT to a block of modulated symbols before mapping them onto subcarriers. This DFT step spreads each symbol's energy across multiple subcarriers, resulting in a single-carrier-like transmission with lower PAPR. For this reason, SC-FDM is also referred to as DFT-s-OFDM.

In SC-FDM, symbol mapping can be either localized or distributed, depending on the arrangement of occupied subcarriers. Localized mapping involves the allocation of consecutive subcarriers, while distributed mapping allocates non-consecutive subcarriers.

SC-FDMA, an uplink multiple access scheme based on SC-FDM, enables different subcarriers to be assigned to multiple users through FDMA. However, it is worth noting that the multiplexing process should ideally be synchronized to ensure efficient resource allocation and minimize inter-user interference, which can pose a challenge in practical implementations.

Both CP-OFDM and SC-FDM utilize frequency domain equalization, which offers computational efficiency. While SC-FDM exhibits a lower PAPR and reduces sensitivity to carrier frequency offset, its transmission bandwidth is generally considered less spectrally efficient compared to CP-OFDM [73]. This relative inefficiency in bandwidth utilization contributes to the preference for CP-OFDM in downlink channels, where high data rates and SE are essential.

However, single-carrier modulation offers advantages, such as lower PAPR compared to OFDM, flexible dynamic frequency allocation, and user orthogonality within the same cell [74]. It also poses certain challenges, including the need for equalizer implementation at the base station (eNodeB, in the case of LTE) on the receiving side, as well as stricter synchronization requirements in the time domain compared to OFDM.

10.2. WiMAX 802.16m (WiMAX-2)

The IEEE and the WiMAX Forum considered 802.16m as a candidate technology for meeting the requirements of 4G networks. WiMAX 802.16m, an amendment to the 802.16-2009 standard, was developed as an evolution of the WiMAX air interface, building upon the earlier 802.16e standard that was explained in Section 9 [75].

WiMAX 802.16m was submitted to ITU for consideration in the 4G mobile systems. Despite competing with LTE Advanced, 802.16m ultimately did not gain recognition as a true 4G mobile system. Its main objective was to develop an air interface that met the requirements of IMT-Advanced next-generation networks and to support the legacy 802.16 OFDMA system at the same time. To meet these requirements, WiMAX-2 introduced features such as higher data rates through multi-carrier transmission, enhanced MIMO capabilities for improved throughput, and a more efficient superframe structure to support higher overall system performance [75]. WiMAX-2 employed OFDM as the underlying waveform and used OFDMA for multiple access, supporting QPSK, 16-QAM, and 64-QAM modulation schemes. It is compatible with both TDD and FDD duplexing modes.

10.3. UMB

UMB was a proposed 4G communication technology designed as the successor to CDMA2000. It was developed under the third generation partnership project 2 (3GPP2) [76] and the standard-setting organization for CDMA2000. UMB intended to utilize OFDMA as its air interface, representing a significant improvement over the CDMA technology used in CDMA2000. However, most vendors and service providers chose to focus their efforts on promoting 3GPP LTE instead. As a result, the development of UMB was halted in 2008, and it was never widely deployed or adopted as a 4G technology.

OFDM was used in UMB as the main modulation scheme. UMB employs sophisticated channelization techniques to achieve high reliability and throughput. It utilizes adaptive coding and modulation alongside synchronous HARQ and turbo coding with a short retransmission latency. The forward link from the base station to the mobile device in UMB implements MIMO technology, which includes a single code word with closed-loop rate and rank adaptation, as well as a multi-code-word with per-layer rate adaptation.

MIMO also features closed-loop precoding and space division multiple access (SDMA). The maximum data rate reaches 260 Mbps in a 20 MHz forward link.

The reverse link from the mobile device to the base station employs quasi-orthogonal transmission, with orthogonal transmission based on OFDMA and non-orthogonal transmission using multiple receive antennas. CDMA is used for the control segment in the reverse link, allowing statistical multiplexing of various control channels, fast access, and request, and it also provides a wideband reference for power control, subband scheduling, and efficient handoff support [77].

In summary, UMB utilized both OFDMA and CDMA in the uplink for the radio link control channel, with OFDMA used for both the uplink and downlink channels. Additionally, UMB employed several advanced multiple-antenna techniques, such as MIMO, spatial multiplexing, SDMA, and beamforming to achieve much higher user data rates.

11. 4.5G

4.5G, also known as LTE-Advanced and introduced in 3GPP Release 10, is a radio interface technology that meets the IMT-Advanced requirements defined by the ITU-R [78]. LTE-Advanced is recognized as the first true 4G technology according to 3GPP standards. The main objectives of 4.5G are outlined as follows [79]:

- To improve the peak data rate in both forward and reverse channels, with a target of 3 Gbps in the downlink and 1.5 Gbps in the uplink.
- To improve the SE, moving from a maximum of 16 bps/Hz in 3GPP Release 8 to 30 bps/Hz in 3GPP Release 10.
- To increase the number of simultaneously active subscribers.
- To provide better performance at cell edges—such as, for example, for downlink 2 × 2 MIMO—at least 2.40 bps/Hz per cell is required.

LTE Advanced includes advanced features, such as carrier aggregation (CA), improved multi-antenna techniques, and support for relay nodes (RN), as outlined in [79]. These features are briefly discussed in Appendix C.

The modulation used in LTE-Advanced is based on 4G technology using OFDM with higher modulation orders, including 256-QAM on the downlink and initially 64-QAM in the uplink [80]. Support for 256-QAM in the uplink was introduced in 3GPP Release 13 as an optional feature, enabling higher data rates under favorable conditions. The use of 256-QAM modulation can increase the peak data rate by approximately 33% compared to 64-QAM.

The latest release to include significant updates for LTE technology is 3GPP Release 16, completed in June 2020, and it is part of the ongoing development within 3GPP. While Release 16 is often referred to as "5G Phase 2" due to its extensive enhancements for 5G networks, it also introduces improvements for LTE, broadly categorized into four key areas, which can be summarized as follows.

Following its initial specification in Release 10, 4.5G (LTE-Advanced and LTE-Advanced Pro) was expanded in Releases 11 to 16 with the addition of advanced capabilities and enhancements. The latest release to include significant updates for LTE technology was 3GPP Release 16, completed in June 2020, and it was part of the ongoing development within 3GPP. While Release 16 is often referred to as "5G Phase 2" due to its extensive enhancements for 5G networks, it also introduced improvements for LTE, broadly categorized into four key areas, which can be summarized as follows:

- Enhanced LTE-WLAN aggregation (eLWA): This feature improves the integration and aggregation of LTE and Wi-Fi networks, leading to an enhanced user experience and more efficient utilization of network resources.

- Support for Unlicensed Spectrum: Release 16 introduces licensed assisted access (LAA) and enhanced LAA (eLAA), enabling LTE operation in unlicensed frequency bands, such as the 5 GHz band. This enables increased network capacity and higher data rates by leveraging unlicensed spectrum resources.
- Internet of Things (IoT) Enhancements: Improvements were made to LTE machine type communication (LTE-MTC or LTE-M) and narrowband IoT (NB-IoT) technologies to better support massive IoT deployments. These enhancements provided better coverage, power efficiency, and lower latency for IoT devices and applications.
- Vehicle-to-everything (V2X) Communications: Advanced support for V2X use cases was introduced, including vehicle-to-vehicle, vehicle-to-infrastructure, and vehicle-to-network communication. This enabled a wide variety of advanced automotive applications and services, such as autonomous driving and smart transportation systems. Overall, 3GPP Release 16 brought significant advancements and new capabilities to LTE, facilitating seamless coexistence and integration with 5G networks. These improvements empower LTE to support a broad spectrum of advanced use cases and services in various industry verticals.

12. 5G

5G, which was introduced in the mid-2010s, aimed to enhance data transfer speeds and reduce latency, enabling the use of new technologies, such as augmented reality, virtual reality, and IoT. According to 3GPP [81], 5G networks were designed to support three main applications [82]:

- Enhanced Mobile Broadband (eMBB).
- Massive Machine Type Communications (mMTC).
- Ultra-Reliable and Low Latency Communications (URLLC).

Additionally, 5G networks are also expected to support enhanced vehicle-to-everything (eV2X) communications, which is considered a crucial service [81].

5G started from 3GPP release 15 (R15). R15, similar to LTE, uses the CP-OFDM as the downlink waveform for new radio (NR); however, in contrast to LTE, CP-OFDM can also be used in the uplink channel in NR. DFT-s-OFDM can be used in the uplink as a complementary waveform with lower PAPR to improve coverage, although it is limited to single-layer transmission [83]. DFT-s-OFDM serves as the underlying modulation technique for SC-FDMA. SC-FDMA is a multiple access scheme employed in the uplink transmission of LTE systems, which utilizes DFT-s-OFDM to provide efficient resource allocation and sharing among multiple users while maintaining the benefits of DFT-s-OFDM. The topic of SC-FDMA is already explained in Section 10.1.

NR supports different carrier frequencies in two possible frequency ranges with different frequency bandwidths, as shown in Table 5 [83]. NR offers enhanced adaptability through its versatile frame structure, which supports variable subcarrier spacing. Subcarrier spacing is the distance between the centers of two consecutive subcarriers, with possible values of 15, 30, 60, 120, and 240 KHz, which are referred to as "multiple numerologies" [83]. NR is compatible with both FDD and TDD modes of operation [84]. The list of the uplink and downlink physical channels in 5G NR, their associated modulation and coding schemes, and the respective roles of these channels in the overall communication process are provided in Appendix D.

R15 uses a waveform very similar to LTE, but there are several candidates proposed in the literature that could achieve the objectives of the 5G system. For example, in eV2X, sensor nodes usually transmit different types of data asynchronously in narrowbands, while OFDM requires different users to be highly synchronized. Other than that there will be a huge interference among the adjacent subbands [24].

This section explores several prominent next-generation waveforms, beginning with an overview of WOLA. It then focuses on two main categories of waveforms and concludes with a comparison of their modulation performance:

- Subband-filtered schemes, including Filtered OFDM (F-OFDM) and Universal filtered multicarrier (UFMC).
- Pulse-shaped schemes, including Filter bank multicarrier (FBMC) and generalized frequency division multiplexing (GFDM).

These waveforms were all potential candidates for use as the waveform of the downlink channel in the 5G cellular systems, which could be used depending on the carrier and user deployments [23].

All five modulations mentioned in the previous paragraph are categorized as modulation techniques in orthogonal multiple access (OMA) in terms of multiplexing techniques. FDMA, TDMA, SDMA, CDMA, and OFDMA are all examples of OMA used through the evolution path from 1G to 4G and even 5G systems. With the rise of 5G and beyond, non-orthogonal multiple access (NOMA) is being widely discussed in the literature as a new class of multiplexing techniques [85,86]. The idea of NOMA is allowing different users to share the same resources. This means, for example, more than one user can be served by the same physical resource block at the same time. This is exactly the opposite of OMA, where the signals of different users should be orthogonal. For example, if one resource block is allocated to a user in OMA, those subcarriers cannot be used for other users at the same time block [87].

Table 5. NR channel bandwidth [83].

Frequency Range	Frequency Range [MHz]	Supported Channel Bandwidth [MHz]
FR1	410–7125	5, 10, 15, 20, 25, 30, 40, 50, 60, 80, 90, 100
FR2	24,250–52,600	50, 100, 200, 400

This paper does not delve into multiplexing techniques, but it is important to note that the idea in 5G and beyond is to use new modulation schemes, such as the ones explained in this section that are categorized under the OMA group or using NOMA for certain purposes (like improvement of resource allocation for better user fairness and SE enhancement). Due to the limitation in the number of orthogonal resource blocks, OMA can support just a limited number of users. This characteristic of OMA limits the capacity and SE of the network, so to support a large number of diverse users and applications, NOMA schemes have gained attention since the advent of 5G networks [24].

12.1. CP-OFDM with WOLA

CP-OFDM with WOLA, also known as WOLA-OFDM, asw considered a potential waveform enhancement for 5G systems and was also discussed for improving spectral characteristics in the LTE downlink channel [59,88]. Figures 15 and 16 illustrate the structure and time-domain processing of WOLA-OFDM, which builds upon the CP-OFDM framework. WOLA introduced extended CPs and postfixes at the boundaries of each symbol, and this was followed by the application of a windowing function at both ends. This smoothed symbol transitions and improved spectral containment. WOLA can be deployed in the transmitter and receiver for different purposes. In the transmitter, WOLA is applied to improve out-of-band suppression, while in the receiver, it reduces users' interference, especially from asynchronous users [89].

As depicted in Figure 16, in WOLA, L samples of the first part of each symbol will be copied and added to the end, while CP + L samples from the end of each symbol are copied and added to the beginning. This means that, if in conventional CP-OFDM, the length of the symbol is equal to N. After these procedures, the size of one symbol in the time domain will be N + CP + 2 × L, which means the time domain symbol is cyclically extended [90]. After the extension, the Nyquist windows, such as the raised-cosine window, the "better than" raised-cosine (BTRC) window, the second-order continuity window (SOCW), or some others, can be applied to improve the performance of the system in terms of BER [91].

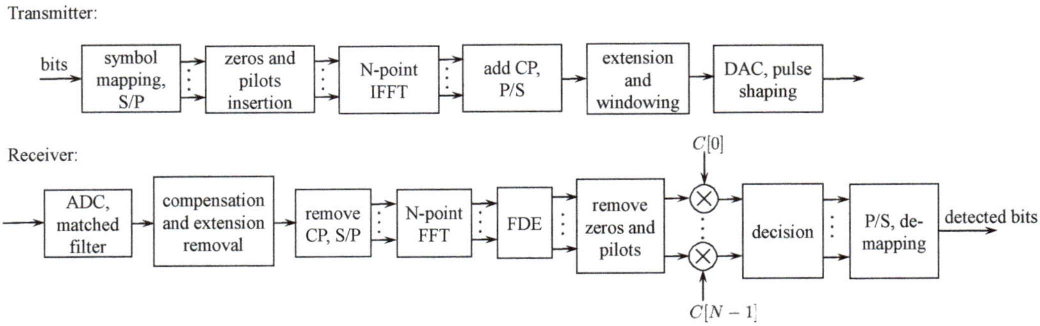

Figure 15. CP-OFDM with a WOLA transmitter and receiver block diagram.

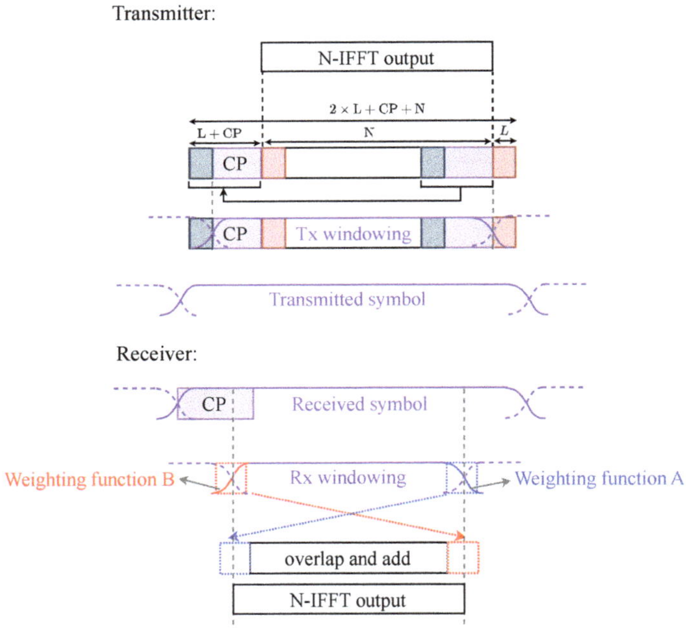

Figure 16. Time domain symbol structure of CP-OFDM with a WOLA in transmitter and receiver. Reproduced from [90].

The WOLA procedures on the transmitter side can be summarized in the following three steps (as shown in Figure 16): (1) extension of the CP-OFDM symbols, (2) applying a window, and (3) overlapping of the consecutive symbols. The overlap step is necessary to maintain the same symbol length as CP-OFDM, but it can also result in collisions between the right side of the previous extended symbol and the left side of the current extended symbol. To avoid collisions, the extended symbols must be separated by the extension length before transmission, resulting in a waste of time and resources. This is the main disadvantage of this waveform [89]. A trade-off arises in the windowing step, as increasing the window length improves out-of-band emissions, and it also increases the probability of collisions and damage to the OFDM data. To reduce the damage, the length of the extension part must be shorter, but this leads to an increase in out-of-band emissions [89].

In CP-OFDM with WOLA, the use of soft-edge pulse shapes (instead of the rectangular pulse shaping filters that are used in conventional CP-OFDM) creates a smooth transition between consecutive symbols [90]. This smooth transition is also achieved by adding a cyclic suffix of length L (which should be greater than the channel delay spread) to the CP-OFDM symbol.

WOLA processing can also be applied on the receiver side, as shown in Figures 15 and 16. The main purpose of receiver WOLA is to reduce interference of asynchronous users, where this interference results from the mismatched FFT capture window [89], by utilizing soft edges. In the WOLA algorithm in the receiver side, weighting functions A and B are applied to the overlapping portions of the window used for circular convolution. The weighting functions are designed to ensure that, when the windowed symbols are summed together, the result is equivalent to the linear convolution of the symbols with the desired filter. The windowed and weighted frames are overlapped and summed to reconstruct the filtered output signal. This process compensates for the overlap between frames, effectively reducing artifacts that would otherwise arise from frame overlapping.

By introducing the weighting function before summing overlapping frames, the WOLA method achieves better spectral containment. This improvement enhances the overall performance and efficiency of the receiver. CP-OFDM with WOLA also has better OOB suppression compared to traditional CP-OFDM; however, it has the drawbacks of a high ACLR and the need for synchronous multiplexing [92]. Despite the benefits of the WOLA processing, this modulation scheme still has the overhead of overlapping adjacent symbols at the edges, similar to traditional CP-OFDM.

12.2. F-OFDM

The development of F-OFDM was driven by the limitations of CP-OFDM in meeting the needs of 5G networks. CP-OFDM requires strict time-synchronization to maintain orthogonality between users, and timing advance (TA) was used in LTE to ensure synchronous arrival of signals from different users. However, as the number of users increases, the overhead due to signaling for TA also increases. This can be problematic in 5G applications, such as mMTC, where numerous devices communicate with one gNodeB simultaneously. Moreover, OFDMA is sensitive to CFO mismatches between different users, which can be an issue in 5G scenarios that aim to provide connectivity in various scenarios, such as machine-to-machine (M2M) and device-to-device (D2D) communications. Therefore, F-OFDM has been proposed in the literature as a candidate waveform for 5G networks [93].

F-OFDM divides the entire frequency band into multiple subbands and applies filters to different subbands in the transmitter [94]. Then, it uses one matched filter per subband to detect the same band in the receiver. From a structural perspective, an F-OFDM waveform is achieved if subband filters are added on top of CP-OFDM, as shown in Figure 17. The bandwidths of the subbands can vary and are not required to be the same. Each subband comprises a different number of subcarriers, but the bandwidth of one subband should be greater than the size of one physical resource block (PRB) in the frequency dimension (as one PRB consists of 12 subcarriers in both LTE and 5G).

Each subband can have its own parameters, such as a specific CP length, TTI, and frequency spacing between subcarriers. This is the primary advantage of F-OFDM, as it adds more flexibility to CP-OFDM by having different subbands and allowing for flexible parameterization of each subband. Each subband, with its unique parameters, can be used for various services and users.

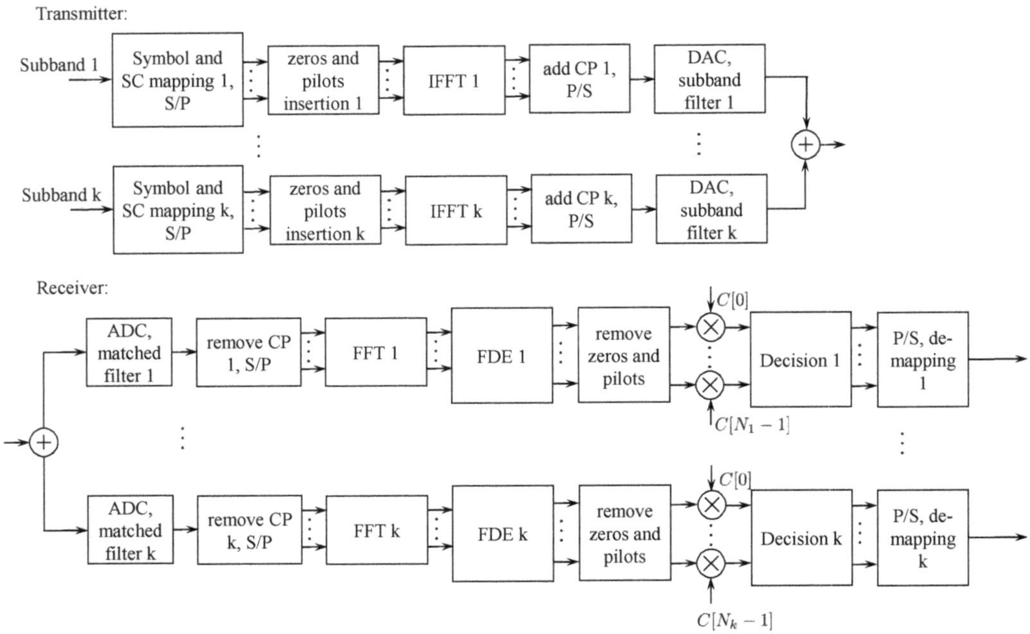

Figure 17. F-OFDM transmitter and receiver.

F-OFDM allows for inter-subband asynchronous transmission (even without the need for synchronization between subcarriers, which is required in CP-OFDM). With subband filtering, the requirement for global synchronization is relaxed. Additionally, a reduction in OOB leakage can be achieved in F-OFDM using well-designed filters. Compared to CP-OFDM, F-OFDM offers better spectrum utilization efficiency due to its minimum level of guard band consumption. Furthermore, within each subband, optimized numerology can be applied to suit the needs of specific types of services [95].

The soft-truncated sinc filter is the most widely used filter for F-OFDM, and it is suitable for many applications with different parameters. Therefore, F-OFDM is highly flexible in terms of frequency multiplexing [24,93]. F-OFDM is an excellent candidate waveform for MIMO communication. It offers all the advantages of CP-OFDM and shows great potential in terms of performance, complexity, cost, and it has an evolutionary path from LTE [95].

Demmer et al. [96] and Gerzaguet et al. [97] both introduced a novel waveform called blocked filtered OFDM (BF-OFDM). F-OFDM and BF-OFDM are both techniques used to improve SE in OFDM systems, but they differ in their filtering approach. F-OFDM involves applying a single filter across the entire OFDM signal to enhance spectral containment and reduce OOB emissions. In contrast, BF-OFDM divides the spectrum into blocks of subcarriers and applies filtering to each block independently, which provides more flexibility in managing interference and spectral allocation across different subbands. While both methods aim to optimize the frequency domain and reduce spectral leakage, BF-OFDM offers more control by allowing separate filtering for different parts of the signal.

12.3. UFMC

UFMC, also known as universal-filtered OFDM or UF-OFDM, shares a concept similar to F-OFDM in terms of filtering a subset of subcarriers. This approach can potentially enhance the SE and mitigate OOB emissions compared to CP-OFDM [98]. UFMC is a combination of two different techniques, F-OFDM and zero-padding OFDM (ZP-OFDM), where ZP-OFDM is equivalent to CP-OFDM while zeros replace the CPs. Unlike CP-OFDM, CP is not officially a part of the structure of UFMC. However, it can still be used to provide extra protection against ISI and enable MIMO transmission [99,100]. Schaich et al. [100] compared UFMC with CP-OFDM in terms of SE and concluded that UFMC outperformed CP-OFDM by approximately 10% in all cases.

The transmitter structure of UFMC, as illustrated in Figure 18, is similar to the structure of F-OFDM. However, the main difference lies in the subband size, which is kept constant over the entire bandwidth. UFMC filtering is executed on every subband, each of which contains multiple subcarriers [101]. In UFMC, the size of each filter in the frequency domain can be equal to the size of a PRB [102]. Thus, each filter is designed to transmit a PRB, and it is a shifted version of the same *prototype* filter, with common filter parameters. Choosing the optimal number of subcarriers per subband is a design optimization problem that needs to balance the baseband processing load and the resulting performance. Increasing the number of subcarriers in a subband can reduce the processing load but can also degrade the performance. Finding the optimal number of subcarriers requires careful consideration of various factors, including the system requirements and available resources.

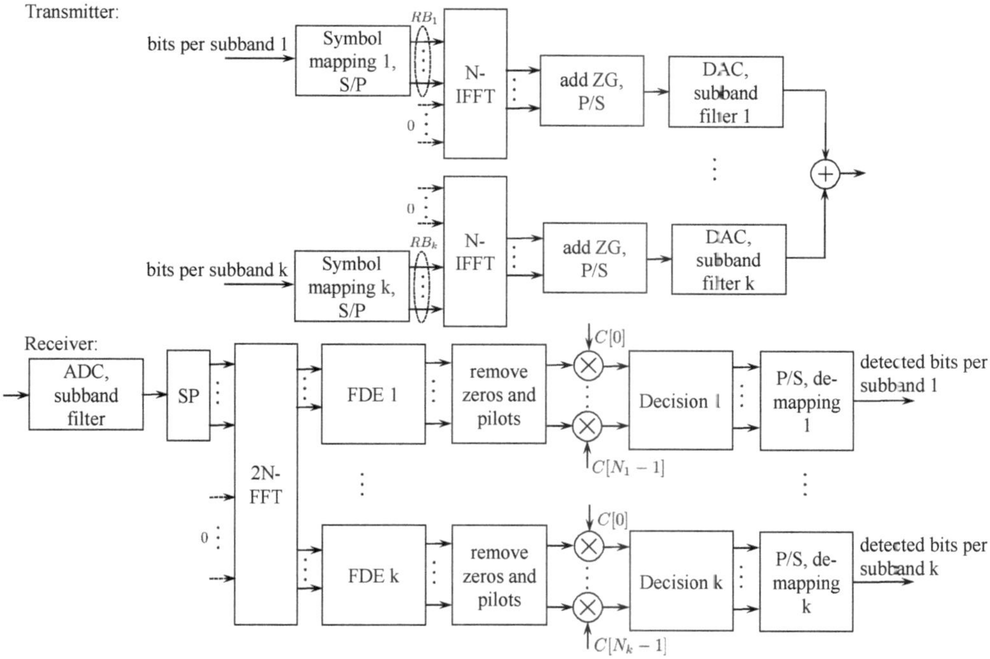

Figure 18. UFMC transmitter and receiver. Partially inspired by [103].

Assuming that N subcarriers are divided into K subbands and each includes $L = N/K$ consecutive subcarriers, then the transmitted signal in UFMC can be expressed as [24]

$$s(n) = \sum_{k=0}^{K-1} s_k(n) * f_k(n), \tag{4}$$

where $f_k(n)$ represents the filter coefficients in the kth subband, and $s_k(n)$ is the OFDM modulated signal over subband k. The expression for $s_k(n)$ is [24]

$$s_k(n) = \sum_{m=0}^{M-1} s_{k,m}(n - m(N+N_g)) = \begin{cases} s_{k,m}(n - m(N+N_g)), & m(N+N_g) \le n \le m(N+N_g) + N - 1 \\ 0, & \text{otherwise} \end{cases}, \quad (5)$$

where N_g is the length of zero padding, m represents number of the symbol blocks, and $s_{k,m}$ denotes the signal at subcarrier k and symbol m. On the receiver side, each symbol interval has the length of $N + N_g$ and will be zero-padded to have a length of $2N$ so that a $2N$-point FFT can be implemented [24].

UFMC is similar to ZP-OFDM in the sense that it appends a zero guard interval between consecutive IFFT symbols to mitigate ISI. At the receiver, a $2N$-point FFT is performed to capture the filter-induced transients, and the relevant N subcarriers corresponding to the original data positions are extracted for detection. UFMC allows all allocated subcarriers to carry data, unlike schemes that restrict transmission to, for example, even-indexed tones. The $2N$-point FFT at the receiver is used solely to capture filter transients and does not imply an increase in the number of data subcarriers.

Several studies in the literature suggest that UFMC has similar OOB emission performance compared to CP-OFDM with WOLA [90]. However, due to the absence of CP in the symbol structure, UFMC may suffer from high ISI. This means less overhead is required by this modulation scheme. Therefore, UFMC is not an optimal choice for channels with high delay spread as it requires the deployment of multi-tap equalizers, which increases the system complexity. Additionally, the larger FFT size required in the receiver for UFMC also increases complexity.

12.4. FBMC

FBMC applies filtering on a subcarrier level [102]. Figure 19 shows a block diagram of a communication system with an N-subcarrier using FBMC. The complex-valued modulated symbols $s_i(n), i = 0, 1, 2, N-1$ come from QAM constellations that are not necessarily identical, with the symbol rate of $1/T$. After upsampling (with the rate of k), each modulated symbol is filtered by a baseband filter known as the prototype filter, with a frequency characteristic of H(f). The N filter output signals are shifted appropriately in the frequency domain and summed up to create the transmitted signal, which is transmitted at the rate of N/T. On the receiver side, matched filtering and downsampling (with the rate of k) are applied. The matched filter is critically sampled when $K = N$, and it is non-critically sampled when $k > N$ [104].

There are different approaches for implementing FBMC, which are mainly categorized into three categories and can be summarized as follows [105].

- Filtered multitone (FMT): FMT uses guard bands to ensure that the subcarriers are separated. This means that conventional frequency division multiplexing (FDM) is used and there is no overlap between subcarriers. However, this reduces bandwidth efficiency. Data symbols are QAM, and the filters used in the transmitter and receiver are a pair of root-Nyquist filters, as shown in Figure 19.
- Multicarrier with Offset QAM/Staggered Modulated Multitone (SMT): SMT uses subcarriers with maximum overlap (minimally spaced subcarriers). Data symbols are offset QAM. If the overlaps are limited to adjacent bands, and the filters in the transmitter and receiver are a pair of root-Nyquist filters, the separation of data symbols at the receiver output is guaranteed [106,107].
- Cosine modulated multitone (CMT): CMT uses pulse amplitude modulated (PAM) data symbols, and subcarriers are maximally overlapped, just like in SMT. Therefore, both SMT and CMT achieve maximum bandwidth efficiency [104,105,108,109].

All three methods require equalization to be implemented after down-sampling in the receiver. SMT, which uses offset-QAM, is a popular recognized potential method for implementing FBMC in 5G systems and beyond. This is why SMT is sometimes referred to as FBMC-OQAM. Because neighboring subcarriers in SMT are overlapped, offset-QAM helps to provide orthogonality to avoid ISI and achieve full capacity. In offset-QAM, the real and imaginary parts of the complex symbols are not transmitted at the same time, and instead the imaginary part of the complex symbols is transmitted with a delay equal to half of the symbol period (as explained in Section 4.2).

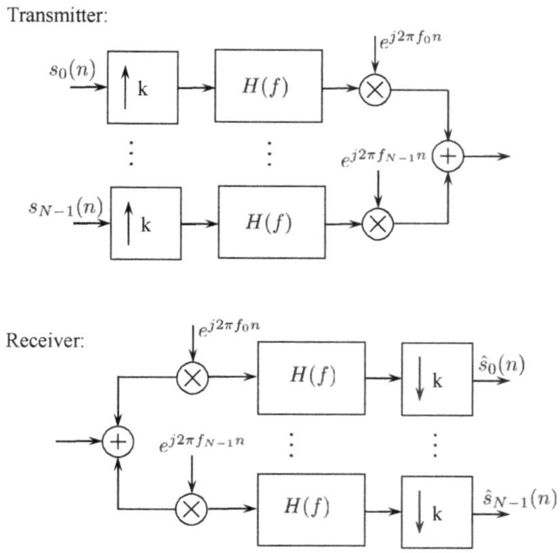

Figure 19. FBMC transmitter and receiver. Reproduced from [104].

Although FBMC uses filter banks to apply filters at the sub-carrier level, it has a higher implementation complexity compared to subband modulation schemes, such as F-OFDM and UFMC. Additionally, the use of offset QAM further increases the complexity. One advantage of FBMC is that it does not require a CP (although it can optionally use it). Filters remove ISI, so it is possible to achieve the same SE as single-carrier schemes without adding overhead due to a CP. Therefore, FBMC has a lower OOB leakage and better bandwidth efficiency compared to CP-OFDM. However, it is not suitable for MIMO schemes like Alamouti space coding [110].

Unlike subband-based waveforms, such as UFMC or F-OFDM, FBMC is not well-suited for short-burst transmissions due to its long filter tails in the time domain. Furthermore, the long filter tails combined with the absence of a CP make FBMC particularly vulnerable to ISI in frequency-selective channels [111]. While FBMC offers excellent spectral confinement and is advantageous for applications like cognitive radio and fragmented spectrum access, it is generally not considered a strong candidate for 5G and beyond due to its lack of orthogonality in the complex domain and the increased complexity associated with filter design and implementation.

Some literature consider UFMC as a generalization of both CP-OFDM and FBMC, based on the granularity of the filtering. CP-OFDM applies filtering over the entire bandwidth, encompassing all subcarriers, whereas FBMC filters each subcarrier individually using well-localized prototype filters. UFMC strikes a balance by applying filtering at the subband level—that is, to groups of subcarriers. Since subband filters are spectrally wider than per-subcarrier filters, they are shorter in the time domain—typically on the order of a

CP length—making UFMC well-suited for short-burst transmissions, which are important in 5G and beyond.

12.5. GFDM

The block diagram of GFDM is illustrated in Figure 20 [112]. Although there are some similarities between GFDM and FBMC (for example, both waveforms are based on pulse shaping, which means that filtering is implemented on the subcarrier level in both waveforms and each subcarrier may have different bandwidth), there are also some differences between them [110,113,114]. The GFDM uses circularly shifted prototype filters (in time and frequency domain) for pulse shaping each subcarrier instead of linear filters used by FBMC, resulting in improved OOB emissions. GFDM also uses a block-based signal structure, allowing for a CP to be added to each block, resulting in low overhead because there is only one CP per block. This means, unlike FBMC, GFDM uses CP insertion to enable low-complexity equalization at the receiver and to provide orthogonality against multipath channels. A tail-biting technique is used to shorten the CP to enhance the SE [112]. GFDM has low PAPR and OOB emission due to filter adjustment, but it has a complex receiver structure that uses matched filters to remove interference.

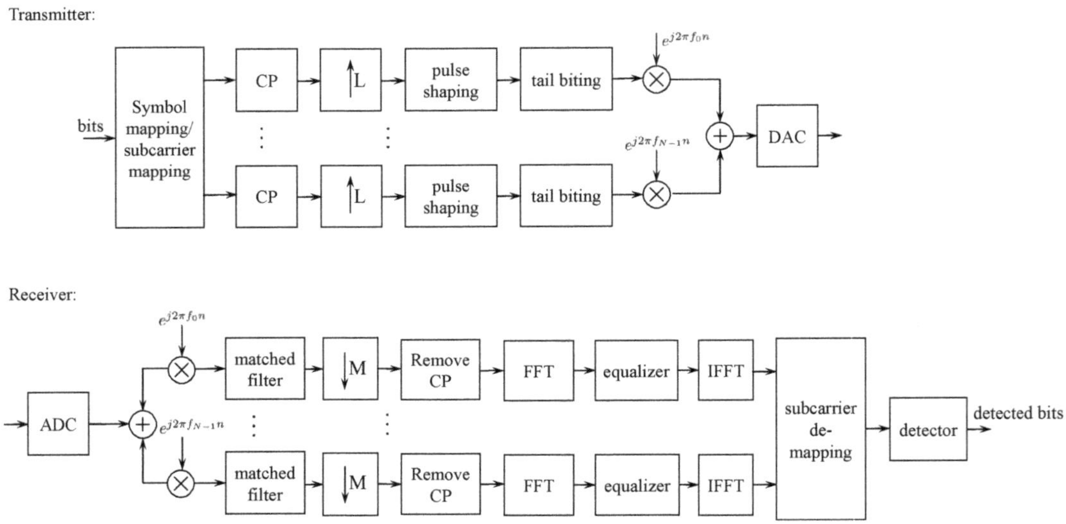

Figure 20. GFDM transmitter and receiver. Here, L denotes the upsampling factor at the transmitter, and M denotes the downsampling factor at the receiver.

In terms of comparison of GFDM with the other schemes, CP-OFDM and SC-FDE may be counted as two special cases of GFDM. GFDM is a parallel realization of SC-FDE, where circularly shifted prototype filters are applied to each subcarrier. In other words, GFDM modulates independent blocks, with each block including multiple subcarriers and subsymbols. Specifically, if there are K different subcarriers and M different subsymbols in each block, then there are $K \times M$ different samples per block in GFDM. GFDM subcarriers are non-orthogonal, which can cause ISI and inter-carrier interference (ICI). However, the adjustment of each filter applied to individual subcarriers, along with the flexibility to choose different signal bandwidths per subcarrier, can help mitigate mutual interference between subcarriers [112]. For example, a matched filter receiver with iterative interference cancellation can achieve the same symbol error rate (SER) performance as OFDM across different channel models [115]. On the other hand, the use of offset-QAM in GFDM makes it unsuitable for MIMO applications. Similar to FBMC, GFDM is well-suited for cognitive radio applications.

12.6. Performance Summary of 5G OFDM-Based Modulation Techniques

Table 6 compares various candidate modulation schemes for 5G and beyond (as previously discussed in Section 12), with CP-OFDM used in LTE (Section 10). The comparison is adapted from the synthesis presented in [116], which consolidates findings from several prior studies, as referenced in that work. All the modulation schemes listed in Table 6 are based on multi-carrier modulation. A comparison between multi-carrier and single-carrier modulation is provided in Appendix E. Although various potential waveform candidates for 5G deployment were considered, by the time 5G rollouts began in 2020, CP-OFDM had emerged as the waveform most widely used [116]. This selection was made due to the compatibility of CP-OFDM with existing infrastructure and its proven performance in 4G LTE systems, while still addressing the key requirements for 5G communications, such as SE and resilience against multipath fading and ISI [117,118].

Table 6. Comparison of 5G modulation candidates. Adapted from [116].

KPI	OFDM	F-OFDM	UFMC	GFDM	FBMC
SE	Lower	Much better	Good	Enhanced	Best
Latency	Low	High	Low	High	High
CP	Used	Used	Not Used	Used	Not Used
OOB	High	Reduced	Reduced	Low	Extremely small
Filter length	Whole-Band	Subband	Subband	Subcarrier	Subcarrier
PAPR	Highest	High	High	Low	Lower than the rest
Computational complexity	Lower	High	Very High	High	High
Introduction of MIMO techniques	Highly flexible	Highly flexible	Flexible	Less flexible	Less flexible
Orthogonality	Orthogonal Subcarriers	Orthogonal Subcarriers	Orthogonal Subcarriers	Non-Orthogonal Subcarriers	Real field orthogonal

13. 6G

The rapid advancements in applications such as extended reality, cloud computing, and smart cities have exposed the limitations of the current 5G network infrastructure. The growing need for higher data rates, lower latency, and enhanced reliability underscores the demand for a more advanced wireless communication solution. The 6G technology has emerged as a response to these challenges, aiming to address the shortcomings of 5G and also enable the next wave of innovations [119]. The cutting-edge technologies, such as terahertz (THz) communications, visible light communications (VLC), ultra-massive multiple-input multiple-output (um-MIMO), hybrid networks, and artificial intelligence (AI)/machine learning (ML) are at the core of the 6G. The strategic integration of these technologies aims to achieve unparalleled performance levels, setting the stage for a range of groundbreaking applications and services [116].

At this stage, 6G is still in the research and development phase, and no single modulation technique has been universally adopted. However, several advanced modulation techniques are being explored for 6G due to their potential to meet the high-performance and diverse requirements of next-generation networks beyond 5G.

Integrated sensing and communication (ISAC) is increasingly viewed as a pivotal technology for advancing 6G wireless networks [120,121]. The intrinsic sensing abilities of ISAC are expected to enable highly accurate sensing services in diverse contexts, such as smart transportation with autonomous vehicles, Wi-Fi sensing for smart homes, V2X interactions in smart cities, and indoor localization, among other applications [122].

Therefore, ISAC will deliver not only communication capabilities, but also the capacity to perceive the surrounding environment, paving the way for a multifunctional network and initiating a paradigm shift in the 6G era [120,123]. Consequently, ISAC has garnered significant research interest recently, encompassing areas, such as waveform design [124], fundamental limitations [125], radio resource management [126], signal processing [127], the coexistence of radar and communication systems [128], and the development of proof-of-concept platforms [129].

Focusing on waveforms, it is crucial to understand that communication and sensing have varying requirements. Communication waveforms are optimized for high data rates, whereas sensing waveforms are ideally characterized by a constant envelope and effective correlation [130]. Thus, developing waveforms that fulfill the requirements of both communication and sensing functions is important for the performance of ISAC systems. Zhou et al. [120] classified ISAC waveform designs into three categories: (1) communication-centric waveform design (CCWD), (2) sensing-centric waveform design (SCWD), and (3) joint waveform optimization and design (JWOD). We expand on this classification by incorporating additional emerging waveforms to provide a more comprehensive review. This extended classification is outlined in Figure 21 and will be explored in detail in this section.

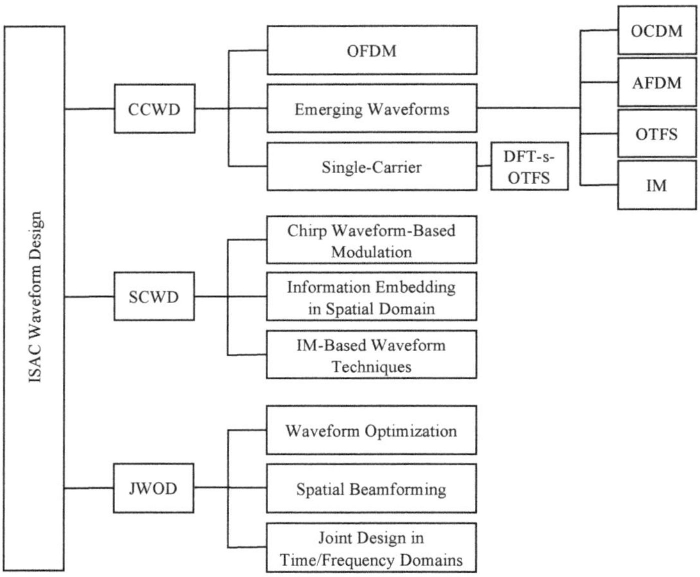

Figure 21. ISAC waveform design classification. Adapted and expanded from Zhou et al. [120].

13.1. CCWD

CCWD focuses on utilizing or adapting traditional communication waveforms to facilitate concurrent sensing capabilities. This approach effectively extracts sensing information from target echoes while ensuring optimal communication performance. While many existing communication waveforms can be applied, Zhou et al. [120] discussed the research progress in three specific categories: OFDM, emerging waveforms, and single carrier.

13.1.1. OFDM

While the fundamental principles of OFDM are described in Section 10, recent advances have extended OFDM for ISAC by incorporating symbol-domain signal processing techniques. A pioneering study [131] introduced a method within OFDM-based ISAC systems that utilizes modulation symbols from both transmission and reception to simplify range and Doppler estimation compared to conventional correlation-based methods. Building on this, subsequent research has highlighted the importance of power allocation and subcarrier configurations in enhancing sensing performance. Adaptive OFDM waveforms have been developed to optimize these parameters, maximizing SE while meeting diverse communication and sensing needs [132,133]. Additionally, techniques such as subcarrier random phase modulation [134] and joint transmit code design [135] have been proposed to mitigate Doppler ambiguity and improve parameter estimation accuracy.

Ensuring communication reliability is essential, with novel approaches such as code-division OFDM [136] and complementary phase code sequences [137], enabling effective signal discrimination from noise. Furthermore, combining OFDM with MIMO technology [138] and full duplex capabilities [139] enhances spatial resolution and reduces self-interference. In general, these advances address challenges, such as range ambiguity, high PAPR, and various interference issues, showcasing the dual functionality of OFDM in modern ISAC systems.

13.1.2. Emerging Waveforms

In addition to classical waveforms, such as single-carrier modulation and OFDM, advanced variants like FBMC and GFDM—detailed in Sections 12.4 and 12.5, respectively—are being actively evaluated for their potential to support ultra-high data rates and improve SE. Beyond these, we delve into four emerging and innovative waveforms: orthogonal chirp division multiplexing (OCDM), affine frequency division multiplexing (AFDM), orthogonal time frequency space (OTFS), and index modulation (IM). These promising waveforms, each offering unique advantages for next-generation communication systems, will be examined in the following paragraphs.

OCDM

The fundamental idea behind OCDM is the simultaneous multiplexing of multiple orthogonal chirp waveforms that share the same bandwidth, enabling overlapping phases and amplitudes. Ouyang and Zhao [140] represent the OCDM system based on OFDM, as shown in Figure 22. The block diagram of a traditional OFDM system is shown if excluding the components highlighted with blue boxes. In the transmitter, the IDFT combines symbols onto parallel subchannels, while at the receiver, the DFT performs the inverse operation to recover the symbols. A single-tap equalizer is then applied to compensate for the subchannels. The discrete-time OCDM signal is given by [140]

$$\mathbf{x} = \mathbf{\Phi}^H \mathbf{s}, \tag{5}$$

where $\mathbf{x}$ is the OCDM symbol vector, $\mathbf{s}$ is a vector of the information symbols, and $\mathbf{\Phi}$ is a discrete Fresnel transform (DFnT) matrix, as defined by [140,141].

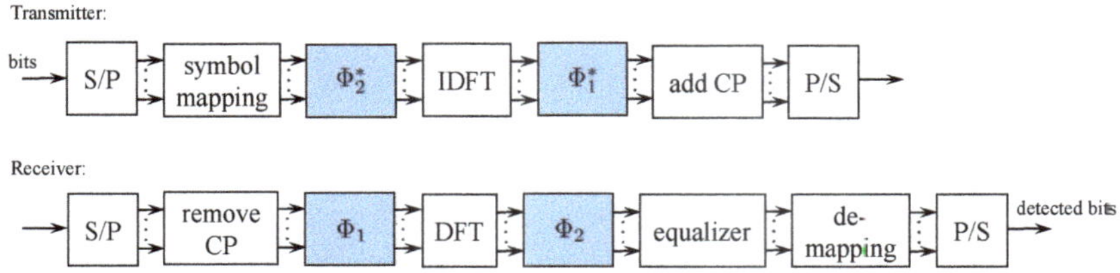

Figure 22. OCDM system schematic diagram based on the OFDM system (including the blue boxes) and a OFDM system schematic diagram (excluding the blue components). Reproduced based on [140].

By utilizing the relationship between the DFT and DFnT outlined by Ouyang and Zhao [140], OCDM can be incorporated into the OFDM system with the added operations shown in the blue boxes. The DFnT can be computed using the DFT in three steps:

1. Multiply by the first quadratic phase $\mathbf{\Phi_1}$.
2. Apply the DFT using the FFT algorithm.
3. Multiply by the second quadratic phase $\mathbf{\Phi_2}$.

Here, $\mathbf{\Phi_1}$ and $\mathbf{\Phi_2}$ are diagonal matrices with diagonal entries, respectively. Using these steps, OCDM can be integrated into the OFDM system with minor modifications [141]. At the transmitter, the IDFT undergoes a three-step process, functioning as the inverse DFnT (IDFnT). On the receiver side, two architectures are introduced by Ouyang and Zhao [140], one based on a structure with a multi-tap TDE or FDE equalizer, and the other based on the conventional FDE structure, allowing for easy adaptation of OFDM channel estimation and equalization schemes to the OCDM system.

Since each chirp covers the entire signal spectrum, with bandwidth controlled by the digital-to-analog converter (DAC) sampling rate, Omar and Ma [142] introduced a method for digitally managing the OCDM spectrum. To reduce the PAPR of OCDM signals, a solution was proposed in [143]. Additionally, to address channel estimation challenges caused by carrier frequency offsets, Zhang et al. [144] suggested using a CP to compensate for these offsets before performing channel estimation [120].

A notable advantage of chirp-based waveforms, such as OCDM, is their ability to support full duplex communication [145]. OCDM leverages a set of orthogonal chirps, which are complex exponentials with linearly varying instantaneous frequencies. By allowing each piece of information to utilize the entire bandwidth, OCDM achieves full diversity in frequency-selective channels, outperforming OFDM. Although the radar images obtained show a slight increase in sidelobe levels compared to those of OFDM, OCDM does not attain full diversity in doubly selective channels [146–149].

Wan et al. [150] propose an innovative ISAC waveform design for mmWave unmanned aerial vehicle (UAV) communications using OCDM, where one subcarrier is dedicated to sensing with frequency-modulated continuous wave (FMCW) radar and the others enhance communication, thereby offering a hardware-efficient, energy-saving solution that outperforms traditional methods like OFDM and OTFS in terms of sensing and communication performance.

AFDM

Recently, a novel chirp-based waveform known as AFDM was introduced [149,151,152]. This approach multiplexes information symbols in the discrete affine Fourier transform (DAFT) domain. AFDM can adapt to varying channel conditions by optimizing its parameters, facilitating the effective separation of all signal paths in the DAFT domain. As a result, AFDM can achieve full diversity, even in doubly selective channels [146].

Bemani et al. [151] represent the AFDM modulation/demodulation block diagram, as shown in Figure 23. The DAFT, which underlies AFDM, is a discrete adaptation of the AFT [152–155]. As shown in Figure 23, inverse DAFT (IDAFT) is used to map $\mathbf{x}$, which is a vector of N QAM symbols, to the time domain. The output of IDAFT is s_n, where $n = 0, \ldots, N-1$, which, in matrix form, can be shown as follows:

$$s = \mathbf{\Lambda}_{c_1}^H \mathbf{F}^H \mathbf{\Lambda}_{c_2}^H \mathbf{x}, \tag{7}$$

where $\mathbf{F}$ is the DFT matrix with entries $e^{-j2\pi mn/N}/\sqrt{N}$, and c_1 and c_2 are defined as the post-chirp and pre-chirp parameters of the AFDM, respectively. $\mathbf{\Lambda}_{c_i} = diag(e^{-j2\pi c_i n^2})$, $n = 0, \ldots, N-1$, and $(.)^H$ denote a Hermitian transpose [146].

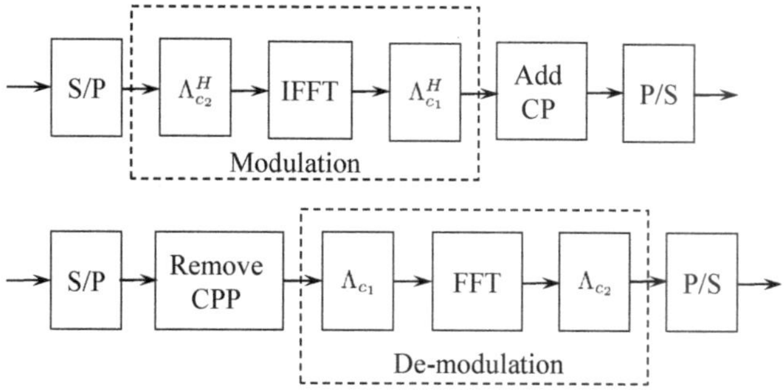

Figure 23. AFDM modulation/demodulation block diagram. Reproduced Figure from [151].

Similar to OFDM, AFDM requires a prefix to emulate a periodic channel response. However, in this case, the prefix utilized is a chirp-periodic prefix (CPP), which is specifically defined by Bemani et al. [151]. After the signal is converted from parallel to serial format and transmitted through the channel, the received samples undergo a serial-to-parallel conversion. Once the CPP is removed, the samples are processed in the DAFT to convert the received time-domain signal back into the frequency domain. The DAFT recovers the transmitted data symbols by transforming the received signal using the appropriate affine scaling and shifting parameters.

OTFS

OTFS modulation is an advanced technology designed for next-generation communication systems beyond 5G, such as 6G. OTFS presents several advantages over OFDM, which has been prevalent in previous generations. These advantages include improved robustness to Doppler shifts, higher data rates, greater flexibility, reduced need for CPs, and a lower PAPR. Among the various techniques for PAPR reduction in OTFS, the classical selected mapping (SLM) method is particularly effective [156]. Due to these improvements, OTFS is regarded as a promising waveform candidate for future high-performance wireless communication systems.

Figure 24 illustrates the block diagram of OTFS modulation [156]. In OTFS, QAM symbols are utilized, and the symbol mapper converts the time-domain input into a 2D delay-Doppler (DD) domain signal, which is represented as $x(k,l)$. This signal is fed into the inverse symplectic finite Fourier transform (ISFFT) block. In OTFS modulation, both the input $x(k,l)$ and output $y(k,l)$, as depicted in Figure 24, are represented in the DD domain. The OTFS block diagram involves two types of FFT: ISFFT and the symplectic finite Fourier transform (SFFT). The ISFFT maps symbols from the DD domain $x(k,l)$ to the time-frequency domain $X(n,m)$, while the SFFT maps from the time-frequency domain $Y(n,m)$ back to the DD domain $y(k,l)$.

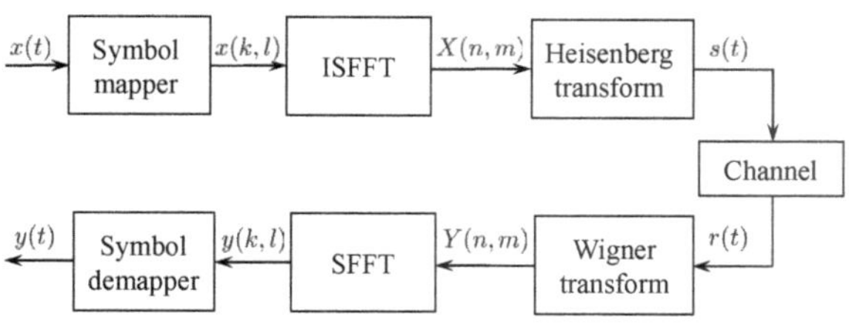

Figure 24. OTFS modulation and demodulation block diagram.

The modulation starts with the input in the DD domain, which is converted into the time-frequency domain. The time-frequency signal is then transformed into the time domain using the Heisenberg transform. The result is the OTFS-modulated signal, which is then transmitted through the channel. Afterward, the signal is converted back to the time-frequency domain using the Wigner transform. To achieve the demodulated output in the DD domain, SFFT is applied. This process completes the OTFS modulation and demodulation cycle [156].

Regarding $x(k,l)$ and $y(k,l)$, the variables k and l refer to the DD domain. For $X(n,m)$ and $Y(n,m)$, the indices n and m indicate the time-frequency domain. Regarding $s(t)$ and $r(t)$, t represents the time domain. The functions x and X are related to the modulated side, while y and Y are associated with the demodulated side [156].

In [157], a system that integrates sensing parameter estimation with communication using OTFS modulation was proposed. Ravitega et al. [158] explored OTFS for estimating target range and velocity. For the IoT applications discussed in [159], OTFS was employed in ISAC systems to enhance reliability and energy efficiency. Additionally, in [160], OTFS signals facilitated roadside units in accurately determining vehicle locations and velocities. OTFS modulation offers notable advantages over OFDM, as highlighted in Table 2 of [120] by Zhou et al., where they compare parameters, such as BER, PAPR, symbol size, stability, and achievable rate, between the two modulation techniques.

IM

IM offers exceptional SE with significantly reduced power losses. This technique enhances data transmission by utilizing indexed resource entities—such as specific timing slots, resonators, subcarriers, and channel states—to convey additional information. The infrastructure required for IM is relatively inexpensive, and it delivers superior throughput, making it a highly desirable modulation method for 6G technology [161].

IM can be further categorized into five subtypes: time domain index modulation (TD-IM), frequency-domain index modulation (FD-IM), spatial domain index modulation (SD-IM), code domain index modulation (CD-IM), and power domain index modulation (PD-IM) [162–164].

- TD-IM: In each data block frame, a small fraction of signal slots is allocated for message transmission, with the corresponding indices used to convey the data. Additionally, combining TD-IM with space-time block coding (STBC) can lead to substantial enhancements compared to non-index modulation (non-IM) methods.
- FD-IM: OFDM is widely used in both 4G and 5G wireless systems and is expected to remain a key technology in 6G due to its high SE [161]. OFDM achieves this using orthogonal subcarriers, each capable of transmitting distinct data, thereby maximizing the utilization of the available channel bandwidth.
  In addition to the conventional use of subcarriers for data transmission, FD-IM introduces an extra layer of data transmission by selecting indices of the activated subcarriers. This technique allows for transmitting additional bits by modulating the index of the active subcarriers rather than their amplitude or phase. FD-IM is also known as subcarrier-index orthogonal frequency division multiplexing (S-OFDM) [161]. Lin et al. [165] provided system models for various IM-based multicarrier systems, including OFDM-IM, GFDM-IM, FBMC-IM, and OTFS-IM.
- SD-IM: This modulation technique utilizes both the spatial domain—by selecting antenna indices—and additional information bits to transmit data, enabling improved SE through antenna index modulation. Unlike traditional MIMO systems, SD-IM reduces synchronization complexity at the transmitter by activating only one antenna per symbol interval, though basic synchronization between transmitter and receiver

is still required. It also reduces inter-antenna electromagnetic interference due to the specific encoding of information in the spatial domain [161,166]. Because SD-IM activates only one transmit antenna per symbol interval and encodes part of the data in the antenna index, it significantly reduces receiver complexity compared to conventional MIMO schemes. By using multiple transmit and receive antennas, SD-IM increases the communication data rate by leveraging spatial diversity and precoding gains, while beamforming at the receiver can further enhance performance.

- CD-IM: In the CD-IM technique, radio frequency (RF) equipment like RF mirrors and electronic switches are employed to enhance communication. This approach, known as media-based modulation, utilizes multiple RF mirrors positioned near the transmitting resonators to direct the RF signal along specific communication paths based on the on/off states of these mirrors or switches. By modifying the RF communication environment, media-based modulation can reorganize the communication channel, resulting in increased transmission speed and efficiency.
- PD-IM: In conventional NOMA operating in the uplink, multiple users share the same time and frequency resources, each being assigned a distinct power level. However, this power-domain multiplexing can limit the SE due to the fixed power allocations. To improve the SE of traditional NOMA, PD-IM is introduced, where the power levels themselves are used to convey index bits, effectively adding a layer of modulation. In power-index modulation multiple access (P-IMMA), the choice of power levels from a predefined set acts as a means of transmitting data, with each power level representing a unique index. This power-based modulation allows for higher data throughput without increasing the required bandwidth. As a result, P-IMMA offers a significant performance gain over conventional NOMA, particularly in terms of BER, by making more efficient use of the available power resources [164].

13.1.3. Single Carrier

In contrast to OFDM waveforms that use multiple subcarriers for parallel data transmission, single carrier signals typically feature a low PAPR [167], ensuring stable power delivery and improved performance of power amplifiers. A key aspect of single carrier waveforms is a spread spectrum, where information is modulated with a pseudorandom sequence to expand the spectrum during transmission, enhancing radar target detection due to excellent autocorrelation properties [168]. This approach has been widely adopted in third-generation technologies (explored in Sections 7–9) for its resilience to interference, low power requirements, confidentiality, and high measurement precision [169].

Recent research has connected sequences to create synthetic spread sequences with superior correlation for ISAC applications [170]. Single-carrier frequency division access offers a solution to the high PAPR associated with multicarrier designs [171], while variations of cyclic prefixed single carrier have also been applied in ISAC [172]. Moreover, strategies such as using linear combinations of communication data for radar detection [173] and phase disturbances to reduce range sidelobes [174], have been explored. A significant area of research in single-carrier modulation for 6G is the exploration of discrete Fourier transform spread orthogonal time-frequency space (DFT-s-OTFS), which we will elaborate in the following.

DFT-s-OTFS

DFT-s-OTFS was recently introduced for THz ISAC to enhance the PAPR and improve robustness against Doppler effects [175–177]. DFT-s-OTFS shows promise for various THz applications. Tarboush et al. [178] analyzed DFT-s-OTFS, its complexity, SE, TTI latency, and resilience to phase noise and THz-specific impairments.

The proposed DFT-s-OTFS system by Hossain et al. [177] is illustrated in the transceiver block diagram in Figure 25. The transmission process begins with the mapping of QAM samples in the DD domain. Subsequently, a Z-point DFT is applied, followed by the ISFFT operation within the same domain. This ISFFT process transforms the symbols into time-frequency samples. A time-frequency modulator then converts these dual data samples into a time-domain waveform suitable for transmission over a wireless channel, effectively utilizing OFDM modulation to map information symbols from the frequency domain to the time domain.

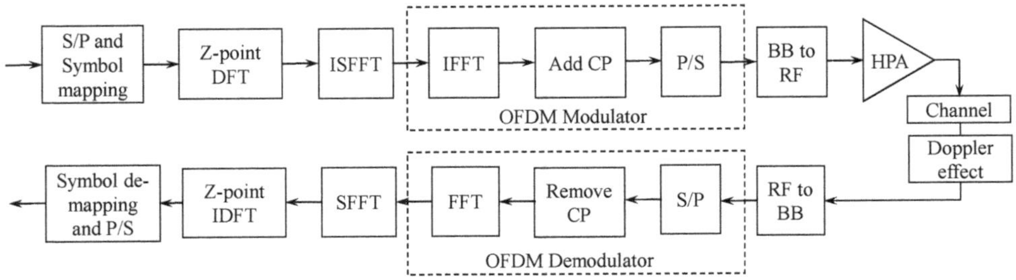

Figure 25. DFT-s-OTFS modulation and demodulation, reproduced based on [177].

Once the time-domain signal is transmitted through a time-varying channel, the receiver performs the reverse operations to retrieve the received signal in the DD domain using the SFFT operation. This is followed by a Z-point IDFT to recover the information symbols. To address the effects of DD spreading, channel equalization is applied after the de-mapping stage. Hossain et al. [177] employed the widely recognized minimum mean square error (MMSE) equalization method for frequency-domain equalization due to its efficient computational requirements. Their simulation results indicate that both the OTFS and the proposed DFT-s-OTFS system attain a gain of 12.8 dB over the OFDM system. In addition, the DFT-s-OTFS system achieves a PAPR reduction of 2.2 dB and 1.8 dB compared to the OFDM and OTFS systems, respectively [177].

The structure of a DFT-s-OTFS data frame includes the same number of symbols as a standard OTFS frame. In single-user uplink scenarios, a selected subset of data symbols undergoes DFT precoding, followed by DD mapping. This mapping constructs the data frame by spreading the symbols and zero-padding the remaining points to form a DD lattice. Subsequently, standard OTFS transmission operations, such as channel coding, are applied to the constructed DFT-s-OTFS frame to ensure reliable and effective signal transmission.

While DFT-s-OTFS offers improved PAPR performance compared to both OTFS and CP-OFDM, it does come with increased complexity for receiver detection and DD channel estimation, particularly under fractional Doppler conditions. Despite this trade-off, the potential for full-time-frequency channel diversity positions DFT-s-OTFS favorably against traditional multiple carrier schemes, highlighting its relevance for future V2X applications in THz-enabled 5G and beyond [178].

13.2. SCWD

SCWD seeks to develop sensing waveforms that facilitate additional communication capability. In contrast to standard communication waveforms that convey modulated data, traditional sensing waveforms focus on gathering information from echoes without transmitting explicit signals. SCWD achieves this by embedding communication data within sensing waveforms while maintaining their primary sensing capabilities. The main approaches to SCWD can be divided into three categories: embedding information using chirp waveforms, spatial domain techniques, and IM [120].

13.2.1. Chirp Waveform-Based Modulation

Chirp waveforms, long used in radar, are now being investigated as ISAC candidates because they can simultaneously embed digital information and support high-resolution sensing [120,179]. Communication symbols (each representing a group of information bits) are mapped to chirp signals by adjusting parameters such as sweep rate, initial phase, or frequency trajectory, so that different chirp patterns represent distinct symbols. While this preserves radar robustness, the main drawback is low SE (resulting in lower data rates for a given bandwidth) compared with conventional communication waveforms [180]. To mitigate this, approaches, such as embedding multiple bits per chirp or employing continuous phase modulation, have been proposed to raise data throughput [181].

Hybrid methods—e.g., combining chirp modulation with phase-shift keying (PSK) or employing sets of orthogonal chirps—aim to balance sensing accuracy with higher communication rates [182,183]. Moreover, linear frequency modulation (LFM) and continuous chirp-based waveforms are being adapted for vehicular ISAC, where controlling range–Doppler coupling is essential for achieving reliable communication alongside precise target detection [184,185].

13.2.2. Information Embedding in the Spatial Domain

Information can be embedded in the spatial domain through techniques such as sidelobe control and array modulation [120]. In an ISAC context, sidelobe control is not only used to improve radar detection, but also to encode communication symbols by exploiting the differences between the mainlobe and sidelobes. By compressing sidelobes, interference with weak targets can be reduced while simultaneously creating distinct sidelobe patterns that carry embedded data [186]. Various studies have shown this by mapping data symbols onto multiple orthogonal sidelobe waveforms transmitted in parallel, thus supporting concurrent sensing and communication [187–189]. However, the achievable data rate is constrained by the radar pulse repetition rate, which directly limits the symbol rate.

Array modulation methods, such as time-modulated arrays (TMAs) and frequency diversity arrays (FDAs), further facilitate joint radar–communication operation. For example, TMAs can temporally switch array elements to embed information without degrading sensing, while FDAs use frequency increments across array elements to provide target detection while embedding low-to-moderate rate data symbols [190,191]. Additionally, techniques like phase-rotational invariance have been utilized to embed information into multi-sensor arrays for concurrent sensing and data exchange [192].

13.2.3. IM-Based Waveform Techniques

IM is an innovative technique that leverages the spatial diversity of antenna arrays, frequency variations, and coding diversity to convey communication symbols without altering the original radar waveform, thereby maintaining radar performance [120]. A prominent implementation is the carrier agile phased array radar (CAESAR), which utilizes frequency and spatial agility to transmit multiple carrier frequencies simultaneously, reducing hardware complexity while enabling effective data transmission [193].

Another application is frequency-hopping MIMO (FH-MIMO) radar, which employs rapid frequency changes among sub-pulses, allowing different data symbols to be represented based on selected pulses and antennas. FH-MIMO incorporates techniques such as Differential PSK to ensure phase continuity and minimize spectral leakage, as well as FH code selection (FHCS) to enhance communication rates [194,195]. Additionally, frequency nulling modulation is used to embed communication information, though it may require larger bandwidths and complicate hardware implementation [196,197].

13.3. JWOD

JWOD focuses on collaboratively enhancing ISAC waveforms through various performance metrics across multiple domains, thereby achieving a favorable balance between sensing and communication functionalities. JWOD provides greater degrees of freedom (DoF) and flexibility, enabling performance improvements without the constraints of traditional waveform structures. This approach encompasses three primary categories that are detailed by Zhou et al. [120]: waveform optimization, spatial beamforming, and joint design in time/frequency domains.

In waveform optimization, various metrics, like the signal-to-interference-plus-noise ratio (SINR), Cramér–Rao bound (CRB), and mutual information (MI), are utilized to maximize performance while maintaining certain constraints [198–201]. Spatial beamforming enhances the directionality of radar and communication signals, facilitating spectrum sharing [202–204]. Lastly, joint design strategies in the time and frequency domains optimize frame structures and signal allocations to effectively meet the requirements of both communication and sensing functionalities [205–207].

As introduced in Section 12, NOMA has emerged as a promising multiplexing technique for future networks. Both power-domain and code-domain NOMA variants have been under consideration for enhancing capacity and user connectivity [208–210].

14. Performance Evaluation of Modulation Schemes

The performance of modulation schemes is critical in determining the efficiency and quality of mobile communication systems. As mobile networks have evolved, so have the modulation techniques, adapting to the growing demand for higher data rates, improved SE, and reduced latency. This section reviews key performance metrics—BER, throughput, SE, and latency—of the key modulation schemes across various mobile generations, highlighting the reuse and adaptation of certain schemes in multiple generations. Table 7 presents a comparative summary of the theoretical or typical values for these metrics under ideal conditions. These values are approximate and may vary based on implementation details, channel characteristics, interference, and system architecture [30].

14.1. Modulation Performance Metrics Across Mobile Generations

In 1G mobile systems like AMPS, modulation schemes were optimized for analog voice transmission [211]. FM modulation, used for voice channels, was resilient to noise and nonlinearities but lacked a defined BER metric due to its analog nature. Instead, its performance was characterized using signal-to-noise ratio (SNR) or signal-to-noise and distortion ratio (SINAD) [212]. FM achieved a low SE (0.2–0.3 bps/Hz) and a latency of 30–50 ms. Direct FSK, used in digital control channels, had low throughput (9.6 kbps), poor SE (0.1–0.2 bps/Hz), and a BER around 3×10^{-3} at 10 dB E_b/N_0 with non-coherent detection [213]. Its low symbol rate led to high latency, making it unsuitable for data services. Fast FSK, an improved version, increased the symbol rate to achieve higher throughput (up to 14.4 kbps), SE (0.3–0.5 bps/Hz), and lower latency (10–30 ms) [212]. However, it still fell short of the requirements for modern digital communications. Overall, 1G systems were voice-centric and limited by low data rates, high latency, and inefficient spectrum use.

As mobile networks evolved to 2G, 2.5G, and 2.75G systems, modulation schemes began to adapt to support both voice and data. Several modulation schemes, such as GMSK, QPSK, OQPSK, $\pi/4$ DQPSK, and 8PSK, were introduced and optimized to meet the data needs of the mobile internet [30]. GMSK, used in standards such as GSM and GPRS, had a moderate BER in high SNR (making it suitable for voice calls but less robust in noisy conditions), throughput ranging from 9.6 to 171 kbps (depending on the specific technology), and a SE of 0.5 bps/Hz [214]. It was ideal for voice services but inefficient for high-speed data, and the latency was low, making it suitable for real-time communication in voice calls.

Table 7. Performance metrics of the modulation techniques used in mobile communication systems Values represent typical performance under standardized moderate SNR conditions. BER values are based on uncoded transmission at 10 dB E_b/N_0.

Modulation Scheme	BER	Throughput (kbps)	SE (bps/Hz)	Latency (ms)	Generation
Direct FSK	3×10^{-3}	9.6	Low (0.1 to 0.2)	High	1G
FM	N/A (Analog)	10	Low (0.2 to 0.3)	Medium: 30–50	1G
Fast FSK	Low ($\sim 4 \times 10^{-6}$)	9.6 to 14.4	Low (0.3 to 0.5)	Low: 10–30	1G
GMSK	Moderate ($\sim 4 \times 10^{-3}$)	9.6 (GSM), Up to 171 (GPRS)	Low (0.5)	Low (10–30)	2G to 2.75G
QPSK	Low ($\sim 4 \times 10^{-6}$)	Up to 384 (GPRS)	Medium (2)	Medium: 30–50	2G to 2.75G 3G and 4G
OQPSK	Low ($\sim 4 \times 10^{-6}$)	Up to 384 (GPRS)	Medium (2)	Medium: 30–50	2G to 2.75G
$\pi/4$ DQPSK	Moderate ($\sim 10^{-4}$)	Up to 384 (GPRS)	Medium (2)	Medium: 30–50	2G to 2.75G
8PSK	Moderate ($\sim 10^{-3}$)	Up to 1200 (EDGE)	High (3)	High: 100–200	2G to 2.75G 3G to 3.75G
BPSK	Low ($\sim 10^{-6}$)	9.6 to 384	Low (0.5 to 1)	Low: 10–20	2G to 2.75G 3G to 3.75G
16QAM	Moderate ($\sim 10^{-3}$)	Up to 1200	Medium (2 to 3)	Medium to high: 50–100	2G to 2.75G 3G to 3.75G
64QAM	High ($\sim 2 \times 10^{-2}$)	1500 to 3000 (LTE)	High (4 to 6)	Medium: 30–50	3G and 4G
OFDM with 256QAM	High ($\sim 10^{-1}$)	Up to 400,000	High (6 to 8)	Low: 10–30	4G And 4.5G 5G
DFT-s-OFDM with 64QAM	High ($\sim 10^{-2}$)	Up to 150,000	High (6 to 8)	Low: 10–30	4G And 4.5G 5G

QPSK, used extensively in pre-3G and beyond, allowed for a higher throughput (up to 384 kbps) and a SE of 2 bps/Hz. Although maintaining a low BER, it provided better performance for data applications, particularly for services such as web browsing and email. The latency was medium and appropriate for basic internet services. OQPSK, which is essentially a variant of QPSK, was also used in 2G and offered similar throughput, SE, and BER as QPSK. However, it introduced fewer errors during phase transitions, making it more robust in some scenarios [214]. $\pi/4$ DQPSK provides a BER of 10^{-4} using differential encoding to enhance robustness against phase uncertainties and fading. Its throughput (up to 384 kbps) and SE (2 bps/Hz) are comparable to those of QPSK and OQPSK, but its non-coherent detection capability offers improved error resilience in mobile environments, making it suitable for applications requiring reliable data transmission.

8PSK, introduced in EDGE, enables higher throughput (up to 1200 kbps) and SE (3 bps/Hz), allowing more data to be transmitted within the same bandwidth. However, its BER is higher (around 10^{-3}) compared to other modulation schemes, making it less robust in noisy environments. With 3 bits per symbol, 8PSK is more prone to errors at low SNR but is well-suited for high-throughput applications. The increased latency makes it less suitable for real-time voice communication but more appropriate for data-intensive services [215].

Several of these schemes, like QPSK and 8PSK, were used across multiple generations, evolving in their applications and performance as mobile technology advanced from 2G to 3G and beyond. The transition to 3G and 4G technologies brought further advancements in modulation techniques, enabling faster data rates, higher SE, and reduced latency. In these generations, QPSK, 16QAM, and 64QAM emerged as dominant modulation schemes, while multicarrier approaches such as OFDM and DFT-s-OFDM were introduced to effectively meet the escalating demands of high-speed data transmission [30].

QPSK, initially used in earlier generations, was later complemented by higher-order modulation schemes such as 16QAM and 64QAM to meet increasing data demands. These constellations were employed in both single-carrier systems prior to LTE and in multi-carrier schemes like OFDM in LTE and beyond [216]. In its early adoption, 16QAM supported throughputs of up to approximately 1200 kbps and SE between 2 to 3 bps/Hz. With a moderate BER of around 10^{-3} and a latency reduced to 50–100 ms, 16QAM offered a balance of data rate and reliability, making it suitable for multimedia services. The superior performance of 16QAM over QPSK and BPSK, as demonstrated by Tan et al. [217], reflects its advantages in coded systems, where FEC and a higher SE enable improved throughput under favorable SNR conditions. 64QAM, widely deployed in LTE, enabled significantly higher throughput (ranging from 1500 to 3000 kbps) and SE between 4 to 6 bps/Hz. It maintained a BER of 2×10^{-2}, requiring higher SNR for reliable operation. Its latency was medium (30–50 ms), making it suitable for real-time applications such as video conferencing and online gaming.

OFDM with 256QAM further increased the throughput (up to 400 Mbps) and SE (6 to 8 bps/Hz) in 4G networks [218]. It provided high BER and ultra-low latency (10–30 ms), ideal for high-demand data applications such as high definition (HD) video streaming and large file transfers. DFT-s-OFDM with 64QAM was introduced in 4G and 5G for more efficient power use, supporting throughput up to 150 Mbps with similar SE and latency as OFDM with 256QAM. 5G systems introduced significant advancements in modulation techniques, enabling ultra-high data rates and low latencies—critical for real-time applications, such as augmented reality (AR), virtual reality (VR), and autonomous vehicles. Techniques like OFDM with 256QAM and DFT-s-OFDM with 64QAM continue to evolve, allowing for near-gigabit data speeds with extremely low latency. Table 7 provides a summary of these details.

14.2. BER Performance Analysis Across Modulation Schemes

Theoretical uncoded BER performance curves for selected modulation schemes were generated over an additive white Gaussian noise (AWGN) channel using MATLAB R2025a, as shown in Figure 26. All digital modulation schemes were modeled using ideal coherent detection unless otherwise stated. FM was excluded from the comparison due to its analog nature and incompatibility with BER analysis. The GMSK curve was approximated using a simplified MSK-based expression, accounting for its Gaussian pulse shaping. For $\pi/4$-DQPSK, a differential detection model was used to better reflect its practical BER characteristics. OFDM and DFT-s-OFDM were modeled per subcarrier under ideal conditions, assuming no inter-carrier interference (ICI), multipath fading, or nonlinear distortions. The simulation results offer a theoretical benchmark for comparing the noise resilience of twelve widely used modulation schemes across mobile communication systems:

- BPSK and QPSK exhibit the best BER performance, achieving values near 10^{-6} at 10 dB E_b/N_0. Their exceptional robustness makes them ideal for low-SNR, interference-prone, or control channel environments. Both are widely used in legacy and modern systems where reliability is critical.
- GMSK exhibits noticeably worse BER performance compared to BPSK and QPSK, with BERs 4×10^{-3} at 10 dB E_b/N_0, primarily due to its approximation via MSK and non-ideal filtering effects. In contrast, $\boldsymbol{\pi/4}$-DQPSK shows improved robustness, reaching BERs near 10^{-4}. While both modulations offer constant-envelope advantages beneficial for nonlinear RF chains in mobile uplinks, their error performance is inferior to coherent schemes and must be supported by coding in practical systems.
- Fast FSK (Binary Frequency Shift Keying, BFSK) achieves a BER of approximately 10^{-6} at 10 dB, matching BPSK in AWGN under ideal conditions. It provides excellent noise resilience and is suitable for robust, low-complexity links. In contrast, Direct FSK (noncoherent orthogonal) shows a much poorer performance, with a BER around

3×10^{-3} at 10 dB, which limits its standalone usage in modern systems without strong channel coding.

- 8PSK offers higher SE, as shown in Table 7, and achieves BERs around 5×10^{-4} at 10 dB E_b/N_0. While its performance is inferior to BPSK and QPSK due to tighter constellation spacing, it remains a viable option in moderate-to-high SNR scenarios when combined with effective FEC.
- 16QAM and 64QAM offer a balance between SE and BER performance. At 10 dB E_b/N_0, 16QAM achieves a BER of approximately 10^{-3}, while 64QAM reaches around 10^{-2}. These modulations are widely used in high-throughput wireless systems, where channel coding, link adaptation, and favorable SNR conditions enable reliable communication despite their higher symbol density.
- 256QAM, while offering very high SE, is highly sensitive to noise due to its dense constellation and small Euclidean distances between symbols. At 10 dB E_b/N_0, it typically exhibits a BER around 10^{-1}, which is unsuitable for uncoded transmission. However, in high-SNR environments, it becomes viable when combined with robust FEC and accurate channel estimation.
- OFDM (256QAM) and DFT-s-OFDM (64QAM) follow the BER trends of their underlying QAM constellations in AWGN. Since multipath fading, Doppler, and ICI are not modeled here, their BER aligns closely with a single-carrier QAM. In real channels, OFDM's PAPR and frequency selectivity may introduce further distinctions.

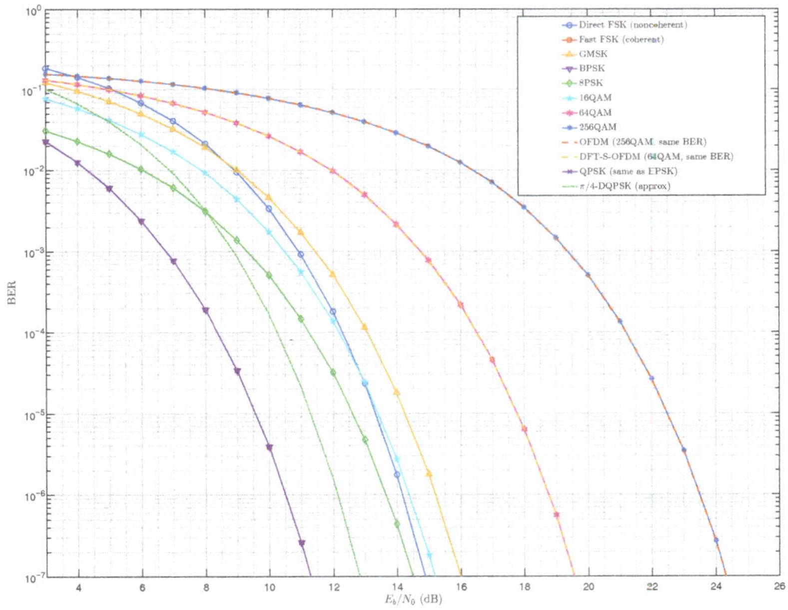

Figure 26. The BER performance of digital modulation schemes in AWGN.

In summary, the results confirm the fundamental trade-off between SE and noise resilience. Lower-order modulations like BPSK and QPSK deliver superior BER at moderate SNRs, while higher-order schemes, such as 64QAM and 256QAM, offer increased throughput with significantly reduced error tolerance. Modern systems mitigate this trade-off through adaptive modulation and coding (AMC), dynamic link adaptation, and MIMO technologies to ensure consistent performance under varying channel conditions.

15. Conclusions

Modulation schemes form the foundation of wireless communications, each offering distinct advantages and inherent trade-offs. This paper provides a comprehensive analysis of the modulation techniques employed across successive generations of mobile cellular networks and their respective standards. It explores the key characteristics, performance metrics, and the trade-offs that define each approach. Figure 27 illustrates a concise overview of the standards, multiple access techniques, and the associated waveforms and/or modulation schemes for each cellular generation from 0G to 6G, as discussed throughout this paper. Additionally, this paper examines how advancements in modulation and waveform technologies have driven the evolution of mobile cellular networks. By analyzing the historical progression of cellular network standards, this paper offers valuable insights into how these networks have become a critical enabler of modern communication.

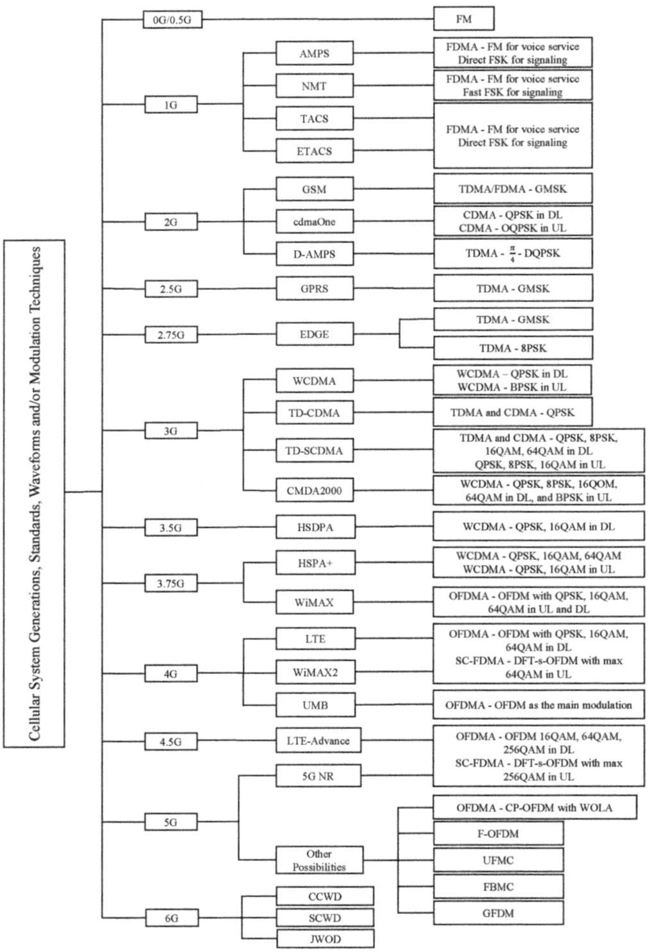

Figure 27. Cellular system generation overview in terms of standard, multiple access, waveform, and/or modulation scheme.

Author Contributions: Conceptualization, F.A.; methodology, F.A.; visualization, F.A.; validation, F.A. and M.S.; formal analysis, F.A.; investigation, F.A.; writing—original draft preparation, F.A.; writing—review and editing, F.A. and M.S.; supervision, F.A. All authors have read and agreed to the published version of the manuscript.

Funding: This research received no external funding.

Conflicts of Interest: The authors declare no conflicts of interest.

Abbreviations

0G	Zero generation
0.5G	Half generation
1G	First generation
2G	Second generation
2.5G	Intermediate 2G
2.75G	Enhanced 2G
3G	Third generation
3.5G	Advanced 3G
3.75G	Advanced 3G and beyond
3GPP2	Third generation partnership project 2
4G	Fourth generation
4.5G	LTE Advanced
5G	Fifth generation
6G	Sixth generation
ACLR	Adjacent channel leakage ratio
ACPR	Adjacent channel power ratio
Ack	Acknowledgment
AFDM	Affine frequency division multiplexing
AI	Artificial intelligence
AMC	Adaptive modulation and coding
AMPS	Advanced mobile phone system
AR	Augmented reality
AWGN	Additive white Gaussian noise
BER	Bit error rate
BF-OFDM	Blocked filtered OFDM
BFSK	Binary frequency shift keying
BPSK	Binary phase shift keying
BTRC	Better than raised-cosine
CA	Carrier aggregation
CAESAR	Carrier agile phased array radar
CATT	China academy of telecommunications technology
CC	Component carrier
CCWD	Communication-centric waveform design
CD-IM	Codedomain index modulation
CDMA	Code division multiple access
cdmaOne	Code division multiple access one
CDPD	Cellular digital packet data
CFO	Carrier frequency offset
CGI	Cell global identity
CMT	Cosine modulated multitone
CP	Cyclic prefix
CP-OFDM	Cyclic-prefix orthogonal frequency division multiplexing
CPP	Chirp-periodic prefix
CQI	Channel quality indicator
CRB	Cramér–Rao bound
CRS	Cell-specific reference signal
CS	Coding scheme

D-AMPS	Digital advanced mobile phone system
DAC	Digital-to-analog converter
DAFT	Discrete affine Fourier transform
DC	Direct current
DD	Delay-Doppler
DeNB	Donor eNodeB
DFE	Decision feedback equalization
DFnT	Discrete Fresnel transform
DFT-s-OFDM	Discrete Fourier transform spread orthogonal frequency division multiplexing
DFT-s-OTFS	Discrete Fourier transform spread orthogonal time-frequency space
DH	Dual-cell HSDPA
DL	Downlink
DMRS	Demodulation reference signal
DoF	Degrees of freedom
DQPSK	Differential quaternary phase shift keying
DS-CDMA	Direct-sequence CDMA
DSSS	Direct sequence spread spectrum
EDGE	Enhanced data rates for GSM evolution
EGPRS	Enhanced GPRS
eLAA	Enhanced LAA
eLWA	Enhanced LTE-WLAN aggregation
eMBB	Enhanced mobile broadband
ETACS	Extended total access communication system
eV2X	Enhanced vehicle-to-everything
F-OFDM	Filtered OFDM
FDAs	Frequency diversity arrays
FD-IM	Frequency-domain index modulation
FEC	Forward error correction
FFSK	Fast frequency shift keying modulation
FH-CDMA	Frequency hopping CDMA
FHCS	FH code selection
FH-MIMO	Frequency-hopping MIMO
FM	Frequency modulation
FMCW	Frequency-modulated continuous wave
FMT	Filtered multitone
FPGA	Field-programmable gate array
FSK	Frequency shift Keying
GFDM	Generalized frequency division multiplexing
GFSK	Gaussian frequency-shift keying
GMSK	Gaussian minimum-shift keying
GSM	Global system for mobile communications
HARQ	Hybrid automatic repeat requests
HD	High definition
HSDPA	Downlink packet access
HSPA	High-speed packet access
HSUPA	High-speed uplink packet access
ICI	Inter-carrier interference
IDAFT	Inverse DAFT
IDFnT	Inverse DFnT
IFFT	Inverse fast Fourier transform
IM	Index modulation
IMT2000	International mobile telecommunications-2000

IMTS	Improved mobile telephone service
IoT	Internet of things
IS-54	Interim standard 54
IS-95	Interim standard 95
IS-136	Interim standard 136
ISFFT	Inverse symplectic finite Fourier transform
ISI	Inter-symbol interference
ITU	International telecommunication union
ITU-R	ITU radiocommunication sector
ITU-T	ITU telecommunication sector
JWOD	Joint waveform optimization and design
LAA	Licensed assisted access
LCR	Low chip rate
LDPC	Low density parity check
LFM	Linear frequency modulation
LTE	Long-term evolution
LTE-MTC	LTE machine type communication
MAC	Medium access control
M2M	Machine-to-machine
MC-CDMA	Multi-carrier CDMA
MF	Matched filter
MIB	Master information block
MIMO	Multiple-input-multiple-output
ML	Machine learning
MLSE	Maximum likelihood sequence estimation
MMSE	Minimum mean square error
mMTC	Massive machine type communications
MTD	Mobile telephony system D (Swedish)
MTS	Mobile telephone service
MUX	Multiplexer
Nack	None-acknowledge
NB-IoT	Narrowband IoT
NMT	Nordic mobile telephone
NR	New radio
NRZ	Non-return-to-zero
NOMA	Non-orthogonal multiple access
OFDMA	Orthogonal frequency division multiple Access
OICF	Iterative clipping and filtering
OMA	Orthogonal multiple access
OLT	Offentlig landmobil telefoni
OQPSK	Offset-quadrature phase shift keying
OTFS	Orthogonal time frequency space
PAM	Pulse amplitude modulated
PAPR	Peak-to-average-power ratio
PBCH	Physical broadcast channel
PD-IM	Power-domain index modulation
PDCCH	Physical downlink control channel
PDSCH	Physical downlink shared channel
PDU	Protocol data unit
PHY	Physical layer
PLL	Phase locked loops
PN	Pseudo-random noise

PRACH	Physical random access channel
PTM	Point-to-multipoint
PTP	Point-to-point
PTT	Push-to-talk
PUSCH	Physical uplink shared channel
PUCCH	Physical uplink control channel
PSK	Phase-shift keying
QAM	Quadrature amplitude modulation
QoS	Quality of service
QPSK	Quadrature phase shift keying
R15	Release 15
RA	Random access
RAM	Random access memory
RF	Radio frequency
RLC	Radio link control
RM	Reed–Muller
RN	Relay nodes
RNC	Radio network controller
RRC	Radio resource control
SC-FDE	Single-carrier systems with frequency-domain equalization
SC-FDMA	Single-carrier frequency-division multiple access
SC-QAM	Single carrier quadrature amplitude modulation
SC-TDE	Single-carrier systems with time-domain equalization
SCWD	Sensing-centric waveform design
SDMA	Space division multiple access
SD-IM	Spatial domain index modulation
SE	Spectral efficiency
SER	Symbol error rate
SIB	System information block
SINAD	Signal-to-noise and distortion ratio
SINR	Signal-to-interference-plus-noise ratio
SFFT	Symplectic finite Fourier transform
SLM	Selected mapping
SMS	Short message service
SMT	Staggered modulated multitone
SNR	Signal-to-noise-ratio
STBC	Space–time block coding
TA	Timing advance
TD-CDMA	Time division CDMA
TD-IM	Time-domain Index modulation
TD-SCDMA	Time division synchronous CDMA
TDD	Time division duplex
TM	Transmission modes
TMAs	Time-modulated arrays
TTI	Time transmission interval
UAV	Unmanned aerial vehicle
UE	User equipment
UMB	Ultra-mobile broadband
UMTS	Universal mobile telecommunications system
UMTS-TDD	UTRA TDD (time division CDMA)
UTRA	UMTS terrestrial radio access
UTRAN	UMTS terrestrial radio access network

URLLC	Ultra-reliable and low latency communications
Uu	E-UTRAN air interface
V2X	Vehicle-to-everything
VFO	Variable frequency oscillator
VR	Virtual reality
WCDMA	Wideband CDMA
WiMAX	Worldwide Interoperability for Microwave Access
WiMAX-2	Worldwide Interoperability for Microwave Access 802.16m
WLL	Wireless local loop
ZF	Zero-forcing
ZP-OFDM	Zero-padding OFDM

Appendix A. Example of Spreading and Despreading Technique in WCDMA

Consider an example with three users, as depicted in Figure A1, each intending to transmit a sequence of three information bits. For example, the first user (UE1) intends to transmit the sequence (1 0 1) after encoding. Each user is assigned a unique orthogonal spreading code, ensuring that the inner product of any two different code sequences is zero.

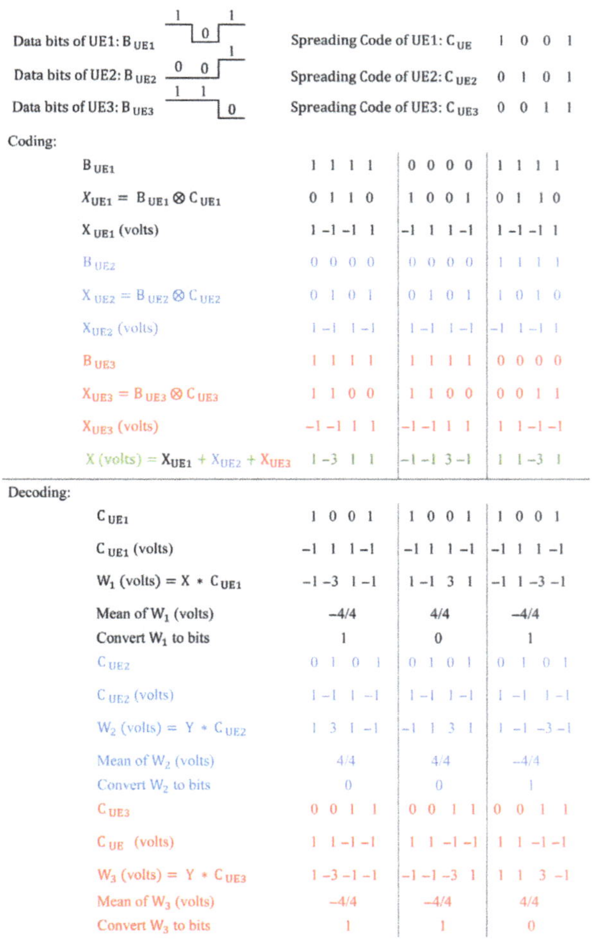

Figure A1. An example of spreading and despreading in WCDMA.

The UE1 is assigned the spreading code (1 0 0 1), which is multiplied by each encoded bit (1 0 1) to produce the spreading output, X_{UE1}. In this example for each encoded binary bit, four digits (called chips) will be transmitted, which means the rate of transmission after spreading increases by four. This process is repeated for all users. Then, all the spreading data should be summed up, of course, after converting bits to voltage (binary 0 means 1 Volt and binary 1 means −1 Volt). The signal after summation is referred to as a composite signal, which will be transmitted through the physical channel after modulation.

On the receiver side, all the users receive the same composite signal, but for decoding, they need to use their unique spread code. For instance, UE1 uses code (1 0 0 1) to decode its information, converting it back to bits for detection, as shown by Figure A1. The advantage of spreading becomes clear when the wrong code is used, as it disrupts the orthogonality of code sequences and only results in zeros after despreading. The spreading codes utilized in WCDMA are based on orthogonal variable spreading factor (OVSF) techniques, which allow the spreading factor to be changed while maintaining orthogonality between different spreading codes [52].

Appendix B. SC-TDE vs. SC-FDE

The use of TDEs has a long history of mitigating ISI in narrowband wireline channels. These techniques were adopted in international CCITT standards (now known as the ITU-T, i.e., the telecommunication sector of the ITU) for dial-up modems. TDEs can also be deployed, in principle, in broadband wireless communications; however, the number of operations per signaling interval grows linearly with the ISI span or, equivalently, with the data rates [219]. An alternative approach using frequency domain equalization with lower computational demands, such as single carrier waveform, is becoming more popular.

IEEE 802.11ad (an amendment to the IEEE 802.11 wireless networking standard) is an example of a SC-FDE implementation in wireless networks. This protocol implements data transmission in the time domain but adds a CP in the transmitter to ensure the channel appears cyclically on the receiver side. With this approach, the receiver can use a single-tab equalizer to perform the signal equalization, which is implemented between fast Fourier transform (FFT) and inverse fast Fourier transform (IFFT), and it can perform the channel equalization in the frequency domain at a lower computational cost than equalization in the time domain. This was shown by Walzman and Schwartz, who explored FDE for the first time in 1973 [220]. Both transmission and detection are performed in the time domain using the SC-FDE waveform.

SC-FDE has a similar structure to the CP-OFDM, as discussed in detail in Section 10.1.1. Both rely on FFT and IFFT; however, in SC-FDE, the IFFT is performed at the receiver instead of the transmitter. One should also note that, in OFDM, FFT and IFFT are used for modulation and demodulation. However, their purpose in SC-FDE is to bring the signal to the frequency domain for equalization and then return it to the time domain for detection.

Another similarity is the block-wise implementation of digital data transmission in both schemes. SC-FDE is a simpler implementation of OFDM to reduce PAPR and the need for highly accurate frequency synchronization. It is also a MIMO-friendly waveform.

The potential of SC-FDE to compete with OFDM was first revealed by Sari et al. in 1995 [221]. Some literature suggests that SC-FDE can perform similarly or better than coded OFDM systems in certain scenarios [219]. One example of SC-FDE is SC-QAM with a CP, in which case further degradation of power SE is expected due to the CP overhead.

Appendix C. LTE Advanced Features

Appendix C.1. CA

CA is a key feature in LTE Advanced that enhances the data rates for individual UEs by allocating multiple carrier frequency bands to a single UE. This mechanism effectively achieves higher data rates while maintaining compatibility with earlier 3GPP releases, such as Release 8 and 9. A component carrier (CC) can have a frequency bandwidth of 1.4, 3, 5, 10, 15, or 20 MHz. According to 3GPP specifications, a maximum of five CC can be aggregated to provide an overall bandwidth of up to 100 MHz (20 MHz $\times$ 5). This technique significantly contributes to the improvement of SE and network capacity in LTE Advanced systems.

It is important to note that the number of aggregated carriers may vary between the uplink and downlink channels; however, the number of uplink carriers cannot be greater than the number of downlink carriers. Moreover, the bandwidths of the aggregated carriers do not have to be identical, which allows for increased flexibility and enhanced spectrum utilization. This feature, known as asymmetric CA, enables network operators to combine carriers of varying bandwidths across different frequency bands, tailoring the network capacity to match the specific traffic demands and channel conditions in a given deployment scenario [79].

In 3GPP Release 11, improvements were made to CA, primarily focusing on enhancing the performance of non-contiguous CA signals. This was achieved by specifying additional UE reference sensitivity requirements for non-contiguous CA signals with different numbers of component carriers. Additionally, 3GPP Release 11 introduced enhancements related to uplink control signaling and intra-band and inter-band CA for TDD-FDD joint operation, as well as enhancements to downlink control information for CA. These updates provided increased flexibility, efficiency, and performance for the evolving LTE-Advanced system.

Appendix C.2. MIMO

LTE Advanced introduces 8 $\times$ 8 MIMO system in the downlink and 4 $\times$ 4 MIMO system in the uplink. In high SNR scenarios, when the channel quality is good, spatial multiplexing is recommended to improve throughput. However, in the low SNR scenarios the other mode of MIMO is suggested, called transmitter diversity, in which multiple antennas transmit the same data stream to the user to reduce the risk of deep path fading.

To manage the adjustment of the MIMO technique, the radio link control (RLC) layer, which is layer two in the UMTS/LTE air interface, is responsible for determining the transmission mode for each user based on parameters such as channel quality. The user is informed of their transmission mode through radio resource control (RRC) signaling. The number of transmission modes (TM) increased from seven in 3GPP Release 8 to nine in 3GPP Release 9, then to ten in 3GPP Release 10, which is 4.5G, with TM1 to TM10 being the available modes. The major differences between the transmission modes include the number of layers (streams or rank); the number of antenna ports used; the type of reference signals, such as the cell-specific reference signal (CRS) or demodulation reference signal (DMRS) introduced in R10; and the precoding type [79].

Appendix C.3. RN

RNs are low-power base stations utilized to enhance coverage and capacity in areas with weak signals or high demand, such as cell edges and hot spots. Each RN is connected to the base station, known as an eNodeB in a 4.5G system, and the eNodeB that serves the RN is referred to as a donor eNodeB (DeNB). The connection between an RN and a DeNB is established through the air interface, which is a modified version of the E-UTRAN air interface (Uu) and is referred to as Un. All the cell resources of the DeNB are shared among users directly served by the DeNB and those served by the RN [79].

Appendix D. 5G Physical Channels and Associated Modulation and Coding

The names of the uplink and downlink physical channels in 5G NR, along with their related modulation and coding schemes, are listed in Table A1 [83]. The following highlights the respective roles of these channels in the overall communication process. These channels are key to the synchronization procedure, where the user equipment (UE) must first establish synchronization with the network to ensure efficient data transmission.

Table A1. NR physical channels, modulations, and channel coding [83].

Physical Channel Name	Modulation	Channel Coding	DL/UL
Physical downlink shared channel (PDSCH)	QPSK, 16QAM, 64QAM, 256QAM	Low density parity check (LDPC) coding	DL
Physical broadcast channel (PBCH)	QPSK	Polar coding	DL
Physical downlink control channel (PDCCH)	QPSK	Polar coding	DL
Physical uplink shared channel (PUSCH)	QPSK, 16QAM, 64QAM, 256QAM, $\pi/2$-BPSK when DFT-s-OFDM is selected	LDPC coding	UL
Physical uplink control channel (PUCCH)	$\pi/2$-BPSK, BPSK, QPSK depending on PUCCH format and information bit size	Reed–Muller (RM) block coding and Polar coding	UL
Physical random access channel (PRACH)	$\pi/2$-BPSK	-	UL

The PCI value determines the scrambling code, allowing the UE to decode the PBCH and PDSCH channels to access cell parameters. The PBCH carries the master information block (MIB), which includes frequency and time domain information. The PDSCH carries the system information block (SIB), which the UE can access using information from the MIB. The UE then reads the PDCCH to obtain scheduling information, after which it can decode dedicated channels such as the PDSCH and PUSCH. The SIB includes all the high-layer cell parameters, allowing the UE to determine which cell is the best to make random access through the PRACH channel [222].

Upon power-on, the user equipment initiates an extensive search for the primary synchronization signal (PSS) and the secondary synchronization signal (SSS) to attain proper synchronization with the network in both time and frequency domains. The PSS serves as a fundamental component for time-domain synchronization, enabling the UE to identify the slot boundaries. Subsequently, the SSS facilitates radio frame synchronization and establishes the cell-specific frame timing, as well as provides essential information about the PCI.

The calculation of the PCI in 5G involves a two-step process: First, the UE recognizes the PSS sequence, which comprises one of three distinct Zadoff–Chu sequences. This recognition allows the UE to derive the PCI within a range of 0 to 2. Next, the UE identifies the SSS sequence, which consists of 336 unique sequences derived from the interleaving of two length-31 binary sequences. The UE can then determine the PCI within a range of 0 to 335. The UE concludes the PCI calculation by combining the PCI values obtained from the PSS and SSS, resulting in a unique physical cell identity ranging from 0 to 1007, as shown by (A1). This is a notable difference compared to LTE, where the PCI ranges from 0 to 503.

$$\text{PCI} = \text{PSS ID} + 3 \times \text{SSS ID}, \tag{A1}$$

where PSS ID has a value between 0 and 2, and SSS ID can take a value between 0 and 335. This means UE should find its cell among 1008 candidates. This is double compared to LTE, in which there were 504 candidates.

Appendix E. Multi-Carrier vs. Single-Carrier Modulation Performance

Table A2 provides a comparison of the modulation schemes discussed in Sections 10 and 12, i.e., the multi-carrier modulation techniques versus single-carrier waveforms explained in Section 4, based on some important key performance indicators and characteristics. It is evident that single-carrier modulations can perform the following:

- Utilize a single carrier frequency for data transmission.
- Are ideal for systems with low data rates and narrowband channels.
- Feature simple transmitter and receiver designs, resulting in lower implementation costs.
- Are more vulnerable to ISI due to multipath fading.

Examples of single-carrier modulations are constant envelope waveforms, such as BPSK and QPSK, SC-TDE, SC-FDE, and SC-FDM waveforms.

However, the multi-carrier modulation techniques are the following:

- Able to transmit data using multiple orthogonal subcarriers within the available bandwidth.
- Are well-suited for high data rate and wideband communication systems.
- Are able to involve more complex transmitter and receiver designs compared to single-carrier modulation.
- Are able to offer enhanced resilience to ISI by distributing data across multiple subcarriers.

Examples of multi-carrier modulations would be CP-OFDM, other OFDM variants such as F-OFDM, UFMC, FBMC, and GFDM as explored in Sections 10 and 12.

Since each modulation technique has its unique features, a practical approach is to combine different waveforms to enhance overall system performance. For example, in the LTE system, CP-OFDM can be used alongside other OFDM-based multicarrier modulations to retain the advantages of CP-OFDM while alleviating its drawbacks, or it can be combined with different waveforms in the 5G system. This idea is known as waveform coexistence, and if two or more modulation schemes can be used together to enhance system performance, it is referred to as green coexistence [88]. The last row of Table A2 shows the waveforms that can coexist.

Table A2. Comparison of single-carrier and multi-carrier waveform characteristics. ✓ = yes; ✗ = no; ≈ /✓ = nearly yes; optional = depends on implementation.

KPI/Characteristic \ Waveform	Constant Envelope	SC-TDE	SC-FDE	SC-FDM	CP-OFDM	CP-OFDM-WOLA	F-OFDM	UFMC	FBMC	GFDM
Used CP	✗	✗	✓	✓	✓	✓	✓	optional	optional	✓
Lower OOB leakage compare to CP-OFDM	✓	✗	≈ /✓	✓	—	✓	✓	✓	✓	✓
Lower PAPR compare to CP-OFDM	✓	✓	✓	✓	—	✓	✗	✓	✗	✓
Good for short burst transmission	✓	✗	✓	✓	✗	✗	✓	✓	✗	✓
Based on pulse shaping (subcarrier-based filtering)	✗	✗	✗	✗	✗	✗	✗	✗	✓	✓
Based on Subband filtering	✗	✗	✗	✗	✗	✗	✓	✓	✗	✗
Complex structure	✗	✓	✗	✗	✗	✗	✓	✓	✓	✓
Allows asynchronous multiple access	✓	✗	✓	✗	✗	✓	✓	✓	✓	✓
Offset-QAM	✗	✗	✗	✗	✗	✗	✗	✗	✓	✓
MIMO friendly	✗	✗	✗	✓	✓	✓	✓	✓	✗	✗
Support dynamic bandwidth allocation	✗	✗	✗	✗	✓	✓	✓	✓	✓	✓
Top three:										
Highest SE	✗	✗	✗	✗	✗	✓	✗	✗	✓	✓
Lowest out-of-band emission	✗	✗	✗	✗	✗	✗	✗	✓	✓	✓
Lowest implementation complexity	✓	✗	✓	✓	✗	✗	✗	✗	✗	✗
Lowest power consumption	✓	✗	✓	✓	✗	✗	✗	✗	✗	✗
Supports spectral coexistence with	✗	✗	✗	✗	UFMC FBMC	✗	CP-OFDM UFMC	FBMC	UFMC	✗

References

1. Wells, J. The evolution of cellular system design. In Proceedings of the 34th IEEE Vehicular Technology Conference, Pittsburgh, PA, USA, 21–23 May 1984; Volume 34, pp. 1–5. [CrossRef]
2. Huff, D.L. AT&T Cellular Technology Review. In Proceedings of the MILCOM 1985—IEEE Military Communications Conference, Boston, MA, USA, 20–23 October 1985; Volume 2, pp. 490–494. [CrossRef]
3. Abu El-Ata, M. Evolution of mobile cellular communication systems. The journey to UMTS. In Proceedings of the Seventeenth National Radio Science Conference, 17th NRSC'2000 (IEEE Cat. No.00EX396), Minufiya, Egypt, 22–24 February 2000; p. INV2/1-INV217. [CrossRef]
4. Shahabuddin, S.; Rahaman, S.; Rehman, F.; Ahmad, I.; Khan, Z. Evolution of Cellular Systems. In *A Comprehensive Guide to 5G Security*; John Wiley & Sons: Hoboken, NJ, USA, 2018; Chapter 1, pp. 1–29. [CrossRef]
5. Akyildiz, I.F.; Nie, S.; Lin, S.C.; Chandrasekaran, M. 5G roadmap: 10 key enabling technologies. *Comput. Netw.* **2016**, *106*, 17–48. [CrossRef]
6. Veijalainen, J.; Rehmat, W. Mobile Communities in Developing Countries. In Proceedings of the 2010 Eleventh International Conference on Mobile Data Management, Kansas City, MO, USA, 23–26 May 2010; pp. 314–319. [CrossRef]
7. Singh, S.K.; Singh, R.; Kumbhani, B. A Review on Cellular Network Infrastructure Evolution: Challenges and Opportunities. In Proceedings of the 2019 14th Conference on Industrial and Information Systems (ICIIS), Kandy, Sri Lanka, 18–20 December 2019; pp. 163–168. [CrossRef]
8. Hansen, F. Modulation techniques for land mobile radio. In Proceedings of the 31st IEEE Vehicular Technology Conference, Tokyo, Japan, 15–18 June 1981; Volume 31, pp. 164–170. [CrossRef]
9. Nagul, S. A review on 5G modulation schemes and their comparisons for future wireless communications. In Proceedings of the 2018 Conference on Signal Processing and Communication Engineering Systems (SPACES), Vijayawada, India, 4–5 January 2018; pp. 72–76. [CrossRef]
10. Bahuguna, A.S.; Kumar, K.; Pundir, Y.P.; Alaknanda; Bijalwan, V. A Review of Various Digital Modulation Schemes Used in Wireless Communications. In *Proceedings of the Integrated Intelligence Enable Networks and Computing*; Singh Mer, K.K., Semwal, V.B., Bijalwan, V., Crespo, R.G., Eds.; Springer: Singapore, 2021; pp. 561–570. [CrossRef]
11. Acharya, S.; Kabiraj, P.; De, D. Comparative analysis of different modulation techniques of LTE network. In Proceedings of the 2015 Third International Conference on Computer, Communication, Control and Information Technology (C3IT), Hooghly, India, 7–8 February 2015; pp. 1–6. [CrossRef]
12. Guneser, M.T.; Sahab, A.S.; Seker, C. Performance analysis of modulation techniques in 5G communication system. *China Commun.* **2022**, *19*, 100–114. [CrossRef]
13. Liu, T. A Review on the 5G Enhanced OFDM Modulation Technique. In Proceedings of the 2023 3rd Asia-Pacific Conference on Communications Technology and Computer Science (ACCTCS), Shenyang, China, 25–27 February 2023; pp. 677–683. [CrossRef]
14. Arabian, F.; Nordin, G.P.; Rice, M. On Polarization Dependent Equalization in 5G mmWave Systems. In Proceedings of the 2020 International Conference on Computing, Networking and Communications (ICNC), Big Island, HI, USA, 17–20 February 2020; pp. 1058–1062. [CrossRef]
15. Arabian, F.; Nordin, G.P.; Rice, M. On the Ungerboeck and Forney Observation Models for Spatial Combining And Their Application to 5G Millimeter-Wave Bands. *IEEE Access* **2021**, *9*, 22214–22231. [CrossRef]
16. Shakhatreh, H.; Sawalmeh, A.; Hayajneh, K.F.; Abdel-Razeq, S.; Malkawi, W.; Al-Fuqaha, A. A Systematic Review of Interference Mitigation Techniques in Current and Future UAV-Assisted Wireless Networks. *IEEE Open J. Commun. Soc.* **2024**, *5*, 2815–2846. [CrossRef]
17. Siddiqui, M.U.A.; Qamar, F.; Ahmed, F.; Nguyen, Q.N.; Hassan, R. Interference Management in 5G and Beyond Network: Requirements, Challenges and Future Directions. *IEEE Access* **2021**, *9*, 68932–68965. [CrossRef]
18. Ezhilarasan, E.; Dinakaran, M. A Review on Mobile Technologies: 3G, 4G and 5G. In Proceedings of the 2017 Second International Conference on Recent Trends and Challenges in Computational Models (ICRTCCM), Tamil Nadu, India, 3–4 February 2017; pp. 369–373. [CrossRef]
19. Fadhil, H.M.; Dawood, Z.O. Evolutionary Perspective of Mobile Communication Technologies. In Proceedings of the 2018 International Conference on Computer and Applications (ICCA), Beirut, Lebanon, 25–26 September 2018; pp. 80–84. [CrossRef]
20. Ud Din Arshad, Q.K.; Kashif, A.U.; Quershi, I.M. A Review on the Evolution of Cellular Technologies. In Proceedings of the 2019 16th International Bhurban Conference on Applied Sciences and Technology (IBCAST), Islamabad, Pakistan, 8–12 January 2019; pp. 989–993. [CrossRef]
21. Alam, M.S.; Siddiqui, S.T.; Qidwai, K.A.; Aftab, A.; Kamal, M.S.; Shahi, F.I. Evolution of Wireless Communication Networks from 5G to 6G: Future Perspective. *Radioelectron. Commun. Syst.* **2023**, *66*, 213–222. [CrossRef]
22. Liu, G.; Cai, B.; Xie, W. Research on 5G Wireless Networks and Evolution. In Proceedings of the 2021 IEEE International Symposium on Broadband Multimedia Systems and Broadcasting (BMSB), Chengdu, China, 4–6 August 2021; pp. 1–5. [CrossRef]

23. Gerzaguet, R.; Bartzoudis, N.; Baltar, L.G.; Berg, V.; Doré, J.B.; Kténas, D.; Font-Bach, O.; Mestre, X.; Payaró, M.; Färber, M.; et al. The 5G candidate waveform race: A comparison of complexity and performance. *EURASIP J. Wirel. Commun. Netw.* **2017**, *2017*, 13. [CrossRef]
24. Cai, Y.; Qin, Z.; Cui, F.; Li, G.Y.; McCann, J.A. Modulation and Multiple Access for 5G Networks. *IEEE Commun. Surv. Tutor.* **2018**, *20*, 629–646. [CrossRef]
25. Khalifa, O.O.; Ahmed, Z.; Esgiar, A.N.; Saeed, R.A.; Abdalla, A.H. Performance Analysis of LTE Codes System Using Various Modulation Techniques. In Proceedings of the 2021 National Computing Colleges Conference (NCCC), Taif, Saudi Arabia, 27–28 March 2021; pp. 1–5. [CrossRef]
26. Parasher, S.; Kumar, S. Effects of Various Digital Modulation Methods on LTE System Over Orthogonal Frequency Division Multiplexed AWGN: A Review. In Proceedings of the 2022 International Conference on Advances in Computing, Communication and Materials (ICACCM), Dehradun, India, 10–11 November 2022; pp. 1–6. [CrossRef]
27. Meraj, M.; Kumar, S. Evolution of Mobile Wireless Technology from 0G to 5G. *Int. J. Comput. Sci. Inf. Technol.* **2015**, *6*, 2545–2551. Available online: https://ijcsit.com/docs/Volume%206/vol6issue03/ijcsit20150603123.pdf (accessed on 1 August 2025).
28. Frenkiel, R.; Schwartz, M. Creating cellular: A history of the AMPS project (1971–1983) [History of Communications]. *IEEE Commun. Mag.* **2010**, *48*, 14–24. [CrossRef]
29. 62–Glossary. *Telecommunications Engineer's Reference Book*; Mazda, F., Ed.; Butterworth-Heinemann: Oxford, UK, 1993; pp. 62-1–62-32. [CrossRef]
30. Rappaport, T.S. *Wireless Communications: Principles and Practice*; Prentice Hall: Upper Saddle River, NJ, USA, 2002. [CrossRef]
31. Fuller, R.; Epler, F. Overview of an A.M.P.S. mobile call processing system. In Proceedings of the 30th IEEE Vehicular Technology Conference, Dearborn, MI, USA, 15–17 September 1980; Volume 30, pp. 505–512. [CrossRef]
32. Wakatsuki, Y.; Takeuchi, H.; Shimizu, G. FSK Demodulating Device. US Patent US4897857A, 30 January 1990. Available online: https://patents.google.com/patent/US4897857A/en (accessed on 1 July 2025).
33. NMT. Nordic Mobile Telephone. Available online: http://www.s-lab.se/nmt/NMT-DOC450_3.pdf (accessed on 1 July 2025).
34. ERICSSON. Work on NMT. Available online: https://www.ericsson.com/en/about-us/history/changing-the-world/the-nordics-take-charge/work-on-nmt (accessed on 1 June 2025).
35. de Buda, R. Fast FSK Signals and Their Demodulation. *Can. Electr. Eng. J.* **1976**, *1*, 28–34. [CrossRef]
36. Mishra, A.R. *Cellular Technologies for Emerging Markets: 2G, 3G, and Beyond*; Wiley: Chichester, UK; Hoboken, NJ, USA, 2010. [CrossRef]
37. Peersman, C.; Cvetkovic, S.; Griffiths, P.; Spear, H. The Global System for Mobile Communications Short Message Service. *IEEE Pers. Commun.* **2000**, *7*, 15–23. [CrossRef]
38. Hirade, K.; Murota, K. A Study of Modulation for Digital Mobile Telephony. In Proceedings of the 29th IEEE Vehicular Technology Conference, Arlington Heights, IL, USA, 27–30 March 1979; Volume 29, pp. 13–19. [CrossRef]
39. Turletti, T.; Bentzen, H.; Tennenhouse, D. Toward the software realization of a GSM base station. *IEEE J. Sel. Areas Commun.* **1999**, *17*, 603–612. [CrossRef]
40. Krouk, E.; Semenov, S. CDMA2000 and Its Evolution. In *Modulation and Coding Techniques in Wireless Communications*; John Wiley & Sons Ltd.: Hoboken, NJ, USA, 2011; pp. 605–654. [CrossRef]
41. Rice, M. *Digital Communications: A Discrete-Time Approach*; Kindle Digital Publishing: Seattle, WA, USA, 2018; ISBN 978-1790588565.
42. Gutierrez, P.; Wigard, J.; Andersen, P.; Damgaard, H.; Mogensen, P. Performance of link adaptation in GPRS networks. In Proceedings of the Vehicular Technology Conference Fall 2000, IEEE VTS Fall VTC2000. 52nd Vehicular Technology Conference (Cat. No.00CH37152), Boston, MA, USA, 24–28 September 2000; Volume 2, pp. 492–499. [CrossRef]
43. Grami, A. *Introduction to Digital Communications*; Elsevier Science Publishing Co Inc. (Academic Press): Boston, MA, USA, 2015. [CrossRef]
44. Kukushkin, A. *Introduction to Mobile Network Engineering: GSM, 3G-WCDMA, LTE and the Road to 5G*; John Wiley & Sons, Ltd. Hoboken, NJ, USA, 2018. [CrossRef]
45. Richardson, K.W. UMTS overview. *Electron. Commun. Eng. J.* **2000**, *12*, 93–100. [CrossRef]
46. Lee, T.L.; Faure, C.; Grandblaise, D. Impact of FDD/TDD co-existence on overall UMTS system performance. In Proceedings of the IEEE VTS 53rd Vehicular Technology Conference, Spring 2001. Proceedings (Cat. No.01CH37202), Rhodes, Greece, 6–9 May 2001; Volume 4, pp. 2655–2659. [CrossRef]
47. Haas, H.; McLaughlin, S.; Povey, G. The effects of interference between the TDD and FDD mode in UMTS at the boundary of 1920 MHz. In Proceedings of the 2000 IEEE Sixth International Symposium on Spread Spectrum Techniques and Applications. ISSTA 2000. Proceedings (Cat. No.00TH8536), Parsippany, NJ, USA, 6–8 September 2000; Volume 2, pp. 486–490. [CrossRef]
48. Boudreau, D.; Caire, G.; Corazza, G.; De Gaudenzi, R.; Gallinaro, G.; Luglio, M.; Lyons, R.; Romero-Garcia, J.; Vernucci, A.; Widmer, H. Wide-band CDMA for the UMTS/IMT-2000 satellite component. *IEEE Trans. Veh. Technol.* **2002**, *51*, 306–33[illegible] [CrossRef]

49. Berrou, C. Error-Correction Coding Method with at Least Two Systematic Convolutional Codings in Parallel, Corresponding Iterative Decoding Method, Decoding Module and Decoder. United States US5446747A, 29 August 1995. Available online: https://patents.google.com/patent/US5446747A/en (accessed on 1 July 2025).
50. Berrou, C.; Glavieux, A.; Thitimajshima, P. Near Shannon limit error-correcting coding and decoding: Turbo-codes. 1. In Proceedings of the ICC '93—IEEE International Conference on Communications, Geneva, Switzerland, 23–26 May 1993; Volume 2, pp. 1064–1070. [CrossRef]
51. Hudec, P.; Polivka, M. High dynamic range multiplier for detection and localization of BPSK WCDMA signals. In Proceedings of the 2005 European Microwave Conference, Paris, France, 4–6 October 2005; Volume 1, p. 4. [CrossRef]
52. Kurjenniemi, J. A Study of TD-CDMA and WCDMA Radio Network Enhancements. Ph.D. Dissertation, University of Jyväskylä, Jyväskylä, Finland, 2005. Available online: https://jyx.jyu.fi/bitstreams/5e9cd435-394b-4f37-9e88-647515b74ab0/download (accessed on 1 July 2025).
53. Haardt, M.; Klein, A.; Koehn, R.; Oestreich, S.; Purat, M.; Sommer, V.; Ulrich, T. The TD-CDMA based UTRA TDD mode. *IEEE J. Sel. Areas Commun.* **2000**, *18*, 1375–1385. [CrossRef]
54. 3GPP. Overview of 3GPP Release 4–Summary of All Release 4 Features v. TSG #26. 2004. Available online: https://www.3gpp.org/ftp/tsg_ran/TSG_RAN/TSGR_26/Docs/PDF/RP-040548.pdf (accessed on 1 July 2025).
55. Rohde & Schwarz. TD-SCDMA, incl TD-SCDMA Enhanced Features Digital Standard for R&S® Signal Generators. Version 15. Available online: https://scdn.rohde-schwarz.com/ur/pws/dl_downloads/dl_common_library/dl_manuals/dl_user_manual/SMxAMU_TD-SCDMA_OperatingManual_en_15.pdf (accessed on 1 July 2025).
56. Prasad, R.; Ojanpera, T. A survey on CDMA: Evolution towards wideband CDMA. In Proceedings of the 1998 IEEE 5th International Symposium on Spread Spectrum Techniques and Applications—Proceedings. Spread Technology to Africa (Cat. No.98TH8333), Sun City, South Africa, 2–4 September 1998; Volume 1, pp. 323–331. [CrossRef]
57. Knisely, D.; Kumar, S.; Laha, S.; Nanda, S. Evolution of wireless data services: IS-95 to cdma2000. *IEEE Commun. Mag.* **1998**, *36*, 140–149. [CrossRef]
58. Sindhushayana, N.; Black, P. Forward link coding and modulation for CDMA2000 1XEV-DO (IS-856). In Proceedings of the 13th IEEE International Symposium on Personal, Indoor and Mobile Radio Communications, Lisboa, Portugal, 15–18 September 2002; Volume 4, pp. 1839–1846. [CrossRef]
59. Qualcomm. 5G Waveform and Multiple Access Techniques. 2015. Available online: https://www.qualcomm.com/media/documents/files/5g-research-on-waveform-and-multiple-access-techniques.pdf (accessed on 1 June 2025).
60. Arabian, F.; Rice, M. On the Performance of Filter Based Equalizers for 16APSK in Aeronautical Telemetry Environment. In Proceedings of the International Telemetering Conference, Glendale, AZ, USA, 5–8 November 2018.
61. Chande, V.; Sun, H.; Vitthaladevuni, P.K.; Hou, J.; Mohanty, B. Performance Analysis of 64-QAM and MIMO in Release 7 WCDMA (HSPA+) Systems. In Proceedings of the 2010 IEEE 71st Vehicular Technology Conference, Taipei, Taiwan, 16–19 May 2010; pp. 1–5. [CrossRef]
62. Khan, M.N.; Ghauri, S. The WiMAX 802.16e physical layer model. In Proceedings of the 2008 IET International Conference on Wireless, Mobile and Multimedia Networks, Beijing, China, 12–15 October 2008; pp. 117–120. Available online: https://digital-library.theiet.org/doi/10.1049/cp:20080159 (accessed on 1 June 2025).
63. Kamali, B. The IEEE 802.16 Standards and the WiMAX Technology. In *AeroMACS: An IEEE 802.16 Standard-Based Technology for the Next Generation of Air Transportation Systems*; Wiley: Hoboken, NJ, USA, 2018; pp. 189–258. [CrossRef]
64. WiMAX Forum™. MobileWiMAX—Part I: A Technical Overview and Performance Evaluation. 2006. Available online: https://wimaxforum.org/news/downloads/mobile_wimax_part1_overview_and_performance.pdf (accessed on 1 June 2025).
65. 3GPP. LTE Release 8 Specifications. Available online: https://www.3gpp.org/specifications (accessed on 1 July 2025).
66. Chang, R.W. Synthesis of Band-Limited Orthogonal Signals for Multichannel Data Transmission. *Bell Syst. Tech. J.* **1966**, *45*, 1775–1796. [CrossRef]
67. LaSorte, N.; Barnes, W.J.; Refai, H.H. The History of Orthogonal Frequency Division Multiplexing. In Proceedings of the IEEE GLOBECOM 2008—2008 IEEE Global Telecommunications Conference, New Orleans, LA, USA, 30 November–4 December 2008; pp. 1–5. [CrossRef]
68. Weinstein, S.; Ebert, P. Data Transmission by Frequency-Division Multiplexing Using the Discrete Fourier Transform. *IEEE Trans. Commun. Technol.* **1971**, *19*, 628–634. [CrossRef]
69. Matic, D.; Coenen, T.; Schoute, F.; Prasad, R. OFDM timing synchronisation: Possibilities and limits to the usage of the cyclic prefix for maximum likelihood estimation. In Proceedings of the Gateway to 21st Century Communications Village. VTC 1999-Fall. IEEE VTS 50th Vehicular Technology Conference (Cat. No.99CH36324), Amsterdam, The Netherlands, 19–22 September 1999; Volume 2, pp. 668–672. [CrossRef]
70. ETSI. LTE; Evolved Universal Terrestrial Radio Access (E-UTRA); Base Station (BS) Radio Transmission and Reception (3GPP TS 36.104 Version 8.10.0 Release 8). Available online: https://www.etsi.org/deliver/etsi_ts/136100_136199/136104/08.10.00_60/ts_136104v081000p.pdf (accessed on 1 June 2025).

71. Anoh, K.; Tanriover, C.; Adebisi, B. On the Optimization of Iterative Clipping and Filtering for PAPR Reduction in OFDM Systems. *IEEE Access* **2017**, *5*, 12004–12013. [CrossRef]
72. Jeon, H.B.; No, J.S.; Shin, D.J. A Low-Complexity SLM Scheme Using Additive Mapping Sequences for PAPR Reduction of OFDM Signals. *IEEE Trans. Broadcast.* **2011**, *57*, 866–875. [CrossRef]
73. Rahmatallah, Y.; Mohan, S. Peak-To-Average Power Ratio Reduction in OFDM Systems: A Survey and Taxonomy. *IEEE Commun. Surv. Tutor.* **2013**, *15*, 1567–1592. [CrossRef]
74. Ciochina, C.; Sari, H. A Review of OFDMA and Single-Carrier FDMA. In Proceedings of the 2010 European Wireless Conference (EW), Lucca, Tuscany, Italy, 12–15 April 2010; pp. 706–710. [CrossRef]
75. Rohde & Schwarz Inc. IEEE 802.16m Technology Introduction. 2010. Available online: https://scdn.rohde-schwarz.com/ur/pws/dl_downloads/dl_application/application_notes/1ma167/1MA167_3e_IEEE_80216m_technology.pdf (accessed on 1 May 2025).
76. 3GPP2. Ultra Mobile Broadband Air Interface Specification. Technical Report C.S0084-001 v2.0, 3GPP2. 2007. Available online: https://link.springer.com/chapter/10.1007/978-0-387-68192-4_12 (accessed on 1 June 2025).
77. Wang, M. Ultra Mobile Broadband Technology Overview. In Proceedings of the 6th Annual Communication Networks and Services Research Conference (cnsr 2008), Halifax, NS, Canada, 5–8 May 2008; pp. 8–9. [CrossRef]
78. 3rd Generation Partnership Project (3GPP). Release 10 Specification. 2014. Available online: https://www.3gpp.org/ftp/Information/WORK_PLAN/Description_Releases/ (accessed on 1 June 2025).
79. 3GPP. LTE-Advanced. Available online: https://www.3gpp.org/img/pdf/lte_advanced_v2.pdf (accessed on 1 May 2025).
80. Takaharu, N. LTE and LTE-advanced: Radio technology aspects for mobile communications. In Proceedings of the 2011 XXXth URSI General Assembly and Scientific Symposium, Istanbul, Turkey, 13–20 August 2011; pp. 1–4. [CrossRef]
81. 3GPP. Study on Scenarios and Requirements for Next Generation Access Technologies. Technical Specification 38.913, 3rd Generation Partnership Project (3GPP), Version 14.2.0. 2017. Available online: https://www.etsi.org/deliver/etsi_tr/138900_138999/138913/14.02.00_60/tr_138913v140200p.pdf (accessed on 1 July 2025).
82. Popovski, P.; Trillingsgaard, K.F.; Simeone, O.; Durisi, G. 5G Wireless Network Slicing for eMBB, URLLC, and mMTC: A Communication-Theoretic View. *IEEE Access* **2018**, *6*, 55765–55779. [CrossRef]
83. 3GPP. Release 15 Description. Technical Report TR 21.915, 3rd Generation Partnership Project (3GPP). 2019. Available online: https://portal.3gpp.org/desktopmodules/Specifications/SpecificationDetails.aspx?specificationId=3389 (accessed on 1 July 2025).
84. Abdelghaffar, M.; Santhappan, T.V.P.; Tokgoz, Y.; Mukkavilli, K.; Ji, T. Subband Full-Duplex Large-Scale Deployed Network Designs and Tradeoffs. *Proc. IEEE* **2024**, *112*, 487–510. [CrossRef]
85. Saito, Y.; Kishiyama, Y.; Benjebbour, A.; Nakamura, T.; Li, A.; Higuchi, K. Non-Orthogonal Multiple Access (NOMA) for Cellular Future Radio Access. In Proceedings of the 2013 IEEE 77th Vehicular Technology Conference (VTC Spring), Dresden, Germany, 2–5 June 2013; pp. 1–5. [CrossRef]
86. Kebede, T.; Wondie, Y.; Steinbrunn, J.; Kassa, H.B.; Kornegay, K.T. Multi-Carrier Waveforms and Multiple Access Strategies in Wireless Networks: Performance, Applications, and Challenges. *IEEE Access* **2022**, *10*, 21120–21140. [CrossRef]
87. Reddy, P.V.; Reddy, S.; Reddy, S.; Sawale, R.D.; Narendar, P.; Duggineni, C.; Valiveti, H.B. Analytical Review on OMA vs. NOMA and Challenges Implementing NOMA. In Proceedings of the 2021 2nd International Conference on Smart Electronics and Communication (ICOSEC), Trichy, Tamil Nadu, India, 7–9 October 2021; pp. 552–556. [CrossRef]
88. Hammoodi, A.; Audah, L.; Taher, M.A. Green Coexistence for 5G Waveform Candidates: A Review. *IEEE Access* **2019**, *7*, 10103–10126. [CrossRef]
89. An, C.; Ryu, H.G. CPW-OFDM(Cyclic Postfix Windowing OFDM) for the B5G (Beyond 5th Generation) Waveform. In Proceedings of the 2018 IEEE 10th Latin-American Conference on Communications (LATINCOM), Buenos Aires, Argentina, 23–25 October 2018; pp. 1–4. [CrossRef]
90. Zayani, R.; Medjahdi, Y.; Shaiek, H.; Roviras, D. WOLA-OFDM: A Potential Candidate for Asynchronous 5G. In Proceedings of the 2016 IEEE Globecom Workshops (GC Wkshps), Washington, DC, USA, 4–8 December 2016; pp. 1–5. [CrossRef]
91. Beaulieu, N.C.; Tan, P. On the effects of receiver windowing on OFDM performance in the presence of carrier frequency offset. *IEEE Trans. Wirel. Commun.* **2007**, *6*, 202–209. [CrossRef]
92. Gerzaguet, R.; Medjahdi, Y.; Demmer, D.; Zayani, R.; Doré, J.B.; Shaiek, H.; Roviras, D. Comparison of promising candidate waveforms for 5G: WOLA-OFDM versus BF-OFDM. In Proceedings of the 2017 International Symposium on Wireless Communication Systems (ISWCS), Bologna, Italy, 28–31 August 2017; pp. 355–359. [CrossRef]
93. Abdoli, J.; Jia, M.; Ma, J. Filtered OFDM: A new waveform for future wireless systems. In Proceedings of the 2015 IEEE 16th International Workshop on Signal Processing Advances in Wireless Communications (SPAWC), Stockholm, Sweden, 28 June–1 July 2015; pp. 66–70. [CrossRef]
94. Farhang-Boroujeny, B. OFDM Versus Filter Bank Multicarrier. *IEEE Signal Process. Mag.* **2011**, *28*, 92–112. [CrossRef]

95. Zhang, X.; Jia, M.; Chen, L.; Ma, J.; Qiu, J. Filtered-OFDM - Enabler for Flexible Waveform in the 5th Generation Cellular Networks. In Proceedings of the 2015 IEEE Global Communications Conference (GLOBECOM), San Diego, CA, USA, 6–10 December 2015; pp. 1–6. [CrossRef]
96. Demmer, D.; Gerzaguet, R.; Doré, J.B.; Le Ruyet, D.; Kténas, D. Block-filtered OFDM: A novel waveform for future wireless technologies. In Proceedings of the 2017 IEEE International Conference on Communications (ICC), Paris, France, 21–25 May 2017; pp. 1–6. [CrossRef]
97. Gerzaguet, R.; Demmer, D.; Doré, J.B.; Ktiénas, D. Block-filtered OFDM: A new promising waveform for multi-service scenarios. In Proceedings of the 2017 IEEE International Conference on Communications (ICC), Paris, France, 21–25 May 2017; pp. 1–6. [CrossRef]
98. Vakilian, V.; Wild, T.; Schaich, F.; ten Brink, S.; Frigon, J.F. Universal-filtered multi-carrier technique for wireless systems beyond LTE. In Proceedings of the 2013 IEEE Globecom Workshops (GC Wkshps), Atlanta, GA, USA, 9–13 December 2013; pp. 223–228. [CrossRef]
99. Bochechka, G.; Tikhvinskiy, V.; Vorozhishchev, I.; Aitmagambetov, A.; Nurgozhin, B. Comparative analysis of UFMC technology in 5G networks. In Proceedings of the 2017 International Siberian Conference on Control and Communications (SIBCON), Nur-Sultan, Kazakhstan, 29–30 June 2017; pp. 1–6. [CrossRef]
100. Schaich, F.; Wild, T.; Chen, Y. Waveform Contenders for 5G - Suitability for Short Packet and Low Latency Transmissions. In Proceedings of the 2014 IEEE 79th Vehicular Technology Conference (VTC Spring), Seoul, Republic of Korea, 18–21 May 2014; pp. 1–5. [CrossRef]
101. Tao, Y.; Liu, L.; Liu, S.; Zhang, Z. A survey: Several technologies of non-orthogonal transmission for 5G. *China Commun.* **2015**, *12*, 1–15. [CrossRef]
102. Schaich, F.; Wild, T. Waveform contenders for 5G—OFDM vs. FBMC vs. UFMC. In Proceedings of the 2014 6th International Symposium on Communications, Control and Signal Processing (ISCCSP), Athens, Greece, 21–23 May 2014; pp. 457–460. [CrossRef]
103. Kongara, G.; He, C.; Yang, L.; Armstrong, J. A Comparison of CP-OFDM, PCC-OFDM and UFMC for 5G Uplink Communications. *IEEE Access* **2019**, *7*, 157574–157594. [CrossRef]
104. Cherubini, G.; Eleftheriou, E.; Olcer, S. Filtered multitone modulation for VDSL. In Proceedings of the Seamless Interconnection for Universal Services. Global Telecommunications Conference. GLOBECOM'99. (Cat. No.99CH37042), Rio de Janeiro, Brazil, 5–9 December 1999; Volume 2, pp. 1139–1144. [CrossRef]
105. Farhang-Boroujeny, B. Filter Bank Multicarrier (FBMC): An Integrated Solution to Spectrum Sensing and Data Transmission in Cognitive Radio Networks. In Proceedings of the PHYDYAS Workshop, Perugia, Italy, 21–24 June 2009. Available online: https://www.slideshare.net/slideshow/phydyas-09-ffilter-bank-multicarrier-fbmc-an-integrated-solution-to-spectrum-sensing-and-data-transmission-in-cognitive-radio-networksarhangboroujeny/15837461 (accessed on 1 July 2025).
106. Sourck, H.S.; Wu, Y.; Bergmans, J.W.M.; Sadri, S.; Farhang-Boroujeny, B. Effect of Carrier Frequency Offset on Offset QAM Multicarrier Filter Bank Systems over Frequency-Selective Channels. In Proceedings of the 2010 IEEE Wireless Communication and Networking Conference, Sydney, Australia, 18–21 April 2010; pp. 1–6. [CrossRef]
107. Farhang-Boroujeny, B. *Signal Processing Techniques for Software Radios*; Lulu Publishing: Raleigh, NC, USA, 2008.
108. Saltzberg, B. Performance of an Efficient Parallel Data Transmission System. *IEEE Trans. Commun. Technol.* **1967**, *15*, 805–811. [CrossRef]
109. Farhang-Boroujeny, B. Multicarrier modulation with blind detection capability using cosine modulated filter banks. *IEEE Trans. Commun.* **2003**, *51*, 2057–2070. [CrossRef]
110. Farhang-Boroujeny, B.; Moradi, H. OFDM Inspired Waveforms for 5G. *IEEE Commun. Surv. Tutor.* **2016**, *18*, 2474–2492. [CrossRef]
111. Abdel-Atty, H.M.; Raslan, W.A.; Khalil, A.T. Evaluation and Analysis of FBMC/OQAM Systems Based on Pulse Shaping Filters. *IEEE Access* **2020**, *8*, 55750–55772. [CrossRef]
112. Fettweis, G.; Krondorf, M.; Bittner, S. GFDM—Generalized Frequency Division Multiplexing. In Proceedings of the VTC Spring 2009–IEEE 69th Vehicular Technology Conference, Barcelona, Spain, 26–29 April 2009; pp. 1–4. [CrossRef]
113. Datta, R.; Fettweis, G.; Koll´r, Z.; Horv´th, P. FBMC and GFDM Interference Cancellation Schemes for Flexible Digital Radio PHY Design. In Proceedings of the 2011 14th Euromicro Conference on Digital System Design, Oulu, Finland, 31 August–2 September 2011; pp. 335–339. [CrossRef]
114. Zeng, Y.; Liang, Y.C.; Chia, M.W.; Pham, T.H. Fast Algorithms for FBMC and GFDM in Dynamic Spectrum Access. In Proceedings of the 2017 IEEE Wireless Communications and Networking Conference (WCNC), San Francisco, CA, USA, 19–22 March 2017; pp. 1–6. [CrossRef]
115. Michailow, N.; Matthé, M.; Gaspar, I.S.; Caldevilla, A.N.; Mendes, L.L.; Festag, A.; Fettweis, G. Generalized Frequency Division Multiplexing for 5th Generation Cellular Networks. *IEEE Trans. Commun.* **2014**, *62*, 3045–3061. [CrossRef]

116. Alhaj, N.A.; Jamlos, M.F.; Manap, S.A.; Abdelsalam, S.; Bakhit, A.A.; Mamat, R.; Jamlos, M.A.; Gismalla, M.S.M.; Hamdan, M. Integration of Hybrid Networks, AI, Ultra Massive-MIMO, THz Frequency, and FBMC Modulation Toward 6G Requirements: A Review. *IEEE Access* **2024**, *12*, 483–513. [CrossRef]
117. Arabian, F.; Rice, M. Polarization Combining and Equalization in 5G Mobile-to-Mobile Systems. *IEEE Access* **2022**, *10*, 45881–45892. [CrossRef]
118. Arabian, F.; Rice, M.; Hu, R.Q. Who's On First In 5G Mobile Networks: Equalizers or Polarization Diversity Combiners? In Proceedings of the 2020 Intermountain Engineering, Technology and Computing (IETC), Orem, UT, USA, 4–5 May 2020; pp. 1–6. [CrossRef]
119. Yaklaf, S.K.A.; Tarmissi, K.S.; Shashoa, N.A.A. 6G Mobile Communications Systems: Requirements, Specifications, Challenges, Applications, and Technologies. In Proceedings of the 2021 IEEE 1st International Maghreb Meeting of the Conference on Sciences and Techniques of Automatic Control and Computer Engineering MI-STA, Tripoli, Libya, 25–27 May 2021; pp. 679–683. [CrossRef]
120. Zhou, W.; Zhang, R.; Chen, G.; Wu, W. Integrated Sensing and Communication Waveform Design: A Survey. *IEEE Open J. Commun. Soc.* **2022**, *3*, 1930–1949. [CrossRef]
121. Saad, W.; Bennis, M.; Chen, M. A Vision of 6G Wireless Systems: Applications, Trends, Technologies, and Open Research Problems. *IEEE Netw.* **2020**, *34*, 134–142. [CrossRef]
122. Zhang, J.A.; Rahman, M.L.; Wu, K.; Huang, X.; Guo, Y.J.; Chen, S.; Yuan, J. Enabling Joint Communication and Radar Sensing in Mobile Networks—A Survey. *IEEE Commun. Surv. Tutor.* **2022**, *24*, 306–345. [CrossRef]
123. Liu, F.; Cui, Y.; Masouros, C.; Xu, J.; Han, T.X.; Eldar, Y.C.; Buzzi, S. Integrated Sensing and Communications: Toward Dual-Functional Wireless Networks for 6G and Beyond. *IEEE J. Sel. Areas Commun.* **2022**, *40*, 1728–1767. [CrossRef]
124. Hassanien, A.; Amin, M.G.; Zhang, Y.D.; Ahmad, F. Signaling strategies for dual-function radar communications: An overview. *IEEE Aerosp. Electron. Syst. Mag.* **2016**, *31*, 36–45. [CrossRef]
125. Liu, A.; Huang, Z.; Li, M.; Wan, Y.; Li, W.; Han, T.X.; Liu, C.; Du, R.; Tan, D.K.P.; Lu, J.; et al. A Survey on Fundamental Limits of Integrated Sensing and Communication. *IEEE Commun. Surv. Tutor.* **2022**, *24*, 994–1034. [CrossRef]
126. Luong, N.C.; Lu, X.; Hoang, D.T.; Niyato, D.; Kim, D.I. Radio Resource Management in Joint Radar and Communication: A Comprehensive Survey. *IEEE Commun. Surv. Tutor.* **2021**, *23*, 780–814. [CrossRef]
127. Zhang, J.A.; Liu, F.; Masouros, C.; Heath, R.W.; Feng, Z.; Zheng, L.; Petropulu, A. An Overview of Signal Processing Techniques for Joint Communication and Radar Sensing. *IEEE J. Sel. Top. Signal Process.* **2021**, *15*, 1295–1315. [CrossRef]
128. Zheng, L.; Lops, M.; Eldar, Y.C.; Wang, X. Radar and Communication Coexistence: An Overview: A Review of Recent Methods. *IEEE Signal Process. Mag.* **2019**, *36*, 85–99. [CrossRef]
129. Kumari, P.; Mezghani, A.; Heath, R.W. JCR70: A Low-Complexity Millimeter-Wave Proof-of-Concept Platform for a Fully-Digital SIMO Joint Communication-Radar. *IEEE Open J. Veh. Technol.* **2021**, *2*, 218–234. [CrossRef]
130. Chiriyath, A.R.; Ragi, S.; Mittelmann, H.D.; Bliss, D.W. Novel Radar Waveform Optimization for a Cooperative Radar-Communications System. *IEEE Trans. Aerosp. Electron. Syst.* **2019**, *55*, 1160–1173. [CrossRef]
131. Sturm, C.; Wiesbeck, W. Waveform Design and Signal Processing Aspects for Fusion of Wireless Communications and Radar Sensing. *Proc. IEEE* **2011**, *99*, 1236–1259. [CrossRef]
132. Liu, Y.; Liao, G.; Xu, J.; Yang, Z.; Zhang, Y. Adaptive OFDM Integrated Radar and Communications Waveform Design Based on Information Theory. *IEEE Commun. Lett.* **2017**, *21*, 2174–2177. [CrossRef]
133. Liu, Y.; Liao, G.; Yang, Z.; Xu, J. Design of integrated radar and communication system based on MIMO-OFDM waveform. *J. Syst. Eng. Electron.* **2017**, *28*, 669–680. [CrossRef]
134. Tigrek, R.F.; De Heij, W.J.A.; Van Genderen, P. OFDM Signals as the Radar Waveform to Solve Doppler Ambiguity. *IEEE Trans. Aerosp. Electron. Syst.* **2012**, *48*, 130–143. [CrossRef]
135. Johnston, J.; Venturino, L.; Grossi, E.; Lops, M.; Wang, X. MIMO OFDM Dual-Function Radar-Communication Under Error Rate and Beampattern Constraints. *IEEE J. Sel. Areas Commun.* **2022**, *40*, 1951–1964. [CrossRef]
136. Chen, X.; Feng, Z.; Wei, Z.; Zhang, P.; Yuan, X. Code-Division OFDM Joint Communication and Sensing System for 6G Machine-Type Communication. *IEEE Internet Things J.* **2021**, *8*, 12093–12105. [CrossRef]
137. Ellinger, J.; Zhang, Z.; Wu, Z.; Wicks, M.C. Dual-Use Multicarrier Waveform for Radar Detection and Communication. *IEEE Trans. Aerosp. Electron. Syst.* **2018**, *54*, 1265–1278. [CrossRef]
138. Temiz, M.; Alsusa, E.; Baidas, M.W. A Dual-Functional Massive MIMO OFDM Communication and Radar Transmitter Architecture. *IEEE Trans. Veh. Technol.* **2020**, *69*, 14974–14988. [CrossRef]
139. Barneto, C.B.; Liyanaarachchi, S.D.; Heino, M.; Riihonen, T.; Valkama, M. Full Duplex Radio/Radar Technology: The Enabler for Advanced Joint Communication and Sensing. *IEEE Wirel. Commun.* **2021**, *28*, 82–88. [CrossRef]
140. Ouyang, X.; Zhao, J. Orthogonal Chirp Division Multiplexing. *IEEE Trans. Commun.* **2016**, *64*, 3946–3957. [CrossRef]

141. Singh, M.K.; Goel, A. Orthogonal Chirp Division Multiplexing: An Emerging Multi-Carrier Modulation Scheme. In *Innovations in Intelligent Computing and Communication*; Panda, M., Dehuri, S., Patra, M.R., Behera, P.K., Tsihrintzis, G.A., Cho, S., Coello Coello, C.A., Eds.; Springer International Publishing: Cham, Switzerland, 2022; pp. 236–246. [CrossRef]
142. Omar, M.S.; Ma, X. Spectrum Design for Orthogonal Chirp Division Multiplexing Transmissions. *IEEE Wirel. Commun. Lett.* **2020**, *9*, 1990–1994. [CrossRef]
143. Lv, X.; Wang, J.; Jiang, Z.; Jiao, W. A novel PAPR reduction method for OCDM-based radar-communication signal. In Proceedings of the 2018 IEEE MTT-S International Microwave Workshop Series on 5G Hardware and System Technologies (IMWS-5G), Dublin, Ireland, 30–31 August 2018; pp. 1–3. [CrossRef]
144. Zhang, R.; Wang, Y.; Ma, X. Channel Estimation for OCDM Transmissions with Carrier Frequency Offset. *IEEE Wirel. Commun. Lett.* **2022**, *11*, 483–487. [CrossRef]
145. Wang, Q.; Kakkavas, A.; Gong, X.; Stirling-Gallacher, R.A. Towards Integrated Sensing and Communications for 6G. In Proceedings of the 2022 2nd IEEE International Symposium on Joint Communications & Sensing (JC&S), Seefeld, Austria, 9–10 March 2022; pp. 1–6. [CrossRef]
146. Ni, Y.; Wang, Z.; Yuan, P.; Huang, Q. An AFDM-Based Integrated Sensing and Communications. In Proceedings of the 2022 International Symposium on Wireless Communication Systems (ISWCS), Hangzhou, China, 19–22 October 2022; pp. 1–6. [CrossRef]
147. Oliveira, L.G.d.; Alabd, M.B.; Nuss, B.; Zwick, T. An OCDM Radar-Communication System. In Proceedings of the 2020 14th European Conference on Antennas and Propagation (EuCAP), Copenhagen, Denmark, 15–20 March 2020; pp. 1–5. [CrossRef]
148. de Oliveira, L.G.; Nuss, B.; Alabd, M.B.; Li, Y.; Yu, L.; Zwick, T. MIMO-OCDM-based Joint Radar Sensing and Communication. In Proceedings of the 2021 15th European Conference on Antennas and Propagation (EuCAP), Düsseldorf, Germany, 22–26 March 2021; pp. 1–5. [CrossRef]
149. Bemani, A.; Ksairi, N.; Kountouris, M. Affine Frequency Division Multiplexing for Next Generation Wireless Communications. *IEEE Trans. Wirel. Commun.* **2023**, *22*, 8214–8229. [CrossRef]
150. Wan, Z.; Gao, Z.; Héliot, F.; Su, Y.; Wu, Z.; Chen, L.; Luo, Q.; Xiao, P. Orthogonal Chirp Division Multiplexing Waveform Design for 6G mmWave UAV Integrated Sensing and Communication. In Proceedings of the 2024 International Wireless Communications and Mobile Computing (IWCMC), Ayia Napa, Cyprus, 27–31 May 2024; pp. 622–627. [CrossRef]
151. Bemani, A.; Cuozzo, G.; Ksairi, N.; Kountouris, M. Affine Frequency Division Multiplexing for Next-Generation Wireless Networks. In Proceedings of the 2021 17th International Symposium on Wireless Communication Systems (ISWCS), Berlin, Germany, 6–9 September 2021; pp. 1–6. [CrossRef]
152. Bemani, A.; Ksairi, N.; Kountouris, M. AFDM: A Full Diversity Next Generation Waveform for High Mobility Communications. In Proceedings of the 2021 IEEE International Conference on Communications Workshops (ICC Workshops), Montreal, QC, Canada, 14–23 June 2021; pp. 1–6. [CrossRef]
153. Pei, S.C.; Ding, J.J. Closed-form discrete fractional and affine Fourier transforms. *IEEE Trans. Signal Process.* **2000**, *48*, 1338–1353. [CrossRef]
154. Pei, S.C.; Ding, J.J. Relations between fractional operations and time-frequency distributions, and their applications. *IEEE Trans. Signal Process.* **2001**, *49*, 1638–1655. [CrossRef]
155. Moshinsky, M.; Quesne, C. Linear Canonical Transformations and Their Unitary Representations. *J. Math. Phys.* **1971**, *12*, 1772–1780. [CrossRef]
156. Chennamsetty, S.; Boddu, S.; Chandhar, P.; Bulusu, K.C. Analysis of PAPR in OTFS Modulation with Classical Selected Mapping Technique. In Proceedings of the 2023 15th International Conference on COMmunication Systems & NETworkS (COMSNETS), Bengaluru, India, 3–8 January 2023; pp. 319–322. [CrossRef]
157. Gaudio, L.; Kobayashi, M.; Caire, G.; Colavolpe, G. On the Effectiveness of OTFS for Joint Radar Parameter Estimation and Communication. *IEEE Trans. Wirel. Commun.* **2020**, *19*, 5951–5965. [CrossRef]
158. Raviteja, P.; Phan, K.T.; Hong, Y.; Viterbo, E. Orthogonal Time Frequency Space (OTFS) Modulation Based Radar System. In Proceedings of the 2019 IEEE Radar Conference (RadarConf), Boston, MA, USA, 22–26 April 2019; pp. 1–6. [CrossRef]
159. Wu, K.; Zhang, J.A.; Huang, X.; Guo, Y.J. OTFS-Based Joint Communication and Sensing for Future Industrial IoT. *IEEE Internet Things J.* **2023**, *10*, 1973–1989. [CrossRef]
160. Yuan, W.; Wei, Z.; Li, S.; Yuan, J.; Ng, D.W.K. Integrated Sensing and Communication-Assisted Orthogonal Time Frequency Space Transmission for Vehicular Networks. *IEEE J. Sel. Top. Signal Process.* **2021**, *15*, 1515–1528. [CrossRef]
161. Patel, A.; Shukla, A.; Bhalani, J. A Comprehensive Survey on 6G Networks: Key Technologies and Challenges. In Proceedings of the 2021 International Conference on Simulation, Automation & Smart Manufacturing (SASM), Mathura, India, 20–21 August 2021; pp. 1–6. [CrossRef]
162. Basar, E. Index modulation techniques for 5G wireless networks. *IEEE Commun. Mag.* **2016**, *54*, 168–175. [CrossRef]
163. Basar, E.; Wen, M.; Mesleh, R.; Di Renzo, M.; Xiao, Y.; Haas, H. Index Modulation Techniques for Next-Generation Wireless Networks. *IEEE Access* **2017**, *5*, 16693–16746. [CrossRef]

164. Li, J.; Dang, S.; Wen, M.; Li, Q.; Chen, Y.; Huang, Y.; Shang, W. Index Modulation Multiple Access for 6G Communications: Principles, Applications, and Challenges. *IEEE Netw.* **2023**, *37*, 52–60. [CrossRef]
165. Lin, X.; Zhang, J.; Liu, Y.; Kim, J. (Eds.) *Fundamentals of 6G Communications and Networking*; Signals and Communication Technology; Springer International Publishing: Cham, Switzerland, 2024. [CrossRef]
166. Datta, T.; Eshwaraiah, H.S.; Chockalingam, A. Generalized Space-and-Frequency Index Modulation. *IEEE Trans. Veh. Technol.* **2016**, *65*, 4911–4924. [CrossRef]
167. Tanahashi, M.; Ochiai, H. On the distribution of instantaneous power in single-carrier signals. *IEEE Trans. Wirel. Commun.* **2010**, *9*, 1207–1215. [CrossRef]
168. Jegbefume, O.; Saquib, M.; Torlak, M. Performance analysis of spread spectrum protocols with adaptive spreading. *IEEE Trans. Wirel. Commun.* **2010**, *9*, 1063–1073. [CrossRef]
169. Jones, W. Principles of mitigation bandwidth with application to spread spectrum communication in the mobile channel. In Proceedings of the ISSSTA'95 International Symposium on Spread Spectrum Techniques and Applications, Mainz, Germany, 25 September 1996; Volume 2, pp. 618–622. [CrossRef]
170. Mizutani, K.; Kohno, R. Inter-vehicle spread spectrum communication and ranging system with concatenated EOE sequence. *IEEE Trans. Intell. Transp. Syst.* **2001**, *2*, 180–191. [CrossRef]
171. Saadia, R.; Khan, N.M. Single Carrier-Frequency Division Multiple Access Radar: Waveform Design and Analysis. *IEEE Access* **2020**, *8*, 35742–35751. [CrossRef]
172. Zeng, Y.; Ma, Y.; Sun, S. Joint Radar-Communication With Cyclic Prefixed Single Carrier Waveforms. *IEEE Trans. Veh. Technol.* **2020**, *69*, 4069–4079. [CrossRef]
173. Cao, N.; Chen, Y.; Gu, X.; Feng, W. Joint Radar-Communication Waveform Designs Using Signals From Multiplexed Users. *IEEE Trans. Commun.* **2020**, *68*, 5216–5227. [CrossRef]
174. Zhou, S.; Liang, X.; Yu, Y.; Liu, H. Joint Radar-Communications Co-Use Waveform Design Using Optimized Phase Perturbation. *IEEE Trans. Aerosp. Electron. Syst.* **2019**, *55*, 1227–1240. [CrossRef]
175. Wu, Y.; Han, C.; Yang, T. DFT-Spread Orthogonal Time Frequency Space Modulation Design for Terahertz Communications. In Proceedings of the 2021 IEEE Global Communications Conference (GLOBECOM), Madrid, Spain, 7–11 December 2021; pp. 1–6. [CrossRef]
176. Wu, Y.; Han, C.; Chen, Z. DFT-Spread Orthogonal Time Frequency Space System With Superimposed Pilots for Terahertz Integrated Sensing and Communication. *IEEE Trans. Wirel. Commun.* **2023**, *22*, 7361–7376. [CrossRef]
177. Hossain, M.N.; Sugiura, Y.; Shimamura, T.; Ryu, H.G. DFT-Spread OTFS Communication System with the Reductions of PAPR and Nonlinear Degradation. *Wirel. Pers. Commun.* **2020**, *115*, 2211–2228. [CrossRef]
178. Tarboush, S.; Sarieddeen, H.; Alouini, M.S.; Al-Naffouri, T.Y. Single- Versus Multicarrier Terahertz-Band Communications: A Comparative Study. *IEEE Open J. Commun. Soc.* **2022**, *3*, 1466–1486. [CrossRef]
179. Bekar, M.; Baker, C.J.; Hoare, E.G.; Gashinova, M. Joint MIMO Radar and Communication System Using a PSK-LFM Waveform With TDM and CDM Approaches. *IEEE Sens. J.* **2021**, *21*, 6115–6124. [CrossRef]
180. Saddik, G.N.; Singh, R.S.; Brown, E.R. Ultra-Wideband Multifunctional Communications/Radar System. *IEEE Trans. Microw Theory Tech.* **2007**, *55*, 1431–1437. [CrossRef]
181. Jiang, M.; Liao, G.; Yang, Z.; Liu, Y.; Chen, Y. Tunable Filter Design for Integrated Radar and Communication Waveforms. *IEEE Commun. Lett.* **2021**, *25*, 570–573. [CrossRef]
182. Hassanien, A.; Amin, M.G.; Zhang, Y.D.; Himed, B. A dual-function MIMO radar-communications system using PSK modulation. In Proceedings of the 2016 24th European Signal Processing Conference (EUSIPCO), Budapest, Hungary, 29 August–2 September 2016; pp. 1613–1617. [CrossRef]
183. Gu, Y.; Zhang, L.; Zhou, Y.; Zhang, Q. Embedding Communication Symbols Into Radar Waveform With Orthogonal FM Scheme. *IEEE Sens. J.* **2018**, *18*, 8709–8719. [CrossRef]
184. Wang, Q.; Xu, S. Vehicle Width Detection Based on Millimeter-Wave LFMCW Radar for Autonomous Driving. In Proceedings of the 2022 IEEE 95th Vehicular Technology Conference: (VTC2022-Spring), Helsinki, Finland, 19–22 June 2022; pp. 1–6. [CrossRef]
185. Yang, C.; Wang, M.; Zheng, L.; Tang, C. Dual Function System With Shared Spectrum Using FMCW. *IEEE Access* **2018**, *6*, 79026–79038. [CrossRef]
186. Hassanien, A.; Amin, M.G.; Zhang, Y.D.; Ahmad, F. Dual-Function Radar-Communications: Information Embedding Using Sidelobe Control and Waveform Diversity. *IEEE Trans. Signal Process.* **2016**, *64*, 2168–2181. [CrossRef]
187. McCormick, P.M.; Blunt, S.D.; Metcalf, J.G. Simultaneous radar and communications emissions from a common aperture, Part I: Theory. In Proceedings of the 2017 IEEE Radar Conference (RadarConf), Seattle, WA, USA, 8–12 May 2017; pp. 1685–16[illegible] [CrossRef]
188. Feng, L.; Cui, G.; Kong, L.; Zhang, X. A RadCom System with Flexible Array Controls. In Proceedings of the 2019 IEEE Radar Conference (RadarConf), Boston, MA, USA, 22–26 April 2019; pp. 1–5. [CrossRef]

189. Yang, J.; Cui, G.; Yu, X.; Kong, L. Dual-Use Signal Design for Radar and Communication via Ambiguity Function Sidelobe Control. *IEEE Trans. Veh. Technol.* **2020**, *69*, 9781–9794. [CrossRef]
190. Euzière, J.; Guinvarc'h, R.; Lesturgie, M.; Uguen, B.; Gillard, R. Dual function radar communication Time-modulated array. In Proceedings of the 2014 International Radar Conference, Lille, France, 13–17 October 2014; pp. 1–4. [CrossRef]
191. Nusenu, S.Y.; Huaizong, S.; Ye, P.; Xuehan, W.; Basit, A. Dual-Function Radar-Communication System Design Via Sidelobe Manipulation Based On FDA Butler Matrix. *IEEE Antennas Wirel. Propag. Lett.* **2019**, *18*, 452–456. [CrossRef]
192. Hassanien, A.; Amin, M.G.; Zhang, Y.D.; Ahmad, F. Dual-function radar-communications using phase-rotational invariance. In Proceedings of the 2015 23rd European Signal Processing Conference (EUSIPCO), Nice, France, 31 August–4 September 2015; pp. 1346–1350. [CrossRef]
193. Huang, T.; Liu, Y.; Ma, D.; Shlezinger, N.; Eldar, Y.C. Multi-Carrier Agile Phased Array Radar. In Proceedings of the 2019 IEEE International Conference on Signal, Information and Data Processing (ICSIDP), Chongqing, China, 11–13 December 2019; pp. 1–6. [CrossRef]
194. Eedara, I.P.; Amin, M.G. Dual Function FH MIMO Radar System with DPSK Signal Embedding. In Proceedings of the 2019 27th European Signal Processing Conference (EUSIPCO), Coruña, Spain, 2–6 September 2019; pp. 1–5. [CrossRef]
195. Baxter, W.; Aboutanios, E.; Hassanien, A. Dual-Function MIMO Radar-Communications Via Frequency-Hopping Code Selection. In Proceedings of the 2018 52nd Asilomar Conference on Signals, Systems, and Computers, Pacific Grove, CA, USA, 28–31 October 2018; pp. 1126–1130. [CrossRef]
196. Cui, G.; Yang, J.; Lu, S.; Yu, X.; Kong, L. Dual-Use Unimodular Sequence Design via Frequency Nulling Modulation. *IEEE Access* **2018**, *6*, 62470–62481. [CrossRef]
197. Wang, X.; Hassanien, A.; Amin, M.G. Dual-Function MIMO Radar Communications System Design Via Sparse Array Optimization. *IEEE Trans. Aerosp. Electron. Syst.* **2019**, *55*, 1213–1226. [CrossRef]
198. Liu, R.; Li, M.; Liu, Q.; Swindlehurst, A.L. Joint Waveform and Filter Designs for STAP-SLP-Based MIMO-DFRC Systems. *IEEE J. Sel. Areas Commun.* **2022**, *40*, 1918–1931. [CrossRef]
199. Cheng, Z.; Liao, B.; Shi, S.; He, Z.; Li, J. Co-Design for Overlaid MIMO Radar and Downlink MISO Communication Systems via Cramér–Rao Bound Minimization. *IEEE Trans. Signal Process.* **2019**, *67*, 6227–6240. [CrossRef]
200. Liu, F.; Liu, Y.F.; Li, A.; Masouros, C.; Eldar, Y.C. Cramér-Rao Bound Optimization for Joint Radar-Communication Beamforming. *IEEE Trans. Signal Process.* **2022**, *70*, 240–253. [CrossRef]
201. Yuan, X.; Feng, Z.; Zhang, J.A.; Ni, W.; Liu, R.P.; Wei, Z.; Xu, C. Spatio-Temporal Power Optimization for MIMO Joint Communication and Radio Sensing Systems With Training Overhead. *IEEE Trans. Veh. Technol.* **2021**, *70*, 514–528. [CrossRef]
202. Fuhrmann, D.R.; San Antonio, G. Transmit beamforming for MIMO radar systems using signal cross-correlation. *IEEE Trans. Aerosp. Electron. Syst.* **2008**, *44*, 171–186. [CrossRef]
203. Sun, S.; Rappaport, T.S.; Heath, R.W.; Nix, A.; Rangan, S. Mimo for millimeter-wave wireless communications: Beamforming, spatial multiplexing, or both? *IEEE Commun. Mag.* **2014**, *52*, 110–121. [CrossRef]
204. Mu, J.; Gong, Y.; Zhang, F.; Cui, Y.; Zheng, F.; Jing, X. Integrated Sensing and Communication-Enabled Predictive Beamforming With Deep Learning in Vehicular Networks. *IEEE Commun. Lett.* **2021**, *25*, 3301–3304. [CrossRef]
205. Kumari, P.; Choi, J.; González-Prelcic, N.; Heath, R.W. IEEE 802.11ad-Based Radar: An Approach to Joint Vehicular Communication-Radar System. *IEEE Trans. Veh. Technol.* **2018**, *67*, 3012–3027. [CrossRef]
206. Zhang, Q.; Sun, H.; Gao, X.; Wang, X.; Feng, Z. Time-Division ISAC Enabled Connected Automated Vehicles Cooperation Algorithm Design and Performance Evaluation. *IEEE J. Sel. Areas Commun.* **2022**, *40*, 2206–2218. [CrossRef]
207. Dokhanchi, S.H.; Mysore, B.S.; Mishra, K.V.; Ottersten, B. A mmWave Automotive Joint Radar-Communications System. *IEEE Trans. Aerosp. Electron. Syst.* **2019**, *55*, 1241–1260. [CrossRef]
208. Liu, Y.; Qin, Z.; Elkashlan, M.; Ding, Z.; Nallanathan, A.; Hanzo, L. Nonorthogonal Multiple Access for 5G and Beyond. *Proc. IEEE* **2017**, *105*, 2347–2381. [CrossRef]
209. Mu, X.; Wang, Z.; Liu, Y. NOMA for Integrating Sensing and Communications Toward 6G: A Multiple Access Perspective. *IEEE Wirel. Commun.* **2024**, *31*, 316–323. [CrossRef]
210. Al-Dweik, A.; Alsusa, E.; Dobre, O.A.; Hamila, R. Multi-Symbol Rate NOMA for Improving Connectivity in 6G Communications Networks. *IEEE Commun. Mag.* **2024**, *62*, 118–123. [CrossRef]
211. Kaur, S.; Dhand, R.; Sharma, P. Evolutionary Trends in Mobile Networks From 1G to 6 G. In Proceedings of the 2023 International Conference on Research Methodologies in Knowledge Management, Artificial Intelligence and Telecommunication Engineering (RMKMATE), Chennai, India, 1–2 November 2023; pp. 1–5. [CrossRef]
212. Lee, W.C. *Mobile Cellular Telecommunications: Analog and Digital Systems*, 2nd ed.; McGraw-Hill Professional: New York, NY, USA, 1995.
213. Proakis, J.G. *Digital Communications*, 4th ed.; McGraw-Hill: New York, NY, USA, 2000.
214. Kwok, Y.K.R.; Lau, V.K. Wireless Internet and Mobile Computing: Interoperability and Performance. In *Wireless Internet and Mobile Computing: Interoperability and Performance*; Wiley: Hoboken, NJ, USA, 2007; pp. 19–45. [CrossRef]

215. Eriksson, M.; Turina, D.; Arslan, H.; Balachandran, K.; Cheng, J.F. System performance with higher level modulation in the GSM/EDGE radio access network. In Proceedings of the GLOBECOM'01, IEEE Global Telecommunications Conference (Cat. No.01CH37270), San Antonio, TX, USA, 25–29 November 2001; Volume 5, pp. 3065–3069. [CrossRef]
216. Shopov, S.; Gurbuz, O.D.; Rebeiz, G.M.; Voinigescu, S.P. A *D*-Band Digital Transmitter with 64-QAM and OFDM Free-Space Constellation Formation. *IEEE J. Solid-State Circuits* **2018**, *53*, 2012–2022. [CrossRef]
217. Tan, M.; Zhang, L.; He, S. The Implementation of 16QAM Modulation and Demodulation and Performance Comparison. In Proceedings of the 2009 5th International Conference on Wireless Communications, Networking and Mobile Computing, Beijing, China, 24–26 September 2009; pp. 1–4. [CrossRef]
218. Halawa, T.N.; Fathy, R.A.; Zekry, A. Performance analysis of LTE-A with 256-QAM. In Proceedings of the 2016 Sixth International Conference on Digital Information Processing and Communications (ICDIPC), Beirut, Lebanon, 21–23 April 2016; pp. 141–145. [CrossRef]
219. Pancaldi, F.; Vitetta, G.M.; Kalbasi, R.; Al-Dhahir, N.; Uysal, M.; Mheidat, H. Single-carrier frequency domain equalization. *IEEE Signal Process. Mag.* **2008**, *25*, 37–56. [CrossRef]
220. Walzman, T.; Schwartz, M. Automatic equalization using the discrete frequency domain. *IEEE Trans. Inf. Theory* **1973**, *19*, 59–68. [CrossRef]
221. Sari, H.; Karam, G.; Jeanclaude, I. Transmission techniques for digital terrestrial TV broadcasting. *IEEE Commun. Mag.* **1995**, *33*, 100–109. [CrossRef]
222. Johnson, C. *5G New Radio in Bullets*; Independently Published: Farnham, UK, 2019.

telecom

MDPI

Review

RF Energy-Harvesting Techniques: Applications, Recent Developments, Challenges, and Future Opportunities

Stella N. Arinze [1], Emenike Raymond Obi [2], Solomon H. Ebenuwa [3] and Augustine O. Nwajana [4,*]

1 Department of Electrical and Electronic Engineering, Enugu State University of Science and Technology, Agbani PMB 01660, Enugu, Nigeria; ndidi.arinze@esut.edu.ng
2 Raysoft AssetAnalytics, Regina, SK S4N 7S1, Canada; ray.obi@raysoft-aa.net
3 School of Computing and Mathematical Sciences, University of Greenwich, Greenwich Campus, London SE10 9LS, UK; s.h.ebenuwa@greenwich.ac.uk
4 School of Engineering, University of Greenwich, Medway Campus, Chatham, Kent ME4 4TB, UK
* Correspondence: a.o.nwajana@greenwich.ac.uk

Abstract: The increasing demand for sustainable and renewable energy solutions has made radio frequency energy harvesting (RFEH) a promising technique for powering low-power electronic devices. RFEH captures ambient RF signals from wireless communication systems, such as mobile networks, Wi-Fi, and broadcasting stations, and converts them into usable electrical energy. This approach offers a viable alternative for battery-dependent and hard-to-recharge applications, including streetlights, outdoor night/security lighting, wireless sensor networks, and biomedical body sensor networks. This article provides a comprehensive review of the RFEH techniques, including state-of-the-art rectenna designs, energy conversion efficiency improvements, and multi-band harvesting systems. We present a detailed analysis of recent advancements in RFEH circuits, impedance matching techniques, and integration with emerging technologies such as the Internet of Things (IoT), 5G, and wireless power transfer (WPT). Additionally, this review identifies existing challenges, including low conversion efficiency, unpredictable energy availability, and design limitations for small-scale and embedded systems. A critical assessment of current research gaps is provided, highlighting areas where further development is required to enhance performance and scalability. Finally, constructive recommendations for future opportunities in RFEH are discussed, focusing on advanced materials, AI-driven adaptive harvesting systems, hybrid energy-harvesting techniques, and novel antenna–rectifier architectures. The insights from this study will serve as a valuable resource for researchers and engineers working towards the realization of self-sustaining, battery-free electronic systems.

Keywords: wireless power transfer; rectenna; internet of things; IoT powering; wireless sensor networks; WSN; AI-based energy optimization

1. Introduction

The rapid proliferation of wireless communication systems, smart devices, and IoT has intensified the demand for sustainable and self-sufficient energy solutions. Traditional energy sources such as batteries and wired power supplies face significant challenges, with a limited lifespan, the need for frequent replacements, and a negative environmental impact. RFEH [1–132] has emerged as a promising solution to power wireless sensors, biomedical devices, smart city infrastructure, and IoT nodes without relying on conventional power sources [1]. RFEH involves capturing ambient RF signals from communication networks, TV broadcasts, Wi-Fi, and cellular base stations and converting them into usable DC power through RF-to-DC conversion [2]. The core components of an RFEH system include a receiving antenna that captures ambient RF signals, a matching network that optimizes power transfer efficiency, a rectifier circuit (rectenna) that converts RF signals into DC voltage, and an energy storage unit that stores harvested energy for later use. Recent advancements in rectenna design, impedance matching techniques, and high-efficiency rectifiers have significantly improved energy-harvesting efficiency [3]. However, challenges such as low power density, signal attenuation, and varying RF energy levels hinder widespread deployment. Furthermore, this review is motivated by the growing need for a consolidated and up-to-date analysis of RFEH developments across communication technologies, particularly considering emerging 6G requirements.

Several studies have explored different approaches to enhancing RFEH performance. Researchers have developed multi-band rectennas that operate at multiple frequency bands to increase the amount of energy harvested. Muhammad et al. [4] proposed a dual-band rectenna for 0.9 GHz and 1.8 GHz Wi-Fi signals, achieving over 70% RF-to-DC conversion efficiency. Recent studies have integrated metamaterials to enhance energy-harvesting efficiency. Chen et al. [5] introduced a metamaterial-based rectenna that significantly improved power conversion by focusing and amplifying incoming RF signals. The integration of simultaneous wireless information and power transfer (SWIPT) enables wireless networks to transmit both power and information. Basnayaka et al. [6] investigated SWIPT-enabled 5G networks, where base stations simultaneously transfer RF energy and data to IoT devices. Machine learning (ML) models have also been applied to optimize energy harvesting by predicting RF energy availability in dynamic environments. Lima et al. [7] designed an AI-based energy optimization framework for self-sustaining IoT networks, increasing overall energy efficiency. While these advancements have improved RF energy harvesting, several research gaps remain. Scalability remains a major concern, particularly in integrating RFEH into large-scale 5G and 6G infrastructures. Interference management poses another challenge in maximizing harvested RF energy without affecting primary communication signals. Energy storage efficiency remains a critical area in need of improvement, with researchers exploring new materials for ultra-low-power energy storage in RFEH systems.

The purpose of this paper is to provide a thorough synthesis of recent innovations in RFEH techniques and architectures. Its objectives include presenting the underlying principles and configurations of modern RFEH systems, analyzing the latest applications across sectors, identifying prevailing challenges and research gaps, and proposing forward-looking research opportunities. The key contributions include a detailed analysis of existing RFEH techniques, including rectenna designs, impedance matching networks, and power management circuits; a review of recent developments in RFEH for IoT, smart cities, and biomedical applications; the identification of current challenges and research gaps in RF energy harvesting; and an exploration of future opportunities, including machine learning-based energy optimization and the role of 6G networks in wireless power transfer. Unlike prior review articles that focus either exclusively on rectifier design or system architecture, this paper uniquely integrates recent developments in AI-driven optimization and SWIPT techniques with practical implementation considerations, offering a broader and more application-oriented perspective. Table 1 summarizes the key differences between this review and the related literature.

Table 1. Key differences between this review and the related literature.

Reference	Coverage of Applications	Review Period	Focus on AI/ML	Inclusion of SWIPT	Comparative Technique Analysis	Future Opportunities Discussed
[8]	IoT sensors	2015–2020	No	No	Partial	Yes
[9]	Biomedical + IoT	2016–2021	Minimal	No	No	Yes
[10]	Biomedical + IoT	2017–2021	Yes	No	No	Yes
[11]	Smart cities + IoT	2017–2021	Yes	No	Partial	Yes
[12]	Biomedical + IoT	2018–2021	Yes	No	No	Yes
This work	Smart cities, biomedical, industrial, WSN, satellite	2019–2025	Yes	Yes	Yes	Yes

The scope of this review is limited to scholarly publications and technologies developed between 2019 and 2025, ensuring relevance and timeliness. Applications in smart cities, healthcare, Industry 4.0, and satellite systems are emphasized, while foundational theory is only discussed as necessary to contextualize recent advances. The main contributions of this paper are as follows:

- A detailed analysis of recent RF energy-harvesting techniques including rectenna designs, impedance matching networks, and power management circuits.
- A comprehensive review of applications in IoT, smart cities, biomedical devices, industry, and satellite communications.
- The identification of current challenges and research gaps in RF energy harvesting.
- The exploration of future research directions emphasizing AI-driven energy optimization and emerging 6G networks.
- An integrative perspective that uniquely combines AI/ML-based approaches and simultaneous wireless information and power transfer (SWIPT) techniques with practical implementation considerations.

The remainder of this paper is structured as follows. Section 2 covers the literature on the fundamental principles and components of RFEH systems. Section 3 discusses various RFEH techniques, including rectenna design and impedance matching methods. Section 4 reviews recent technological advancements and applications in different fields. Section 5 highlights challenges and limitations. Section 6 presents future opportunities in RFEH. Section 7 concludes the article with key takeaways and recommendations for future research. By addressing these critical aspects, this article aims to advance research into RFEH and facilitate the development of more efficient wireless power solutions for next-generation networks and IoT ecosystems.

2. Literature Collection Methodology

This review is grounded in the systematic collection of recent and high-quality literature to ensure its comprehensiveness and relevance to current trends in RFEH. A meticulous search was conducted across major scholarly databases, including IEEE Xplore, Elsevier (ScienceDirect), SpringerLink, MDPI, Wiley Online Library, Taylor & Francis, and ACM Digital Library. These platforms were selected for their broad coverage of peer-reviewed journal articles and conference proceedings in areas like electrical engineering, communication systems, energy harvesting, and embedded systems. To identify relevant studies, a combination of advanced Boolean search strategies and domain-specific keywords was employed. Search queries typically included phrases such as "RF energy harvesting", "wireless power transfer", "rectenna design", "ambient energy scavenging", "SWIPT", "AI-based energy optimization", "metamaterial antenna", "impedance matching", and "5G/6G energy harvesting." For example, a search query like "RF energy harvesting and 6G and rectifier" returned targeted results focused on recent innovations in rectenna circuits within next-generation wireless systems. Additional filtering ensured that only publications between 2019 and 2025 were retained for relevance and timeliness. The inclusion criteria required that articles be published in peer-reviewed journals or reputable conference proceedings and focus on technical developments, applications, challenges, or future opportunities in RFEH. Studies were included if they provided experimental, simulation-based, or validated theoretical results. Publications were excluded if they appeared before 2019, lacked technical depth, or offered purely speculative commentary without substantial analysis or modelling.

In total, 167 articles were initially gathered. After removing duplicates, screening abstracts, and conducting full-text evaluations, 132 articles were selected for final inclusion. They covered the use of RFEH in diverse domains such as smart cities, biomedical devices, industrial IoT, satellite communication, and AI-integrated systems. The distribution of selected publications across the years is shown in Figure 1, which illustrates that steady growth is evident from 2021, with 2024 marking the highest publication volume, indicative of maturing interest and investment in RF energy-harvesting technologies. The data shows a clear trajectory of increasing innovation and scholarly focus in this domain.

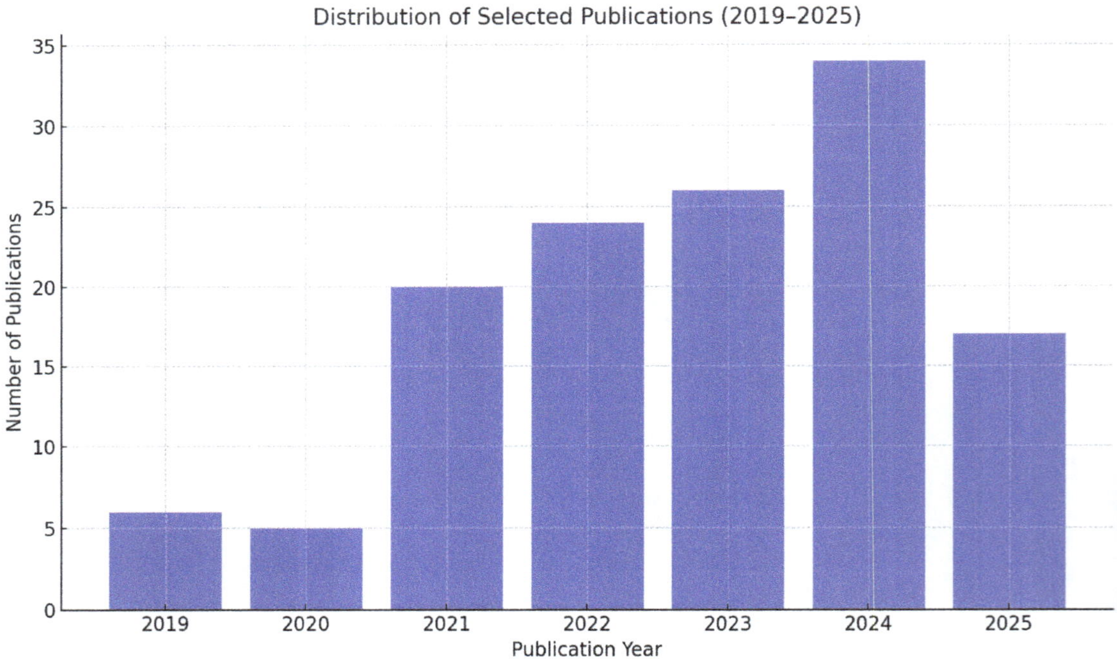

Figure 1. Distribution of selected publications on RF energy harvesting.

To further enhance this review's analytical depth, the 132 selected articles were categorized based on the primary RF energy-harvesting techniques or themes they addressed. These include single-band and multi-band rectenna designs, SWIPT-enabled architectures, AI/ML-assisted energy optimization, and hybrid energy-harvesting systems. Table 2 summarizes the number of publications per category.

Table 2. Categorization of selected RF energy-harvesting studies.

Category	Number of Articles
Single-band rectenna designs	22
Multi-band/multi-source energy harvesters	20
SWIPT-enabled systems	18
AI/ML-assisted energy optimization	15
Hybrid energy harvesting (RF + solar/etc.)	12
Biomedical applications	10
Smart cities/urban infrastructure	13
Industrial IoT/WSN	22

The analysis shows that improving rectenna designs is still the focus of most studies. However, more recent research is starting to explore the use of AI and combinations of RF with other energy sources. These emerging methods are likely to become very important in building more advanced and efficient energy-harvesting systems for future technologies like 6G and smart cities.

3. Fundamentals of RF Energy Harvesting

RF energy harvesting is a transformative technology that enables the collection and conversion of ambient electromagnetic waves into usable electrical energy, allowing wireless electronic devices to operate without relying on traditional battery power sources. This approach is particularly beneficial for powering low-energy applications such as wireless sensor networks, IoT devices, and biomedical implants, thereby reducing dependence on conventional energy storage solutions. By capturing and converting RF signals from sources like mobile networks, Wi-Fi, television broadcasts, and radio stations, RFEH presents a promising alternative for sustainable and autonomous power solutions. A comprehensive understanding of the fundamental principles, core components, available energy sources, and key performance metrics of RFEH systems is essential for optimizing their design and implementation.

3.1. Principles of RF Energy Harvesting

RFEH is based on the principle of capturing electromagnetic waves emitted from ambient RF sources, such as cellular networks, Wi-Fi routers, and broadcasting stations, and converting them into usable direct-current (DC) power. This process involves three fundamental stages: energy capture, conversion, and storage. In the first stage, a receiving antenna captures ambient RF waves from surrounding sources. These captured signals are then fed into a rectifier circuit (also known as a rectenna) that converts the alternating-current (AC) RF signals into DC voltage. Finally, the harvested energy is stored in an energy storage component, such as a battery or supercapacitor, ensuring a continuous and stable power supply for low-power electronic devices [13]. Figure 2 illustrates the RFEH process. The antenna collects RF signals from the environment and transfers them to the RF rectifier, which comprises a rectifying diode, a DC-pass filter, and a terminal load. The impedance matching network (IMN) plays a crucial role in ensuring maximum power transfer from the antenna to the rectifier [14]. The rectifying diode converts the RF signal into DC, while the DC-pass filter removes high-frequency harmonics to ensure a smooth DC output. The DC-pass capacitor filter further regulates the output impedance and enhances the stability of the generated DC power.

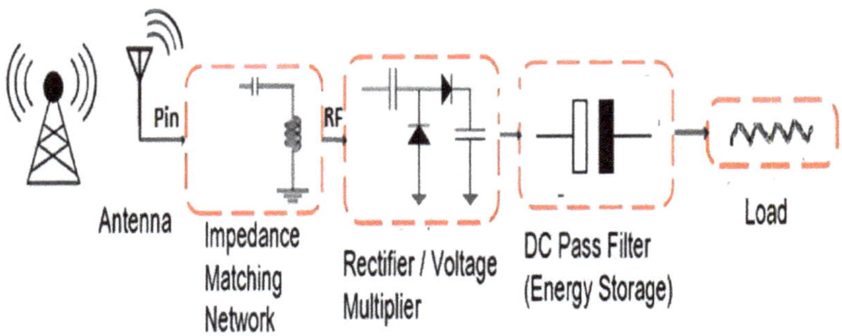

Figure 2. RF energy-harvesting process.

The efficiency of RFEH is influenced by factors such as the strength and frequency of the ambient RF signals, the design and orientation of the antenna, the effectiveness of the rectifier circuit, proper impedance matching, variations in RF signal availability due to interference, distance from the source, and obstacles.

3.2. Components of an RF Energy-Harvesting System

An RFEH system comprises several interconnected components, each playing a critical role in determining overall efficiency. The major components include the antenna, matching network, rectifier circuit, and energy storage unit.

3.2.1. Antenna

The antenna serves as the primary interface for capturing ambient RF signals, making it a crucial component of RFEH systems. The effectiveness of energy harvesting largely depends on the antenna's ability to efficiently collect RF signals from the surrounding environment. Antennas designed for RFEH must exhibit characteristics such as high efficiency, wideband or multi-band operation, and impedance matching with the rectifier to maximize power transfer. Different antenna designs, including dipole, patch, fractal antennas, and planar inverted-F antennas (PIFAs), as shown in Figure 3, have been explored to optimize energy reception. Dipole antennas are among the simplest and most widely used designs due to their omnidirectional radiation pattern and ease of fabrication [8,15]. They operate efficiently over narrow frequency ranges, making them suitable for harvesting RF energy from known, fixed-frequency sources such as broadcast transmitters and cellular base stations. Microstrip patch antennas are compact, lightweight, and suitable for integration into wearable and embedded systems. They typically operate at a single frequency, but with modifications such as slotted designs or stacked configurations, they can achieve multi-band or broadband operation, allowing them to capture RF signals from multiple sources simultaneously [16,17]. Fractal antennas are designed using self-replicating geometric patterns, enabling them to operate efficiently over multiple frequency bands while maintaining a compact size. This property makes them particularly useful for RFEH in urban environments, where multiple RF sources exist at varying frequencies. Monopole antennas and PIFAs offer compact designs and are commonly used in mobile and IoT applications. Their ability to be embedded in small devices while maintaining reasonable efficiency makes them attractive for RFEH in wearable electronics and wireless sensor networks [18,19].

Recent advancements have introduced metamaterial-based and reconfigurable antennas, which utilize artificial electromagnetic structures to enhance signal reception and dynamically adapt to varying RF environments. These antennas can be tuned to optimize performance based on the availability of RF signals, improving the overall efficiency of energy-harvesting systems [20]. Traditional antennas are often designed for a specific frequency band, limiting their ability to harvest RF energy from diverse sources. Multi-band and broadband antennas address this limitation by capturing energy from multiple frequency bands simultaneously, significantly improving the total harvested power. These antennas are particularly beneficial in urban environments, where RF signals from cellular networks (e.g., 4G, 5G), Wi-Fi, Bluetooth, and broadcast stations coexist across various frequency ranges.

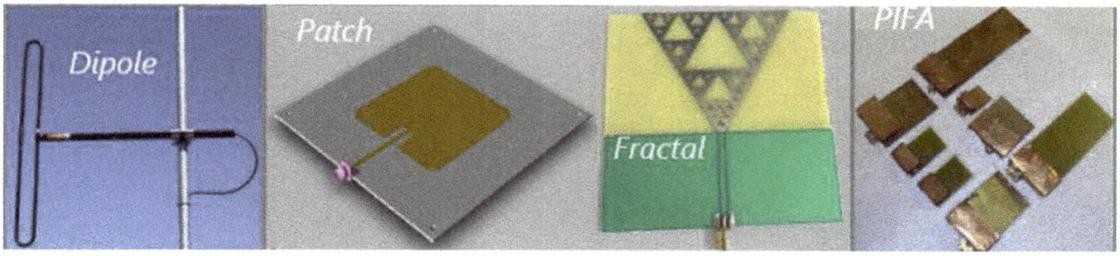

Figure 3. Antenna designs [21].

The choice of an antenna for RFEH depends on factors such as operating frequency, efficiency, gain, polarization, and impedance matching with the rectifier circuit. Antenna design for RFEH must prioritize high efficiency due to the typically low power levels of ambient RF signals. Multi-band or broadband antennas enhance energy capture by covering various RF sources, while compact and flexible designs are essential for applications like wearable electronics and IoT devices [22]. Directional antennas improve harvesting from specific sources, whereas omnidirectional antennas capture signals from multiple directions. Proper impedance matching with the rectifier is crucial in minimizing reflection losses and optimizing power transfer. Considering these factors ensures higher efficiency, supporting the development of self-sustaining wireless devices for smart cities, IoT networks, and biomedical applications. Table 3 provides a comparison of different antenna designs, highlighting their frequency ranges, circuits, and efficiency for RFEH systems.

Table 3 presents a detailed comparison of representative antenna structures used in RFEH systems, emphasizing not only their efficiency but also critical design attributes such as physical size, gain, and sensitivity. These parameters play a vital role in the performance and practicality of RF energy harvesters. Antenna size directly influences both gain and sensitivity. Larger antennas, such as logarithmic spiral or tapered slot designs, typically provide higher gain, improving energy capture from weak signals. However, these designs are less suitable for space-constrained or wearable applications due to their footprint. In contrast, compact antennas, such as those based on meander or fractal structures, offer miniaturization benefits but often exhibit lower gain, requiring optimized matching circuits or rectifiers to maintain acceptable sensitivity [28]. Sensitivity, expressed as the minimum input power level (in dBm) at which the rectenna begins effective energy conversion, is a critical metric for evaluating antenna suitability in low-RF environments. Systems designed for ambient RF environments, particularly in rural or shielded settings, benefit from antennas with higher sensitivity and low activation thresholds. There is an inherent trade-off between size and efficiency. While miniaturized designs are ideal for embedded or implantable applications, they often suffer from reduced gain and narrower bandwidth [29]. This necessitates advanced rectifier topologies and power management strategies to compensate for the lower energy input. Moreover, antennas developed using flexible or wearable materials (e.g., printed or textile-based dipoles) offer exciting prospects for biomedical and consumer electronics but may compromise performance under mechanical stress. Hence, material properties should be considered in conjunction with electrical characteristics for real-world deployments.

Table 3. Comparison among different antennas for RFEH systems [9,23–27].

Antenna Type and Frequency	Antenna Structure	Size (mm)	Antenna Gain (dBi)	Circuit Type and Frequency	Efficiency (%)	Input Power (dBm)	Minimum Sensitivity (dBm)
Logarithmic Spiral Antenna (2–18 GHz)		60 × 60	3.2	Half-Wave Rectifier (2.4 GHz)	20	+4	−18
Linear Tapered Slot Antenna (1.85/2.4 GHz)		80 × 40	5.1	Differential Rectifier (1.85 GHz)	13	−15	−22
Concentric Square Patch (2.4/5.5 GHz)		45 × 45	4.0 @ 2.4 GHz	Full-Wave Rectifier	36% @ 2.4 GHz, 5% @ 5.5 GHz	0	−16
Slotted Patch Antenna (2.45 GHz)		35 × 30	5.3	Half-Wave Cockcroft–Walton	68	+5	−21
Meander Antenna (434 MHz)		45 × 25	1.2	Voltage Doubler	20	−30	−35
Fractal Antenna (900 MHz/ 2.45 GHz)		25 × 25	2.5	Doubler/ Full-wave	30 @ −10 dBm	−10	−20
Printed Dipole (2.45 GHz)	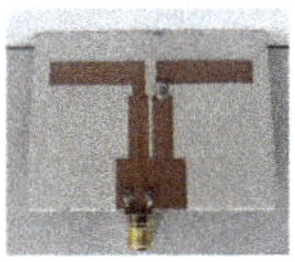	60 × 20	2.0	Full-Wave Rectifier	33	−5	−18

3.2.2. Matching Network

A matching network is a circuit designed to adjust the impedance between two components, typically between an antenna and a rectifier, to ensure efficient power transfer. It plays a critical role in maximizing energy transfer by minimizing signal reflection and ensuring that the maximum possible RF power is delivered from the antenna to the rectifier. Proper impedance matching enhances the overall efficiency of the energy conversion process and significantly improves the performance of RFEH systems [21,30]. Traditional matching networks utilize passive components such as inductors and capacitors configured in various L, T, or π-network topologies, as shown in Figure 4. An L-network consists of two reactive components (inductors and/or capacitors) arranged in either a series-parallel or parallel-series configuration. There are two basic types of L-networks: the series L configuration, where an inductor is placed in series with the signal path and a capacitor is placed in shunt, and the shunt L configuration, where the positions of the inductor and capacitor are reversed. L-networks are simple to design and are typically used for narrowband applications, where the load impedance is either greater or less than the source impedance [31]. A π-network consists of two capacitors connected in parallel (shunt) with the transmission line, while an inductor is placed in series. This configuration forms a structure resembling the Greek letter "π."

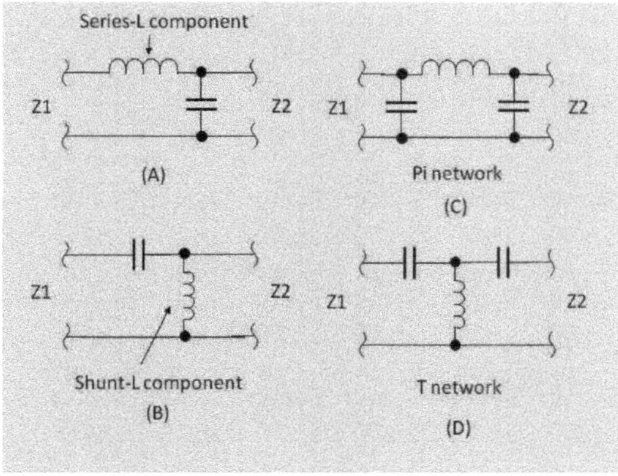

Figure 4. L–C impedance matching networks. (**A**) Series L component. (**B**) Shunt L component. (**C**) Pi-network. (**D**) T-network.

The π-network provides a broader bandwidth compared to L-networks and is effective in high-frequency applications requiring better selectivity and attenuation of unwanted signals. Additionally, it allows for greater flexibility in impedance transformation, making it suitable for dynamic RF environments. A T-network consists of two inductors, connected in series with the transmission path, and a capacitor placed in a shunt to ground at the centre, forming a structure that resembles the letter "T." T-networks are commonly used in wideband applications and provide higher flexibility in impedance matching over a broad frequency range. They also help in filtering out unwanted high-frequency components while maintaining efficient impedance transformation [32]. The selection of an appropriate matching network depends on the application requirements, frequency range, and impedance characteristics of the antenna and rectifier circuit. While L-networks offer simplicity and efficiency for narrowband applications, π- and T-networks provide improved bandwidth performance and greater adaptability for dynamic RF harvesting scenarios. Recent advancements have introduced tunable and reconfigurable matching networks that employ varactors, MEMS switches, and machine learning algorithms to dynamically adjust impedance in real time, further optimizing energy-harvesting efficiency

3.2.3. Rectifier

The rectifier is a crucial component in RFEH, responsible for converting RF signals into DC power. Its efficiency directly impacts the overall energy conversion performance of the system. The rectenna, an integrated structure combining a receiving antenna and a rectifier, as shown in Figure 5, is a key advancement in RFEH technology. Various rectifier configurations, including single-stage, voltage doubler, and multi-stage rectifiers, as shown in Figure 6, have been explored to optimize conversion efficiency. Schottky diodes are widely used in rectifier circuits due to their low forward voltage drop and high switching speed, making them suitable for low-power applications [33,34]. Figure 6a shows a single-diode half-wave rectifier, also known as an envelope detector. This circuit consists of a series diode followed by a shunt capacitor and operates by allowing only one half-cycle of the alternating-current (AC) signal to pass while blocking the other half. However, half-wave rectifiers introduce significant voltage ripples in the rectified output, necessitating additional filtering to reduce harmonic distortions and stabilize the DC voltage. Figure 6b presents a voltage doubler, also referred to as a single-stage voltage multiplier [35]. This configuration functions as a full-wave rectifier, ensuring more stable DC output compared to the half-wave rectifier. The circuit consists of a one-stage clamper with a pumping capacitor at the input, a shunt diode, a series-rectifying diode, and a shunt capacitor at the output. The output capacitor acts as a low-pass filter to smooth out variations in the rectified DC voltage. This setup results in a higher average DC voltage, making it more efficient for energy-harvesting applications. Figure 6c illustrates a multi-stage voltage multiplier, which is essentially an extended version of the voltage doubler. It further increases the output voltage using multiple capacitor–diode stages. Each stage builds upon the voltage of the previous stage, and the maximum output voltage is determined by the total number of stages [36,37]. However, the primary drawback of multi-stage rectifiers is their decreasing overall efficiency. As more stages are added, cumulative power losses occur due to the inherent inefficiencies of each stage, making the design a trade-off between output voltage and energy conversion efficiency.

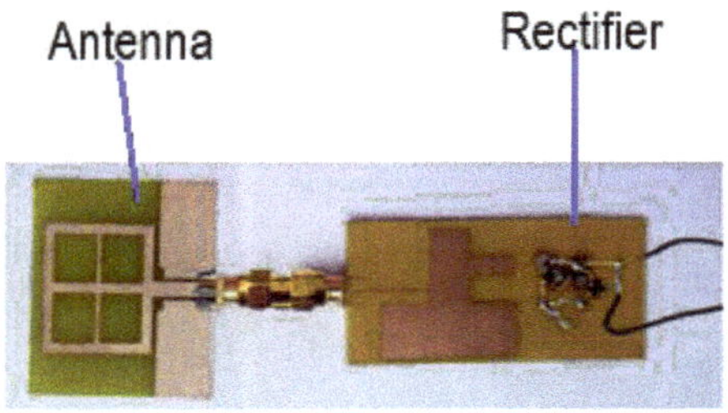

Figure 5. Rectenna [38].

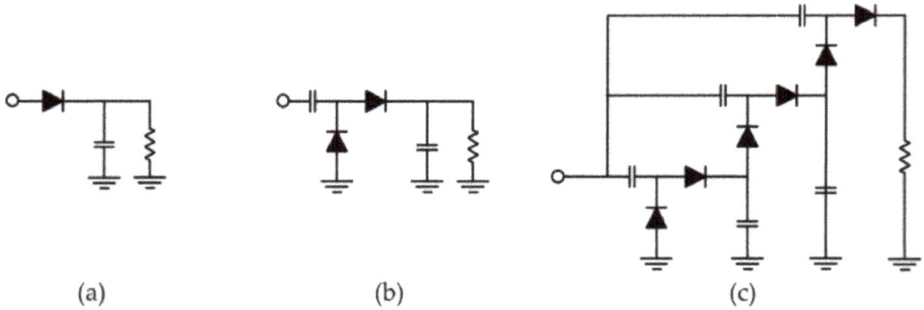

Figure 6. Rectifier topologies. (**a**) Single diode (single stage). (**b**) Voltage doubler. (**c**) Voltage multiplier (multi-stage) [35].

The choice of rectifier topology significantly impacts the efficiency of RF-to-DC conversion. In scenarios where ample AC power is available, multi-stage rectifiers can substantially enhance DC voltage output. However, the reduction in overall power efficiency must be carefully considered when designing RF energy-harvesting systems for practical applications [39,40].

3.2.4. Energy Storage Unit

The energy storage unit plays a crucial role in RFEH systems by ensuring that there is a stable and continuous power supply to electronic devices. Since ambient RF energy is highly intermittent and varies in intensity depending on the surrounding environment, an efficient storage mechanism is needed to accumulate and regulate the harvested energy before it can be utilized. The two primary storage solutions in RF energy-harvesting applications are supercapacitors and rechargeable batteries, each with distinct characteristics suited for different use cases. Supercapacitors are widely used for their ability to charge and discharge rapidly, making them ideal for applications requiring frequent energy bursts. They have a longer lifespan compared to conventional batteries, as they can withstand millions of charge–discharge cycles without significant degradation [41]. However, supercapacitors have lower energy density, meaning they store less energy per unit volume compared to batteries. This makes them more suitable for low-power, short-duration applications such as wireless sensor nodes, IoT devices, and energy buffering in transient-powered systems [42,43]. Rechargeable batteries, such as lithium-ion and lithium-polymer batteries, offer significantly higher energy density, allowing them to store more energy for prolonged operation. This makes them ideal for applications where continuous power is required, such as medical implants, remote monitoring systems, and wearable electronics. However, batteries have longer charging times and a limited number of charge–discharge cycles, which can impact their long-term sustainability. Advances in battery technology, including solid-state batteries and nanomaterial-based electrodes have improved their efficiency, making them a viable option for RF energy-harvesting applications that demand extended operational periods [42,43].

A combination of supercapacitors and batteries is often used in hybrid energy storage systems to balance short-term energy fluctuations and long-term energy supply [44]. Supercapacitors can handle sudden power surges and provide immediate energy when needed, while batteries ensure sustained operation over extended durations [2]. Table 4 provides a clearer understanding of their roles in RF energy-harvesting systems and also helps determine the most appropriate storage solution based on specific application requirements.

Table 4. Comparative evaluation of supercapacitors and batteries [45].

	Supercapacitor	Battery
Recharge cycle lifetime	$> 10^6$ cycles	$< 10^3$ cycles
Fastest charging time	sec~min	Hours
Fastest discharging time	<a few min	0.3~3 h
Self-discharge rate	30%	5%
Voltage	0–2.7 V	3.7–4.2 V
Energy density (Wh/kg)	low (0.8–10)	high (20–150)
Power density (W/kg)	high (500–400)	low (50–300)
Charging circuit	simple	Complex

3.3. RF Energy Sources and Availability

The efficiency of RFEH systems is intricately linked to the availability and strength of ambient RF signals, which serve as the primary energy source. These RF energy sources are broadly categorized into intentional and unintentional emitters. Intentional sources involve dedicated RF power transmitters specifically deployed for wireless power transfer applications, providing a consistent and controllable energy supply. Unintentional sources encompass ambient RF emissions from existing communication infrastructures, including cellular networks, television broadcasts, and Wi-Fi routers, which, while ubiquitous, offer variable energy levels depending on the environment [46,47]. The ambient RF energy landscape varies significantly across different environments. In densely populated urban areas, the proliferation of communication devices and broadcasting systems results in higher RF power densities. Conversely, rural or remote locations, with fewer such devices, exhibit lower RF energy availability. Studies have indicated that RF power density can range from approximately 0.1 μW/cm^2 in rural settings to over 10 μW/cm^2 in urban centres [8].

To effectively harness ambient RF energy, especially given its fluctuating nature, researchers have proposed adaptive techniques such as dynamic spectrum sensing and intelligent frequency selection. These methods enable energy-harvesting devices to identify and utilize the most energy-rich frequencies available in real time, thereby optimizing the energy capture process. Such adaptive strategies are crucial for enhancing the viability of RF energy harvesting in environments where RF energy availability is inconsistent [48,49]. Visual representations of RF energy availability, such as heatmaps, offer valuable insights into the distribution of ambient RF energy across various environments. Several studies have conducted measurements and developed these visualizations to assess RF power density. A study conducted in Montreal, for instance, measured ambient RF power levels along major streets and highways, providing critical data for understanding RF energy distribution in urban settings. The findings shown in Figure 7 reinforce the feasibility of RF energy harvesting in densely populated commercial districts, where multiple cellular networks operate at higher power levels. However, the study also highlights significant variations in RF power levels across different frequency bands. Notably, GSM and LTE bands exhibit stronger RF signals compared to wireless fidelity (Wi-Fi) and Digital Television (DTV), indicating that RF energy-harvesting systems should prioritize cellular bands for optimal efficiency rather than relying on weaker signals [50,51].

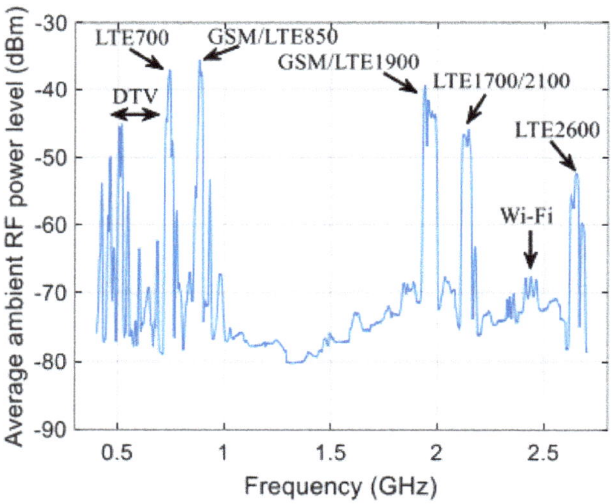

Figure 7. Ambient RF power levels [50].

In contrast, RF power levels in rural areas tend to be significantly lower due to fewer base stations and lower communication activity. A study conducted in Geidam Town, Yobe State, Nigeria, measured RF power densities around mobile telephony base stations, reporting values ranging from 0.3017 mW/m^2 to 1.265 mW/m^2 at distances between 10 m and 100 m from the stations. These values are substantially lower than those recorded in urban environments, reinforcing the challenge of RF energy harvesting in rural areas. The disparity in RF power levels between urban and rural regions underscores the need for adaptive and efficient energy-harvesting techniques tailored to different environments [52,53].

3.4. Performance Metrics

Evaluating the performance of RFEH systems relies on key metrics such as RF-to-DC conversion efficiency, sensitivity, and power density. RF-to-DC conversion efficiency measures the proportion of captured RF energy that is successfully converted into usable DC power. Optimizing impedance matching networks, rectifier circuits, and energy storage components is essential for improving this efficiency. Recent advancements have led to significant improvements in rectenna performance. As shown in Figure 8, a rectifier circuit achieved a peak efficiency of 88% at an input power of 14 dBm while maintaining over 60% efficiency within the input power range of 3 dBm to 16 dBm at 2.4 GHz. Similarly, at 5.8 GHz, it attained a maximum efficiency of 81% at 10 dBm input power, sustaining efficiency above 60% for input power levels between 7 dBm and 12 dBm [54]. Another study reported a rectenna system with a conversion efficiency of 66.7% at 2.4 GHz under an input power of 5 dBm [38]. Additionally, a dual-band rectenna design demonstrated peak conversion efficiencies of 72% at 1.8 GHz and 65% at 2.45 GHz [55]. These findings highlight ongoing advancements in rectenna design to enhance RF energy-harvesting capabilities.

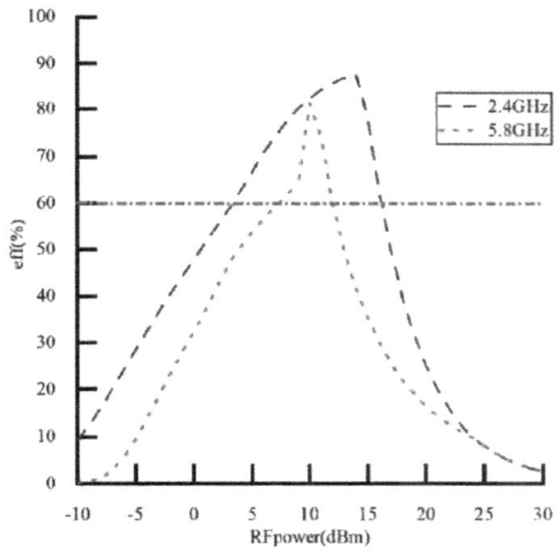

Figure 8. Performance metrics of RFEH system for RF-to-DC conversion efficiency [54].

Sensitivity defines the minimum RF power level required for an energy-harvesting system to function effectively. Devices with higher sensitivity can operate in low-power RF environments, making them more versatile. For instance, a 2.4 GHz RFEH CMOS chip demonstrated a sensitivity of −19 dBm, achieving a peak efficiency of 51% and a power dynamic range of 24 dB, as reported in [56]. Power density refers to the amount of energy harvested per unit area of the receiving antenna, directly impacting the system's overall energy collection potential. Maximizing power density requires the optimization of antenna gain, the selection of suitable operating frequencies, and the strategic positioning of rectennas to maximize exposure to RF sources. Studies indicate that harvested power levels vary significantly based on signal strength and distance from RF sources, reinforcing the need for adaptive energy-harvesting strategies [8].

4. RF Energy-Harvesting Techniques

RFEH encompasses various techniques that enable the conversion of ambient RF signals into usable electrical power. These techniques differ in frequency range, circuit complexity, and integration with other energy sources. This section discusses key approaches, including single-band vs. multi-band harvesting, narrowband vs. broadband harvesting, passive vs. active systems, and hybrid energy-harvesting methods that integrate RF energy with solar, vibration, and thermal sources [57].

4.1. Single-Band vs. Multi-Band RF Energy Harvesting

RFEH systems are classified based on their ability to capture energy from either a single frequency band or multiple frequency bands. Single-band systems are designed to operate at a specific frequency, such as 2.4 GHz or 5.8 GHz, which are commonly used in Wi-Fi, industrial, and ISM (Industrial, Scientific, and Medical) band applications. These systems optimize impedance matching networks and rectifier circuits to maximize efficiency at a single frequency, ensuring minimal losses. Typical antennas used for single-band RREH are microstrip patch antennas, monopole antennas, and dipole antennas. They leverage the principle of resonance, whereby the antenna and matching circuit are finely tuned to a specific frequency, achieving maximum power transfer. The high-Q (quality factor) design enables narrowband operation with minimal signal loss, enhancing the energy harvested per unit of RF power. Studies have demonstrated high conversion efficiencies in single-band designs. A rectifier designed for 2.4 GHz operation, shown in Figure 9a, achieved a conversion efficiency of 62.4% at an input power of −10 dBm and a conversion efficiency of 31.1% at an input power of −20 dBm, incorporating a voltage booster circuit to enhance performance at low power levels [58]. Another optimized rectifying circuit for RF energy scavenging at 2.45 GHz reached a conversion efficiency of 31.16% at an input power of 20 dBm, delivering an output voltage of 10.32 V [59]. These advancements highlight the progress made in enhancing rectifier circuit efficiencies, particularly under conditions with varying input power. However, single-band systems are limited in their applicability as they rely on a single RF source, making them less effective in environments where RF energy levels fluctuate. In addition, their narrow frequency targeting makes them susceptible to detuning due to temperature variations or material inconsistencies, which can degrade performance. These systems are best suited for predictable environments with stable RF sources, where maximum efficiency at one frequency outweighs adaptability.

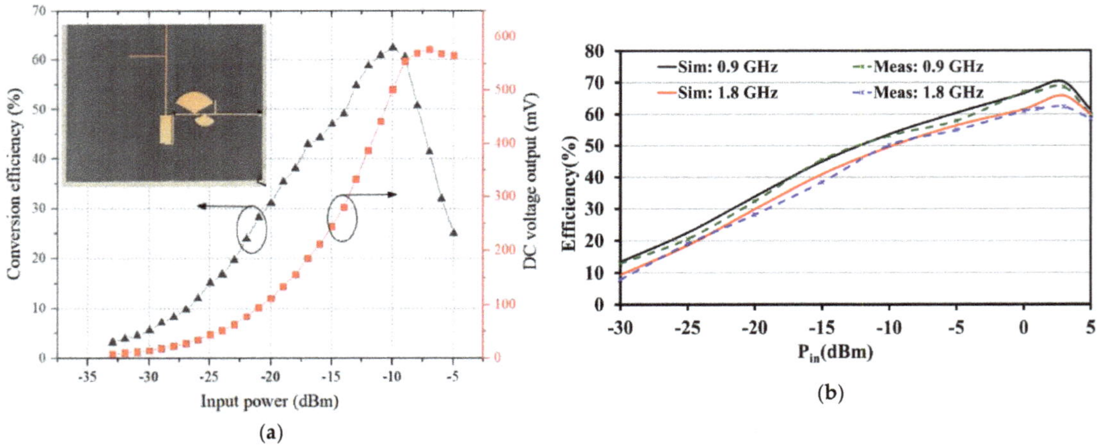

Figure 9. RF energy-harvesting system. (**a**) Single band [58]. (**b**) Multiple bands [62].

In contrast, multi-band RFEH systems operate across multiple frequency bands, allowing them to harvest energy from diverse RF sources such as GSM at 900 MHz and 1.8 GHz, LTE ranging from 700 MHz to 2.6 GHz, and Wi-Fi at 2.4 GHz and 5 GHz. The typical antennas used for multi-band RF energy harvesting are multi-band microstrip patch antennas, PIFAs, and fractal antennas. The underlying design principle in multi-band systems involves creating antennas and rectifiers that resonate at multiple frequencies. This can be achieved through multi-resonant structures, nested antenna elements, or fractal geometries that support various harmonic modes. Recent studies have demonstrated high-efficiency multi-band rectennas. Maher et al. [34] reported that a dual-band rectenna operating at 1.8 GHz and 2.45 GHz achieved measured peak conversion efficiencies of 72% at 1.8 GHz (input power of 4 dbm) and 65% at 2.45 GHz (input power of 3 dbm), as shown in Figure 9b [60]. Another study designed a triple-band rectenna operating at 915 MHz, 1.8 GHz, and 2.1 GHz frequency bands. The proposed rectenna achieved efficiencies greater than 52% when harvesting RF energy from outdoor sources. The design incorporates a modified E-shaped patch antenna and an optimized impedance matching network to enhance performance across the targeted frequency bands [61]. While multi-band RF energy harvesters offer greater flexibility, particularly in urban and industrial environments where RF signals vary across multiple frequencies, they also introduce additional design complexity. Efficiency variations across different frequency bands require careful impedance matching and rectifier optimization to ensure stable energy conversion across multiple bands. Each added frequency band increases the challenge of minimizing cross-band interference and achieving compact form factors, particularly when integrating tunable matching circuits or broadband rectifiers. This complexity can impact size, cost, and integration with existing wireless sensor networks. The integration of broadband antennas and tunable matching networks is a promising approach to address these challenges.

A comparison of the two approaches reveals key trade-offs. Single-band RFEH systems offer high efficiency within the targeted frequency band and a simpler circuit design, making them suitable for dedicated power transfer applications such as RFID and wireless sensors operating in a fixed-frequency environment. However, they are ineffective in scenarios where the targeted RF source is unavailable. Multi-band RFEH, on the other hand, enables energy collection from multiple sources, making it more adaptable to dynamic RF environments. Nevertheless, the trade-off lies in the difficulty of maintaining consistent performance across a wide range of frequencies, often requiring multi-stage rectification circuits or switchable matching networks, which can increase power losses and reduce net efficiency. This advantage comes at the cost of increased design complexity, as multi-band rectennas require sophisticated impedance matching networks to maintain efficiency across different frequencies.

4.2. Narrowband vs. Broadband RF Energy Harvesting

RFEH systems can be categorized based on their operational bandwidth into narrowband and broadband systems. Narrowband systems are engineered to capture and convert RF energy at specific, predetermined frequencies (a few MHz around the centre frequency), optimizing components such as impedance matching networks and rectifiers to achieve high efficiency at these targeted frequencies. This design approach ensures minimal energy loss and is particularly effective in environments where the RF source operates consistently at a known frequency. High-Q microstrip patch antennas, helical antennas, and loop antennas are the typical antennas used for narrowband RF energy-harvesting systems. The high-Q nature of these antennas means they are resonant over a very narrow frequency band, which leads to increased voltage across the rectifier and thus higher conversion efficiency. This narrow resonance also allows for simplified impedance matching circuits, which are easier to design and fabricate. However, the primary limitation of narrowband systems lies in their restricted frequency range, which renders them less effective in dynamic environments where RF energy sources vary in frequency. In contrast, broadband RFEH systems are designed to operate over a wide range of frequencies (from hundreds of MHz to GHz), enabling them to capture energy from multiple RF sources simultaneously. This capability is particularly advantageous in urban and industrial settings, where numerous RF signals are present across various frequencies. Vivaldi antennas, spiral antennas, log-periodic dipole arrays, bowtie antennas and ultra-wideband patch antennas are the typical antennas used for broadband RFEH systems. These antennas are inherently frequency-independent or support multiple resonant modes, allowing them to respond to a broad spectrum of incoming signals. Their geometrical designs are often based on self-similarity or logarithmic scaling, which enables effective energy reception across wide frequency bands.

A notable example of broadband design was presented by the authors in [60] where a quad-band RF energy harvester was developed to operate within the 0.8 GHz to 2.6 GHz frequency range. The proposed system incorporated a self-complementary log-periodic antenna and an improved impedance matching network, achieving a DC rectification efficiency of approximately 35% at an input power of −15 dBm, which went up to 52% at −20 dBm. These results underscore the potential of broadband systems to effectively harness ambient RF energy across multiple frequency bands. The performance gains seen in this system were attributed not only to the antenna design but also to a dynamic impedance matching network capable of adapting to varying signal strengths and frequencies, ensuring a relatively stable DC output even under fluctuating RF conditions. Despite their versatility, broadband systems face challenges related to design complexity. Achieving efficient impedance matching across a wide frequency spectrum is inherently more complex than in narrowband systems. Additionally, the integration of broadband antennas and rectifiers requires meticulous design to maintain consistent performance across all targeted frequencies. This often involves trade-offs such as using wideband matching techniques like multisection transformers or active tuning mechanisms, which can increase circuit size, power consumption, or component count. Maintaining a flat response across the frequency range is difficult, and performance may drop significantly at frequency extremes. A comprehensive review by the authors in [13,63] highlights various impedance matching network configurations and their impact on the performance of broadband RF energy harvesters, emphasizing the need for advanced design methodologies to address these challenges. The choice between narrowband and broadband RF energy-harvesting systems involves a trade-off between efficiency at specific frequencies and versatility across multiple frequencies. Narrowband systems offer high efficiency and simpler design but are limited to specific RF sources, making them suitable for applications with known, stable frequency environments. Conversely, broadband systems provide greater adaptability in capturing energy from diverse RF sources but require more complex designs to ensure consistent performance across a wide frequency range [64,65]. In scenarios such as applications of wireless sensor networks, mobile health monitoring, and smart city deployments, the ability of broadband systems to harvest energy opportunistically from multiple sources may outweigh their slightly lower peak efficiencies, particularly when energy availability is unpredictable. The decision between these approaches should be guided by the specific application requirements and the characteristics of the RF environment in which the system will operate.

4.3. Passive vs. Active RF Energy Harvesting

RFEH systems can be broadly categorized into two types: passive and active. Passive systems rely solely on ambient RF energy present in the environment, capturing and converting it into usable electrical power without the aid of external power sources. In contrast, active systems incorporate external power sources to enhance their energy-harvesting capabilities, often employing amplifiers or other active components to boost performance. Passive RF energy harvesters capture ambient RF signals emitted from various sources such as mobile phones, Wi-Fi routers, and broadcasting stations. These systems typically consist of an antenna to capture RF energy, an impedance matching circuit to maximize power transfer, a rectifier to convert AC signals into DC signals, and an energy storage unit. Typical antennas used for passive RF energy harvesting are dipole, simple patch, and omnidirectional antennas. The energy captured in passive systems is often in the microwatt range, making ultra-low-power circuit design essential. The careful tuning of the impedance matching network is critical to ensure maximum voltage transfer to the rectifier. Energy storage units such as supercapacitors or thin-film batteries are used to accumulate harvested energy for intermittent power delivery. One notable design improvement in passive systems is the incorporation of harmonic harvesting loops, which can enhance rectifying efficiency by over 25% compared to conventional rectifiers. This improvement allows for more efficient energy harvesting from ambient RF signals within the 825 MHz to 2.5 GHz range [57]. These loops operate by capturing and recycling harmonic components of the RF signal, which would otherwise be dissipated, thereby increasing the total harvested energy. They also help stabilize the output voltage by reducing spectral leakage.

Active RFEH systems utilize external power sources to enhance their energy-harvesting capabilities. By integrating amplifiers with the antenna feed line, these systems can significantly improve the dynamic range and efficiency of energy harvesting. For instance, active wearable energy-harvesting systems have demonstrated substantial bandwidth and gain improvements, achieving an active antenna gain of 24 $\pm$ 2.5 dB for frequencies ranging from 200 MHz to 900 MHz. This enhancement is particularly beneficial for wearable sensors in medical and sports applications, where consistent and reliable energy harvesting is crucial. The inclusion of low-noise amplifiers (LNAs) or power amplifiers (PAs) in the signal chain compensates for weak ambient signals and enables operation even in RF-sparse environments. This also allows active systems to respond dynamically to changes in the RF environment by tuning or steering the antenna pattern. Typical antennas used for active RF energy harvesting are tunable/reconfigurable microstrip, electronically steered antennas. These antennas are capable of frequency agility or beamforming, enabling the system to selectively focus on the strongest RF sources and dynamically adapt to environmental changes. This adaptability extends the operational range and improves link reliability, especially in mobile or obstructed scenarios. The choice between passive and active RF energy-harvesting systems depends on the specific application requirements and environmental conditions. Passive systems are advantageous in scenarios where simplicity, low cost, and maintenance-free operation are prioritized. They are suitable for powering low-energy devices such as wireless sensors and IoT devices in environments with sufficient ambient RF energy. However, their performance is limited by the availability and strength of ambient RF signals. To compensate for low input power levels, passive systems often employ highly efficient rectifier topologies, such as voltage doublers or Schottky diode-based circuits, and require careful PCB layout design to minimize losses. Active systems, while more complex and requiring an external power source, offer enhanced performance and reliability [49,58,66]. They are ideal for applications demanding higher power levels or operating in environments with low ambient RF energy. The integration of active components allows for the better control and optimization of the RFEH process, making them suitable for critical applications such as medical implants and wearable electronics. For example, closed-loop control circuits in active systems can dynamically adjust load impedance or antenna parameters based on real-time input, improving conversion efficiency and enabling smart energy management.

4.4. Hybrid Energy Harvesting

Hybrid energy-harvesting systems are designed to capture and convert energy from multiple ambient sources, such as solar radiation, mechanical vibrations, and thermal gradients, into electrical power. By integrating different energy conversion mechanisms, these systems enhance overall efficiency and reliability, particularly in environments where a single energy source may be inconsistent or insufficient. The multi-modal design of hybrid harvesters allows them to operate across a wider range of environmental conditions, increasing their duty cycle and enabling continuous power delivery, even when one energy source becomes unavailable. This robustness is critical in off-grid or intermittently powered applications. The integration of solar and vibration-based energy harvesting has been explored as a means of capturing energy from both light and mechanical motion. A notable example is the development of hybrid harvesters that combine photovoltaic (PV) cells with piezoelectric, electromagnetic, or electrostatic mechanisms. This approach allows for simultaneous energy conversion from sunlight and mechanical oscillations, resulting in improved performance over standalone energy-harvesting systems. For instance, PV cells generate electricity from photons, while piezoelectric components exploit strain-induced charge separation to harvest energy from vibration or motion. Electromagnetic harvesters use relative motion between magnets and coils to induce current, and electrostatic systems rely on variable capacitance. Their joint operation can offer complementary power outputs—steady outputs from solar energy, and pulsed outputs from vibrations. Recent studies have demonstrated that such hybrid systems can significantly enhance energy output, making them suitable for self-powered devices in remote locations [67–69]. Hybrid systems that merge solar and thermal energy harvesting leverage the synergy between photovoltaic cells and thermoelectric generators (TEGs). Photovoltaic thermal hybrid solar collectors (PVT collectors) exemplify this approach by integrating solar cells with thermal collectors to convert solar radiation into both electrical and thermal energy. This combination increases overall efficiency by utilizing waste heat from photovoltaic cells to generate additional power. Traditional PV cells suffer from thermal losses, which reduce electrical conversion efficiency. By harvesting this waste heat using TEGs, PVT systems increase total energy yield. The TEGs exploit the Seebeck effect, where voltage is generated across two different conductors in the presence of a temperature gradient, to recover otherwise lost thermal energy. Research on PVT collectors has shown that they provide a higher energy yield compared to standalone photovoltaic or thermal systems, making them an attractive option for renewable energy applications. The integration of vibration and thermal energy harvesting has also been investigated to take advantage of both piezoelectric and thermoelectric effects [70,71]. Piezoelectric materials convert mechanical vibrations into electricity, while thermoelectric materials generate power based on temperature differences. Researchers have developed hybrid systems that combine these two mechanisms, resulting in increased power generation compared to single-source energy harvesters. These hybrid configurations are particularly suitable for environments like industrial plants or vehicles, where both mechanical vibrations and thermal gradients naturally coexist. Efficient coupling between the two subsystems is critical to avoid energy loss and to allow simultaneous operation without signal interference or thermal degradation. Such systems have demonstrated improved energy capture efficiency and have potential applications in wireless sensor networks and wearable electronics. For example, a hybrid piezo-thermoelectric energy harvester embedded in a shoe sole could generate electricity from both walking motion and foot-sole temperature, enabling the continuous monitoring of gait or health parameters without requiring battery replacement.

The typical antennas used for hybrid RFEH are flexible, transparent, printable, conformal antennas. These advanced antenna designs support the integration of RFEH into hybrid platforms by ensuring mechanical adaptability, esthetic compatibility (e.g., wearables), and low fabrication costs. Their material flexibility allows them to be embedded in curved or irregular surfaces, such as clothing, medical implants, or building materials. Hybrid energy-harvesting systems offer multiple advantages, including enhanced energy capture, achieved by tapping into different ambient energy sources; improved overall efficiency due to the integration of various conversion mechanisms; and greater adaptability for diverse applications such as remote sensors and medical implants. However, the design and implementation of hybrid energy harvesters present certain challenges, including the increased complexity in ensuring compatibility among different energy-harvesting mechanisms, potential increases in size and weight, and higher production costs due to the use of multiple materials and technologies. The integration process demands careful electrical isolation, thermal management, and power conditioning to avoid mutual interference and to optimize overall output. System miniaturization and unified control circuitry remain key areas of ongoing research. Despite these challenges, hybrid energy-harvesting systems represent a promising approach to sustainable and reliable power generation. Ongoing research and technological advancements continue to refine these systems, making them more efficient and cost-effective for real-world applications. Innovative fabrication techniques such as inkjet printing, MEMS integration, and 3D packaging are enabling more compact, scalable, and customizable hybrid harvesters tailored to specific deployment scenarios. Table 5 shows the comparison of the various RFEH techniques described in Section 4.

Table 5. Comparison of RFEH techniques.

Aspect	Single-Band	Multi-Band	Narrowband	Broadband	Passive	Active	Hybrid
Operating Frequency	Single fixed frequency (e.g., 2.4 GHz, 5.8 GHz)	Multiple distinct frequencies (e.g., 900 MHz, 1.8 GHz, 2.4 GHz)	Very narrow frequency range around centre frequency	Wide range (hundreds of MHz to several GHz)	Ambient RF signals only	Ambient RF signals + assisted with amplifiers	Combination of RF, solar, vibration, and thermal energy sources
Typical Antennas Used	Microstrip patch, dipole, monopole	Fractal antennas, multi-band patch antennas, PIFAs	High-Q microstrip patch, helical, loop antennas	Vivaldi, spiral, LPDA, bowtie, ultra-wideband (UWB) patch	Dipole, simple patch, omnidirectional antennas	Reconfigurable microstrip antennas, electronically steered antennas	Flexible, transparent, conformal, printable antennas
Design Complexity	Simple and optimized	High (requires separate tuning or broadband matching)	Relatively simple	Complex (requires broadband matching and rectification)	Simple	Moderate to high (additional power circuitry needed)	High (multiple harvesting units and integration circuits)
Conversion Efficiency	High at target frequency	Varies across bands, needs optimization	Very high at centre frequency	Moderate to high (depending on design)	Moderate (depends on ambient signal strength)	Higher than passive systems under weak RF conditions	Varies based on environmental energy source combination
Adaptability to RF Environments	Low (only effective at specific frequency)	High (can capture different signals)	Low (best in controlled frequency environment)	Very high (suitable for dynamic environments)	Depends on environment	More adaptable due to power amplification	Very high (suitable for dynamic and harsh environments)
Power Output	Depends on presence of single RF source	Generally better (due to harvesting from multiple sources)	Stable if RF source is constant	Fluctuates depending on wideband ambient RF availability	Moderate	Improved and stabilized by amplifiers	Typically higher due to multiple energy sources being combined
Typical Applications	RFID tags, dedicated wireless sensors	Smart cities, industrial IoT, dynamic wireless systems	IoT devices in controlled environments (e.g., factories)	Urban RF harvesting, smart sensors, IoT devices in dense areas	Low-power IoT nodes, environmental sensors	Medical implants, wearable electronics	Remote sensing, smart agriculture, biomedical implants
Challenges	Limited flexibility to changing environments	Complexity in circuit and impedance matching	Limited to known RF source environments	Difficult impedance matching and efficiency management	Limited by ambient RF power density	Complexity, size, and energy cost of amplification	Integration issues, size, cost, and weight
Typical Antenna Gain (dB)	2–4 dB	4–8 dB	5–10 dB	8–15 dB	2–3 dB	5–10 dB	3–8 dB
Circuit Performance (Efficiency)	85–90%	75–85%	90–95%	60–80%	30–50%	60–80%	70–85%
Size (cm^2)	20–40 cm^2	50–100 cm^2	10–20 cm^2	100–200 cm^2	5–20 cm^2	20–50 cm^2	50–150 cm^2
Energy Efficiency	85–90%	70–80%	90–95%	60–75%	30–50%	60–80%	70–85%

5. Recent Developments in RF Energy Harvesting

Recent advancements in RFEH have focused on improving efficiency, scalability, and adaptability in diverse applications. The field has witnessed significant breakthroughs in rectenna design, impedance matching circuits, metamaterial integration, AI-based optimization, and SWIPT systems. These developments have driven the expansion of RF energy-harvesting technology for applications such as the IoT, smart cities, and wireless sensor networks. These developments are particularly crucial in addressing the limitations of traditional passive and active systems, as discussed in Section 4.3. For instance, AI-optimized rectenna designs can dynamically adjust to weak ambient RF environments, enhancing performance without relying entirely on external power sources.

5.1. Advances in Rectenna Design and Optimization

Rectennas, integral to RFEH systems, have seen significant advancements aimed at enhancing their efficiency, compactness, and multi-frequency capabilities. A notable development is the dual-band, dual-polarized rectenna operating efficiently at 2.4 GHz and 5.8 GHz. This design achieves a conversion efficiency exceeding 30% at 3.6 GHz for input ranges from -10 to 6.5 dBm, and one over 20% at 5.8 GHz for input ranges from -13 to 3 dBm [72]. The incorporation of dual polarization enhances energy capture from various orientations, making it suitable for powering IoT devices in environments with diverse RF signals. Similarly, a dual-band rectenna that operates efficiently at 2.4 GHz and 5.8 GHz was developed, demonstrating a high conversion efficiency of over 70% at -10 dBm input power [54]. The design incorporates a novel impedance matching network that adapts to different frequency bands, ensuring optimized power conversion efficiency across Wi-Fi and ISM bands. Another significant innovation is the development of a triple-band rectenna designed for energy harvesting at 915 MHz, 1.8 GHz, and 2.45 GHz. This rectenna integrates a high-gain antenna with a voltage doubler rectifier, enabling efficient energy scavenging from multiple RF sources. Such designs are particularly beneficial for IoT applications within smart office environments, where multiple communication standards are prevalent [54]. These recent multi-band and adaptive designs represent a notable evolution over conventional passive systems that are often limited to narrow frequency bands and fixed impedance characteristics. The added flexibility improves harvesting capability, even in environments with fluctuating signal strength. The emergence of flexible and wearable electronics has driven the development of rectennas using novel materials. Researchers have reported highly flexible graphene-film-based rectennas capable of wireless energy harvesting [73]. These rectennas maintain high performance even under significant bending, making them ideal for integration into wearable devices and flexible electronics. The integration of compact rectifier circuits using Schottky diodes and GaN-based transistors has also contributed to higher efficiency at low RF input power levels. Compared to traditional rigid structures in passive and hybrid systems (as discussed in Section 4.4), flexible rectennas offer a significant advantage in applications requiring conformability and durability under mechanical stress, such as biomedical wearables.

5.2. Impedance Matching and Power Management Circuits

Impedance matching is critical for maximizing RF energy transfer from the antenna to the rectifier. At a fundamental level, proper matching ensures that the antenna's complex impedance is the complex conjugate of the rectifier input, thereby minimizing reflection ($|\Gamma| \rightarrow 0$) and maximizing power delivered to the load. Recent studies have introduced adaptive impedance matching techniques that adjust in real time to variations in RF power levels and frequency shifts. A dynamic impedance matching circuit based on varactor diodes was proposed to enhance energy transfer across varying RF environments. It employs a three-state tuning scheme that adapts to variations in the RF-to-DC converter's input impedance, resulting in enhanced power extraction efficiency [74]. The varactor approach exploits the voltage-dependent capacitance of the diode to shift the network's resonant frequency on the fly; its main strength is rapid, continuous tuning, but drawbacks include added insertion loss and potential non-linearity at high RF power. The circuit adaptively adjusts the impedance based on the incident RF power, resulting in an efficiency improvement of up to 30% compared to traditional fixed impedance matching designs. Another approach involves integrating AI-based adaptive impedance matching circuits, which utilize machine learning models such as Convolutional Neural Networks (CNNs) and Long Short-Term Memory (LSTM) networks to predict optimal impedance settings for different ambient RF conditions [75]. These data-driven techniques excel in highly dynamic, multi-band environments where analytical tuning rules break down; however, they require on-chip processing or periodic cloud computation, adding silicon area and quiescent power overhead. In power management, the latest RF harvesting circuits incorporate ultra-low-power DC-DC converters to optimize the voltage levels required for energy storage and utilization. Synchronous buck/boost topologies with sub-μA quiescent currents are now common, ensuring that the converter's consumption does not negate the harvested energy. A novel maximum power point tracking (MPPT) circuit was introduced, which dynamically adjusts the rectifier's load resistance to extract the maximum available power from the ambient RF signals [76]. While MPPT boosts net energy by 10–25% in fluctuating fields, it introduces control-loop latency and demands the precise sensing of rectifier voltage/current, which can be challenging at μW power levels. An equivalent circuit model has been developed that closely aligns with the mechanical model of piezoelectric energy harvesters, facilitating improved impedance matching and power management [77]. This cross-domain modelling unifies electrical and mechanical perspectives, but its practical adoption hinges on the accurate parameter extraction of real-world piezoelectric materials, which can vary with age and temperature. These improvements enhance the overall efficiency of RFEH systems in dynamic environments.

5.3. Metamaterial-Based RF Energy Harvesting

Metamaterials are engineered materials with customized electromagnetic properties that are widely applied to enhance the efficiency of RFEH systems. By leveraging artificial resonant structures, metamaterial-based energy harvesters can concentrate RF waves and significantly increase the harvested power. A recent breakthrough introduced a metamaterial absorber-enhanced rectenna, which improved power conversion efficiency by over 40% compared to conventional designs [78]. The structure utilized a polarization-insensitive metamaterial, ensuring efficient energy capture across multiple angles of incidence. Another study proposed a wideband metamaterial-integrated rectenna capable of harvesting energy from 900 MHz to 6 GHz, demonstrating a conversion efficiency of 68% across all bands [79]. This broadband operation is attributed to multi-resonant unit cells that overlap spectrally, allowing concurrent energy capture across heterogeneous RF sources such as Wi-Fi, cellular, and radar systems. Moreover, the introduction of reconfigurable metamaterials has enabled tunable RF energy harvesting. A smart metamaterial array was developed to dynamically adjust its resonance frequency based on the surrounding RF spectrum, optimizing energy capture in real time [80]. Such tunability is often achieved using embedded varactors, MEMS switches, or tunable substrates like liquid crystal layers, enabling frequency agility without significant structural modification. From a practical standpoint, metamaterial-based harvesters offer benefits such as miniaturization, angular stability, and selective filtering. However, limitations include the narrow intrinsic bandwidth of unit cells and added fabrication complexity due to active reconfiguration. This innovation presents a compact, wide-angle, polarization-insensitive metamaterial harvester capable of efficiently harvesting electromagnetic energy in the S, C, X, and Ku bands. These advancements have significantly improved the adaptability and performance of RFEH systems in urban and industrial environments.

5.4. AI and Machine Learning in RF Energy-Harvesting Optimization

The integration of AI and machine learning has revolutionized RFEH by enabling adaptive and intelligent power management. AI-driven algorithms are now used to optimize energy-harvesting efficiency based on predictive RF energy availability. One of the major applications of AI in RF energy harvesting is spectrum prediction. A machine learning-based model was developed to analyze real-time RF spectrum data and predict the optimal frequency bands for energy harvesting [81]. The model improved energy-harvesting efficiency by 25% by dynamically switching between different frequency bands. Another key innovation is the use of reinforcement learning for adaptive impedance matching and power management. A deep reinforcement learning algorithm was designed to adjust rectifier circuit parameters based on real-time energy-harvesting conditions, achieving significant efficiency gains over conventional methods [82]. Such algorithms commonly use reward functions that maximize output voltage or power while minimizing response time. They are trained in simulated environments that mimic stochastic RF conditions, such as mobile or urban scenarios. AI has also been applied in power distribution and storage optimization. A neural network-based energy management system was introduced to predict energy demand and distribute harvested energy efficiently across multiple IoT devices [83]. In addition, federated learning frameworks have emerged to enhance distributed energy management by allowing local AI models on edge devices to collaboratively improve global energy-harvesting strategies without centralized data sharing. These AI-driven solutions contribute to the automation and efficiency enhancement of RF energy-harvesting networks.

5.5. Simultaneous Wireless Information and Power Transfer

Simultaneous wireless information and power transfer (SWIPT) is an emerging technology that enables the concurrent transmission of energy and data over the same RF spectrum. This approach is highly beneficial for next-generation wireless networks and IoT applications. A recent study developed a SWIPT-enabled rectenna that operates at 2.45 GHz and achieves an energy-harvesting efficiency of 65% while simultaneously decoding data signals with minimal interference [84]. The system employs power-splitting and time-switching architectures to optimize both power transfer and data communication. Another innovative SWIPT approach utilizes MIMO (Multiple-Input Multiple-Output) techniques to enhance energy and data transfer efficiency. A SWIPT-MIMO system was proposed for 5G networks, demonstrating a 35% improvement in power transfer efficiency compared to conventional SWIPT designs [80]. The system dynamically allocates power and information-carrying signals based on the user's energy demand. Recent SWIPT systems also leverage hybrid relaying protocols and cooperative communication strategies to improve energy fairness and spectrum efficiency across multi-user environments. The integration of SWIPT with AI-driven optimization further enhances performance. A predictive AI model was integrated into a SWIPT-enabled IoT network to optimize resource allocation, leading to a 30% improvement in energy-harvesting efficiency. This model incorporates real-time contextual awareness of user mobility and signal fading, enabling the development intelligent modulation and coding schemes for dual-function transmission. Additionally, advanced SWIPT channel models have been developed to simultaneously enhance the Age of Information (AoI) and energy-harvesting timeliness. This was achieved through an online adaptive packet transmission strategy based on the Robbins–Monro algorithm, which dynamically adjusts transmission parameters to maintain optimal system performance in fluctuating environments. These innovations reinforce the status of SWIPT as a pivotal technology for the future of wireless communication and energy-harvesting applications [85].

6. Applications of RF Energy Harvesting

RFEH has emerged as a transformative solution that enables electronic devices to operate independently by capturing ambient electromagnetic energy and converting it into usable electrical power. Its significance spans multiple industries, including smart cities, biomedical engineering, industrial IoT, agriculture, and even space communications, as shown in Figure 10. Recent advancements have further accelerated its integration into real-world applications, paving the way for sustainable, maintenance-free, and efficient device networks.

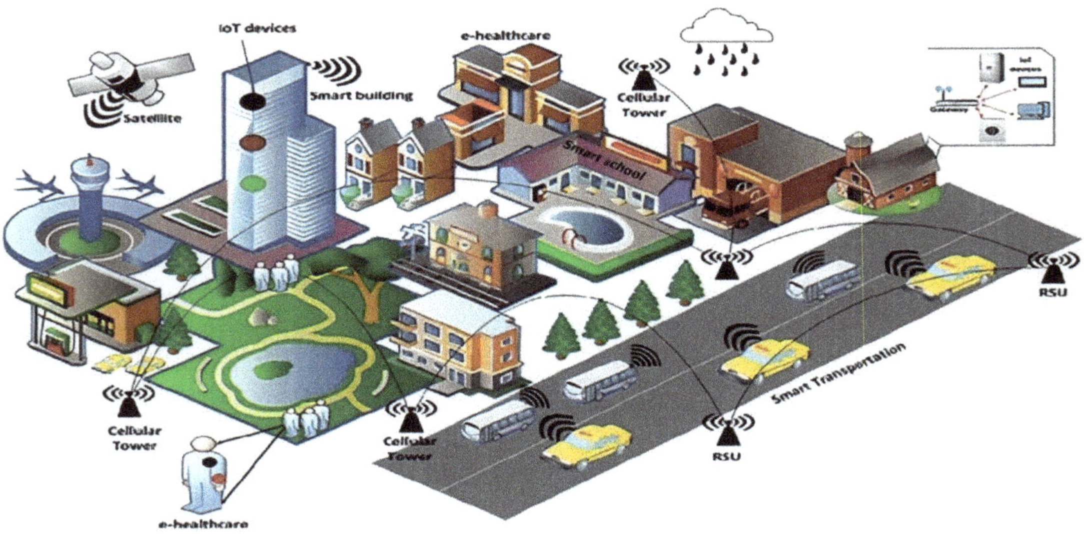

Figure 10. RFEH applications [86].

6.1. Smart Cities

In smart city infrastructure, RF energy harvesting plays a critical role in achieving sustainable urban development. By embedding RF energy harvesters into streetlights, security cameras, and IoT sensors, cities can establish large-scale wireless networks without relying on extensive cabling or frequent battery replacements. A recent study demonstrated the deployment of RF-powered streetlights capable of continuous operation by harvesting ambient energy from nearby 5G base stations, significantly reducing maintenance costs and enhancing environmental sustainability [87]. For instance, Seoul, South Korea, has initiated the installation of multifunctional smart poles that integrate street lighting, Wi-Fi access points, and CCTV cameras. These poles are designed to support various smart city applications, including environmental monitoring and public safety enhancements. While the primary power source is the electrical grid, the integration of energy-efficient technologies and the potential for incorporating energy-harvesting methods are being explored to further reduce energy consumption and maintenance costs [88]. Security cameras and environmental sensors equipped with RFEH capabilities can operate autonomously, improving urban surveillance and environmental monitoring while minimizing human intervention. This approach lowers operational expenses and supports the deployment of the vast sensor networks necessary for real-time data collection within smart city infrastructures. EnOcean GmbH has developed battery-less wireless sensors that harvest energy from their environment, such as motion, light, and temperature differences. These sensors have been implemented in various smart building projects across Europe, enabling maintenance-free operation and reducing electronic waste. Their applications include occupancy detection, temperature monitoring, and energy management, contributing to the efficiency of smart city infrastructures [89]. Notably, security surveillance systems powered by RF harvesters have achieved uninterrupted functionality, even during grid failures, thereby strengthening public safety initiatives [90]. In Milan, Italy, research initiatives have focused on deploying autonomous RF-powered security cameras in metro stations. These systems are designed to operate without external power sources, maintaining real-time video feeds under varying electromagnetic conditions. This implementation aims to enhance public safety while reducing infrastructure costs associated with power supply and maintenance [91]. Furthermore, recent developments have introduced self-powered environmental sensors designed to monitor air quality, traffic flow, and energy consumption. These sensors, enabled by RF energy-harvesting technologies, form the foundation of data-driven decision-making in modern smart cities, allowing for optimized urban management and sustainable growth [92,93]. Singapore's Cooperative and Unified Smart Traffic System (CRUISE) exemplifies the integration of AI with real-time data from sensors to optimize traffic flow and enhance pedestrian safety. While CRUISE primarily relies on data from various sensors, the potential incorporation of energy-harvesting technologies could further improve the sustainability and maintenance efficiency of such smart traffic systems [94].

6.2. Biomedical and Wearable Devices

The biomedical field has witnessed remarkable innovation through the integration of RF energy harvesting into wearable and implantable medical devices. Wearable health-monitoring devices, such as continuous glucose monitors and cardiac patches, have incorporated flexible RF harvesters capable of collecting energy from ambient Wi-Fi and cellular signals to extend battery life or even achieve full autonomy [95]. Studies have reported the development of skin-conformal harvesters that maintain operational efficiency during motion, bending, and stretching, making them ideal for body sensor networks (BSNs) [96]. A notable example is the InfiniWolf smart bracelet, which combines thermal and solar energy harvesting to achieve self-sustainability. This device demonstrates the feasibility of integrating multiple energy-harvesting methods in wearable electronics [44]. In implantable applications, RFEH addresses a critical challenge by enabling pacemakers, cochlear implants, and neural stimulators to operate without repeated surgical battery replacements. On-chip micro-rectennas have been designed to power implants from ambient RF sources, achieving sufficient energy delivery while minimizing tissue heating and maximizing biocompatibility [97]. Researchers in [30] reported the practical demonstration of an RF-powered implantable stimulator, tested in vivo, which maintained wireless charging efficiency above 70% when under tissue layers. Their findings support the feasibility of RF harvesting in powering next-generation medical implants with clinical reliability. These breakthroughs significantly improve patient outcomes, reduce healthcare costs, and open new avenues for personalized medicine.

6.3. Industrial IoT

In industrial environments, the integration of RFEH into remote sensors has significantly advanced predictive maintenance strategies and smart factory operations. Wireless sensors equipped with RF harvesters can continuously monitor key parameters such as machine vibration, temperature, humidity, environmental conditions, and structural integrity without relying on wired connections or frequent battery replacements. This capability enables real-time data collection and analysis, resulting in improved operational efficiency, reduced downtime, and cost savings [98]. For instance, a study demonstrated the deployment of RF-powered low-energy sensor nodes for predictive maintenance in electromagnetically harsh industrial environments. The system utilized 2.45 GHz RF energy to power sensor nodes monitoring machinery conditions, effectively eliminating the need for battery replacements and enhancing maintenance efficiency [53]. Integrating RF energy harvesters into sensors placed on rotating machinery or in inaccessible locations facilitates continuous health monitoring without the logistical challenges associated with traditional battery-powered systems. This proactive approach supports predictive maintenance strategies, minimizing the risk of sudden equipment failures and the costly unplanned downtime that accompanies them [99]. Moreover, EnOcean's energy-harvesting technology has been adopted in industrial automation to create maintenance-free wireless sensors that operate without batteries. These sensors harvest energy from their environment, drawing on sources such as motion, light, or temperature differences, and are used in various applications, including the monitoring of equipment status and environmental conditions in industrial settings [89]. Recent research demonstrated that RF-powered predictive maintenance systems are capable of detecting mechanical faults approximately 25% earlier than conventional battery-operated sensors, leading to a substantial reduction in equipment downtime and maintenance costs [100]. In support of this, a comprehensive survey by Ibrahim et al. (2022) highlighted the performance determinants of RF energy-harvesting systems in industrial applications, emphasizing their role in enhancing predictive maintenance by providing continuous power to sensor networks, thereby improving fault detection and reducing maintenance costs [13]. Moreover, the scalability of RFEH technologies enables the deployment of thousands of wireless sensors across expansive industrial sites, paving the way for the realization of Industry 4.0, where machine-to-machine communication, real-time monitoring, and advanced data analytics drive unprecedented levels of industrial automation and efficiency.

6.4. Wireless Sensor Networks (WSN) and Smart Agriculture

The agricultural sector has greatly benefited from RFEH, particularly in the deployment of wireless sensor networks (WSNs) for smart farming. Sensors distributed across vast farmlands monitor soil moisture, nutrient levels, weather conditions, and pest activity, allowing farmers to make precise decisions that optimize crop yield and resource utilization. RF energy harvesters power these sensors continuously, eliminating the need for costly battery replacements and reducing the ecological footprint [101]. For instance, one study demonstrated the use of aerial drones equipped with RF energy-harvesting sensor tags for autonomous agricultural monitoring. These battery-less sensors harvested energy from drone-emitted RF signals to measure environmental parameters like temperature and humidity, enabling efficient data collection over large farmlands [102]. Recent research demonstrated that RF-powered agricultural sensors enabled a 30% improvement in irrigation efficiency by providing real-time soil moisture data, contributing to water conservation and sustainable farming practices. Additionally, they proposed a hierarchical clustering-based dynamic data fusion algorithm for WSNs in smart agriculture. This approach enhanced energy efficiency and event detection precision, ensuring the reliable operation of RF-powered sensors in monitoring various environmental parameters [103]. Hence, integrating RFEH with machine learning models allows for dynamic adaptation to environmental changes, ensuring the long-term reliability and operational resilience of agricultural WSNs.

6.5. Space and Satellite Communications

In space applications, RFEH offers novel approaches to powering satellites, space probes, and communication devices. Satellite systems can harvest energy from Earth-based RF transmissions or cosmic microwave background radiation to supplement their onboard energy storage, extending mission lifespan. A notable advancement includes the development of RF-harvesting modules that are integrated into CubeSats, providing an alternative energy source for low-Earth-orbit operations without increasing payload weight [104]. Furthermore, the concept of space-based solar power involves collecting solar energy in space, converting it into RF signals, and beaming it down to Earth, where ground-based harvesters can capture and convert it into usable electricity. This futuristic application is actively being researched as a solution for global energy sustainability [105]. Such innovations not only support long-term space exploration but also lay the groundwork for a sustainable extraterrestrial energy economy. A significant milestone in this field was achieved by the California Institute of Technology (Caltech) in 2023. Their Space Solar Power Demonstrator (SSPD-1) successfully demonstrated wireless power transfer in space using the Microwave Array for Power-transfer Low-orbit Experiment (MAPLE) technique. This experiment marked the first time that power was wirelessly transmitted in space and detected on Earth, validating the feasibility of space-based solar power systems [106].

7. Challenges and Limitations

While RFEH presents a promising solution for powering low-energy devices and supporting sustainable wireless technologies, several technical and practical challenges remain that hinder its widespread adoption. These limitations span from energy conversion inefficiencies to regulatory concerns, requiring continued research and innovation.

7.1. Low RF-to-DC Conversion Efficiency

One of the primary challenges facing RFEH systems is the inherently low RF-to-DC conversion efficiency, especially at low input power levels. While modern rectenna designs have achieved efficiencies exceeding 80% under optimal conditions, this level of performance typically only occurs at relatively high input powers or specific frequency bands. Under ambient RF conditions where power levels are often below −20 dBm, the efficiency drops significantly, often falling below 30% [107]. Recent studies have emphasized the importance of optimizing impedance matching, rectifier design, and power management circuits to mitigate these losses, yet achieving consistently high efficiency across diverse and unpredictable RF environments remains a major hurdle [10]. This inefficiency is primarily caused by the poor sensitivity of conventional Schottky diodes at ultra-low input powers and their increased threshold voltage, which limits the rectification of weak signals. Furthermore, the non-linear behaviour of rectifying diodes under low power inputs complicates circuit design, requiring innovative materials and circuit topologies to sustain performance [14,34]. To address these issues, researchers are exploring the use of zero-threshold MOSFET-based rectifiers, differential-drive rectifier topologies, and advanced multi-stage voltage multipliers to maintain effective rectification, even under sub-microwatt input power levels. Additionally, the integration of wideband impedance matching networks and self-tuning rectifiers using varactors or MEMS-based components has shown promise in dynamically adapting to fluctuating RF power levels, thereby enhancing conversion efficiency across real-world environments.

7.2. Unpredictable RF Energy Availability

The unpredictable and dynamic nature of ambient RF energy sources poses another significant limitation. In urban areas, RF power densities are relatively high due to dense cellular, Wi-Fi, and broadcast infrastructures. However, in rural, suburban, or industrial environments, RF energy availability fluctuates greatly and may drop to levels insufficient for practical energy harvesting [36]. A recent field measurement campaign indicated that RF power densities can vary by two to three orders of magnitude within the same urban area depending on time, location, and user density [108]. This variability makes it challenging to design harvesters that can reliably operate under all environmental conditions without supplemental energy storage or hybrid systems. Additionally, the sporadic availability of RF signals raises concerns about the stability of critical applications such as biomedical implants or industrial IoT sensors, where continuous power supply is mandatory. These fluctuations are caused by factors such as dynamic user behaviour, time-varying transmission loads, and physical obstructions that cause multipath fading and shadowing. As a result, designing effective harvesters requires practitioners to capture ambient energy when available but also store it efficiently during surplus conditions. Potential technical solutions include integrating advanced energy storage devices (like supercapacitors or micro-batteries) that can buffer energy during periods of high RF availability. Another approach is the use of hybrid energy-harvesting systems that combine RF with other renewable sources such as solar or vibrational energy, thus smoothing out the intermittency associated with any single source. Furthermore, adaptive algorithms that predict RF energy availability based on historical data can be implemented to dynamically adjust power management and load conditions, ensuring more reliable operation in environments with highly variable RF signals.

7.3. Antenna Size and Design Constraints

Another critical challenge lies in the physical design of antennas optimized for RFEH. Efficient RF energy capture demands high-gain, wideband, or multi-band antennas; however, developing these features often conflicts with the requirements of miniaturization, flexibility, or wearable applications [109]. For instance, at lower frequencies such as 900 MHz, the physical size of an efficient antenna becomes impractically large for compact devices. On the other hand, scaling down antenna size using miniaturization techniques often results in reduced radiation efficiency, a narrow bandwidth, and lower gain [110]. Flexible and wearable applications further complicate the design, as antenna performance can degrade due to bending, stretching, or varying proximity to the human body. These design constraints are primarily caused by the trade-offs between antenna size, resonant frequency, and radiation performance. The wavelength of lower-frequency RF signals directly dictates larger antenna dimensions, which are incompatible with the objectives of compactness or a wearable form. In wearable scenarios, body-induced detuning and absorption can also significantly impair antenna function. To address these issues, researchers have proposed several technical solutions. One approach involves using metamaterial-inspired antennas or fractal geometries that maintain performance while having a smaller physical size. These structures allow multi-band operation and improved gain in a miniaturized footprint. Additionally, reconfigurable and tunable antennas that can dynamically adapt to changing frequency and loading conditions are being developed to sustain efficient harvesting across a range of environments. Material innovation also plays a key role. The adoption of novel conductive materials such as silver nanowires, graphene films, or conductive polymers can improve flexibility while maintaining electrical performance. Moreover, embedding antennas within stretchable substrates or incorporating 3D printing techniques allows better integration into wearable electronics without sacrificing performance [110]. Ongoing research is also exploring hybrid antenna systems that combine compact omnidirectional receivers with directional high-gain components for dynamic orientation control, ensuring effective energy capture, even during movement or deformation. New materials such as conductive textiles and graphene composites have shown promise, but they introduce additional manufacturing complexities and costs [111].

7.4. Signal Interference and Electromagnetic Compatibility

RF energy harvesters operate in environments crowded with numerous communication signals. In such settings, electromagnetic compatibility (EMC) and signal interference become serious issues. Harvesters may unintentionally pick up noise or unwanted signals, degrading the efficiency of energy capture and potentially causing harmful interference to nearby sensitive communication systems [112]. Studies have reported that rectifiers and matching circuits can experience significant energy conversion degradation when exposed to wideband noise. Moreover, the presence of harmonics generated by the rectification process can interfere with nearby RF communication systems if proper harmonic filtering is not implemented [113]. The root causes of these issues include the broadband nature of ambient RF sources, the non-linear behaviour of rectification circuits, and insufficient electromagnetic isolation between components. Rectifiers operating over multiple frequency bands tend to lack selectivity, making them prone to harvesting irrelevant or noisy signals. Additionally, poorly matched or untuned circuits may reflect the transmission of energy into the environment, further increasing electromagnetic pollution [112]. Technical solutions to mitigate these issues include the integration of bandpass or notch filters that selectively allow the attainment of desired frequencies while attenuating out-of-band noise and harmonics. Shielded enclosures or multilayer PCB designs with ground planes can also be employed to improve isolation and minimize cross-talk. Moreover, adaptive interference cancellation techniques based on software-defined radio (SDR) and digital signal processing (DSP) have been proposed to dynamically suppress unwanted signals in real time. These methods enable rectifier circuits to adapt their behaviour based on the spectral environment. Recent developments also explore using reconfigurable impedance matching networks to reduce sensitivity to broadband interference and optimize energy extraction from clean, narrowband sources. By incorporating machine learning algorithms, these networks can autonomously detect and respond to spectral anomalies that would otherwise degrade performance [113]. The management of this coexistence between energy-harvesting devices and traditional RF communication infrastructures without introducing mutual interference remains an open research problem, requiring sophisticated filtering, shielding, and adaptive control techniques.

7.5. Regulatory and Standardization Issues

Regulatory constraints and the lack of standardized frameworks for RFEH further hinder its large-scale deployment. Current wireless communication regulations, such as those implemented by the Federal Communications Commission (FCC) and the International Telecommunication Union (ITU), primarily focus on communication and safety, not on energy-harvesting considerations [114]. As a result, many RF energy-harvesting initiatives must operate within strict power emission limits, limiting the amount of RF energy available for collection. One major issue is the absence of frequency bands specifically allocated for energy harvesting, which forces systems to rely on ambient RF signals not intended for power transfer. This creates unpredictability in energy availability and complicates legal compliance when deploying intentional RF power sources. Furthermore, the use of dedicated RF transmitters for power delivery risks exceeding Specific Absorption Rate (SAR) limits, raising health and safety concerns. To address these challenges, researchers and industry bodies have begun proposing new regulatory models that include power transfer allowances within existing spectrum allocations. This includes low-power dedicated RF transmitters governed by duty-cycle constraints or time-division multiplexing to minimize interference with traditional communication services. For example, frequency-sharing protocols have been proposed that allow the coordinated use of a spectrum by communication and energy-harvesting devices, subject to adaptive control based on real-time spectral occupancy [114]. Additionally, the coexistence of dedicated RF power transmitters with conventional communication networks raises concerns about spectrum congestion, fairness, and interference. These coexistence issues have spurred work on dynamic spectrum access (DSA) and cognitive radio-based energy-harvesting systems, which allow harvesters to intelligently detect and utilize idle spectrum without disrupting licenced users. In parallel, policy discussions are emerging around enabling unlicensed low-power RF power transmission zones in dense IoT environments, similar to Wi-Fi and Bluetooth regulations. Recent efforts have begun exploring standardization for wireless power transfer and energy-harvesting systems, particularly in 5G and IoT contexts, but comprehensive regulatory guidelines remain lacking. Standardization bodies such as IEEE P2668 and ETSI are beginning to draft interoperability standards and performance benchmarks for energy-harvesting modules, including guidelines for rectenna efficiency, output power levels, and safe integration with IoT networks. However, these are still in the early stages and require broader consensus among stakeholders [115]. Without proper frameworks, the deployment of dedicated energy-harvesting infrastructures or collaborative energy-harvesting from existing networks faces significant uncertainty and legal barriers.

7.6. Real-World Performance and Implementation Challenges

While laboratory-scale advancements in RFEH have shown promising efficiencies and system improvements, real-world deployments reveal additional complexities that impact practical performance. Field studies indicate that ambient RF energy levels are often significantly lower and more variable than those reported in controlled environments, which affects the continuous power supply to energy-harvesting devices. For example, variations in user density, environmental obstacles, and temporal factors such as time of day or weather conditions can cause substantial fluctuations in harvested energy, challenging the reliability of autonomous devices [8]. Moreover, implementation trials exhibit highlighted integration issues when embedding RF harvesters within commercial IoT devices or wearable systems. These include challenges related to antenna placement, device form factor constraints, and interference from other onboard electronic devices, which often degrade the expected energy capture and conversion efficiency [14]. The thermal management and long-term durability of novel materials like graphene-based rectennas under real operating conditions also remain open concerns [16].

In industrial and biomedical applications, maintaining stable operation amidst highly dynamic electromagnetic environments demands the use of hybrid energy-harvesting approaches and advanced power management algorithms to buffer against intermittent RF sources [19]. Several pilot deployments underscore the necessity of combining RFEH with auxiliary power sources or energy storage components to guarantee uninterrupted device functionality. Furthermore, the regulatory landscape complicates large-scale implementation, with field deployments sometimes restricted by emission limits and spectrum use policies that vary significantly across regions [22]. Collaborative efforts are increasingly required to harmonize these standards to enable the development of viable commercial solutions. These real-world insights emphasize the need for further research focusing on deployment-driven design, robust system integration, and adaptive energy management strategies that can cope with the unpredictability and heterogeneity of ambient RF environments. Addressing these challenges will be critical to transitioning RFEH from experimental setups to reliable, mass-market applications.

8. Future Opportunities and Research Directions

The future of RFEH promises to bring about transformative changes across diverse industries, from wireless communication and smart cities to biomedical applications and space technology. As emerging demands for autonomous, maintenance-free devices continue to grow, significant research opportunities are unfolding to advance RFEH technologies. Key areas for future development include innovations in materials, reconfigurable systems, integration with next-generation networks, AI-driven energy management, and strategies for enabling widespread deployment and commercialization.

8.1. Advanced Materials for High-Efficiency Rectennas

One of the most promising future directions in RFEH is the development of advanced materials to improve rectenna efficiency. Traditional rectenna designs, typically based on conventional metallic conductors and semiconductor diodes, face efficiency bottlenecks due to material limitations. Recent research has explored the use of two-dimensional materials such as graphene and molybdenum disulfide (MoS_2) for ultrafast, low-threshold Schottky diodes, resulting in significant improvements in RF-to-DC conversion efficiency. Future research should explore scalable fabrication techniques such as chemical vapour deposition (CVD) for synthesizing high-quality graphene films with controlled doping and defect minimization to optimize carrier transport properties. Studies have demonstrated that graphene-based flexible rectennas can maintain high conversion efficiency, even when subjected to mechanical deformation, making them ideal for wearable and biomedical applications [116]. Implementing conformal rectenna structures based on flexible substrates such as polyimide or PDMS, using printable conductive materials like silver nanowires or graphene inks, would support the development of miniaturized and stretchable designs. These should be experimentally validated under mechanical strain conditions (e.g., bending or stretching) to assess real-world performance reliability [34]. Moreover, the exploration of metamaterials and meta-surfaces, integrated into rectenna structures, has opened new avenues for enhanced energy capture and impedance matching across wide frequency ranges [117,118]. A focused research agenda should include the design and simulation of tunable meta-surfaces using embedded MEMS or varactor-based elements, enabling real-time frequency agility in dynamic RF environments. These reconfigurable structures can be modelled using full-wave electromagnetic solvers and optimized through machine learning algorithms to tailor performance to specific use cases [79]. Research efforts focusing on nanomaterial-enhanced substrates and tunable smart surfaces are expected to drive the development of a new generation of rectennas with record-breaking efficiency and flexibility. Additionally, developing standardized testing protocols for evaluating rectenna performance across a range of frequencies, input power levels, and mechanical deformation scenarios will support the benchmarking and commercialization of next-generation rectenna technologies [5].

8.2. Reconfigurable and Adaptive RF Harvesting Systems

Conventional RF energy harvesters are typically designed for fixed-frequency operation and static environmental conditions. However, real-world RF environments are dynamic, with fluctuating signal strengths, frequencies, and interference patterns. Reconfigurable and adaptive RFEH systems, capable of dynamically adjusting their operating parameters in response to environmental changes, represent a critical future direction.

Recent works have proposed reconfigurable antennas and tunable matching networks controlled by varactor diodes and MEMS switches, allowing dynamic frequency adaptation and optimal power transfer [119]. Furthermore, the concept of cognitive RFEH, where harvesters can sense the RF spectrum and opportunistically adapt to available energy sources, is gaining momentum. Adaptive systems could also leverage machine learning models to predict optimal operating parameters, enhancing overall system resilience and energy yield in heterogeneous environments [120]. An example of such a system is the implementation of a frequency-agile rectenna integrated with a spectrum-sensing module, capable of detecting spectrum occupancy and retuning its antenna and impedance matching circuit accordingly using a closed-loop feedback controller. This allows the system to adapt in real time to multi-band RF sources such as 5G base stations, Wi-Fi, and Bluetooth and ensures optimal power transfer, even in highly variable spectral environments [121]. Furthermore, research prototypes have demonstrated the feasibility of using reinforcement learning algorithms, such as Deep Q-Networks (DQN), to optimize harvesting policies based on real-time signal strength and usage statistics from surrounding RF bands. This integration of AI enables autonomous decision-making in selecting the best frequencies and configurations to maximize energy harvesting under complex urban scenarios. Future roadmaps could involve the development of integrated RF harvesting modules with built-in AI chips (e.g., edge TPU) for low-power spectrum analytics, coupled with reconfigurable meta-surface antennas that can electronically steer or reshape their radiation patterns in response to environmental feedback, enabling the development of highly adaptive, location-aware harvesting architectures [122].

8.3. Integration with 5G, 6G, and Beyond

The rollout of 5G networks and the development of 6G technologies present unprecedented opportunities for RF energy harvesting. With the proliferation of small cells, massive MIMO antennas, and dense heterogeneous networks operating at millimetre-wave (mmWave) and terahertz (THz) frequencies, ambient RF energy will become more abundant and diverse. Future RFEH systems must be designed to exploit the high-frequency, high-density RF environment of 5G and 6G networks. Research has highlighted the potential of harvesting from mmWave bands using specialized rectennas optimized for high-frequency capture [123]. Furthermore, 6G systems are expected to embrace SWIPT more fully, where devices will simultaneously receive data and energy from the same carrier waves [124,125]. Future research should prioritize the co-design of high-frequency rectennas and beamforming protocols tailored to mmWave and THz energy-harvesting scenarios, where dynamic energy beam steering from base stations can be synchronized with user equipment for optimal power reception [40]. Moreover, the detailed modelling and prototyping of SWIPT-enabled transceivers using power-splitting or time-switching architectures will be crucial in determining optimal trade-offs between energy-harvesting efficiency and communication performance in real-world 6G deployments. A phased roadmap for implementation could involve the initial lab-scale demonstration of SWIPT at sub-6 GHz frequencies, followed by mmWave integration in small-cell testbeds, and eventual deployment in 6G pilot networks featuring edge intelligence for real-time energy scheduling [85]. Integrating RF energy harvesting into 5G/6G base stations, IoT devices, and even user terminals could dramatically reduce the energy footprint of future networks, aligning with sustainability goals such as net-zero emissions.

8.4. AI-Based Energy-Harvesting Management Systems

Artificial intelligence (AI) and machine learning (ML) are set to revolutionize the way RF energy-harvesting systems operate. Intelligent energy management systems, powered by AI, can dynamically optimize the allocation, storage, and distribution of harvested energy based on real-time network conditions and user demands. Predictive models using deep reinforcement learning (DRL) have been proposed to adaptively control matching circuits, select optimal harvesting frequencies, and manage energy storage to maximize system uptime and efficiency [126,127]. Moreover, federated learning techniques are emerging as a solution for distributed RFEH networks, allowing devices to collaboratively learn optimal harvesting strategies without centralized control, preserving data privacy. AI-based optimization also supports the realization of cognitive radio energy harvesting, where devices not only sense and adapt to RF environments but intelligently forecast RF spectrum usage patterns to maximize energy capture [11,128]. Recent works have introduced implementation frameworks where DRL agents are deployed on edge devices using TensorFlow Lite or PyTorch 2.0 Mobile to minimize computational overhead while enabling real-time decision-making in dynamic RF environments. Additionally, federated learning prototypes have been demonstrated on embedded platforms like Raspberry Pi and NVIDIA Jetson Nano for decentralized energy prediction tasks in IoT clusters. To address power constraints, ongoing research is focusing on quantized and pruned neural networks tailored to microcontrollers such as STM32 or ESP32, enabling AI-driven harvesting with sub-milliwatt power consumption. These lightweight models can be trained on cloud servers and deployed on nodes for inference, offering a feasible roadmap for integrating intelligence into ultra-low-power harvesting systems [129]. Future research is likely to explore lightweight AI models specifically designed for ultra-low-power energy-harvesting systems.

8.5. Large-Scale Deployment and Commercialization Prospects

Although tremendous progress has been made in RFEH at the laboratory scale, transitioning to widespread commercial deployment remains a major challenge. One of the primary hurdles is standardizing the performance metrics and testing protocols for RF energy harvesters across different frequency bands and application scenarios. Collaborative efforts between academia, industry stakeholders, and regulatory bodies are essential to develop global standards that facilitate interoperability and commercialization [130]. Emerging market sectors such as smart homes, smart cities, healthcare monitoring, and environmental sensing represent the initial targets for the large-scale deployment of RF-powered devices. Implementation roadmaps are increasingly focusing on phased pilot deployments in smart city environments, leveraging existing municipal IoT infrastructures to validate RF harvesting technologies under real-world conditions. Testbeds integrating RF harvesters with common IoT platforms such as LoRaWAN and NB-IoT are being proposed to evaluate system interoperability and scalability [128,131].

Innovative business models will also be needed, such as energy-as-a-service platforms where energy harvested from ambient RF sources is monetized or shared across IoT ecosystems [131,132]. Furthermore, hybrid energy-harvesting solutions that integrate RF with solar, vibration, and thermal energy harvesting are expected to gain traction, ensuring a reliable and continuous power supply, even with fluctuating RF availability. Future commercialization strategies include developing modular RF harvesting units with plug-and-play capabilities, enabling easy integration into consumer electronics and industrial sensors. Additionally, industry consortia are exploring certification programmes to guarantee device performance and compatibility, thus building consumer and enterprise trust in RF-powered products [68]. Future research must address cost optimization, miniaturization, and seamless integration into consumer products to unlock the full commercial potential of RFEH.

9. Conclusions

This paper has provided a comprehensive review of RFEH technologies, covering the fundamental principles, system architectures, recent technological developments, applications, challenges, and future opportunities. Key advancements have been recorded in rectenna design, impedance matching networks, metamaterial-assisted harvesting, and AI-driven energy optimization, resulting in improved RF-to-DC conversion efficiencies and more robust multi-band operation. The applications explored across smart cities, biomedical devices, industrial IoT, smart agriculture, and satellite communications highlight the growing significance of RFEH in enabling self-sustaining, maintenance-free electronic systems. However, the review also identified persistent challenges, including low ambient RF power density, unpredictable energy availability, antenna size limitations, electromagnetic interference, and regulatory constraints, that must be addressed to realize the full potential of this technology. The prospects for RFEH are highly promising, especially as the world moves toward the large-scale deployment of 5G, 6G, and future communication systems. Emerging technologies such as reconfigurable energy-harvesting systems, advanced materials for high-efficiency rectennas, AI-enhanced resource management frameworks, and hybrid energy-harvesting solutions are expected to drive further breakthroughs. To fully harness these opportunities, future research should focus on developing comprehensive implementation roadmaps that encompass pilot-scale deployment, standardization protocols, and cost-effective manufacturing processes. There is a critical need for multidisciplinary approaches combining material science, RF engineering, machine learning, and regulatory policy to design adaptive, environment-aware RF harvesting systems capable of operating efficiently under real-world conditions. The exploration of AI-driven predictive energy management and federated learning in distributed harvester networks can optimize energy capture and consumption dynamically. Additionally, hybrid harvesting architectures, integrating RF with solar, thermal, and vibrational energy sources, will improve reliability and broaden application scenarios. As research and development efforts continue to bridge existing performance gaps, RF energy harvesting is poised to become a cornerstone technology for self-powered IoT ecosystems, resilient smart infrastructures, and sustainable wireless communication networks. Further investigation into large-scale commercialization models, including energy-as-a-service platforms and modular plug-and-play harvesting units, will be vital to accelerate adoption across diverse markets. Moreover, the establishment of global regulatory frameworks, addressing spectrum usage, emission limits, and interoperability standards, is essential to remove current legal and operational uncertainties. By addressing current technical and regulatory barriers, RFEH can unlock a future where millions of devices operate autonomously, contributing to a more intelligent, connected, and energy-efficient world.

Author Contributions: S.N.A.: conceptualization, data curation, validation, writing—original draft. E.R.O.: investigation, formal analysis, writing—review & editing. S.H.E.: visualization, resources, writing—review & editing. A.O.N.: methodology, project administration, funding acquisition, supervision. All authors have read and agreed to the published version of the manuscript.

Funding: This research received no external funding.

Data Availability Statement: No new data were created or analyzed in this study.

Conflicts of Interest: Author Emenike Raymond Obi was employed by the company Raysoft AssetAnalytics. The remaining authors declare that the research was conducted in the absence of any commercial or financial relationships that could be construed as a potential conflict of interest.

References

1. Moloudian, G.; Hosseinifard, M.; Kumar, S.; Simorangkir, R.B.V.B.; Buckley, J.L.; Song, C.; Fantoni, G.; O'Flynn, B. RF Energy Harvesting Techniques for Battery-less Wireless Sensing, Industry 4.0 and Internet of Things: A Review. *IEEE Sens. J.* **2024**, *24*, 5732–5745. [CrossRef]
2. López, O.L.A.; Alves, H.; Souza, R.D.; Montejo-Sánchez, S.; Fernández, E.M.G.; Latva-aho, M. Massive Wireless Energy Transfer: Enabling Sustainable IoT Toward 6G Era. *IEEE Internet Things J.* **2021**, *8*, 8816–8835. [CrossRef]
3. Niotaki, K.; Carvalho, N.B.; Georgiadis, A.; Gu, X.; Hemour, S.; Wu, K.; Matos, D.; Belo, D.; Pereira, R.; Figueiredo, R.; et al. energy harvesting and wireless power transfer for energy autonomous wireless devices and RFIDs. *IEEE J. Microw.* **2023**, *3*, 763–782. [CrossRef]
4. Muhammad, S.; Tiang, J.J.; Wong, S.K.; Smida, A.; Ghayoula, R.; Iqbal, A. A Dual-Band Ambient Energy Harvesting Rectenna Design for Wireless Power Communications. *IEEE Access* **2021**, *9*, 99944–99953. [CrossRef]
5. Chen, D.; Li, R.; Xu, J.; Li, D.; Fei, C.; Yang, Y. Recent progress and development of radio frequency energy harvesting devices and circuits. *Nano Energy* **2023**, *117*, 108845. [CrossRef]
6. Basnayaka, C.M.W.; Jayakody, D.N.K.; Perera, T.D.P.; Beko, M. DataAge: Age of Information in SWIPT-Driven Short Packet IoT Wireless Communications. *IEEE Internet Things J.* **2024**, *11*, 26984–26999. [CrossRef]
7. Lima, L.S.C.; Trevisoli, R.; Suyama, R.; Casella, I.R.S.; Doria, R.T.; Capovilla, C.E. Optimizing RF Energy Harvesting Systems for IoT Applications using Reinforcement Learning. In Proceedings of the 2023 Symposium on Internet of Things (SIoT), São Paulo, Brazil, 25–27 October 2023; pp. 1–5. [CrossRef]
8. Sherazi, H.H.R.; Zorbas, D.; O'Flynn, B. A Comprehensive Survey on RF Energy Harvesting: Applications and Performance Determinants. *Sensors* **2022**, *22*, 2990. [CrossRef] [PubMed]
9. Ali, A.; Shaukat, H.; Bibi, S.; Altabey, W.A.; Noori, M.; Kouritem, S.A. Recent progress in energy harvesting systems for wearable technology. *Energy Strateg. Rev.* **2023**, *49*, 101124. [CrossRef]
10. Wang, R.; Qi, Y.; Lavasani, H.M. A Highly Efficient CMOS Rectifier for Ultra-Low-Power Ambient RF Energy Harvesting. In Proceedings of the IEEE International Midwest Symposium on Circuits and Systems (MWSCAS), Lansing, MI, USA, 9–11 August 2021; pp. 212–215. [CrossRef]
11. Ahmed, Q.W.; Garg, S.; Rai, A.; Ramachandran, M.; Jhanjhi, N.Z.; Masud, M.; Baz, M. AI-Based Resource Allocation Techniques in Wireless Sensor Internet of Things Networks in Energy Efficiency with Data Optimization. *Electronics* **2022**, *11*, 2071. [CrossRef]
12. Eunice, A.; Carvalho, N.B. Energy Harvesting Mechanisms in a Smart City—A Review. *Smart Cities* **2021**, *4*, 476–498. [CrossRef]
13. Ibrahim, H.H.; Singh, M.J.; Al-Bawri, S.S.; Ibrahim, S.K.; Islam, M.T.; Alzamil, A.; Islam, M.S. Radio Frequency Energy Harvesting Technologies: A Comprehensive Review on Designing, Methodologies, and Potential Applications. *Sensors* **2022**, *22*, 4144. [CrossRef] [PubMed]
14. Muhammad, S.; Tiang, J.J.; Wong, S.K.; Rambe, A.H.; Adam, I.; Smida, A.; Waly, M.I.; Iqbal, A.; Abubakar, A.S.; Mohd Yasin, M.N. Harvesting Systems for RF Energy: Trends, Challenges, Techniques, and Tradeoffs. *Electronics* **2022**, *11*, 959. [CrossRef]
15. Rao, A.S.; Aziz, A.; Aljaloud, K.; Qureshi, M.A.; Muhammad, A.; Rafique, A.; Hussain, R. Concomitance of radio frequency energy harvesting and wearable devices: A review of rectenna designs. *Int. J. RF Microw. Comput.-Aided Eng.* **2022**, *32*, e23536. [CrossRef]
16. Odiamenhi, M.; Jahanbakhsh Basherlou, H.; Hwang See, C.; Ojaroudi Parchin, N.; Goh, K.; Yu, H. Advancements and Challenges in Antenna Design and Rectifying Circuits for Radio Frequency Energy Harvesting. *Sensors* **2024**, *24*, 6804. [CrossRef] [PubMed]
17. Rashwan, Y.R.; El-Hameed, A.S.A.; Dousoky, G.M.; Tammam, E. Design of an Efficient 2.45 GHz Antenna Array for RF Energy Harvesting. *SVU-Int. J. Eng. Sci. Appl.* **2024**, *5*, 175–183. [CrossRef]
18. Behera, B.; Mishra, S.; Alsharif, M.H.; Uthansakul, P.; Uthansakul, M. A metasurface-inspired printed monopole antenna for 5G and RF energy harvesting application. *Eng. Sci. Technol. Int. J.* **2024**, *51*, 101638. [CrossRef]
19. Ojha, S.S.; Sharma, J.K.; Dhakad, B.; Kumar, S.; Sharma, N.; Pandey, A.K.; Hasnain, S.M.; Kumar, S.; Kumar, R. A dual ultra-wideband rectenna with a compact conical antenna for RF energy harvesting from S and C bands. *Results Eng.* **2024**, *22*, 102279. [CrossRef]
20. Behera, B.R.; Mishra, S.K.; Alsharif, M.H.; Jahid, A. Reconfigurable Antennas for RF Energy Harvesting Application: Current Trends, Challenges, and Solutions from Design Perspective. *Electronics* **2023**, *12*, 2723. [CrossRef]
21. Caltech. In a First, Caltech's Space Solar Power Demonstrator Wirelessly Transmits Power in Space. 2023. Available online: https://www.caltech.edu/about/news/in-a-first-caltechs-space-solar-power-demonstrator-wirelessly-transmits-power-in-space (accessed on 11 June 2025).
22. Liu, X.; Li, M.; Zhang, Y.; Zhu, X.; Guan, Y. Self-Powered wireless sensor node based on RF energy harvesting and management combined design. *Energy Convers. Manag.* **2023**, *292*, 117393. [CrossRef]
23. Nwalike, E.D.; Ibrahim, K.A.; Crawley, F.; Qin, Q.; Luk, P.; Luo, Z. Harnessing Energy for Wearables: A Review of Radio Frequency Energy Harvesting Technologies. *Energies* **2023**, *16*, 5711. [CrossRef]

24. Patel, R.; Upadhyaya, T.; Desai, A.; Palandoken, M. Low Profile Multiband Meander Antenna for LTE/WiMAX/WLAN and INSAT-C Application. *AEU-Int. J. Electron. Commun.* **2019**, *102*, 90–98. [CrossRef]
25. Patanvariya, D.; Chatterjee, A. Planar nested square fractal patch antenna for multi-band wireless applications. *Microsyst. Technol.* **2023**, *29*, 347–357. [CrossRef]
26. Abdulkawi, W.M.; Sheta, A.F.; Elshafiey, I.; Alkanhal, M.A. Design of Low-Profile Single- and Dual-Band Antennas for IoT Applications. *Electronics* **2021**, *10*, 2766. [CrossRef]
27. Cumbajin, M.; Sánchez, P.; Pillajo, D.; Gordón, C. RF Energy Harvesting System Based on Spiral Logarithmic Dipole Rectenna Array. In *CSEI: International Conference on Computer Science, Electronics and Industrial Engineering (CSEI)*; Garcia, M.V., Gordón-Gallegos, C., Eds.; Springer: Cham, Switzerland, 2023; Volume 678. [CrossRef]
28. Chattha, H.; Huang, Y.; Ishfaq, M.; Boyes, S. A Comprehensive Parametric Study of Planar Inverted-F Antenna. *Wirel. Eng. Technol.* **2019**, *3*, 1–11. [CrossRef]
29. Kwiatkowski, E.; Estrada, J.A.; López-Yela, A.; Popović, Z. Broadband RF Energy-Harvesting Arrays. *Proc. IEEE* **2022**, *110*, 74–88. [CrossRef]
30. Magno, M.; Wang, X.; Eggimann, M.; Cavigelli, L.; Benini, L. InfiniWolf: Energy Efficient Smart Bracelet for Edge Computing with Dual Source Energy Harvesting. In Proceedings of the 2020 Design, Automation & Test in Europe Conference & Exhibition (DATE), Grenoble, France, 9–13 March 2020. [CrossRef]
31. Yunus, M.; Mahdi, A.R.; Tan, Y.; Maulana, M.F.; Nurmantris, D.A.; Munir, A. Utilization of LC Circuit as Impedance Matching for Spiral Resonator-based Planar Antenna. In Proceedings of the 2024 Photonics & Electromagnetics Research Symposium (PIERS), Chengdu, China, 21–25 April 2024; pp. 1–5. [CrossRef]
32. Van Ieperen, A.; Derammelaere, S.; Minnaert, B. Coupling-independent capacitive wireless power transfer with one transmitter and multiple receivers using frequency bifurcation. *IEEE Open J. Power Electron.* **2024**, *5*, 891–901. [CrossRef]
33. Eidaks, J.; Kusnins, R.; Babajans, R.; Cirjulina, D.; Semenjako, J.; Litvinenko, A. Fast and Accurate Approach to RF-DC Conversion Efficiency Estimation for Multi-Tone Signals. *Sensors* **2022**, *22*, 787. [CrossRef] [PubMed]
34. Chew, Z.J.; Kuang, Y.; Ruan, T.; Zhu, M. Energy Harvesting in Smart Cities. In *Handbook of Smart Cities*; Augusto, J., Ed.; Springer: Cham, Switzerland, 2020. [CrossRef]
35. Abdul, Q.; Antoniades Marco, A.; Photos, V.; Sym, N. Voltage-Doubler RF-to-DC Rectifiers for Ambient RF Energy Harvesting and Wireless Power Transfer Systems. In *Recent Wireless Power Transfer Technologies*; IntechOpen: London, UK, 2019. [CrossRef]
36. Jocqué, J.; Sennesael, J.; Verhaevert, J.; Van Torre, P.; Schreurs, D.; Rogier, H. Rectifier Optimization for Low-Power RF Energy Harvesting Tags. In Proceedings of the 2024 IEEE International Symposium on Antennas and Propagation and INC/USNC-URSI Radio Science Meeting (AP-S/INC-USNC-URSI), Firenze, Italy, 14–19 July 2024; pp. 17–18. [CrossRef]
37. Zhang, X.; Grajal, J.; López-Vallejo, M.; McVay, E.; Palacios, T. Opportunities and Challenges of Ambient Radio-Frequency Energy Harvesting. *Joule* **2020**, *4*, 1148–1152. [CrossRef]
38. Dey, A.B.; Semwal, N.; Arif, W. Design of a compact and efficient 2.4 GHz rectenna system for energy harvesting. *J. Electromagn. Waves Appl.* **2022**, *36*, 1850–1868. [CrossRef]
39. Ahmed, A.E.; Abdullah, K.; Habaebi, M.H.; Ramli, H.A.; Asnawi, A.L. RF Energy Harvesting Wireless Networks: Challenges, and Opportunities. *Indones. J. Electr. Eng. Inform. (IJEEI)* **2021**, *9*, 101–113. [CrossRef]
40. Alhashimi, H.F.; Hindia, M.N.; Dimyati, K.; Hanafi, E.B.; Alden, F.Z.; Qamar, F.; Nguyen, Q.N. Survey on AI-Enabled Resource Management for 6G Heterogeneous Networks: Recent Research, Challenges, and Future Trends. *Comput. Mater. Contin.* **2025**, *83*, 3585–3622. [CrossRef]
41. Zhang, J.; Gu, M.; Chen, X. Supercapacitors for renewable energy applications: A review. *Micro Nano Eng.* **2023**, *21*, 100229. [CrossRef]
42. Gopi, C.V.M.; Ramesh, R. Review of battery-supercapacitor hybrid energy storage systems for electric vehicles. *Results Eng.* **2024**, *24*, 103598. [CrossRef]
43. Şahin, M.E.; Blaabjerg, F.; Sangwongwanich, A. A Comprehensive Review on Supercapacitor Applications and Developments. *Energies* **2022**, *15*, 674. [CrossRef]
44. A*STAR Cooperative and Unified Smart Traffic System (CRUISE). 2025. Available online: https://www.a-star.edu.sg/enterprise/our-stories/connectivity-trade-logistics/cooperative-and-unified-smart-traffic-system-(cruise) (accessed on 11 June 2025).
45. Muntather, A.; Aasheesh, S.; Hemalatha, S.; Kavitha, P.; Gambhire, G.M.; Pankaj, R.; Pardeshi, B.P. Comparative Analysis of Supercapacitors vs. Batteries. *E3S Web Conf.* **2024**, *591*, 01010. [CrossRef]
46. Szut, J.; Piątek, P.; Pauluk, M. RF Energy Harvesting. *Energies* **2024**, *17*, 1204. [CrossRef]
47. Alcaraz López, O.L.; Suto, K. RF Energy Harvesting and Wireless Power Transfer for IoT. *Sensors* **2024**, *24*, 7567. [CrossRef] [PubMed]
48. Mushtaq, M.U.; Venter, H.; Singh, A.; Owais, M. Advances in Energy Harvesting for Sustainable Wireless Sensor Network Challenges and Opportunities. *Hardware* **2025**, *3*, 1. [CrossRef]

49. Sarma, M.P.; Sarma, K.K.; Barman, P. An Optimized and Highly Efficient Low Power RF Energy Harvesting System with Current Boost Technique Designed using 45 nm Technology. *Appl. Energy* **2023**, *350*, 121756. [CrossRef]
50. Gu, X.; Grauwin, L.; Dousset, D.; Hemour, S.; Wu, K. Dynamic Ambient RF Energy Density Measurements of Montreal for Battery-Free IoT Sensor Network Planning. *IEEE Internet Things J.* **2021**, *81*, 13209–13221. [CrossRef]
51. Xu, Z.; Khalifa, A.; Mittal, A.; Nasrollahpourmotlaghzanjani, M.; Etienne-Cummings, R.; Sun, N.X.; Cash, S.S.; Shrivastava, A. Analysis and Design Methodology of RF Energy Harvesting Rectifier Circuit for Ultra-Low Power Applications. *IEEE Open J. Circuits Syst.* **2022**, *3*, 82–96. [CrossRef] [PubMed]
52. Geidam, H.A.; Kassim, S.O. Evaluation of Radiofrequency Radiation Emissions from Mobile Telephony Base Stations In Geidam Town, Yobe State. *IOSR J. Electr. Electron. Eng. (IOSR-JEEE)* **2021**, *16*, 12–17.
53. Jin, P.; Fu, J.; Wang, F.; Zhang, Y.; Wang, P.; Liu, X.; Jiao, Y.; Li, H.; Chen, Y.; Ma, Y.; et al. A flexible, stretchable system for simultaneous acoustic energy transfer and communication. *Sci. Adv.* **2021**, *7*, eabg2507. [CrossRef]
54. Deng, X.; Yang, P.; Chen, S.; Ren, W. Design of a 2.4 & 5.8 GHz Efficient Circularly Polarized Rectenna for Wireless Power Transfer Applications. *Electronics* **2023**, *12*, 2645. [CrossRef]
55. Maher, R.; Allam, A.; Kanaya, H.; Abdelrahman, A.B. Dualband rectenna for RF energy harvesting using metamaterial reflect array and novel matching technique. *AEU-Int. J. Electron. Commun.* **2024**, *173*, 155020. [CrossRef]
56. Yan, J.R.; Huang, Y.W.; Lai, W.J.; Liao, J.H.; Lin, C.C.; Liao, Y.T. A 2.4 GHz, −19 dBm Sensitivity RF Energy Harvesting CMOS Chip with 51% Peak Efficiency and 24 dB Power Dynamic Range. In Proceedings of the 2024 IEEE Radio Frequency Integrated Circuits Symposium (RFIC), Washington, DC, USA, 16–18 June 2024; pp. 43–46. [CrossRef]
57. Khan, N.U.; Khan, F.U.; Farina, M.; Merla, A. RF energy harvesters for wireless sensors, state of the art, future prospects and challenges: A review. *Phys. Eng. Sci. Med.* **2024**, *47*, 385–401. [CrossRef] [PubMed]
58. Du, J.; Wang, R.; Zheng, P. A 2.4 GHz High-Efficiency Rectifier Circuit for Ambient Low Electromagnetic Power Harvesting. *Sensors* **2024**, *24*, 6854. [CrossRef]
59. Abd Razak, I.S.; Hamid, Z. Optimization of Rectifying Circuit for RF Energy Scavenging. *ANP J. Soc. Sci. Humanit.* **2021**, *2*, 60–67. [CrossRef]
60. Roy, S.; Tiang, J.-J.; Roslee, M.B.; Ahmed, M.T.; Kouzani, A.Z.; Mahmud, M.A.P. Design of a Highly Efficient Wideband Multi-Frequency Ambient RF Energy Harvester. *Sensors* **2022**, *22*, 424. [CrossRef]
61. Boursianis, A.D.; Papadopoulou, M.S.; Koulouridis, S.; Rocca, P.; Georgiadis, A.; Tentzeris, M.M.; Goudos, S.K. Triple-Band Single-Layer Rectenna for Outdoor RF Energy Harvesting Applications. *Sensors* **2021**, *21*, 3460. [CrossRef]
62. Antonio Estrada, J.; Kwiatkowski, E.; Lopez-Yela, A.; Borgonos-Garcia, M.; Segovia-Vargas, D.; Barton, T.; Popovic, Z. RF-Harvesting Tightly Coupled Rectenna Array Tee-Shirt with Greater Than Octave Bandwidth. *IEEE Trans. Microw. Theory Tech.* **2020**, *68*, 3908–3919. [CrossRef]
63. Liu, H.; Fu, H.; Sun, L.; Lee, C.; Yeatman, E.M. Hybrid energy harvesting technology: From materials, structural design, system integration to applications. *Renew. Sustain. Energy Rev.* **2021**, *137*, 110473. [CrossRef]
64. Sharma, P.; Singh, A.K. A survey on RF energy harvesting techniques for lifetime enhancement of wireless sensor networks. *Sustain. Comput. Inform. Syst.* **2023**, *37*, 100836. [CrossRef]
65. Nosrati, M.; Rezaei, P.; Danaie, M.; Khalilpour, J. A broadband integrated rectenna for microwave radio energy harvesting using an integrated hybrid sandwich power divider. *E-Prime-Adv. Electr. Eng. Electron. Energy* **2025**, *11*, 100891. [CrossRef]
66. Alex-Amor, A.; Palomares-Caballero, A.; Fernández-González, M.J.; Padilla, P.; Marcos, D.; Sierra-Castañer, M.; Esteban, J. RF Energy Harvesting System Based on an Archimedean Spiral Antenna for Low-Power Sensor Applications. *Sensors* **2019**, *19*, 1318. [CrossRef] [PubMed]
67. Rahman, M.; Chakravarty, U. Design and Analysis of a Hybrid Solar and Vibration Energy Harvester. *Energy Harvest. Syst.* **2019**, *6*, 39–55. [CrossRef]
68. Liu, X.; Yang, J.; Yang, T.; Gao, J.; Li, J. Design of Radio Frequency Energy Harvesting System. In *Advances in Intelligent Information Hiding and Multimedia Signal Processing*; Smart Innovation, Systems and Technologies, Pan, J.S., Li, J., Namsrai, O.E., Meng, Z., Savić, M., Eds.; Springer: Singapore, 2021; Volume 211. [CrossRef]
69. Basim, M.; Khan, D.; Ain, Q.U.; Shehzad, K.; Shah, S.A.; Jang, B.; Pu, Y.; Yoo, J.; Kim, J.; Lee, K. A Highly Efficient RF-DC Converter for Energy Harvesting Applications Using a Threshold Voltage Cancellation Scheme. *Sensors* **2022**, *22*, 2659. [CrossRef]
70. Kim, S.; Shin, J.; Kim, H.; Lee, D.; Park, J.; Baik, J.M.; Kim, S.Y.; Kang, C.; Choi, W.; Song, H.; et al. A synergetic effect of piezoelectric energy harvester to enhance thermoelectric Power: An effective hybrid energy harvesting method. *Energy Convers. Manag.* **2023**, *298*, 117774. [CrossRef]
71. Ren, R.; Sun, H.; Zhang, H. A hybrid solar and RF energy harvester for applications of self-Powered wireless sensor nodes. *J. Electromagn. Waves Appl.* **2024**, *38*, 2049–2066. [CrossRef]
72. Bairappaka, S.K.; Ghosh, A.; Halimi, M.A.; Roy, B. A Dual-Band Dual-Polarized Rectenna for Efficient RF Energy Harvesting in Battery-Less IoT Devices With Broad Power Range. *Int. J. Commun. Syst.* **2025**, *38*, e6103. [CrossRef]

73. Aboualalaa, M.; Mansour, I.; Pokharel, R.K. Energy Harvesting Rectenna Using High-Gain Triple-Band Antenna for Powering Internet-of-Things (IoT) Devices in a Smart Office. *IEEE Trans. Instrum. Meas.* **2023**, *72*, 2001312. [CrossRef]
74. Zhang, J.; Wang, Y.; Song, R.; Kou, Z.; He, D. Highly Flexible Graphene-Film-Based Rectenna for Wireless Energy Harvesting. *Energy Environ. Mater.* **2024**, *7*, e12548. [CrossRef]
75. Mohan, A.; Sahoo, A.K.; Mondal, S. A tunable impedance matching strategy for RF energy harvesting systems. *Analog Integr. Circuits Signal Process* **2022**, *113*, 287–294. [CrossRef]
76. Zhang, H.; Wang, Y.; Park, H.K.; Sung, T.H. Design and investigation of small-scale long-distance RF energy harvesting system for wireless charging using CNN, LSTM, and reinforcement learning. *Front. Phys.* **2024**, *12*, 1337421. [CrossRef]
77. Chen, M.-C.; Sun, T.-W.; Tsai, T.-H. Dual-Domain Maximum Power Tracking for Multi-Input RF Energy Harvesting with a Reconfigurable Rectifier Array. *Energies* **2022**, *15*, 2068. [CrossRef]
78. Yao, M.; Li, J.; Niu, Y. Adaptive impedance matching for power management circuit for a piezoelectric energy harvester on the bridge. *Sens. Actuators A Phys.* **2021**, *331*, 112986. [CrossRef]
79. Clayton, F.; Sinhara, S.; Grija, T.; Jiangfeng, Z. High efficiency ambient RF energy harvesting by a metamaterial perfect absorber. *Opt. Mater. Express* **2022**, *12*, 1242–1250.
80. Lee, W.; Choi, S.-I.; Kim, H.-I.; Hwang, S.; Jeon, S.; Yoon, Y.-K. Metamaterial-Integrated High-Gain Rectenna for RF Sensing and Energy Harvesting Applications. *Sensors* **2021**, *21*, 6580. [CrossRef]
81. Ullah, N.; Islam, M.S.; Hoque, A.; Alzamil, A.; Soliman, M.S.; Islam, M.T. Design and development of a compact, wide-angle metamaterial electromagnetic energy harvester with multiband functionality and polarization-insensitive features. *Sci. Rep.* **2024**, *14*, 19000. [CrossRef]
82. Aygül, M.A.; Çırpan, H.A.; Arslan, H. Machine learning-based spectrum occupancy prediction: A comprehensive survey. *Front. Commun. Netw.* **2025**, *6*, 1482698. [CrossRef]
83. Sawaguchi, S. Lightweight Highly-Adaptive Reinforcement Learning Method for Energy-Harvesting IoT Applications. Ph.D. Thesis, Université Grenoble Alpes, Saint-Martin-d'Hères, France, 2021.
84. Lv, M.; Xu, E. Deep Learning on Energy Harvesting IoT Devices: Survey and Future Challenges. *IEEE Access* **2022**, *10*, 124999–125014. [CrossRef]
85. Nguyen, X.; Kha, H.H. Energy-Spectral Efficiency Trade-Offs in Full-Duplex MU-MIMO Cloud-RANs with SWIPT. *Wirel. Commun. Mob. Comput.* **2021**, *2021*, 6678792. [CrossRef]
86. Rafiq, I.; Mahmood, A.; Razzaq, S.; Jafri, S.H.M.; Aziz, I. IoT applications and challenges in smart cities and services. *J. Eng.* **2023**, *2023*, e12262. [CrossRef]
87. Wang, Z.; He, J.; Liang, X.; Ge, L. A novel SWIPT channel model and jointly age optimization in SWIPT-enabled access networks based on age of energy harvesting. *Sci. Rep.* **2025**, *15*, 10377. [CrossRef]
88. Aslam, S.; Aung, P.P.; Rafsanjani, A.S.; Majeed, A.P. Machine learning applications in energy systems: Current trends, challenges and research directions. *Energy Inform.* **2025**, *8*, 62. [CrossRef]
89. National Center for Mobility Management. Seoul Is Replacing Street Lamps with Smart Poles that Offer Electricity and WiFi. 2021. Available online: https://nationalcenterformobilitymanagement.org/news/seoul-is-replacing-street-lamps-with-smart-poles-that-offer-electricity-and-wifi/ (accessed on 11 June 2025).
90. Mouapi, A. Radiofrequency Energy Harvesting Systems for Internet of Things Applications: A Comprehensive Overview of Design Issues. *Sensors* **2022**, *22*, 8088. [CrossRef]
91. EnOcean. Energy Harvesting. 2023. Available online: https://www.enocean.com/en/technology/energy-harvesting/ (accessed on 11 June 2025).
92. Mishra, P.; Singh, G. Energy Management Systems in Sustainable Smart Cities Based on the Internet of Energy: A Technical Review. *Energies* **2023**, *16*, 6903. [CrossRef]
93. Raghav, K.S.; Bansal, D. RF Energy Harvesting for Self-Sustained IoT Applications: Challenges and Opportunities. In Proceedings of the 2023 IEEE 9th World Forum on Internet of Things (WF-IoT), Aveiro, Portugal, 12–27 October 2023; pp. 1–2. [CrossRef]
94. Politecnico di Milano. Luca Mottola-Politecnico di Milano. 2022. Available online: https://onlineservices.polimi.it/manifesti/manifesti/controller/ricerche/RicercaPerDocentiPublic.do?evn_downloadcv=evento&id_cv=1222978 (accessed on 11 June 2025).
95. Ali, A.; Eid, R.; Manaseer, D.E.; AbuJaber, H.K.; Ware, A. Dual-Band 802.11 RF Energy Harvesting Optimization for IoT Devices with Improved Patch Antenna Design and Impedance Matching. *Sensors* **2025**, *25*, 1055. [CrossRef]
96. Shuvo, M.M.; Titirsha, T.; Amin, N.; Islam, S.K. Energy Harvesting in Implantable and Wearable Medical Devices for Enduring Precision Healthcare. *Energies* **2022**, *15*, 7495. [CrossRef]
97. Biswas, S.; Lee, S.W.; Lee, Y.; Choi, J.; Chen, J.; Yang, X.; Du, Y.; Falcone, N.; Lee, M.; Kim, H.; et al. Emerging Energy Harvesters in Flexible Bioelectronics: From Wearable Devices to Biomedical Innovations. *Small Sci.* **2024**, *4*, 2300148. [CrossRef]
98. Nagaveni, S.; Hunasigidad, P.; Pathak, D.; Dutta, A. On-Chip Configurable RF Energy Harvester for Biomedical Implantable Devices. *IEEE Trans. Circuits Syst. I Regul. Pap.* **2024**, *71*, 5030–5039. [CrossRef]

99. Shokoor, F.; Shafik, W. Harvesting energy overview for sustainable wireless sensor networks. *J. Smart Cities Soc.* **2023**, *2*, 165–180. [CrossRef]
100. Yahya Alkhalaf, H.; Yazed Ahmad, M.; Ramiah, H. Self-Sustainable Biomedical Devices Powered by RF Energy: A Review. *Sensors* **2022**, *22*, 6371. [CrossRef]
101. Oliveira, L.F.P.D.; Morais, F.J.D.O.; Manera, L.T. Development of an energy harvesting system based on a thermoelectric generator for use in online predictive maintenance systems of industrial electric motors. *Sustain. Energy Technol. Assess.* **2023**, *60*, 103572. [CrossRef]
102. Paolini, G.; Guermandi, M.; Masotti, D.; Shanawani, M.; Benassi, F.; Benini, L.; Costanzo, A. RF-Powered Low-Energy Sensor Nodes for Predictive Maintenance in Electromagnetically Harsh Industrial Environments. *Sensors* **2021**, *21*, 386. [CrossRef] [PubMed]
103. León Ávila, B.Y.; García Vázquez, C.A.; Pérez Baluja, O.; Cotfas, D.T.; Cotfas, P.A. Energy harvesting techniques for wireless sensor networks: A systematic literature review. *Energy Strategy Rev.* **2025**, *57*, 101617. [CrossRef]
104. Srinivasan, D.; Kiran, A.; Parameswari, S.; Vellaichamy, J. Energy efficient hierarchical clustering based dynamic data fusion algorithm for wireless sensor networks in smart agriculture. *Sci. Rep.* **2025**, *15*, 7207. [CrossRef]
105. Pratik, J.; Aqsa, S. RF System Advancements for Satellite and Space Communications: Integrating AI with High-Frequency and Hybrid Systems. *Int. J. Sci. Res. Eng. Manag.* **2025**, *8*, 1–7.
106. Kudyba, P.S.; Sun, H. Autonomous Agricultural Monitoring with Aerial Drones and RF Energy-Harvesting Sensor Tags. *arXiv* **2025**. Available online: https://arxiv.org/abs/2502.16028 (accessed on 11 June 2025).
107. Kang, H.; Kim, H.; Hong, J.; Zhang, R.; Lee, M.; Hong, T. Harnessing sunlight beyond earth: Sustainable vision of space-based solar power systems in smart grid. *Renew. Sustain. Energy Rev.* **2024**, *202*, 114644. [CrossRef]
108. Kamuda, K.; Klepacki, D.; Sabat, W.; Kuryło, K.; Skoczylas, M. Efficiency Measurements of Energy Harvesting from Electromagnetic Environment for Selected General Purpose Telecommunication Systems. *Electronics* **2024**, *13*, 3111. [CrossRef]
109. Obaideen, K.; Albasha, L.; Iqbal, U.; Mir, H. Wireless power transfer: Applications, challenges, barriers, and the role of AI in achieving sustainable development goals—A bibliometric analysis. *Energy Strateg. Rev.* **2024**, *53*, 101376. [CrossRef]
110. Koohestani, M.; Tissier, J.; Latrach, M. A miniaturized printed rectenna for wireless RF energy harvesting around 2.45 GHz. *AEU-Int. J. Electron. Commun.* **2020**, *127*, 153478. [CrossRef]
111. Ullah, M.A.; Keshavarz, R.; Abolhasan, M.; Lipman, J.; Esselle, K.P.; Shariati, N. A Review on Antenna Technologies for Ambient RF Energy Harvesting and Wireless Power Transfer: Designs, Challenges and Applications. *IEEE Access* **2022**, *10*, 17231–17267. [CrossRef]
112. Riaz, A.; Khan, S.; Arslan, T. Design and Modelling of Graphene-Based Flexible 5G Antenna for Next-Generation Wearable Head Imaging Systems. *Micromachines* **2023**, *14*, 610. [CrossRef]
113. Parveen, N.; Abdullah, K.; Badron, K.; Javed, Y.; Khan, Z.I. Coexistence in Wireless Networks: Challenges and Opportunities. *Telecom* **2025**, *6*, 23. [CrossRef]
114. Moloudian, G.; O'Flynn, B.; Buckley, J.L. A Rectenna Design With Harmonic Suppression for RF Energy Harvesting Applications. In Proceedings of the 2023 17th European Conference on Antennas and Propagation (EuCAP), Florence, Italy, 26–31 March 2023; pp. 1–5. [CrossRef]
115. Liu, Z.; Li, T.; Li, S.; Mi, C.C. Advancements and challenges in wireless power transfer: A comprehensive review. *Nexus* **2024**, *1*, 100014. [CrossRef]
116. Van Mulders, J.; Delabie, D.; Lecluyse, C.; Buyle, C.; Callebaut, G.; De Strycker, L. Wireless Power Transfer: Systems, Circuits, Standards, and Use Cases. *Sensors* **2022**, *22*, 5573. [CrossRef]
117. Sun, X.; Zhang, J.; Wang, W.; He, D. A Wearable Dual-Band Magnetoelectric Dipole Rectenna for Radio Frequency Energy Harvesting. *Electronics* **2025**, *14*, 1314. [CrossRef]
118. Alsharif, M.H.; Kim, S.; Kuruoğlu, N. Energy Harvesting Techniques for Wireless Sensor Networks/Radio-Frequency Identification: A Review. *Symmetry* **2019**, *11*, 865. [CrossRef]
119. Ghaneizadeh, A.; Gavriilidis, P.; Joodaki, M.; Alexandropoulos, G.C. Metasurface Energy Harvesters: State-of-the-Art Designs and Their Potential for Energy Sustainable Reconfigurable Intelligent Surfaces. *IEEE Access* **2024**, *12*, 160464–160494. [CrossRef]
120. Alhamad, R.; Boujemâa, H. Simultaneously transmitting and reflecting reconfigurable intelligent surfaces (STAR-RIS) with energy harvesting and adaptive power. *Signal Image Video Process* **2024**, *18*, 1155–1160. [CrossRef]
121. Kayed, S.I.; Elsheakh, D.N.; Mohamed, H.A.; Shawkey, H.A. Multiband Microstrip Rectenna Using ZnO-Based Planar Schottky Diode for RF Energy Harvesting Applications. *Micromachines* **2023**, *14*, 1006. [CrossRef] [PubMed]
122. Wang, J.; Yu, H. RF energy harvesting schemes for intelligent reflecting surface-aided cognitive radio sensor networks. *Sci. Rep.* **2022**, *12*, 22462. [CrossRef]
123. Worka, C.E.; Khan, F.A.; Ahmed, Q.Z.; Sureephong, P.; Alade, T. Reconfigurable Intelligent Surface (RIS)-Assisted Non-Terrestrial Network (NTN)-Based 6G Communications: A Contemporary Survey. *Sensors* **2024**, *24*, 6958. [CrossRef]

124. Surender, D.; Halimi, M.A.; Khan, T.; Talukdar, F.A.; Rengarajan, S.R. 5G/Millimeter-Wave Rectenna Systems for Radio-Frequency Energy Harvesting/Wireless Power Transmission Applications: An overview. *IEEE Antennas Propag. Mag.* **2023**, *65*, 57–76. [CrossRef]
125. Sharma, S.; Popli, R.; Singh, S.; Chhabra, G.; Saini, G.S.; Singh, M.; Sandhu, A.; Sharma, A.; Kumar, R. The Role of 6G Technologies in Advancing Smart City Applications: Opportunities and Challenges. *Sustainability* **2024**, *16*, 7039. [CrossRef]
126. Chataut, R.; Nankya, M.; Akl, R. 6G Networks and the AI Revolution—Exploring Technologies, Applications, and Emerging Challenges. *Sensors* **2024**, *24*, 1888. [CrossRef]
127. Ramalingam, L.; Mariappan, S.; Parameswaran, P.; Rajendran, J.; Nitesh, R.S.; Kumar, N.; Nathan, A.; Yarman, B.S. The Advancement of Radio Frequency Energy Harvesters (RFEHs) as a Revolutionary Approach for Solving Energy Crisis in Wireless Communication Devices: A Review. *IEEE Access* **2021**, *9*, 106107–106139. [CrossRef]
128. Alhussien, N.; Aaron Gulliver, T. Toward AI-Enabled Green 6G Networks: A Resource Management Perspective. *IEEE Access* **2024**, *12*, 132972–132995. [CrossRef]
129. Song, C.; Wang, L.; Lu, P.; Zhang, C.; Chen, Z.; Zheng, X.; He, Y.; Goussetis, G.; Vandenbosch, G.A.E.; Huang, Y. Highly Efficient Wideband mmWave Rectennas for Wireless Power Transfer System with Low-Cost Multi-Node Tracking Capability. *IEEE Trans. Antennas Propag.* **2023**, *71*, 8773–8787. [CrossRef]
130. Ezzeddine, Z.; Khalil, A.; Zeddini, B.; Ouslimani, H.H. A Survey on Green Enablers: A Study on the Energy Efficiency of AI-Based 5G Networks. *Sensors* **2024**, *24*, 4609. [CrossRef] [PubMed]
131. Ali, A.; Shaukat, H.; Elahi, H.; Taimur, S.; Manan, M.Q.; Altabey, W.A.; Kouritem, S.A.; Noori, M. Advancements in energy harvesting techniques for sustainable IoT devices. *Results Eng.* **2025**, *26*, 104820. [CrossRef]
132. Safaei, B.; Peiravian, M.; Siamaki, M. Eco-Friendly IoT: Leveraging Energy Harvesting for a Sustainable Future. *IEEE Sens. Rev.* **2025**, *2*, 32–75. [CrossRef]

telecom

MDPI

Review

Supporting Global Communications of 6G Networks Using AI, Digital Twin, Hybrid and Integrated Networks, and Cloud: Features, Challenges, and Recommendations

Shaymaa Ayad Mohammed [1], Sallar S. Murad [2,3,*], Havot J. Albeyboni [4], Mohammad Dehghani Soltani [5], Reham A. Ahmed [6], Rozin Badeel [7] and Ping Chen [8]

1 Department of Computer Engineering, University of Technology, Baghdad 10066, Iraq; shaimaa.a.mohammed@uotechnology.edu.iq
2 Department of Computing, College of Computing and Informatics, Universiti Tenaga Nasional (UNITEN), Kajang 43000, Malaysia
3 Research and Development Division, BEAMBRIDGE LIMITED, London EC3N 2LB, UK
4 Department of Electrical and Computer Engineering, University of Duhok, Duhok P.O. Box 78, Iraq; havot.albeyboni@uod.ac
5 ESWIN, 4500 Parkway, Whiteley, Fareham PO15 7AZ, UK; msoltan2024@gmail.com or mohammad.soltani@eswincomputing.com
6 Faculty of Technology Management and Business, Universiti Tun Hussein Onn Malaysia (UTHM), Batu Pahat 86400, Johor, Malaysia; rehamabdul.ahmed@gmail.com
7 Department of Network, Parallel & Distributed Computing, University Putra Malaysia, Seri Kembangan 43400, Malaysia; rozinbabdal1987@gmail.com
8 Department of Educational Studies, University Putra Malaysia, Seri Kembangan 43400, Malaysia; chenn.ping06@gmail.com
* Correspondence: sallarmurad@gmail.com

Abstract: The commercial deployment of fifth generation (5G) mobile communication networks has begun, bringing with it novel offerings, improved user activities, and a variety of opportunities for different types of organizations. However, there still exist several challenges to implementing 5G technology. Sixth generation (6G) wireless communication technology development has begun on a worldwide scale in response to these challenges. Even though there have been many discussions on this topic in the past, many questions remain unanswered in the present literature. The article provides a comprehensive overview of 6G, including the common understanding of the concept, as well as its technical requirements and potential applications. A comprehensive analysis of the 6G network design, potential uses, and key elements are covered. This research article delineates future study topics and unresolved challenges to stimulate an ongoing global discourse. This analysis and content of this study supports the use of different applications and services that will benefit the community in the near future using the 6G technology. Subsequently, recommendations for each problem are provided, offering solutions to unresolved difficulties where functionalities are anticipated to improve, hence enhancing the overall user experience.

Keywords: 6G network; 6G application; 6G future technology

1. Introduction

Due to the swift advancement of communication applications, communication technologies are experiencing successive revolutionary transformations. So far, the progress of cellular mobile communication systems has experienced five generations of development. Each subsequent generation of wireless communication systems, starting from the first generation (1G) analogue systems to the 5G online systems, progressively includes greater frequencies, wider bandwidths, and faster data rates.

The 5G base units are designed to utilize advanced technologies such as massive multiple-input multiple-output (MIMO) [1], mmWave, and ultra-dense networking (UDN) [2]. These innovations allow the ground stations to accommodate up to 64 transceiver links with a greater number of antenna components. At present, there are fully developed commercialized 5G infrastructure solutions equipped having 128 antennas [3]. HUAWEI has taken the lead in introducing a huge MIMO base station with 384 antennas. Furthermore, 5G can facilitate augmented reality (AR) and virtual reality (VR). Despite the substantial advancements it brings beyond 4G connectivity technology, 5G possesses certain limitations. Presently, there exist apps and solutions that demand superior communication reliability surpassing the potential of 5G. These include worldwide coverage, transfer of data at extremely high rates, minimal delay, densely connected networks, precise positioning, highly trustworthy and secure connections, low power consumption, high energy effectiveness, and pervasive creativity. Therefore, the limitations and challenges of the 5G technology are part of the motivation behind the need for higher and faster level of connectivity such as the 6G.

For low transmission speed, the end-to-end (E2E) latency must be below 1 millisecond, and for high speed, the latency ought to be in the microsecond range. Furthermore, to achieve high-precision placement, the exactness of the placement must be enhanced to attain centimetre-level accuracy in outdoor environments and sub-centimetre-level accuracy in indoor environments. Greater dependability is needed in a variety of innovative scenarios, including vehicle-to-everything (V2X) and mobile data centres. Energy consumption is a significant concern for many different applications; hence, it is necessary to decrease power spending and boost network energy efficiency by a factor of 100. Furthermore, a significant multitude of sophisticated applications necessitate communication systems to possess a heightened level of intellect.

The progression of wireless communication systems follows a recurring pattern with a duration of around ten years. Currently, the study on 6G is at its initial phase of investigation, and multiple nations and standards organizations worldwide have declared their intentions to perform an investigation on 6G such as Finland [4], China [5], Japan [6], Germany [7], USA [8], Korea [9], and the EU [10].

The world's first major 6G research initiative was initiated in 2018 by the Finnish government. In the U.S., 6G development was suggested in 2019, and the terahertz (THz) spectrum was made available for research by the Federal Communications Commission (FCC). The 6G vision for Japan was unveiled in April 2020 by the Ministry of Internal Affairs and Telecommunications. The 6G timetable for South Korea was announced in January 2020, with commercialization anticipated by 2028. The European Union has established the 6G Smart Networks and Solutions Business Organization to promote and develop 6G networks and products.

All of the aforementioned plans include vigorous 6G preparatory work. For instance, in September 2019, an initial 6G white paper [4] was published by Finland's 6G flagship agency. The article outlined the concept of "ubiquitous wireless intelligence" and focused on the main drivers, obstacles, and associated scientific concerns of 6G. Additionally, following that, a number of white papers were published [11] that addressed many facets of 6G, including networking, ML uses, enterprise, edge intellect, protection, and more. In a white paper published in [12], the authors examined the transition from 5G to 6G and predicted the essential components that will make up 6G. Furthermore, a number of companies have conducted preliminary 6G investigations, including UNISOC [13], China Unicom [14], HUAWEI [15], and China Mobile [16].

In a study conducted by NTT DOCOMO [17], they explored the future development of 5G and the necessary components, applications, and innovations for 6G. They highlighted that the portable communication system undergoes significant shifts every decade, and the process of generating new market worth in the mobile communication industry takes approximately 20 years. In a statement made in [18], Samsung predicted that 6G technology would offer customers an unparalleled experience by enabling hyper-connectivity between people and all other entities. In addition, other alliances within the communication industry have also undertaken preliminary efforts in the development of 6G technology. The Next Generation Mobile Networks alliance conducted an analysis of the key elements that influence the development of 6G. These variables include social aspirations, market desires, and priorities. The alliance also explored the transition from 5G to 6G, focusing on the viewpoint of a 6G vision [19].

There are technical objectives can be obtained by using the 6G system, summarized as follows:

(a) Data transfer rates of one petabit per second or more, or one hundred times faster than 5G. We anticipate peak data rates of up to 10 Tb/s for some unique use cases, including THz wireless backhaul and fronthaul (x-haul);
(b) One gigabit per second (Gb/s) as experienced by users, ten times faster than 5G. In certain cases, such indoor hotspots, it is anticipated to offer a data rate of up to 10 Gb/s as perceived by the user;
(c) A delay of 10–100 μs across the air and elevated mobility exceeding 1000 km/h. This will ensure a satisfactory Quality of Experience for scenarios such as hyper-high-speed rail and aviation systems;
(d) Tenfold the connectivity frequency of 5G. This will accommodate up to 10 units per km^2 and an area traffic capacity of up to 1 $Gb/s/m^2$ for situations such as hotspots;
(e) A spectrum effectiveness of 5–10 times and an energy effectiveness of 10–100 times compared to 5G.

Figure 1 shows some of the capabilities of the 6G networks.

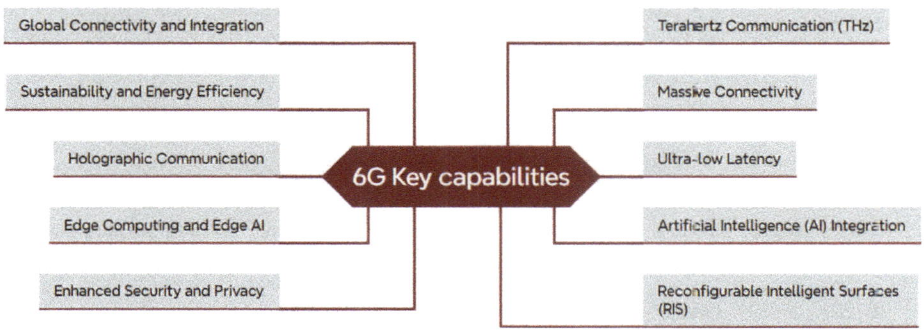

Figure 1. Sixth generation key capabilities.

1.1. Summary

Table 1 provides a summary of the shortcomings and constraints of 5G, which are expected to be addressed and handled in coming 6G communication networks. This analysis is based on an examination of the shortcomings between the features of 5G and the anticipated needs of the next decade.

Table 1. Concise constraints and difficulties encountered in the 5G technology.

	Industrial Application	Requirement	Performance Indicator	Expected Challenges
Coverage	Marine and satellite communications	Cover sea and remote area including rural	Ocean coverage: 5% Land coverage: 20%	Coverage: space-air-ground-sea
Data transmission	Ultra-high-definition video	Very high-speed transmission	Rate: <20 Gbps User data rate: ~100 Mbps	Tbps level peak data rate. User date rate: 1–10 Gbps
Low latency	Auto drive, high production level	High speed, and low latency	Delay: <1 ms but low speed	Sub-second (<1 ms)
Solid connection	Crowded malls, auto production lines, stations	Super dense equipment and population	10^6 devices/km^2	Density: 10^8 device/km^2
High placement (accuracy)	UMV positioning and navigation, indoor positioning	Positioning of indoor or outdoor	Indoor: ~1 m Outdoor: ~10 m	Indoor centimetre level positioning. Outdoor metre level positioning
Reliable/safe	V2X, wireless data, telemedicine, internet	Super safe	99%	99%
Power use/energy	Intermediate altitude communications, internet of things	Reduce the consumption, increase efficiency	Energy efficiency: 10^7 bit/J	Network energy enhancement: 10^9 bit/J
Universal intelligent	AI applications, sensing technologies, digital twins	Support various applications	Low	High

As seen in the table above, the use of 6G networks will provide a wide coverage that could reach far areas and cities including rural areas. Given the swift progress of the wireless communications sector, it is imperative to address the constraints of 5G and advance the invention of 6G. Utilizing a collection of current proactive 6G studies, we present a thorough analysis and overview of 6G. This study seeks to provide a precise description of 6G, encompassing the prevailing knowledge of 6G, and to thoroughly examine the latest advancements in 6G challenges in different aspects are discussed and explored. Recommendations for different issues are presented in detail. Related literature is presented showing the existing studies and their publication data that hold the necessary knowledge that could support the 6G systems.

1.2. Recent Studies and Insights

Supporting technologies including artificial intelligence (AI), cloud, and/or the digital twin technology (DT) are expected to contribute to the development of the coming 6G networks. Table 2 shows recent studies regarding the use of these technologies.

Table 2. Recent studies showing the use of supporting technologies for the 5G/6G networks.

Technology	Ref.	Source (Journal)	Year	Contribution
Digital twin	[20]	IEEE Internet of Things	2024	The aim of this study is to improve computational performance and network security in 6G Internet of Vehicles (IoV) environments. To attain this, this study proposes an AI enhanced DT framework incorporating an advanced feature engineering module comprising stacked sparse autoencoders (ssAE) for dimension reduction wherein the AutoFS and AutoCM specialize in an online learning module for accurate attack detection. With the proposed DT framework, the performance of attack detection increases significantly, while simultaneously reducing system latency, energy consumption and RAM usage while increasing packet delivery rates in dynamic 6G IoV networks.
	[21]	Sensors	2022	The objective of this study is to develop a DT for NB-IoT wireless communication in an industrial indoor environment, and to analyze its performance when operating within real world conditions. To achieve this, simulations are conducted using Wireless InSite software to model and analyze critical communication metrics like throughput, SNR and received power. This study revealed that industrial environments reduce NB-IoT performance, with throughput 21% to 32% below peak rates. The results provide insights for improving communication efficiency in such settings.
	[22]	IEEE Access	2023	This study seeks to formulate a DT model aimed at unobtrusive patient respiration monitoring in healthcare 4.0. For this purpose, Wi-Fi Carrier State Information (CSI) is used and signal processing approaches like Elliptic Filters, Principal Component Analysis (PCA) to denoise raw data of the signal as well as Machine Learning (ML) algorithms for binary and multi-class classification such as Neural-Networks. To improve patient monitoring and decision support in healthcare settings, the study contributes to improving data processing and classification accuracy.
	[23]	IEEE Wireless Communications	2024	This study focused on the generative AI-enabled DT for wireless networks and propose a hierarchical design approach based on DT considering generative AI to handle challenges in 6G network architecture. This is accomplished using generative AI models such as Transformers and GANs to create message-level and policy-level DTs of the core network control plane. Through admission control and resource allocation, the study improves the stability and reliability of the network and proposes a scalable 6G network DTs-based framework.
	[24]	IEEE Conference	2024	The purpose of this work is to develop a wireless multi-hop network management system that continuously examines the network context in order to obtain network performance insight and suggest reconfiguration. To achieve this, the system combines simulations within the Click Router architecture, graph-based algorithms using the LEMON C++ graph library, and a theoretical link model to construct a Digital Twin Network (DTN) for optimizing Bluetooth Mesh networks. The DTN based system is demonstrated using testbeds which results in improved network resilience and management, especially for traffic scheduling in IoT constrained devices.

Table 2. *Cont.*

Technology	Ref.	Source (Journal)	Year	Contribution
Digital twin	[25]	Sensors	2022	The objective of this study is to enhance the positioning accuracy of UWB technology in indoor environments by developing a DT based indoor positioning system that addresses challenges posed by NLOS and LOS indoor obstructions. This is achieved using Slime Mould Algorithm (SMA) for optimizing position of anchor point. Employing a simulated model of anchor placement with Slime Mold Algorithm and addresses every network's error utilization through the implementation of neural networks. This approach enhances UWB indoor positioning systems accuracy by improving anchor point positioning and handling errors in LOS and NLOS situations.
AI	[26]	IEEE Internet of Things	2022	The objective of this study is Reducing total MAC layer latency and improving reliability in Beyond 5G (B5G) networks. To achieve this, a Reinforcement Learning (RL)-enabled MAC scheduler is applied, using the UCB1 algorithm. The study successfully lowers MAC layer latency and improves reliability in B5G networks by optimizing resource allocation and queuing for heterogeneous traffic.
	[27]	IEEE BITS the Information Theory Magazine	2022	This objective of this study is to demonstrate how ML can be used to bypass traditional channel modelling to optimize wireless communication system design. This is achieved through the use of Deep Neural Networks (DNNs) and end-to-end training to directly optimize system-level objectives. The study enhances system performance by optimizing RIS reflection coefficients, distributing source coding for massive MIMO and mmWave initial alignment, and solving data-driven optimization problems.
	[28]	IEEE Communications Standards Magazine	2023	This paper attempts to solve bandwidth, privacy, data security, and inconsistent internet speed for Federated Learning (FL) for the Internet of Things (IoT). This is achieved by introducing a new framework based on 6G technology called super wireless over-the-air FL. Using interference resistant radio waves, the framework provides privacy protection for enhancing AI performance, ensuring that IoT system is smarter, faster and safer.
	[29]	IEEE Access	2024	The aim of this work is to determine the optimal beamforming angles for transmitters and receivers in 5G and 6G wireless communication systems while reducing computation complexity. To achieve this RL algorithms were applied such as Q Learning, SARSA, Double Q Learning, and Expected SARSA. With enhanced data throughput and avoiding interference, the study effectively reduces complexity and improves channel capacity.
	[30]	IEEE Networks	2022	The objective of this study is improving the quality of service (QoS) decision-making by preventing traffic congestion in Mobile Networks (MNs). This is achieved by applying Convolutional Neural Networks (CNN) which directly maps the condition of the MN to optimized admission control strategies. The model uses real world data from a telecom operator to enhance decision-making processes to help network operators make better informed more intelligent and timely decision to reduce congestion and improve QoS of user.

Table 2. *Cont.*

Technology	Ref.	Source (Journal)	Year	Contribution
AI	[31]	MDPI Electronics	2023	The aim of this study is to enhance 6G native AI training latency and task accuracy by enhancing the quality of AI services (QoAIS) in 6G native AI wireless networks. To achieve this, the study utilize G-TSRA and NSG-TSRA heuristic algorithms. This improved accuracy and reduction in latency in 6G native AI wireless networks by resolving the issue of optimizing task scheduling and resource allocation in AI training services.
Cloud	[32]	IEEE	2023	The objective of this work is to improve the flexibility in resource management and adaptability to the changing service requirements in the 5G/6G networks. Integration of Software Defined Networking (SDN) and cloud virtualization approaches in the context of Multi-access Edge Computing (MEC) architecture is used to achieve this. Thus, allowing rapid adaptation to critical user requirements and ensure service continuity for the expected user mobility in MEC environments enables service delivery and low latency in time critical services.
	[33]	CMES	2024	The objective of this study is to optimize the performance of FL in 5G/6G networks by considering issues related to non-IID data, low client engagement and heterogeneous device clients. This is achieved by applying the adaptive server selection FedAdaSS algorithm, which utilize cloud computing resources. The study effectively reduces communication overhead by dynamically selecting the best server for each round of training, and random reshuffling on client sides to mitigate performance loss due to low participation in FL processes.
	[34]	Future Generation Computer Systems	2023	This study aims at reducing service request delays and achieving low cost in 6G cloud-edge systems. To achieve this, the Service Deployment with Service Rise and Fall (SD-SRF) algorithm is developed for offline service arrangement based on a greedy approach, and the Proximal Policy Optimization for Multi-layer Service Deployment (PPO-MSD) algorithm is used on an online service deployment model as Markov Decision Process (MDP). The study effectively reduces service delays and improves cost efficiency in 6G networks by optimizing service deployment in multi-layer edge networks.
	[35]	ICTACT Journal on Microelectronics	2022	The aim of this study is to eliminate the interference in 5G ultra-wideband communication antennas and enhance performance in the cloud computing networks. This is achieved by proposing a novel Ultra-Wideband Communication Antenna (UWBCA) design with a 'double square' girder structure. This results in reducing latency, and improving spectral efficiency, data rates and connection density over existing antenna models.
	[36]	Sensors	2023	This study aims to improve eHealth IoT system performance by integrating 5G network slicing, cloud computing, and fog computing. This is achieved using an IoT–fog–cloud architecture queuing model analysis employing performance metrics. Through resource allocation optimization with virtual BBU and fog nodes, the study improves throughput, QoS, and minimizes latency in 5G eHealth applications.

Table 2. *Cont.*

Technology	Ref.	Source (Journal)	Year	Contribution
Cloud	[37]	Sensors	2023	This study aims to enhance energy efficiency and data management for 6G surveillance systems. This is accomplished by utilizing a clustering algorithm with ZigBee energy optimization to minimize power usage and extend the lifetime of WSN. The reduction in network consumption by integration with the AODVjr algorithm, as well as embedded cloud computing providing data processing improvement through reduced latency and better storage, are accomplished. The proposed solution increases system reliability, enhances data security and effectiveness of multi-scale monitoring in 6G environment.

1.3. Paper Organization

The rest of the study is organized as follows: Section 2 presents the goals of the 6G globally with future applications and components. In Section 3, open issues in some applications are discussed and analyzed then supported with recommendations for solving each challenge that could enhance the functionalities of the system and overcome certain issues. Finally, conclusions are written in Section 4.

2. Worldwide 6G Goals and Expectations

Research manuscripts reporting large data sets that are deposited in a publicly available database should specify where the data have been deposited and provide the relevant accession numbers. If the accession numbers have not yet been obtained at the time of submission, please state that they will be provided during review. They must be provided prior to publication.

It is anticipated that 6G will be distinct from 5G in a number of relevant respects. When it comes to the demands of applications, 5G expanded the "Mobile Internet" that was available in 4G to the "Internet of Everything". After 5G, 6G will strive to improve mobile Internet and Internet of Everything (IoE), but it will also deeply combine these technologies with artificial intelligence and big data in order to achieve smart IoE.

In terms of technological specifications, 6G will aim for broader coverage, enhanced data speeds, increased connectivity, ultra-low latency, exceptional location precision, merging of communication and detecting, greater intellect, improved security, and superior replacement compared to 5G. It is important to recognize that 6G will transcend mere communication. The practical use and technological demands prompt us to conceptualize the nature of 6G.

The 6G concept has been presented in several publications [38–47]. Specifically, in [38], the authors lay out the cutting-edge 5G tech and point out why 6G research is essential. They expected mobile ultra-broadband, super Internet of Things (IoT) [48], and AI to be three main components of 6G, which will take into account the present and future growth of wireless communications. They then go into the essential technologies needed to make each part a reality. Machine learning methods [49] show great promise as AI candidates, symbiotic radio and satellite-assisted communications can be utilized to accomplish super IoT, and THz communications can be utilized to provide mobile ultra-broadband.

The usages of 6G can be integrated with many technologies including AI, IoT, and mobile broadband, Figure 2. Wireless transmissions of terabits per second are possible with mobile ultra-broadband, which can support far and rural areas including villages and small communities. With super IoT, the present Internet of Things can have better connectivity capabilities and coverage. AI has the potential to intelligently design and improve wireless networks in the years to come.

Figure 2. Expected uses of 6G in the community within various areas: AI, IoT, and mobile ultra-broadband.

The writers devoted their attention on studying intelligent 6G wireless networks and multi-Tb/s in [39], with an eye toward the years 2030 and beyond. They laid out a 6G vision, talked about potential uses, and highlighted 6G's main features. The study [40] examined the primary factors influencing 6G, which arise from the problems and operational constraints of 5G, with the technology-driven paradigm shift and growth of wireless networks.

As the global installation of 5G systems accelerates, it is imperative to anticipate the evolution towards above 5G systems. This shift is primarily motivated by arising societal trends that demand entirely automated and smart services, bolstered by extra reality and tangible communications. To meet the rigorous demands of these prospective applications, characterized by data-driven needs and necessitating highly low delay, ultra-reliable, rapid, and seamless wireless connectivity, research initiatives are presently concentrating on an evolving roadmap for sixth generation (6G) networks, anticipated to induce revolutionary shifts in this domain.

Reference [41] elucidates several important enabling innovations for 6G, anticipated to transform the foundational architectures of cellular networks and deliver a variety of homogeneous artificial intelligence-enhanced services, encompassing spread communications, management, computing, detecting, and energy, from the centre to the end nodes. The study [42] articulated a comprehensive, future-oriented perspective that delineates the principles of a 6G infrastructure. We believe that 6G will not only include the search of additional spectrum in high-frequency groups but will instead represent a convergence of emerging technology trends propelled by innovative underlying services. Figure 3 presents an overview of the Essential foundations for 6G.

Establishing an idea of potential communications is crucial to guide that research. In [43], scientists endeavoured to delineate the comprehensive landscape of communication requirements and technology within the context of 6G. The 6G will be refined and economically adapted for the novel applications presented in 5G, facilitating their widespread deployment. Simultaneously, it will facilitate novel applications that we are currently unable to conceive or articulate comprehensively. The proliferation of mobile cellular into various sectors, initiated by the advent of affordable IoT technologies in 4G and ultra-reliable low latency Industrial Internet of Things (IIoT) in 5G, will persist, expanding in both scope and depth in 6G.

Figure 3. Essential foundations for 6G.

Figure 4 Present the Corresponding Analytical Instruments for 6G and concepts pertinent to many open topics. Using different tools for the 6G system will definitely support the community and provide higher quality to all users.

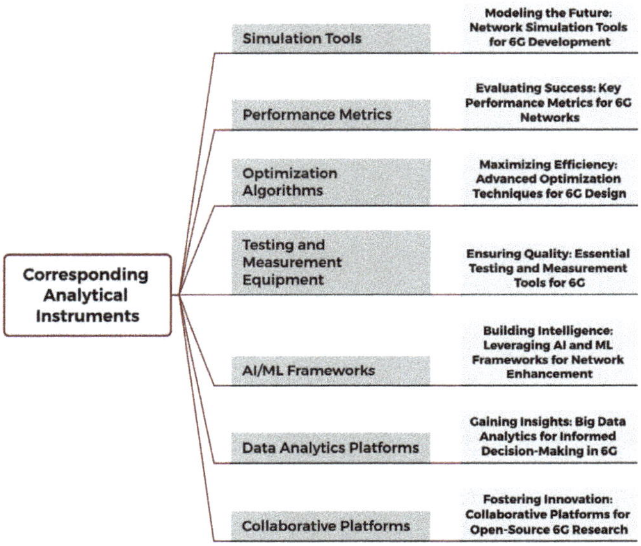

Figure 4. Corresponding Analytical Instruments for 6G.

The swift progression of AI/ML innovation and its efficacy in addressing challenges across various sectors indicates a 6G system that will fundamentally leverage these new capabilities to enhance productivity by more effectively responding to the working context. Emerging topics are expected to influence the needs and advancements of 6G systems, including:

(a) Novel man–machine interactions established by a consortium of various local equipment functioning collaboratively;
(b) Pervasive global computation dispersed across numerous local devices and the cloud;

(c) Integration of multi-sensory information to generate multi-verse maps and innovative mixed-reality experiences; and
(d) Accurate sensing and actuation for the manipulation of the tangible environment.

The above points are summarized in Figure 5, which shows expectations and components. As shown, many devices will be able to function in higher performance and without interruption. This means the operation of services in the community will be enhanced.

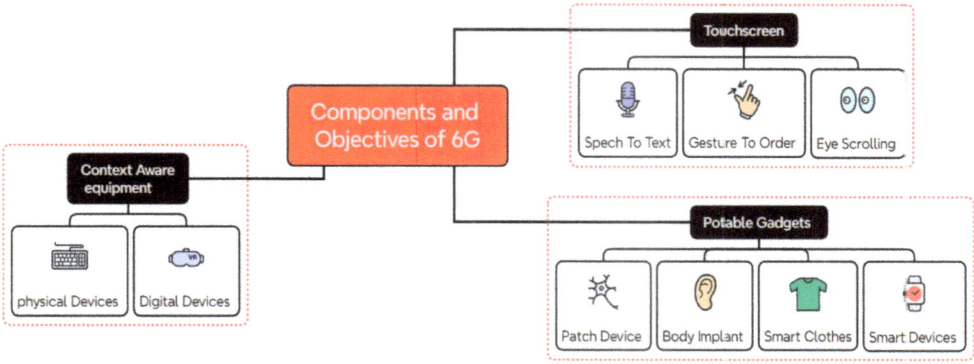

Figure 5. Components and expectations for using smartphones and tablets with man–machine interfaces in the community.

Within the next 5 to 10 years, we will initiate the employment of gadgets that individuals can employ to connect to the network. Although smartphones and tablets will persist, we are likely to encounter novel human–machine designs that will significantly enhance our ability to access and manage information [43,47]. Limited assumptions may be established regarding the existing enabling capabilities associated with 6G innovations, described as follows:

(a) Portable gadgets, including earphones and those integrated into attire, will become prevalent, while skin overlays and bio-implants could also gain acceptance. We may grow dependent on novel cognitive detectors to operate machines. We will possess numerous electronic devices that will function harmoniously together, offering smooth and user-friendly layouts.
(b) Touchscreen input will become obsolete. Instead, we will use gestures and verbal communication with our devices as standard practice and possibly use AI personal assistant for it.
(c) The equipment we utilize will be entirely context-aware, and the network will progressively enhance its ability to anticipate our requirements. The integration of context awareness with advanced human–machine gateways will enhance our interactions with both the tangible and virtual domains, rendering them less confusing and more productive.

The computational requirements for these electronics will probably not be entirely contained within the gadgets alone due to constraints related to their size and battery life [43]. Instead, they may need to depend on locally accessible computer assets to do tasks, beyond the edge-cloud [41]. Telecommunications are going to play a crucial part in the future man–machine interface [50,51]. As clients, we may anticipate that the following may occur:

(a) Contemporary self-driving idea vehicles will be accessible to the general public by the 2030s. They will predominantly operate autonomously, but a remote driver or rider may still be required to assume control under specific circumstances. This will significantly augment the amount of time given for us to access data from the internet in the forms of enhanced amusement, enriched communications, or schooling. Automobiles will substantially increase data consumption: (i) real-time uploads of automobile sensor information to the network, (ii) downloads of high-resolution visualizations, and (iii) immediate inter-vehicle connectivity will occur.

(b) A substantial deployment of wireless cameras as sensors will occur. Advancements in AI and machine vision, along with its ability to identify individuals and things, will render the camera a ubiquitous sensor applicable in all contexts. Concerns about privacy will be mitigated by restricting access to data and masking material. Additionally, radio and other sensing approaches, such as acoustics, will be employed to collect environmental data.

(c) Sophisticated methodologies will be employed in security screening protocols to eradicate protection queues. A multifaceted approach utilizing many sensing modalities will be employed to check individuals as they navigate through congested regions, rather than solely at entry points. Radio sensing will be a crucial element in accomplishing this, bolstered by upcoming communication technologies.

(d) Digital currencies and credentials may emerge as standard, with transactions in both physical and digital realms executed via the multitude of gadgets at our disposal. The forthcoming network must ensure the security and privacy essential for this shift.A multitude of household utility robots will augment the vacuum cleaners and lawnmowers currently in use. These may manifest as a collective of mini robots collaborating to do duties. The robots will be outfitted with video cameras transmitting to a nearby computing servers for immediate execution. Consequently, we will observe a rise in the quantity of products and augmented resource demands within our home networks.

(e) Healthcare will undergo significant transformation, featuring continuous observation of critical metrics for both healthy individuals and patients via various portable gadgets. Health surveillance will encompass internal gadgets that interact with external wearables, which subsequently transmit information to the web.

Figure 6 shows the network capabilities and possibilities for the man–machine applications including applications and components that support different purposes in the community. Some applications that will be supported by the 6G system are maps and localizations, which means the rural areas will benefit from the services with the wide and strong transmission within the right settings. Other wireless technologies can be used together with the 6G network as hybrid systems including WiFi, light fidelity (LiFi), and/or other broadband systems.

The transition to Industry 4.0 and the initial phase of wireless-enabled automation will occur prior to the 2030s. Fifth-generation systems offering ultra-reliable minimal latency connections will enable real-time computing in the cloud computing, including task scheduling in cloud environment [52]. Nevertheless, applications in industry necessitating significantly more strict demands for wireless connectivity will need 6G. For example, holographic virtual presence is going to be standard for professional and everyday interactions. It will be feasible to create the illusion of being in a specific position while actually residing in another, such as looking to be in the office while being situated in a vehicle. We will develop technologies that integrate current face expressions with an avatar that appears in the virtual illustration of any genuine environment. There will be extensive utilization of mobile robot groups and drones [53,54] across multiple sectors, including tourism, healthcare, warehousing, and parcel delivery. Active digital twins in the digital age, featuring progressively precise and synchronized refreshes of the physical environment, will serve as a crucial platform for enhancing cognitivity among humans.

From the aforementioned view of the future, we may derive different key use cases, including augmented reality, telepresence, security monitoring and surveillance, and more. This encompasses the capabilities that 5G will facilitate, alongside widespread adoption anticipated within the 6G period, utilizing novel technology, in addition to a range of new use cases made possible by these 6G innovations.

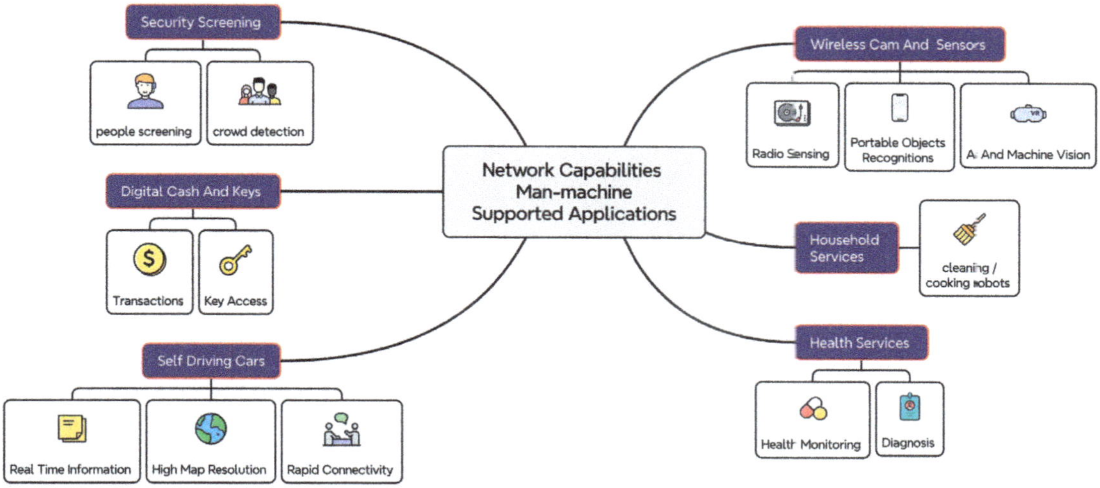

Figure 6. Network support for the man–machine applications and services for the community.

In 2030 as well as thereafter, the rapid advancement of informatization and digitalization will create a new realm of digital twins [55], where each tangible object possesses an online counterpart, as seen in Figure 7. The virtual realm creates a conduit for the effortless exchange of information and intelligence among individuals, between humans and objects, and even among objects themselves within the living creature. The digital realm simulates and forecasts the live entity, precisely mirroring and anticipating its actual condition. The modelling and anticipation of biological entities from the virtual realm may also be employed to pre-emptively intervene in the operations of a living entity or the physical world, thereby mitigating risks and calamities in preparation.

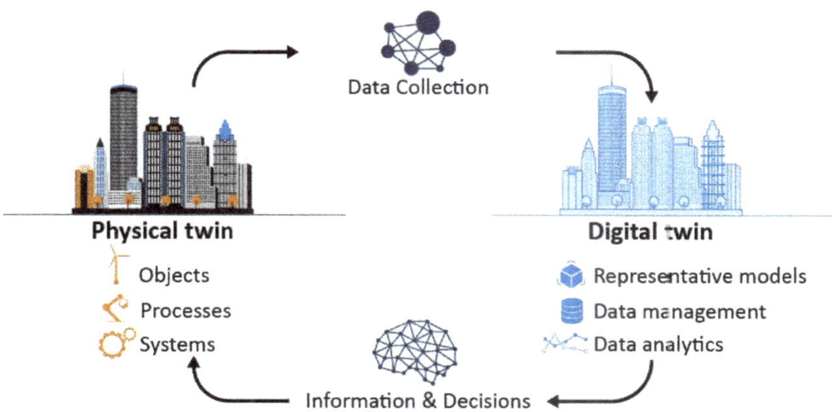

Figure 7. Illustration of the digital twin.

Detailed explanations of anticipated needs pertaining to the vision will be provided in the paragraphs that follow. First, 6G should be a fully merged network that covers more ground and has better reception in more places. This includes both land and satellite connectivity as well as short-range communication among devices.

Second, 6G is anticipated to operate at greater frequencies in order to attain a broader bandwidth, including mmWave, Terahertz, visible light, and others. With a maximum transfer rate of 1 Tb/s and a user encounter data rate of 10 Gb/s, 6G may achieve a data rate that is 10 to 100 times higher than 5G. Further improvement of reusing frequency productivity is possible with 6G's dynamic frequency-sharing solution. Third, 6G is a smart, tailored network. When 6G and AI come together, personal mobile communication will become a reality. Instead of a standard function-centralized system, there will be a new three-centralized model: user-centred, data-centred, and completely content-centred. Fourth, the 6G system will incorporate operational security into the architecture or use an internal security mechanism. Cyberspace safety can be achieved with the support of 6G's self-awareness, real-time flexible assessment, and evolved threat and trust assessment, which are made possible by trust and safety systems. Fifthly, interactions, directions, and detection will all be rolled into 6G. Satellite positioning and localization systems, radar detection systems, and global telecommunication systems are all part of 6G. With its software-defined aerial access networks and backbone systems, 6G will have an accessible design that allows for quick and sophisticated construction as well as flexible activation of network functionalities. Lastly, 6G has the potential to produce vast amounts of data via the Internet of Everything. It may additionally integrate with cutting-edge technologies like edge computing, cloud computation, artificial intelligence, blockchain, and more. With 6G, we can finally achieve entire genius and cluster intellect, which is a kind of group collective intelligence. We can also finally establish a pervasive smarter mobile Community.

3. Challenges and Related Recommendation

More investigation into numerous topics and avenues of study is necessary before the 6G goal of global reach, complete applications, solid security, all perceptions, and all electronics can be realized. Furthermore, there is an immediate need for innovations in 6G telecommunication concepts. In this section, many challenges will be discussed and analyzed. Furthermore, recommendations and features will be suggested for the discussed challenges.

3.1. Fundamental Theories Challenges

Few aspects are included in this section including channel research, information theory of the electromagnetic, and baseband processing including the unified, iterative, hardware generation, and architecture. When discussing Unique Network Research, there are typically four stages to conventional channel research: measuring the channel, estimating its parameters, analyzing its characteristics, and finally, modelling it. Several restrictions are imposed by this passive method of channel recognition. It takes an extensive amount of time, assets, and effort to perform the channel assessment. Also, channel assessments will never be able to capture every possible frequency range or circumstance in the real world. Channel parameter evaluation is complicated and plagued by massive amounts of data. The channel characteristics can only be examined at predetermined frequencies and in predetermined circumstances; thus, the complex link among novel qualities and frequencies/scenarios cannot be thoroughly explored. Last but not least, conventional channel modelling cannot foretell the future channel properties of frequency ranges or scenarios for which no data are now available.

Various channel models, such as 6G established pervasive channel demonstration [56], AI-based forecasting 6G channel demonstrating [57,58], scenario flexible channel modelling, and RIS medium modelling, will need to switch from passively acknowledging channels to proactively controlling them for 6G [59]. Various methods for optimizing antenna–channel interaction are shown in Figure 8. Two methodologies have been employed for channel interaction, distinguished by their level of sophistication. The primary research avenues for AI-based strategies in optimizing antenna and channel compounds involve either the straightforward modification of certain high-level station attributes, such as antenna/beam/tilt choice, or the additional immediate alteration of antenna arrangements according to channel information. AI is anticipated to be especially beneficial for the ideal pattern creation technique illustrated in Figure 8, since a flexible construction of the radiating architecture entails a substantial improvement search space.

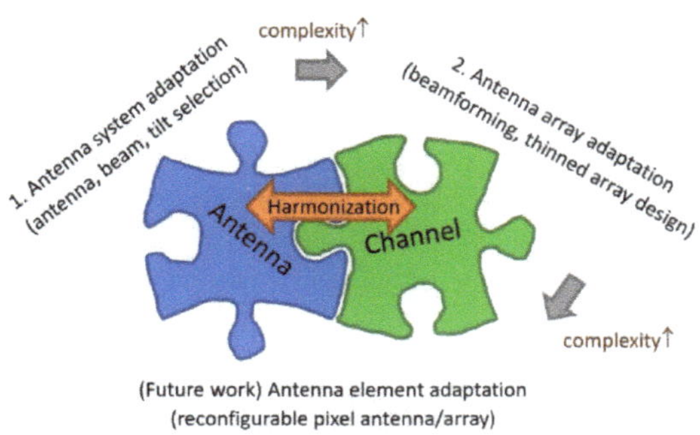

Figure 8. Various methods for optimizing antenna–channel interaction. The grey arrows represent the complexity direction (increasing), while the orange arrow represents a two-way direction between the antenna and the channel.

When it comes to ML-based channel description and forecasting, there are two main areas where the techniques shine. In the first, it is commonly used to find the locating function among the surroundings and the channel characteristics or to immediately mimic the assessment data (Figure 9a,b), respectively. In the second, it is more common to acquire knowledge about the past behaviour of channels and use that information to forecast or extrapolate future channels (Figure 9c).

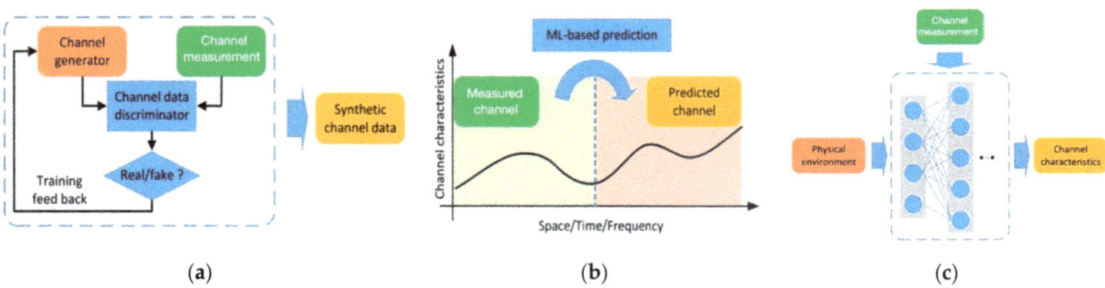

Figure 9. Fundamental concepts of ML-based channel (**a**) steps included in a simulation, (**b**) modelling of prediction using ML, and (**c**) forecasting.

One might come across it in the most recent research: most ML-based channel forecasting techniques are built on top of artificial neural networks (ANNs), which have varied network topologies but both offline training and online executions with minimal computing overhead. As a result, creating effective channel prediction algorithms presents several formidable obstacles related to the type of the ANN, and the input/output type that should be used for training.

Conversely, the conceptual underpinnings of wireless communications rest on the theory of information and electromagnetic (EM) theory. The (EM theory) describes how electromagnetic waves are created and how they travel through space. When dealing with constrained resources like power and bandwidth, the information theory lays out a framework for precise and effective data transmission.

The continuous-space EM field distribution can be derived from the EM theory, which is based on uninterrupted space and time. It is not possible to determine the channel capability only from electromagnetic theory. The channel capability in finite space can be determined using information theory; nevertheless, in an ongoing space, it is not possible. Transmitter theory and wireless transmission channel modelling theory link the electromagnetic theory and information theory. Specifically, the wireless transmission channel connects electromagnetic theory and information theory; it starts at antennas and conveys data via electromagnetic pulses.

Keep in mind that as 6G core innovations advance, there will be issues that none of the aforementioned theories can fully address, creating an opportunity and a threat for a joint application of the four theories. The quantity of users, base stations, relays, and/or RISs, for instance, keeps going up as 6G wireless communication networks grow from covering only a local area on land to covering the entire planet through space, air, ground, and sea interconnected networks.

In mobile networks like low earth orbit (LEO) satellite, UAV, and vehicle networks, the locations of both base stations and users are always shifting. In order to aid in system design, 6G wireless communication networks are evolving from discontinuous to ongoing spaces, necessitating the collecting of channel strength information (CStrI) and its calculation regardless of location in the continuous area. The aforementioned four theories will inevitably be integrated in this scenario. Furthermore, as the dimensions and elements of transmitters in ultra-massive MIMO expand, the Tx/Rx transmitters become increasingly interconnected with their surroundings. Therefore, it is essential to integrate the theories of transmitters and wireless transmission channels in order to describe Tx/Rx antennas and the underlying wireless transmission channel. There are additional requirements for channel characterization, antenna architecture, and continuous-space channel potential computation due to the transformation of antenna grids from isolated to ongoing windows and the decreasing antenna unit separation in holographic MIMO.

Once again, all four of these theories must be considered together. In conclusion, 6G wireless networks offer fresh uses and practical specifications that surpass the reach of any one theory when compared to 5G. Traditional theoretical frameworks have hit roadblocks in their advancement. To lay the theoretical groundwork for 6G wireless communication networks and facilitate fresh findings, it is crucial to investigate the merging of electromagnetic theory and information theory, namely, EM information theory [60,61].

In terms of baseband processing, two distinct varieties can be described: unified and iterative. In communication systems, channel coding has always been crucial for the unifying thanks to its error-correcting potential. Sixth-generation communication systems must include unified implementations of improved coding schemes and simplified decoding algorithms to accommodate dynamic and varied circumstances. In order to satisfy the strict criteria of 6G, it is often necessary to look at the architecture and execution of the entire baseband processor, rather than just refining one channel coding unit. In contrast, iterative receptions have a longer history of being considered to have better system throughput and connection dependability than the current state of distinct baseband processing techniques [62]. The channel estimation, MIMO detection system, network decoder, NOMA detector, and origin decoder perform recurrent soft information swaps with each other [63–66].

The heightened utilization of every module due to repetitions may impose significant delay and computational overhead on the system. Since the majority of modules focus on addressing the greatest a posteriori estimate problem, they can be integrated into a cohesive Bayesian system by leveraging their identical features, hence enhancing the efficiency of the iterative receiver. Given the serial nature of the signal circulation, one alternative method is to consider many modules collectively, treating the end result as the target for joint processing, including joint detection and decoding [64].

Figure 10 illustrates an example [65] of the instance of sparse code multiple access (SCMA) encoding for $J = 6$ and $K = 4$, employing quadrature phase shift keying (QPSK) modulation, with a modulation magnitude of 2 for QPSK. Every user is allocated two assets based on the codebook selected. The J codewords are, thereafter, transferred using K supplies. Each SCMA codeword takes only a portion of the supply. The whole configuration of the SCMA might be depicted by a $K \times J$ factor graph matrix. A component that is greater than zero in the k-th line and j-th field of the factor graph matrix indicates that the asset k is allocated to user j. Every asset is utilized by N_u users, and the communication from each user is conveyed by N_r assets.

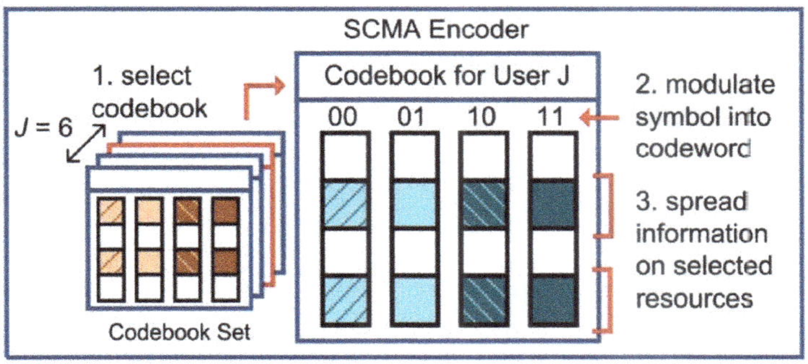

Figure 10. Example of SCMA encoding procedure.

Reliable unified equipment design and accompanying implementations are crucial for meeting many realistic purposes based on cyclic or cooperative baseband processing. In pre-6G telecommunication networks, the layout and construction of each module are segregated, and the absence of system-level design considerations leads to increased hardware asset expenditures. To enhance the adaptability, compatibility, and hardware effectiveness of 6G baseband handling executions, the VLSI-DSP technique [67], which includes retiming, bending, and unfolding, may be employed to create a solid design featuring fixed handling aspects, improved network interconnections, and iterative timing timetables.

The intricate and varied uses of 6G networks of communication necessitate specialized electronic architecture automated solutions for baseband handling, which can reduce the entry obstacles for circuit designs across numerous scenarios and fulfil client requirements. An effectively developed automation tool can autonomously produce the circuit architecture according to the customer's quality specifications. Utilizing the unified Bayesian system, the majority of the computation modules may be defined to enhance the auto-generation technique, hence establishing a design space for optimizing hardware attributes [68,69]. Figure 11 illustrates the automated development of hardware architectures, it shows a hardware architectures auto generation where the F represents the latency and throughput while I, II, and III refer to a different type of the polar encoders. Standardization of polar codes for use in improved mobile broadband management channels has piqued the interest of other potential users. For the purpose of automating the generation of Verilog HDL files for the hardware implementation of $F_{I(N,M)}$, $F_{II(N,M)}$, and $F_{III(N,H)}$, a Python compiler called the Polar Compiler was employed in the study [68]. The compiler files are available on GitHub [70]. AI can assist in the discovery of design spaces with complex needs. It can also assist in identifying the compensatory parameters induced by estimation, quantization methodologies, and iterative techniques in the automated hardware design process [71].

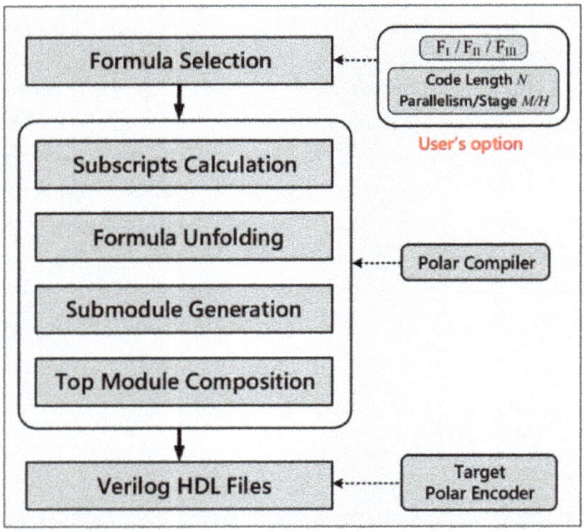

Figure 11. Auto-generation of hardware architectures for target polar encoders [68].

3.2. Future Research Challenges

6G networks approaching challenges by using hybrid RF–Optical networks are discussed in this section in addition to space–air–ground–sear (SAGS) networks, integrated sensing and computing networks, digital twins' networks, and AI-enabled networks.

3.2.1. Hybrid RF–Optical Networks

One exciting new direction for 6G is the integration of systems operating on many frequency channels—including RF and the entire optical wireless bands—to create diverse and mixed networks capable of supporting all of these bands. This will help bring about the entire spectrum vision of 6G. Indoor, outdoor V2V, open space, and submarine settings are only a few examples of the many uses for RF–optical diverse mixed networks [72,73].

However, there are a lot of obstacles to overcome on the road to RF–optical mixed networks and systems. First, RF–optical mixed-signal networks share many of the same difficulties as other types of mixed networks, such as those that span space, air, ground, and sea. These include but are not limited to the following: the development of transfer protocols for the network, traffic balancing [74], synchronization of mixed networks, distribution of resources and enhancement of energy efficiency [75], distribution of spectrum [76], and distribution of power and connection access points [77].

Because of its low energy usage and high attainable data speeds, white LEDs have increased the energy efficiency of data transmission and are now used for both lighting and communications. Consequently, LiFi approaches have a wide range of possible uses, including as anti-jamming in military settings, vulnerable civilian contexts like hospitals and airplanes, and broadband indoor communication. One way to increase the system's performance is by allowing wireless networks with various technologies to collaborate, which is called a mixed wireless network. The variations in fading channels, transmission losses, and accessible resources among networks allow for these improvements. Such improvements, however, are contingent upon successfully completing the enormous task of creating resource distribution algorithms that distribute bandwidth and power among diverse networks in order to meet varying operation demands. One way to improve the energy efficiency of RF communication systems is to use numerous radio gateways.

6G networks can be utilized for the support of smart buildings together with other technologies. Hybrid LiFi/WiFi networks have been used in smart buildings to assist social distancing [78,79], where vehicle access is managed and controlled after data exchange at the entrance at the gate of the building using the mixed wireless network. The presented approach can be assisted and improved using 6G networks, especially when using other services such as cloud servers [52], and other solutions. However, the challenge will be providing the proper infrastructure and power for big buildings and high-density environments. When designing RF and optical communication networks, it is important to think about transceivers with unique characteristics. For example, users face significant difficulties with frequent handovers (HOs) [80–83] due to the diversity of access points when they roam. Several studies have proposed different methods for HO for hybrid LiFi networks. However, since LiFi networks used to provide higher data rates than WiFi, when mixing 6G networks with other networks such as LiFi or any optical network, new challenges will occur including the mobility and access point assignment due to the high data rates not only from the LiFi network but also from the 6G networks, which makes the criteria of data rates, capacity, or network speed are not the main aspects of allocating the network to the user. This remains unknown since LiFi and 6G networks are still under development and are not yet popular for the end-user.

Additionally, uplink OWCs still have significant limits owing to the low energy of terminal equipment, the significant impact of mobility on them, and the interference that these uplink OWCs cause with lighting. Lastly, it is impossible to overlook the mixed systems' impact on system security [84].

Other problems that were studied and investigated in hybrid LiFi networks include optimization of network throughput and data rate provided to the user mainly in load balancing and selecting the best possible access point [74,85–87]. On the other hand, machine learning has been used in such networks for problem solving and network optimization [29,88–90]. Various simulation results were obtained by the existing studies including increasing the network throughput and/or data rate, reducing the switching/outage probability and complexity. Optimizing LiFi hybrid network performance is essential for user quality in indoor scenarios and in other applications such as vehicle communication.

3.2.2. Space–Air–Ground–Sear (SAGS) Networks and Integrated Sensing and Computing Networks

Realizing the dream of 6G worldwide coverage will require the SAGS integrated network, which is both a trend and a critical technology. There are numerous technical and theoretical concerns with network design, building, preservation, and optimization that need to be addressed before the SAGS integrated network can be considered fully operational. Developing and researching communication systems and networks relies on first establishing communication channels. A prevalent channel structure [73] must take into account how to incorporate distinct channel properties in satellite [91], unmanned aerial vehicle (UAV) [92], ocean [93,94], and ground scenarios, among others, due to the significant difficulties posed by different frequency bands and scenarios when measuring and predicting channels [95].

Satellite networking stands apart due to its extensive support coverage qualities and the fact that it is a proven and operational technology with a lengthy history. Several fields have begun to rely on communications via satellite in recent years, including the broadcasting industry, directions, and earth monitoring. There is a lot to explore in using next-gen domestic wireless communication infrastructure, which means that satellite networks have to meet greater standards for things like bandwidth, QoS, and ubiquitous connection if integrated with 6G networks. Predicting the fundamental genuine transmission channel properties requires a generic, precise, and low-complexity approach, which is essential for the development and efficiency assessment of communications satellites [96].

In addition to dependable control and non-payload communication links, several emerging UAV uses, like distant sensing and aerial snapshots, necessitate a significant transmission data rate for payload connectivity [97]. The vast range and enormous capacity of existing and upcoming 6G networks make it a viable option to use future cellular networks to meet UAV communication requirements. Visions of future 6G wireless communication network deployments point to UAVs acting as base stations for mobile devices as a key component in expanding network reach [39]. Figure 12 shows some tools that enables the UAV system.

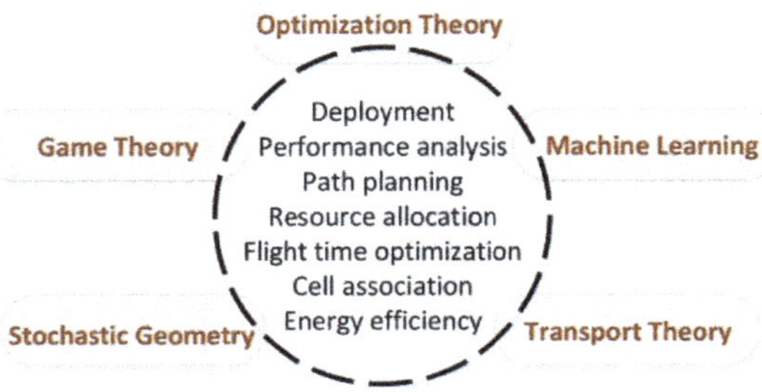

Figure 12. UAV communication system tools.

An attractive option for providing ubiquitous access from above to ground user equipment in some situations and locations (such as hotspot regions, big public spaces, and isolated places without network service) is the installation of aerial BSs [39]. The worldwide Internet of Things or global reach via 6G SAGS connected networks can be realized by integrating UAVs with other critical innovations pertaining to satellite, terrestrial, and ocean telecommunications [98]. Multipath elements (MPEs) produced by ground-level landscapes and structures are the root source of small-scale channel fading.

Figure 13a shows a conventional UAV-to-ground channel situation that incorporates all conceivable forms of MPCs. Two possible paths for the broadcast data to reach are the LoS route and the multibounce route. To improve the accuracy of the UAV-to-ground channel simulation while decreasing computational complexity. In Figure 13b, we can see a simplified representation of the network structure for UAV-to-ship connectivity in maritime settings. A transceiver is attached to the UAV side, while a receiver is attached to the ship side, where SB refers to a single bounce, while MB is a multi-bounce. It is possible to place several antennas on each side.

While sub-6 GHz is the most common frequency for UAV-to-ship communication, mmWave bands are also an option for seamless integration with future 6G networks [99]. Ocean surface roughness increases reflections and causes fading of multiple paths as a result of water fluctuation [100]. Figure 13c shows that there are numerous possible uses for UAVs and 6G wireless channels across various frequency bands. There is a striking disparity between the sounders and features of each channel.

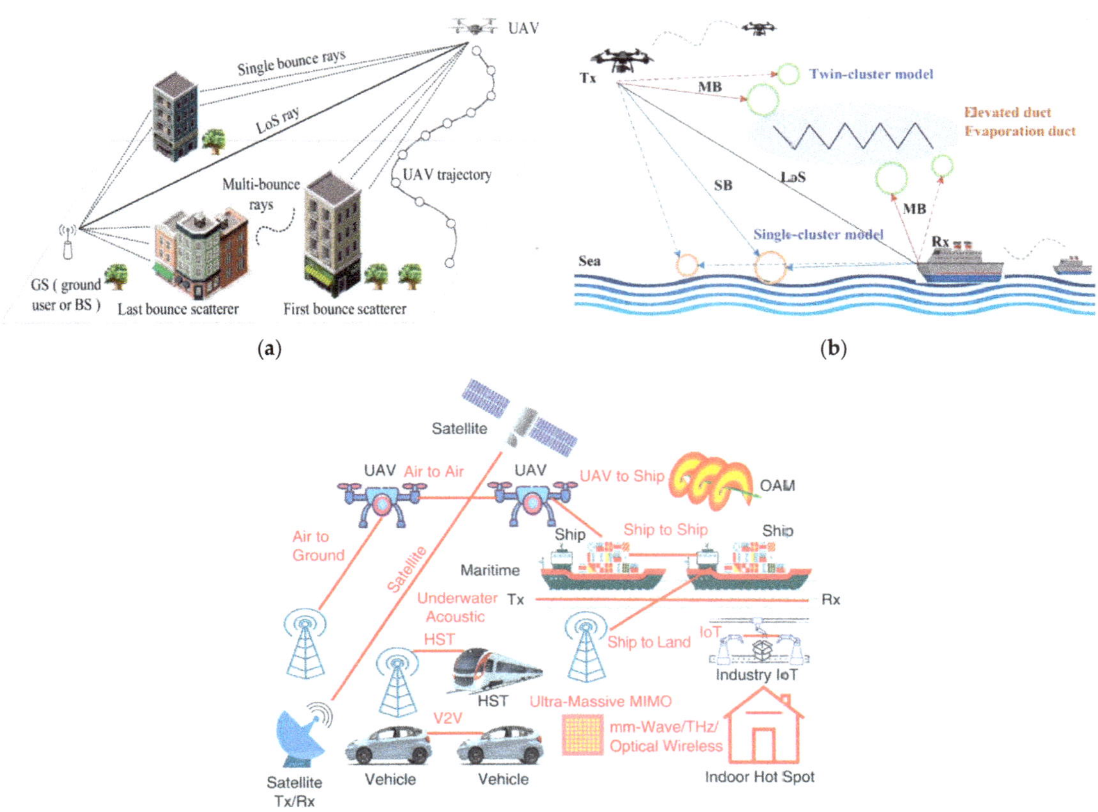

Figure 13. Wireless and broadband applications in different scenarios. (**a**) UAV-to-ground communication in urban and rural areas, illustrating line-of-sight (LOS) and multi-bounce rays with a UAV trajectory over diverse building environments, (**b**) UAV-based communication over marine areas, depicting single-cluster and twin-cluster models, with evaporation duct effects and various signal propagation paths, and (**c**) an overview of various 6G transmission and reception scenarios, including UAV-to-UAV, UAV-to-ship, ship-to-land, underwater acoustic communication, vehicle-to-vehicle (V2V), ultra-massive MIMO, mmWave/THz/optical wireless, and IoT applications for industry and indoor hotspots.

In addition, there is an immediate need for creativity in the following areas of the SAGS unified network: structure layout, movement administration, network procedure, resource placement, routing tactics, EE improvement, and beyond. To ensure that users have access to secure and dependable communication solutions, it is important to plan a network layout that is both effective and secure [101]. It is essential that the interconnected networks have efficient, secure, and transparent authentication mechanisms [102]. Advanced mobility control systems are necessary to ensure smooth mobility control between diverse and homogeneous networks, taking into account various forms of movement in different settings [46]. When planning network protocols, resource allocation, and routing techniques, it is important to keep in mind the SAGS combined network's flexible architecture and elevated movement features [103]. Preservation and improvement of SAGS-based networks necessitates taking into account the power provision of the connectivity base as well as the burden of UAVs and other equipment, all of which place high demands on the total EE of the system.

Simultaneously, it is essential to evaluate how to achieve a balance between enhancing network efficiency and controlling expenses [104]. Furthermore, the application of AI, deep learning, and other smart technologies to improve network structure and enhance entire network reliability [105] is a significant contemporary concern. The current network designs are inadequate for delivering customized network solutions tailored to various task kinds in SAGS. Furthermore, they are unable to effectively address numerous issues in SAGS, including diverse network separation, elevated network latency, sporadic interruptions, and imbalanced network traffic. Edge-cloud processing technology and networks might be employed to resolve the aforementioned issues, facilitating smart networking, and minimizing latencies. Various tasks operating within SAGS and their scheduling [52] may possess distinct network performance needs. SAGS cannot uniformly fulfil the requirements of all tasks. The user quality may be compromised, resulting in an imbalanced network load while addressing a more critical demand.

Moreover, equipment roaming, and an inadequate connection setting may result in sporadic network outages and extended network reaction delay. The current network topologies are inadequate for SAGS in this complex environment. The conventional TCP/IP design is incapable of addressing link interruptions and exhibits drawbacks such as suboptimal bandwidth usage and inadequate security for the network. The information hub system [106], constrained by the extent and volume of data within SAGS, will result in a mass of material namespaces. This will result in significant latency cost for material extraction, hence extending network latency. Moreover, current network topologies do not address the issue of mixed network cooperation, resulting in suboptimal resource use, and proactive propagation is frequently neglected.

3.2.3. Cloud-Edge Computing

Edge and cloud computing can be utilized to improve, secure, the advances the SAGS systems. The advancement of the web, IoT, and cloud computing has enabled diverse innovative wireless gateway devices to connect to the network, offering sophisticated and practical solutions for individuals. The notion of mobile-edge computing (MEC) was established. MEC units are to be deployed at the network's edge to service individuals, with edges referring to random places along the path between the data provider and the cloud computing core [107]. Every edge-computing component possesses computational, networking, and storage assets that can aid the cloud centre in managing activities. To enhance service efficiency, it is essential to develop an innovative network structure to accommodate the demanding SAGS.

There are numerous tasks in SAGS. Adhering solely to the conventional cloud-computing model will eventually result in significant strain on the cloud-computing core, possibly leading to network congestion and paralysis, so impairing the general efficiency of the entire network. The process of collecting, interpreting, analyzing, and mining data is a collaborative effort between edge computing and conventional cloud computing. Therefore, merging edge computing will add significant advantages to the system. The edge-computing terminals are multi-purpose and can store user data among their many other functions. In order to expedite the reaction time to the demand, the working result and the information to be forwarded can be saved locally in the information storage table. To some degree, the network's accessibility is enhanced since data packets can be swiftly routed in the event of a disruption or malfunction. They are able to compute and extract the needs for task-oriented networking. Not only do edge-computing components manage the network and provide interface services, but they are also tasked with communicating with cloud-computing canters and other realms of the network. It receives feedback data and tasks from users and transfers it to a cloud-based data centre. Latency decrease, balance of load, and enhanced network security are some of the benefits that can be achieved through the use of edge computing [108].

As seen in Figure 14 [109], the four sector networks are represented by four separate but interconnected processors that make up the cloud computing core. The central data centre in the cloud is able to exert control over SAGS on a worldwide scale by collecting data from the edge computing terminals in every network region. The use of SAGS supports far and wide connections including rural, maritime, and highlands.

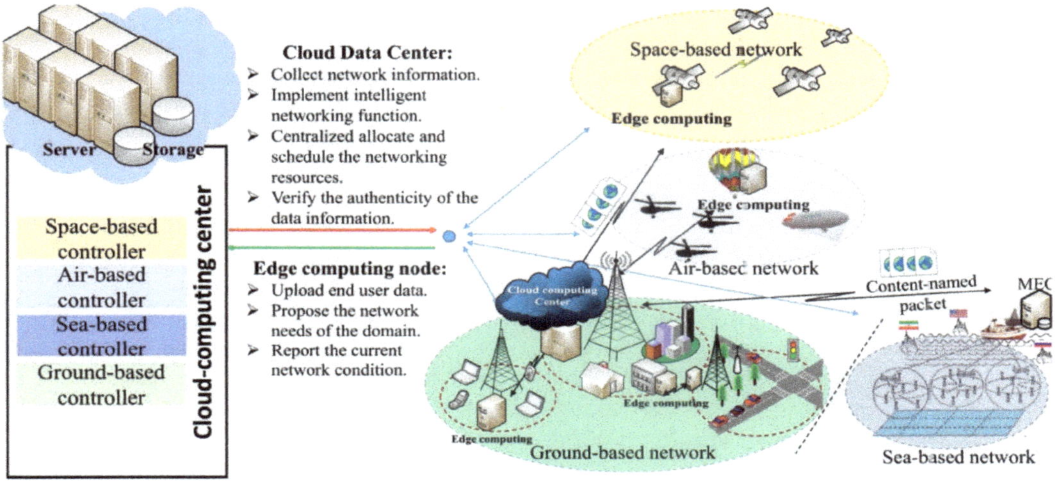

Figure 14. Illustration of cloud and edge computing in a SAGS system.

Through high-level convergence and collaborative augmentation, the 6G network will make possible the combined use of mobile communication, smart awareness, and processing power activities. The innovative extensive network known as the interconnected network of communication, detecting, and processing was created by combining all three of these technologies. Waveform layout and processing of signals [110], sensory wireless systems [111], integrated sensing and communication (ISAC) network-based detecting [112], calculating power offloading [113], partnership of multi-layer calculating power supplies [113,114], and cloud-edge gateway asset distribution for the multi-layer widespread computing system are some of the primary areas of present investigation in the field of information system architecture and computer networks.

Cloud/fog computing have made application offloading a popular subject. The decision-making process can be centralized (as is the case in major existing works) or decentralized, with a single or multiple user devices (UEs), and every app can be loaded in three different ways: coarse-grained application level, fine-grained task level, or percentage. Then, it can be loaded into the cloud/fog [115]. It is common practice to think about task allocation and segmentation in a single-UE scenario, with each job having to decide whether to offload based on specific requirements. For optimal productivity in a multi-UE environment, it is important to distribute assets (such as power and bandwidth) that are divided among the fog cluster and the UEs.

By implementing computing solutions close to mobile consumers through the fog computing model, the end-to-end latency that users encounter can be greatly minimized. This lowered latency is helpful for fresh applications that are sensitive to delays, like electronic healthcare, real-time management, and automobile communication [116], which can only withstand latencies of several milliseconds [117]. Limitations on future networks are largely due to their substantial use of energy. Energy is a major factor in increasing running costs and negatively impacting the environment. Due to the fact that base stations consume the vast majority of energy in cellular systems, implementing smart small base station (SBS) execution techniques can greatly decrease the consumption of energy [118].

There are two main approaches to managing SBS operations: centralization and distribution. Within the former, an SBS functions independently by virtue of its intelligent self-organizing characteristics. On the other hand, in a centralized management system, data are pooled in the centralized baseband unit (BBU) for optimized network-aware treatment from BSs and other assisting components. Centralized BBU pools have the capacity to make decisions that affect the entire network and are responsible for managing all of its components. Centralized management has a larger cost than dispersed management, but it improves system performance due to its judgments being more informed and certain. Hence, network units must be in perfect sync to keep QoS adequate before starting the On/Off and traffic offloading procedures [119].

According to the required quality of service demands, such as power and delay, processing workloads can be offloaded from edge units to the cloud or vice versa, much as traffic offloading in cellular networks [120]. That is to say, computing duties can be handled locally by the edge equipment or remotely by the cloud via the MBS's backhaul connections [121]. Nevertheless, the load on servers in the cloud, communication supplies, and backhaul lines will naturally grow when jobs are pushed to the main cloud. Planned task loading in stacked cloud-fog architectures can also add significant latency due to increased communication latency [122]. Hence, the effects of job loading on the computational and communication nodes must be considered.

From a computational standpoint, executing complicated activities in the central cloud represents a significant advancement in enhancing computing possibilities for humans and computers [123]. Nonetheless, the continually rising quantity of connected equipment within the realms of the IoTs, intelligent homes, and driverless driving will ultimately overwhelm or potentially incapacitate servers in the cloud. Therefore, it is imperative to filter data to alleviate the strain on cloud and network assets and to enhance the quality of user experience, particularly concerning end-to-end latency [124]. Numerous obstacles remain for the combined networks for connectivity, detecting, and computing, particularly in the advancement of ISAC innovations, connected sensing technologies, and processing power network solutions facilitated by the combination of these disciplines. The accessibility and main networks must be able to communicate, sense, and compute in order for the linked sensing with additional computation, sensing, and connectivity to be realized. As a result, specifications for the processing and transceiver parts, network architecture, media access control, navigation procedure, and placement of resources are proposed.

Collaboration sensing across multiple nodes in a network and the need for individualized data collection are further factors to think about. Lastly, research into methods of providing extensive advance knowledge for the best scheduling selection of decentralized computational power via the network sensing feature is still required based on the multi-layer ubiquitous computing system. As a result, using exchanged global computing resources to conduct tailored extraction of features and data synthesis on sensor data is difficult. In addition, it would be fascinating to investigate in the future ways to increase the power of ubiquitous computing by improving communication quality.

3.2.4. Digital Twins Networks

The implementation of 6G's all-digital concept relies heavily on digital twin modern technology. One positive aspect is that new digital twin applications, such as digital twin body area network [46] and digital twin city [125], will be made possible by the 6G network, which will be an improvement over 5G in terms of performance. By contrast, a more protected, productive, smart, and visually represented 6G network can be realized more quickly with the help of digital twin innovation applied to communication networks. This is achieved via the use of real-time linking and interactivity between the two networks.

Significant network necessities, such as holo-graphic system virtual-real communication visualization, entire life cycle administration, and real-time closed-loop management [126], are inherent to the network virtual twin concept but also provide a number of challenges when trying to put it into practice. Initially, the 6G network will have extensive network components interconnected through intricate network architectures. Consequently, modelling a genuine actual network in real time presents a significant problem.

The wireless route digital twin is an essential component of the system digital twin. A precise, real-time scenario-adaptive channel framework must effectively define the transmission setting in immediate terms and forecast upcoming alterations in wireless channels. Furthermore, due to the discrepancies in the practical execution and functionalities of hardware from various manufacturers inside the system, it is imperative to focus on the suitability of network machinery during data gathering, handling, and modelling [127]. Moreover, the extensive 6G network presents significant issues with gathering information, retrieval, administration, and analysis. It is essential to investigate the computational construction of complex network structure links in extensive systems [127].

3.2.5. AI-Enabled Networks

Network AI is a crucial enabling technology in the 6G era for AI-enabled smart wireless telecommunication networks. It offers a whole AI ecosystem within the network itself, including AI facilities AI process logic, data and design solutions, etc. [128]. The field of network AI remains in its infancy, which means that there is a great deal of unanswered questions. This is despite the fact that AI innovations are advancing at a quick pace and that the previously described fresh approaches may make an impact. To begin, studies on artificial intelligence in wireless networks rely heavily on data. Consequently, the most pressing issue is figuring out methods to set up a shared data set for investigations as well as how to gather and utilize data in wireless networks. Paying close attention to the functioning of AI products is also essential when implementing AI to the network. Improved AI functionality may be obtained by carefully selecting AI algorithms and allocating network assets.

An internet-based network design method that adjusts to network fluctuations and modelling the association across AI outcomes and network setup are the biggest obstacles [129]. It is critical to investigate ways to enhance the EE and decrease energy usage as well as expenses prior to the widespread implementation of AI services [130]. This is because more sophisticated and precise AI systems tend to use a lot of resources and have substantial negative environmental effects.

If present trends continue, the increasing computing load of deep learning may soon limit its use in various fields, rendering the attainment of crucial benchmark goals unreachable. Deep learning becoming computationally limited would not be unprecedented in history. Productivity was constrained by the processing resources accessible even when Frank Rosenblatt created the initial neural networks. Thanks to acceleration from switching to specially designed equipment and a desire to dedicate additional assets to enhance productivity, many functional limitations have been loosened in the last decade. As the amount of data sets increases, the amount of deep learning parameters also has to expand, which can lead to overparameterization. Computational demands increase by a minimum the square of the data sets in the overparameterized case, while the expense of training a deep learning structure grows with a combination of the total amount of variables and the quantity of data sets. When it comes to deep learning systems, this exponential expansion is not nearly quick enough to boost ability. In order to see a linear improvement in achievement, the volume of data used for training typically needs to rise at a faster-than-linear rate. Linear regression provides a good example of a well-developed mathematical acquiring concept (and, by extension, a 1-layer by layer neural system with linear activations) that can be used to comprehend the connection among deep learning system variables, data, and computational demands.

One possible approach to reducing the computational load on deep learning systems is enhancing data curation. More specifically, instead of using more data, models might be trained with improved data to decrease processing requirements [131]. Although higher-quality data would ideally reduce computing expenses, it may still have an unclear impact on overall prices due to the high expense of data curation (e.g., higher human annotator expenses).

One alternative to deal with deep learning's increasing computing load is to switch to different forms of machine learning, some of which may be just as unexplored or undervalued. Even though "expert" approaches are often cheaper in terms of computing, their efficiency will eventually halt if the experts involved are unable to identify all of the elements that a model with greater flexibility could test. Awareness of known items (e.g., automobiles) [132] and techniques encouraged by biology (e.g., for learning neural controller structures [133]) are two areas in which such approaches are currently outperforming deep learning models.

By locating calculations in energy-efficient data canters, we may lessen the environmental effect of every calculation. Many factors contribute to the efficiency of such data canters, including (i) their location in regions with typical cooling and (ii) their architecture, which allows for more efficient use of the cooling systems that are required. Methods like smart load allocation, flexible voltage expanding, and server downsizing are all examples of how server administration can improve data centre efficiency [134].

The carbon footprint of deep learning systems is further affected by the energy that powers servers. For instance, according to Google's 2021 sustainability reveal, the tech giant bought sufficient clean energy to power all of its workplaces and data canters [135]. This indicates that carbon production can vary substantially depending on the region of hosting [136].

All the challenges that were explained in this section are summarized in Table 3.

Intelligent slice management solutions for 5G and beyond networks have been the subject of few research that have taken AI's potential into account. The concept of network slicing allows for the coexistence of numerous separate virtual networks, or slices, on top of a single physical network. Network slicing has many benefits [137]. First, network slicing allows for multiple tenancy, or the sharing of physical network infrastructure, via multiplexing virtual networks. Second, network slicing allows for the creation of customized slices for different service types with different quality of service (QoS) requirements. This allows for service differentiation and guarantees service level agreements (SLAs) for each service type. Third, network slicing increases the flexibility and adaptability in network management. It reduces capital expense in network deployment and operation. In order to keep network operation performance high, fulfil new key performance indicators (KPIs), such as dependability, energy consumption, and data quality, and support various and heterogeneous services, the study [138] suggested an efficient, intelligent network slicing framework.

Table 3. Summary of future research challenges of supporting technologies for the 6G networks.

Hybrid RF–Optical Networks	Space–Air–Ground–Sear (SAGS) Networks	Integrated Sensing and Computing Networks	Cloud-Edge Computing	Digital Twins Networks	AI-Enabled Networks
1. mixed-signal and interference, 2. transfer protocols development, 3. traffic and load balancing, 4. data synchronization, 5. distribution of resources, 6. energy efficiency enhancement, 7. distribution of power, 8. connection access points, 9. access point selection, 10. handover and blockages, 11. infrastructure and power for buildings, 12. mobility and user speed.	1. network design, 2. building and development, 3. preservation and maintenance, 4. optimization and upgrading, 5. channel structure and topology, 6. coverage and hardware capabilities, 7. environment and weather, 8. adaptability and scalability, 9. installation of aerial BSs, 10. accuracy of the UAV-to-ground channel, 11. UAV-to-ship connectivity settings, 12. mobility, 13. single and multi-bounce, 14. reflections and fading of multiple paths, 15. structure layout, 16. network procedure, 17. resource placement and routing tactics, 18. elevated network latency, 19. sporadic interruptions, 20. imbalanced network traffic, 21. information hub and data privacy, 22. data collection.	1. computational complexity, 2. transmission data rate requirements, 3. range of transmission, 4. base stations and network reach, 5. LOS route and multi-bounce route, 6. hardware requirements, 7. authentication mechanisms, 8. user data and privacy, 9. components management, 10. data collection and management, 11. communication, detecting, and processing, 12. waveform layout and processing of signals, 13. integrated sensing and communication detection, 14. power supplies, 15. systems' architectures.	1. computing core location, 2. data provider reliability, 3. security and data retrieval, 4. storage and capacity, 5. network structure, 6. cloud-computing model, 7. network congestion and delay, 8. collecting, interpreting, analyzing, and mining data, 9. user data and privacy, 10. network speed, 11. gateway asset distribution, 12. decision-making process, 13. task allocation, assets distribution, and segmentation, 14. use of energy, 15. servers and load distribution, 16. task loading and scheduling, 17. media access control.	1. computational complexity, 2. user data and privacy, 3. power supplies, 4. running costs, 5. systems' architectures, 6. user interaction modelling, 7. network speed, 8. media access control, 9. body area network requirements, 10. performance optimization, 11. visualization, 12. closed-loop management, 13. life cycle administration, 14. practical execution and functionalities of hardware, 15. data gathering, handling, and modelling, 16. information analysis.	1. resource usage, 2. processing techniques, 3. AI model, 4. sensory wireless systems, 5. decision-making process, 6. running costs, 7. systems' architectures, 8. transmission setting and network setup, 9. information analysis, 10. shared data set policies, 11. AI algorithms, 12. assets allocation, 13. negative environmental effects, 14. computing load of deep learning, 15. computational demands, 16. new model training, 17. high quantity of data sets, 18. data types, 19. data curation expenses, 20. energy-efficient in data centres.

While achieving high-speed communication technologies, the goal of constructing green, low-carbon 6G networks is to minimize environmental effect. Sixth-generation networks may greatly mitigate their impact on climate change by lowering their energy usage and carbon emissions. Maximizing the usage of renewable energy in network base stations and data centres is a primary goal in the design and implementation of 6G network equipment and infrastructure. This will improve energy efficiency and decrease waste. The research [139] offered a model for a 6G wireless environmentally friendly network design. This design's standout qualities are its ability to schedule tasks in an energy-efficient manner, its intelligent communication nodes, and the thorough separation of the user and control planes. Ensures energy efficiency and sustainability in 6G networks by discussing the difficulties and opportunities in [140].

4. Conclusions

6G networks will enhance worldwide communication solutions by introducing novel application cases, offering innovative technology knowledge, and fostering growth in the economy. We have rigorously evaluated the most recent approaches presented in a substantial corpus of pertinent literature, emphasizing the related advancements and obstacles. We have deliberated on the corresponding concept for 6G. The fundamental difficulties confronting 6G research and the relevant research paths have been examined via the lenses of basic research, sustainable networks, and the essential technologies produced to promote the envisioned 6G framework.

In summary, 6G studies and regulation continue to encounter several unresolved problems. This study has elucidated the anticipated attractive characteristics of 6G, and it is anticipated that it has offered renewed inspiration and encouragement for the community's 6G development efforts. The discussion and analysis that were presented in this study is expected to benefit the community in the near future especially the developers and researchers.

The following contributions have been achieved in this study:

1. Overview of the 6G networks, its features, characteristics, goals and objectives;
2. Recent studies and insights of the supporting technologies including AI, digital twin, hybrid and integrated networks, and cloud;
3. Review and discussion of the challenges of the supporting technologies;
4. Challenges and recommendations for the challenges have been presented and discussed;
5. Future research challenges have been presented and summarized.

Future directions for the 6G networks are expected to strengthen the technology and provide new solutions, including integrating 6G network with the supporting technologies in various environments and applications. Another promising avenue for future research in 6G networks lies in AI-driven studies, encompassing the development of advanced models and innovative applications tailored to enhance network intelligence, efficiency, and adaptability. AI algorithm performance comparison is also another future direction that will provide further insights.

Author Contributions: Conceptualization, S.S.M. and R.B.; methodology, H.J.A. and S.A.M.; validation, P.C., R.A.A. and M.D.S.; formal analysis, S.S.M. and S.A.M.; investigation, M.D.S.; resources, H.J.A. and R.A.A.; data curation, P.C.; writing—original draft preparation, S.S.M. and S.A.M.; writing—review and editing, H.J.A., P.C. and R.B.; visualization, R.B. and P.C.; supervision, M.D.S.; project administration, S.A.M. and S.S.M. All authors have read and agreed to the published version of the manuscript.

Funding: This research received no external funding.

Conflicts of Interest: Author Sallar S. Murad is employed by the company BEAMBRIDGE LIMITED, and author Mohammad D. Soltani is employed by the company ESWIN. The remaining authors declare that their research was conducted independently and without any commercial or financial relationships that could be construed as potential conflicts of interest.

Abbreviations

The following abbreviations are used in this manuscript:

1G	The first generation
5G	The fifth generation
6G	The sixth generation
AI	Artificial intelligence
ANN	Artificial neural networks
AR	Augmented reality
B5G	Beyond 5G
BBU	Baseband unit
CNN	Convolutional neural networks
CStrI	Channel strength information
CSI	Carrier state information
DNN	Deep neural networks
DT	The digital twin technology
DTN	Digital twin network
E2E	End-to-end
EM	Electromagnetic
FCC	The federal communications commission
FL	Federated learning
Gb/s	Gigabit per second
HOs	Handovers
IIoT	Industrial internet of things
IoT	Internet of things
IoV	Internet of vehicles
ISAC	Integrated sensing and communication
LEO	Low earth orbit
LiFi	Light fidelity
MEC	Multi-access edge computing
MEC	Mobile-edge computing
MDP	Markov decision process
ML	Machine learning
MIMO	Multiple-input multiple-output
MNs	Mobile networks
MPEs	Multipath elements
PCA	Principal component analysis
PPO-MSD	Proximal policy optimization for multi-layer service deployment
QoAIS	Quality of AI services
QoS	Quality of service
QPSK	Quadrature phase shift keying
RL	Reinforcement learning
SAGS	Space-air-ground-sear
SBS	Small base station
SCMA	Sparse code multiple access
SD-SRF	Service deployment with service rise and fall
SMA	Slime mould algorithm
ssAE	Stacked sparse autoencoders
THz	Terahertz

UAV	Unmanned aerial vehicle
UDN	Ultra-dense networking
UE	User devices
UWBCA	Ultra-wideband communication antenna
VR	Virtual reality
V2X	Vehicle-to-everything

References

1. Lu, L.; Swindlehurst, A.L.; Ashikhmin, A.; Zhang, R. An overview of massive MIMO: Benefits and challenges. *IEEE J. Sel. Top. Signal Process.* **2014**, *8*, 742–758. [CrossRef]
2. Kamel, M.; Hamouda, W.; Youssef, A. Ultra-Dense Networks: A Survey. *IEEE Commun. Surv. Tutor.* **2016**, *18*, 2522–2545. [CrossRef]
3. 384 Antenna Elements! Huaweis Third Generation Massive MIMO Continues to Lead the Way. Available online: https://www.sohu.com/a/525177315_110683 (accessed on 14 August 2023). (In Chinese).
4. Key Drivers and Research Challenges for 6G Ubiquitous Wireless Intelligence, 6G Flagship, Oulu, Finland, White Paper. Available online: https://www.6gflagship.com/key-drivers-and-research-challenges-for-6g-ubiquitous-wireless-intelligence/ (accessed on 20 May 2025).
5. Key R&D Projects 2019: Wideband Communications and New Networks. Available online: https://service.most.gov.cn/u/cms/static/201812/12164952skqa.pdf (accessed on 19 July 2023).
6. Research and Development on Satellite-Terrestrial Integration Technology for Beyond 5G. Available online: https://www2.nict.go.jp/spacelab/en/pj_stit.html (accessed on 20 July 2023).
7. 6G SENTINEL—The Next Generation of Mobile Communications. Available online: https://www.fraunhofer.de/en/research/lighthouse-projects-fraunhofer-initiatives/fraunhofer-lighthouse-projects/6g-sentinel.html (accessed on 20 May 2025).
8. RINGS. Available online: https://www.nsf.gov/funding/opportunities/rings-vo-resilient-intelligent-nextg-systems-virtual-organization/506034/nsf22-590 (accessed on 20 May 2025).
9. KT and Hanwha Systems Are Jointly Developing 6G Quantum Cryptography Technology. Available online: https://www.kedglobal.com/cn/6g/newsView/ked202207120033 (accessed on 8 October 2023).
10. 6GStart: Starting the Sustainable 6G SNS Initiative for Europe. Available online: https://5g-ppp.eu/6gstart/ (accessed on 3 October 2023).
11. 6G White Paper. Available online: https://www.oulu.fi/6gflagship/6g-white-papers (accessed on 11 April 2023).
12. Rohde & Schwarz, 5G Evolution—On the Path to 6G: Expanding the Frontiers of Wireless Communications, White Paper. Available online: https://www.microwavejournal.com/articles/34454-g-evolution-on-the-path-to-6g-expanding-the-frontiers-of-wireless-communications (accessed on 20 May 2025).
13. 6G: Unbounded, with AI, UNISOC, Shanghai, China, White Paper. Available online: http://unisocdata.oss-cn-shanghai.aliyuncs.com/other/20201126/202011261130506G (accessed on 3 November 2024).
14. 6G White Paper, China Unicom, Beijing, China, White Paper. Available online: https://www.chinaunicom.com.hk/en/esg/reports/csr2023.pdf (accessed on 20 May 2025).
15. Tong, W.; Zhu, P. *6G, The Next Horizon: From Connected People and Things to Connected Intelligence*, 1st ed.; Cambridge University Press: Cambridge, UK, 2021.
16. 6G Wireless Endogenous AI Architecture and Technology, China Mobile, Beijing, China, White Paper. Available online: https://mp.weixin.qq.com/s/xh-gau4p5pn3YYaHRephAw (accessed on 8 August 2024).
17. NTTDOCOMOINC. 5G Evolution and 6G, White Paper. Available online: https://www.docomo.ne.jp/english/binary/pdf/corporate/technology/whitepaper_6g/DOCOMO_6G_White_PaperEN_20200124.pdf (accessed on 20 May 2025).
18. Samsung Research, 6G the Next Hyper-Connected Experience for All, White Paper. Available online: https://cdn.codeground.org/nsr/downloads/researchareas/6G%20Vision.pdf (accessed on 20 May 2025).
19. Next Generation Mobile Networks Alliance. 6G Drivers and Vision. Available online: https://www.ngmn.org/work-programme/ngmn-6g-drivers-and-vision.html (accessed on 20 May 2025).
20. Yigit, Y.; Maglaras, L.; Buchanan, W.J.; Canberk, B.; Shin, H.; Duong, T.Q. AI-Enhanced Digital Twin Framework for Cyber-Resilient 6G Internet-of-Vehicles Networks. *IEEE Internet Things J.* **2024**, *11*, 36168–36181. [CrossRef]
21. Dangana, M.; Ansari, S.; Asad, S.M.; Hussain, S.; Imran, M.A. Towards the Digital Twin (DT) of Narrow-Band Internet of Things (NBIoT) Wireless Communication in Industrial Indoor Environment. *Sensors* **2022**, *22*, 9039. [CrossRef]
22. Khan, S.; Alzaabi, A.; Iqbal, Z.; Ratnarajah, T.; Arslan, T. A Novel Digital Twin (DT) Model Based on WiFi CSI, Signal Processing and Machine Learning for Patient Respiration Monitoring and Decision-Support. *IEEE Access* **2023**, *11*, 103554–103568. [CrossRef]

23. Tao, Z.; Xu, W.; Huang, Y.; Wang, X.; You, X. Wireless Network Digital Twin for 6G: Generative AI as a Key Enabler. *IEEE Wirel. Commun.* **2024**, *31*, 24–31. [CrossRef]
24. Wieme, J.; Baert, M.; Hoebeke, J. Digital Twin Network for dynamic management of a Bluetooth Mesh Network. In Proceedings of the NOMS 2024—2024 IEEE Network Operations and Management Symposium, Seoul, Republic of Korea, 6–10 May 2024; pp. 1–3.
25. Lou, P.; Zhao, Q.; Zhang, X.; Li, D.; Hu, J. Indoor Positioning System with UWB Based on a Digital Twin. *Sensors* **2022**, *22*, 5936. [CrossRef]
26. Nauman, A.; Nguyen, T.N.; Qadri, Y.A.; Nain, Z.; Cengiz, K.; Kim, S.W. Artificial Intelligence in Beyond 5G and 6G Reliable Communications. *IEEE Internet Things Mag.* **2022**, *5*, 73–78. [CrossRef]
27. Yu, W.; Sohrabi, F.; Jiang, T. Role of Deep Learning in Wireless Communications. *IEEE BITS Inf. Theory Mag.* **2022**, *2*, 56–72. [CrossRef]
28. Pei, J.; Li, S.; Yu, Z.; Ho, L.; Liu, W.; Wang, L. Federated Learning Encounters 6G Wireless Communication in the Scenario of Internet of Things. *IEEE Commun. Stand. Mag.* **2023**, *7*, 94–100. [CrossRef]
29. Geranmayeh, P.; Grass, E. Machine Learning Based Beam Selection for Maximizing Wireless Network Capacity. *IEEE Access* **2024**, *12*, 45176–45186. [CrossRef]
30. Luo, G.; Yuan, Q.; Li, J.; Wang, S.; Yang, F. Artificial Intelligence Powered Mobile Networks: From Cognition to Decision. *IEEE Netw.* **2022**, *36*, 136–144. [CrossRef]
31. Chen, T.; Deng, J.; Tang, Q.; Liu, G. Optimization of Quality of AI Service in 6G Native AI Wireless Networks. *Electronics* **2023**, *12*, 3306. [CrossRef]
32. Seliuchenko, M.; Beshley, M.; Shpur, O.; Seliuchenko, N.; Klymash, M.; Beshley, H. Software-Defined Multi-Access Edge/Cloud Computing for 5G/6G Time-Critical Services. In Proceedings of the 2023 17th International Conference on the Experience of Designing and Application of CAD Systems (CADSM), Jaroslaw, Poland, 22–25 February 2023; IEEE: Piscataway, NJ, USA, 2023; pp. 1–4. [CrossRef]
33. Xu, Y.; Zhao, B.; Zhou, H.; Su, J. FedAdaSS: Federated Learning with Adaptive Parameter Server Selection Based on Elastic Cloud Resources. *Comput. Model. Eng. Sci.* **2024**, *141*, 609–629. [CrossRef]
34. Wang, L.; Liu, A.; Xiong, N.N.; Zhang, S.; Wang, T.; Dong, M. SD-SRF: An Intelligent Service Deployment Scheme for Serverless-operated Cloud-Edge Computing in 6G Networks. *Future Gener. Comput. Syst.* **2024**, *151*, 242–259. [CrossRef]
35. Logeshwaran, J.; Ramesh, G.; Logeshwaran, J.; Aravindarajan, V.; Thachil, F. Eliminate the Interference In 5G Ultra-Wide Band Communication Antennas in Cloud Computing Networks. *ICTACT J. Microelectron.* **2022**, *8*, 1338–1344. [CrossRef]
36. AlQahtani, S.A. An Evaluation of e-Health Service Performance through the Integration of 5G IoT, Fog, and Cloud Computing. *Sensors* **2023**, *23*, 5006. [CrossRef]
37. Yang, H.; Zhou, H.; Liu, Z.; Deng, X. Energy Optimization of Wireless Sensor Embedded Cloud Computing Data Monitoring System in 6G Environment. *Sensors* **2023**, *23*, 1013. [CrossRef]
38. Zhang, L.; Liang, Y.-C.; Niyato, D. 6G Visions: Mobile Ultra-Broadband, Super Internet-of-Things, and Artificial Intelligence. *China Commun.* **2019**, *16*, 1–14. [CrossRef]
39. Zhang, Z.; Xiao, Y.; Ma, Z.; Xiao, M.; Ding, Z.; Lei, X.; Karagiannidis, G.K.; Fan, P. 6G Wireless Networks: Vision, Requirements, Architecture, and Key Technologies. *IEEE Veh. Technol. Mag.* **2019**, *14*, 28–41. [CrossRef]
40. Zong, B.; Fan, C.; Wang, X.; Duan, X.; Wang, B.; Wang, J. 6G Technologies: Key Drivers, Core Requirements, System Architectures, and Enabling Technologies. *IEEE Veh. Technol. Mag.* **2019**, *14*, 18–27. [CrossRef]
41. Bariah, L.; Mohjazi, L.; Muhaidat, S.; Sofotasios, P.C.; Kurt, G.K.; Yanikomeroglu, H.; Dobre, O.A. A prospective look: Key enabling technologies, applications and open research topics in 6G networks. *IEEE Access* **2020**, *8*, 174792–174820. [CrossRef]
42. Saad, W.; Bennis, M.; Chen, M. A Vision of 6G Wireless Systems: Applications, Trends, Technologies, and Open Research Problems. *IEEE Netw* **2020**, *34*, 134–142. [CrossRef]
43. Viswanathan, H.; Mogensen, P.E. Communications in the 6G Era. *IEEE Access* **2020**, *8*, 57063–57074. [CrossRef]
44. Liu, G.; Huang, Y.; Dong, J.; Jin, J.; Wang, Q.; Li, N. Vision, requirements and network architecture of 6G mobile network beyond 2030. *China Commun.* **2020**, *17*, 92–104. [CrossRef]
45. Chen, S.; Liang, Y.C.; Sun, S.; Kang, S.; Cheng, W.; Peng, M. Vision, Requirements, and Technology Trend of 6G: How to Tackle the Challenges of System Coverage, Capacity, User Data-Rate and Movement Speed. *IEEE Wirel. Commun.* **2020**, *27*, 218–228. [CrossRef]
46. You, X.; Wang, C.X.; Huang, J.; Gao, X.; Zhang, Z.; Wang, M.; Huang, Y.; Zhang, C.; Jiang, Y.; Wang, J.; et al. Towards 6G wireless communication networks: Vision, enabling technologies, and new paradigm shifts. *Sci. China Inf. Sci.* **2021**, *64*, 110301. [CrossRef]
47. Jiang, W.; Han, B.; Habibi, M.A.; Schotten, H.D. The road towards 6G: A comprehensive survey. *IEEE Open J. Commun. Soc.* **2021**, *2*, 334–366. [CrossRef]
48. Albahri, O.S.; Alamleh, A.; Al-Quraishi, T.; Thakkar, R. Smart Real-Time IoT mHealth-based Conceptual Framework for Healthcare Services Provision during Network Failures. *Appl. Data Sci. Anal.* **2023**, *2023*, 110–117. [CrossRef]

49. Khaleel, Y.L.; Habeeb, M.A.; Albahri, A.S.; Al-Quraishi, T.; Albahri, O.S.; Alamoodi, A.H. Network and cybersecurity applications of defense in adversarial attacks: A state-of-the-art using machine learning and deep learning methods. *J. Intell. Syst.* **2024**, *33*, 20240153. [CrossRef]
50. Kreutz, D.; Ramos, F.M.V.; Verissimo, P.E.; Rothenberg, C.E.; Azodolmolky, S.; Uhlig, S. Software-Defined Networking: A Comprehensive Survey. *Proc. IEEE* **2015**, *103*, 14–76. [CrossRef]
51. Tariq, F.; Khandaker, M.R.A.; Wong, K.-K.; Imran, M.A.; Bennis, M.; Debbah, M. A Speculative Study on 6G. *IEEE Wirel. Commun.* **2020**, *27*, 118–125. [CrossRef]
52. Murad, S.S.; Badeel, R.; Alsandi, N.S.A.; Ahmed, R.F.A.R.A.; Muhammed, A.; Derahman, M. Optimized Min-Min Task Scheduling Algorithm for Scientific Workflows In A Cloud Environment. *J. Theor. Appl. Inf. Technol.* **2022**, *100*, 480–506.
53. Murad, S.S.; Badeel, R.; Ahmed, R.A.; Yussof, S. Using Drones and Robots for Social Distancing: Literature Review, Challenges and Issues. In Proceedings of the 2024 Panhellenic Conference on Electronics and Telecommunications, PACET 2024—Proceedings, Thessaloniki, Greece, 28–29 March 2024; Institute of Electrical and Electronics Engineers Inc.: Athens, Greece, 2024. [CrossRef]
54. Țoța, P.; Vaida, M.F.; Chiorean, L. Mesh networks with interconnected Li-fi and Wi-Fi communications applied to telepresence robots for medical emergency situations. In Proceedings of the 2020 8th E-Health and Bioengineering Conference, EHB 2020, Iasi, Romania, 29–30 October 2020. [CrossRef]
55. Singh, M.; Fuenmayor, E.; Hinchy, E.P.; Qiao, Y.; Murray, N.; Devine, D. Digital twin: Origin to future. *Appl. Syst. Innov.* **2021**, *4*, 36. [CrossRef]
56. Wang, C.X.; Lv, Z.; Gao, X.; You, X.; Hao, Y.; Haas, H. Pervasive Wireless Channel Modeling Theory and Applications to 6G GBSMs for All Frequency Bands and All Scenarios. *IEEE Trans. Veh. Technol.* **2022**, *71*, 9159–9173. [CrossRef]
57. Huang, C.; He, R.; Ai, B.; Molisch, A.F.; Lau, B.K.; Haneda, K.; Liu, B.; Wang, C.X.; Yang, M.; Oestges, C.; et al. Artificial Intelligence Enabled Radio Propagation for Communications-Part I: Channel Characterization and Antenna-Channel Optimization. *IEEE Trans. Antennas. Propag.* **2022**, *70*, 3939–3954. [CrossRef]
58. Huang, C.; He, R.; Ai, B.; Molisch, A.F.; Lau, B.K.; Haneda, K.; Liu, B.; Wang, C.-X.; Yang, M.; Oestges, C.; et al. Artificial Intelligence Enabled Radio Propagation for Communications-Part II: Scenario Identification and Channel Modeling. *IEEE Trans. Antennas Propag.* **2022**, *70*, 3955–3969. [CrossRef]
59. Sun, Y.; Wang, C.X.; Huang, J.; Wang, J. A 3D Non-Stationary Channel Model for 6G Wireless Systems Employing Intelligent Reflecting Surfaces with Practical Phase Shifts. *IEEE Trans. Cogn. Commun. Netw.* **2021**, *7*, 496–510. [CrossRef]
60. Migliore, M.D. On electromagnetics and information theory. *IEEE Trans. Antennas Propag.* **2008**, *56*, 3188–3200. [CrossRef]
61. Migliore, M.D. Blending electromagnetic and information theory in antenna synthesis. In Proceedings of the 2019 International Conference on Electromagnetics in Advanced Applications (ICEAA), Granada, Spain, 9–13 September 2019; pp. 1383–1386.
62. Hochwald, B.M.; Brink, S.T. Achieving near-capacity on a multiple-antenna channel. *IEEE Trans. Commun.* **2003**, *51*, 389–399. [CrossRef]
63. Shen, Y.; Zhou, W.; Huang, Y.; Zhang, Z.; You, X.; Zhang, C. Fast Iterative Soft-Output List Decoding of Polar Codes. *IEEE Trans. Signal Process.* **2022**, *70*, 1361–1376. [CrossRef]
64. Deng, X.; Sha, J.; Zhou, X.; Fu, Y.; Zhang, Z.; You, X.; Zhang, C. Joint Detection and Decoding of Polar-Coded OFDM-IDMA Systems. *IEEE Trans. Circuits Syst. I Regul. Pap.* **2019**, *66*, 4005–4017. [CrossRef]
65. Sun, W.C.; Su, Y.C.; Ueng, Y.L.; Yang, C.H. An LDPC-Coded SCMA receiver with multi-user iterative detection and decoding. *IEEE Trans. Circuits Syst. I Regul. Pap.* **2019**, *66*, 3571–3584. [CrossRef]
66. Sun, W.C.; Wu, W.H.; Yang, C.H.; Ueng, Y.L. An iterative detection and decoding receiver for LDPC-coded MIMO systems. *IEEE Trans. Circuits Syst. I Regul. Pap.* **2015**, *62*, 2512–2522. [CrossRef]
67. Parhi, K.K. VLSI Digital Signal Processing Systems. 1999. Available online: http://www.bestbookbuys.com (accessed on 1 October 2024).
68. Zhong, Z.; Gross, W.J.; Zhang, Z.; You, X.; Zhang, C. Polar compiler: Auto-generator of hardware architectures for polar encoders. *IEEE Trans. Circuits Syst. I Regul. Pap.* **2020**, *67*, 2091–2102. [CrossRef]
69. Ji, C.; Shen, Y.; Zhang, Z.; You, X.; Zhang, C. Autogeneration of Pipelined Belief Propagation Polar Decoders. *IEEE Trans. Very Large Scale Integr. VLSI Syst.* **2020**, *28*, 1703–1716. [CrossRef]
70. The Polar Compiler. Available online: https://github.com/zzwshuai/The-Polar-Compiler (accessed on 25 June 2024).
71. You, X.; Zhang, C.; Tan, X.; Jin, S.; Wu, H. AI for 5G: Research Directions and Paradigms. Available online: http://arxiv.org/abs/1807.08671 (accessed on 10 July 2024).
72. Chowdhury, M.Z.; Hasan, M.K.; Shahjalal, M.; Hossan, M.T.; Jang, Y.M. Optical Wireless Hybrid Networks: Trends, Opportunities, Challenges, and Research Directions. *IEEE Commun. Surv. Tutor.* **2020**, *22*, 930–966. [CrossRef]
73. Badeel, R. A Review on LiFi Network Research: Open Issues, Applications and Future Directions. *Appl. Sci.* **2021**, *11*, 1111[illegible] [CrossRef]
74. Murad, S.S.; Yussof, S.; Hashim, W.; Badeel, R. Three-Phase Handover Management and Access Point Transition Scheme for Dynamic Load Balancing in Hybrid LiFi/WiFi Networks. *Sensors* **2022**, *22*, 7583. [CrossRef]

75. Kashef, M.; Ismail, M.; Abdallah, M.; Qaraqe, K.A.; Serpedin, E. Energy efficient resource allocation for mixed RF/VLC heterogeneous wireless networks. *IEEE J. Sel. Areas Commun.* **2016**, *34*, 883–893. [CrossRef]
76. Kim, S. Hybrid RF/VLC Network Spectrum Allocation Scheme Using Bargaining Solutions. *IEEE Access* **2022**, *10*, 20019–20028. [CrossRef]
77. Aboagye, S.; Ngatched, T.M.N.; Dobre, O.A.; Ibrahim, A. Joint Access Point Assignment and Power Allocation in Multi-Tier Hybrid RF/VLC HetNets. *IEEE Trans. Wirel. Commun.* **2021**, *20*, 6329–6342. [CrossRef]
78. Murad, S.S.; Yussof, S.; Badeel, R. Wireless Technologies for Social Distancing in The Time Of COVID-19: Literature Review, Open Issues, and Limitations. *Sensors* **2022**, *22*, 2313. [CrossRef]
79. Murad, S.S.; Yussof, S.; Badeel, R.; Hashim, W. A Novel Social Distancing Approach for Limiting the Number of Vehicles in Smart Buildings Using LiFi Hybrid—Network. *Int. J. Environ. Res. Public Health* **2023**, *20*, 3438. [CrossRef] [PubMed]
80. Wu, X.; Haas, H. Handover Skipping for LiFi. *IEEE ACCESS* **2019**, *7*, 38369–38378. [CrossRef]
81. Wu, X.; OBrien, D.C.; Deng, X.; Linnartz, J.P.M.G. Smart Handover for Hybrid LiFi and WiFi Networks. *IEEE Trans. Wirel. Commun.* **2020**, *19*, 8211–8219. [CrossRef]
82. Palas, M.R.; Islam, R.; Roy, P.; Razzaque, A.; Alsanad, A.; AlQahtani, S.A.; Hassan, M.M. Multi-criteria handover mobility management in 5G cellular network. *Comput. Commun.* **2021**, *174*, 81–91. [CrossRef]
83. Badeel, R.; Subramaniam, S.K.; Muhammed, A.; Hanapi, Z.M. A Multicriteria Decision-Making Framework for Access Point Selection in Hybrid LiFi/WiFi Networks Using Integrated AHP–VIKOR Technique. *Sensors* **2023**, *23*, 1312. [CrossRef] [PubMed]
84. Pan, G.; Ye, J.; Ding, Z. Secure Hybrid VLC-RF Systems with Light Energy Harvesting. *IEEE Trans. Commun.* **2017**, *65*, 4348–4359. [CrossRef]
85. Farrag, M.; Shamim, M.Z.; Usman, M.; Hussein, H.S. Load Balancing Scheme in Hybrid WiGig/LiFi Network. *IEEE Access* **2020**, *8*, 222429–222438. [CrossRef]
86. Ahmad, R.; Soltani, M.D.; Safari, M.; Srivastava, A. Reinforcement learning-based near-optimal load balancing for heterogeneous LiFi WiFi network. *IEEE Syst. J.* **2021**, *16*, 3084–3095. [CrossRef]
87. Murad, S.S.; Yussof, S.; Hashim, W.; Badeel, R. Card-Flipping Decision-Making Technique for Handover Skipping and Access Point Assignment: A Novel Approach for Hybrid LiFi Networks. *IEEE Access* **2024**, *12*, 146635–146667. [CrossRef]
88. Wu, X.; OBrien, D.C. A Novel Machine Learning-Based Handover Scheme for Hybrid LiFi and WiFi Networks. In Proceedings of the 2020 IEEE Globecom Workshops, GC Wkshps 2020—Proceedings, Taipei, Taiwan, 7–11 December 2020; IEEE: Piscataway, NJ, USA, 2020; pp. 1–5. [CrossRef]
89. Odabaşı, A.S.; İşçi, O.; Rodoplu, V. Machine Learning Based Seamless Vertical Handoff Mechanism for Hybrid Li-Fi/Wi-Fi Networks. In Proceedings of the 2022 International Conference on INnovations in Intelligent SysTems and Applications (INISTA), Biarritz, France, 8–12 August 2022; IEEE: Piscataway, NJ, USA, 2022; pp. 1–6. [CrossRef]
90. Wang, J.; Jiang, C.; Zhang, H.; Zhang, X.; Leung, V.C.M.; Hanzo, L. Learning-aided network association for hybrid indoor LiFi-WiFi systems. *IEEE Trans Veh. Technol.* **2017**, *67*, 3561–3574. [CrossRef]
91. Bai, L.; Wang, C.X.; Goussetis, G.; Wu, S.; Zhu, Q.; Zhou, W.; Aggoune, E.H. Channel modeling for satellite communication channels at q-band in high latitude. *IEEE Access* **2019**, *7*, 137691–137703. [CrossRef]
92. Chang, H.; Wang, C.-X.; Liu, Y.; Huang, J.; Sun, J.; Zhang, W.; Gao, X. A Novel Nonstationary 6G UAV-to-Ground Wireless Channel Model with 3-D Arbitrary Trajectory Changes. *IEEE Internet Things J.* **2021**, *8*, 9865–9877. [CrossRef]
93. Liu, Y.; Wang, C.X.; Chang, H.; He, Y.; Bian, J. A Novel Non-Stationary 6G UAV Channel Model for Maritime Communications. *IEEE J. Sel. Areas Commun.* **2021**, *39*, 2992–3005. [CrossRef]
94. He, Y.; Wang, C.-X.; Chang, H.; Huang, J.; Sun, J.; Zhang, W. A Novel 3D Non-Stationary Maritime Wireless Channel Model. *IEEE Trans. Commun.* **2022**, *70*, 2102–2116. [CrossRef]
95. Wang, C.X.; Huang, J.; Wang, H.; Gao, X.; You, X.; Hao, Y. 6G Wireless Channel Measurements and Models: Trends and Challenges. *IEEE Veh. Technol. Mag.* **2020**, *15*, 22–32. [CrossRef]
96. Molisch, A.F. *Wireless Communications*; John Wiley & Sons: Hoboken, NJ, USA, 2012.
97. Zeng, Y.; Zhang, R.; Lim, T.J. Wireless communications with unmanned aerial vehicles: Opportunities and challenges. *IEEE Commun. Mag.* **2016**, *54*, 36–42. [CrossRef]
98. Li, B.; Fei, Z.; Zhang, Y. UAV communications for 5G and beyond: Recent advances and future trends. *IEEE Internet Things J.* **2019**, *6*, 2241–2263. [CrossRef]
99. Zhu, L.; Zhang, J.; Xiao, Z.; Cao, X.; Xia, X.G.; Schober, R. Millimeter-Wave Full-Duplex UAV Relay: Joint Positioning, Beamforming, and Power Control. *IEEE J. Sel. Areas Commun.* **2020**, *38*, 2057–2073. [CrossRef]
100. Meng, Y.S.; Lee, Y.H. Measurements and characterizations of air-to-ground channel over sea surface at C-band with low airborne altitudes. *IEEE Trans. Veh. Technol.* **2011**, *60*, 1943–1948. [CrossRef]
101. Li, X.; Feng, W.; Chen, Y.; Wang, C.X.; Ge, N. Maritime Coverage Enhancement Using UAVs Coordinated with Hybrid Satellite-Terrestrial Networks. *IEEE Trans. Commun.* **2020**, *68*, 2355–2369. [CrossRef]

102. Liu, D.; Wu, H.; Ni, J.; Shen, X. Efficient and Anonymous Authentication with Succinct Multi-Subscription Credential in SAGVN. *IEEE Trans. Intell. Transp. Syst.* **2022**, *23*, 2863–2873. [CrossRef]
103. Li, D.; Wu, S.; Jiao, J.; Zhang, N.; Zhang, Q. Age-Oriented Transmission Protocol Design in Space-Air-Ground Integrated Networks. *IEEE Trans. Wirel. Commun.* **2022**, *21*, 5573–5585. [CrossRef]
104. Yan, S.; Qi, L.; Peng, M. User Access Mode Selection in Satellite-Aerial Based Emergency Communication Networks. In Proceedings of the ICC Workshops: 2018 IEEE International Conference on Communications Workshops, Kansas City, MO, USA, 20–24 May 2018; IEEE: Piscataway, NJ, USA, 2018; p. 486.
105. Kato, N.; Fadlullah, Z.M.; Tang, F.; Mao, B.; Tani, S.; Okamura, A. Optimizing Space-Air-Ground Integrated Networks by Artificial Intelligence. *IEEE Wirel. Commun.* **2019**, *26*, 140–147. [CrossRef]
106. Yang, P.; Li, Z.; Yang, P.; Dong, Y. Information-centric mobile ad hoc networks and content routing: A survey. *Ad Hoc Netw.* **2017**, *58*, 255–268. [CrossRef]
107. Zhou, Z.; Feng, J.; Tan, L.; He, Y.; Gong, J. An air-ground integration approach for mobile edge computing in IoT. *IEEE Commun. Mag.* **2018**, *56*, 40–47. [CrossRef]
108. Abbas, N.; Zhang, Y.; Taherkordi, A.; Skeie, T. Mobile Edge Computing: A Survey. *IEEE Internet Things J.* **2018**, *5*, 450–465. [CrossRef]
109. Liu, J.; Du, X.; Cui, J.; Pan, M.; Wei, D. Task-oriented intelligent networking architecture for the space-air-ground-aqua integrated network. *IEEE Internet Things J.* **2020**, *7*, 5345–5358. [CrossRef]
110. Sturm, C.; Wiesbeck, W. Waveform design and signal processing aspects for fusion of wireless communications and radar sensing. *Proc. IEEE* **2011**, *99*, 1236–1259. [CrossRef]
111. Zhang, A.; Rahman, M.L.; Huang, X.; Guo, Y.J.; Chen, S.; Heath, R.W. Perceptive Mobile Networks: Cellular Networks with Radio Vision via Joint Communication and Radar Sensing. *IEEE Veh. Technol. Mag.* **2021**, *16*, 20–30. [CrossRef]
112. Shi, Q.; Liu, L.; Zhang, S.; Cui, S. Device-Free Sensing in OFDM Cellular Network. *IEEE J. Sel. Areas Commun.* **2022**, *40*, 1838–1853. [CrossRef]
113. Ghosh, S.; Mukherjee, A.; Ghosh, S.K.; Buyya, R. Mobi-iost: Mobility-aware cloud-fog-edge-iot collaborative framework for time-critical applications. *IEEE Trans. Netw. Sci. Eng.* **2019**, *7*, 2271–2285. [CrossRef]
114. Alnoman, A.; Anpalagan, A. Computing-Aware Base Station Sleeping Mechanism in H-CRAN-Cloud-Edge Networks. *IEEE Trans. Cloud Comput.* **2021**, *9*, 958–967. [CrossRef]
115. Du, J.; Zhao, L.; Feng, J.; Chu, X. Computation Offloading and Resource Allocation in Mixed Fog/Cloud Computing Systems with Min-Max Fairness Guarantee. *IEEE Trans. Commun.* **2018**, *66*, 1594–1608. [CrossRef]
116. Shah, S.A.A.; Ahmed, E.; Imran, M.; Zeadally, S. 5G for Vehicular Communications. *IEEE Commun. Mag.* **2018**, *56*, 111–117. [CrossRef]
117. Shih, Y.Y.; Chung, W.H.; Pang, A.C.; Chiu, T.C.; Wei, H.Y. Enabling Low-Latency Applications in Fog-Radio Access Networks. *IEEE Netw.* **2017**, *31*, 52–58. [CrossRef]
118. Niu, Z.; Guo, X.; Zhou, S.; Kumar, P.R. Characterizing energy-delay tradeoff in hyper-cellular networks with base station sleeping control. *IEEE J. Sel. Areas Commun.* **2015**, *33*, 641–650. [CrossRef]
119. Oo, T.Z.; Tran, N.H.; Saad, W.; Niyato, D.; Han, Z.; Hong, C.S. Offloading in HetNet: A Coordination of Interference Mitigation, User Association, and Resource Allocation. *IEEE Trans. Mob. Comput.* **2017**, *16*, 2276–2291. [CrossRef]
120. Meng, X.; Wang, W.; Zhang, Z. Delay-Constrained Hybrid Computation Offloading with Cloud and Fog Computing. *IEEE Access* **2017**, *5*, 21355–21367. [CrossRef]
121. Cui, Y.; Jiang, D. Analysis and Optimization of Caching and Multicasting in Large-Scale Cache-Enabled Heterogeneous Wireless Networks. *IEEE Trans. Wirel. Commun.* **2017**, *16*, 250–264. [CrossRef]
122. Lyu, X.; Tian, H.; Jiang, L.; Vinel, A.; Maharjan, S.; Gjessing, S. Selective Offloading in Mobile Edge Computing for the Green Internet of Things. *IEEE Netw.* **2018**, *32*, 54–60. [CrossRef]
123. Li, J.; Huang, L.; Zhou, Y.; He, S.; Ming, Z. Computation partitioning for mobile cloud computing in a big data environment. *IEEE Trans. Ind. Inf.* **2017**, *13*, 2009–2018. [CrossRef]
124. Brogi, A.; Forti, S. QoS-aware deployment of IoT applications through the fog. *IEEE Internet Things J.* **2017**, *4*, 1185–119 [CrossRef]
125. Ivanov, S.; Nikolskaya, K.; Radchenko, G.; Sokolinsky, L.; Zymbler, M. Digital Twin of City: Concept Overview. In *2020 Glob Smart Industry Conference (GloSIC)*; IEEE: Piscataway, NJ, USA, 2020.
126. Sun, T.; Zhou, C.; Duan, X.; Lu, L.; Chen, D.; Yang, H.; Zhu, Y.; Liu, C.; Li, Q.; Wang, X.; et al. Digital Twin Network (DTN Concepts, Architecture, and Key Technologies. *Zidonghua Xuebao/Acta Autom. Sin.* **2021**, *47*, 569–582. [CrossRef]
127. Zhu, Y.; Chen, D.; Zhou, C.; Lu, L.; Duan, X. A knowledge graph based construction method for Digital Twin Network. Proceedings of the 2021 IEEE 1st International Conference on Digital Twins and Parallel Intelligence, DTPI 2021, Beijing, Chi 15 July–15 August 2021; pp. 362–365. [CrossRef]

128. 6GANA TG1, Network AI concept and terminology. Available online: https://www.6g-ana.com/About aspx?ClassID=29 (accessed on 15 October 2024).
129. He, Y.; Ren, J.; Yu, G.; Cai, Y. Optimizing the Learning Performance in Mobile Augmented Reality Systems with CNN. *IEEE Trans. Wirel. Commun.* **2020**, *19*, 5333–5344. [CrossRef]
130. Thompson, N.; Greenewald, K.; Lee, K.; Manso, G.F. The Computational Limits of Deep Learning. In The Computational Limits of Deep Learning. *arXiv* **2021**, arXiv:2007.05558.
131. Bender, E.M.; Gebru, T.; McMillan-Major, A.; Shmitchell, S. On the dangers of stochastic parrots: Can language models be too big? In *FAccT 2021—Proceedings of the 2021 ACM Conference on Fairness, Accountability, and Transparency*; Association for Computing Machinery, Inc.: New York, NY, USA, 2021; pp. 610–623. [CrossRef]
132. He, T.; Soatto, S. Mono3D++: Monocular 3D Vehicle Detection with Two-Scale 3D Hypotheses and Task Priors. Available online: www.aaai.org (accessed on 10 October 2024).
133. Lechner, M.; Hasani, R.; Amini, A.; Henzinger, T.A.; Rus, D.; Grosu, R. Neural circuit policies enabling auditable autonomy. *Nat. Mach. Intell.* **2020**, *2*, 642–652. [CrossRef]
134. Masanet, E.; Shehabi, A.; Lei, N.; Smith, S.; Koomey, J. Recalibrating global data center energy-use estimates. *Science* **2020**, *367*, 984–986. [CrossRef]
135. Google. Google Environmental Report 2022—Gstatic.com. Available online: https://sustainability.google/reports/google-2022-environmental-report/ (accessed on 20 May 2025).
136. Patterson, D.; Gonzalez, J.; Le, Q.; Liang, C.; Munguia, L.-M.; Rothchild, D.; So, D.; Texier, M.; Dean, J. Carbon Emissions and Large Neural Network Training. *arXiv* **2021**, arXiv:2104.10350.
137. Shen, X.; Gao, J.; Wu, W.; Lyu, K.; Li, M.; Zhuang, W. AI-assisted network-slicing based next-generation wireless networks. *IEEE Open J. Veh. Technol.* **2020**, *1*, 45–66. [CrossRef]
138. Abdellatif, A.A.; Abo-Eleneen, A.; Mohamed, A.; Erbad, A.; Navkar, N.V.; Guizani, M. Intelligent-Slicing: An AI-Assisted Network Slicing Framework for 5G-and-Beyond Networks. *IEEE Trans. Netw. Serv. Manag.* **2023**, *20*, 1024–1039. [CrossRef]
139. Li, L. Research on future 6G green wireless networks. *Green Technol. Sustain.* **2025**, *3*, 100156. [CrossRef]
140. Tyagi, A.K.; Tiwari, S.; Gupta, S.; Mishra, A.K. Energy Efficiency and Sustainability in 6G Networks. In *6G—Enabled Technologies for Next Generation*; Wiley: Hoboken, NJ, USA, 2025; pp. 168–186. [CrossRef]

MDPI AG
Grosspeteranlage 5
4052 Basel
Switzerland
Tel.: +41 61 683 77 34

Telecom Editorial Office
E-mail: telecom@mdpi.com
www.mdpi.com/journal/telecom

www.ingramcontent.com/pod-product-compliance
Lightning Source LLC
Chambersburg PA
CBHW061615220926
51850CB00005B/512